T0342565

INTRODUCTION TO TYPE-2 FUZZY LOGIC CONTROL

IEEE Press
445 Hoes Lane
Piscataway, NJ 08854

INTRODUCTION TO TYPE-2 FUZZY LOGIC CONTROL

THEORY AND APPLICATIONS

Jerry M. Mendel
Hani Hagras
Woei-Wan Tan
William W. Melek
Hao Ying

IEEE Press Series on Computational Intelligence
David B. Fogel, Series Editor

WILEY

For general information on our other products and services or for technical support, please contact our Customer Care Department within the United States at (800) 762-2974, outside the United States at (317) 572-3993 or fax (317) 572-4002.

Wiley also publishes its books in a variety of electronic formats. Some content that appears in print may not be available in electronic formats. For more information about Wiley products, visit our web site at www.wiley.com.

Library of Congress Cataloging-in-Publication Data:

Mendel, Jerry M., 1938–
Introduction to type-2 fuzzy logic control : theory and applications / Jerry M. Mendel, Hani Hagras, Woei-Wan Tan, William W. Melek, Hao Ying.
pages cm
Includes bibliographical references and index.
ISBN 978-1-118-27839-0 (cloth)
1. Automatic control. 2. Fuzzy systems. I. Hagras, Hani. II. Tan, Woei-Wan. III. Melek, William W. IV. Ying, Hao, 1958- V. Title.
TJ217.5.M46 2014
629.8′95633–dc23

2014000084

Printed in the United States of America

10 9 8 7 6 5 4 3 2 1

To
the memory of
Ebrahim Mamdani (1943–2010)
Founder of Fuzzy Logic Control

CONTENTS

PREFACE

When Lotfi Zadeh invented fuzzy sets in 1965, he never dreamt that the field in which they would be most widely used would arguably be the one that became the most hostile to the concept of fuzziness, namely control. Perhaps this was because the word "fuzzy" in Western civilization does not have a positive connotation and suggests an abandonment of mathematical rigor, one of the cornerstones of control. Perhaps it was because some famous mathematical probabilists (incorrectly) claimed that there was no difference between a fuzzy set and subjective probability. Perhaps it was because for almost a decade, until the 1974 seminal paper by Prof. Ebrahim Mamdani, who founded the field of fuzzy logic control and to whose memory our book is dedicated, there were no substantial real-world applications for fuzzy sets. Or, perhaps, it was because after the founding of this field many exaggerated claims were made by the fuzzy logic control community that flew in the face of mathematical rigor and did not pay attention to the same metrics that were and still are the cornerstones for control and cannot be ignored.

Now, 40 years after Mamdani's seminal paper, fuzzy logic control using regular (i.e., type-1) fuzzy sets and logic has been extensively studied, applied to practical problems, and is very widely used in many real-world applications. It can and has been studied with the same level of mathematical rigor that control theorists are accustomed to, and is now considered a matured field; however, it still has some shortcomings. Its major shortcoming (in the opinions of the authors of this book) goes back to one of the earliest criticisms made about a type-1 fuzzy set, namely the unfuzziness of its membership function, that is, the word "fuzzy" has the connotation of being uncertain. But how can this connotation be captured by a membership function that is completely certain?

Importantly, in 1975 Zadeh introduced more general kinds of fuzzy sets in which their membership function grades are themselves fuzzy. The two most widely studied of these are *interval-valued fuzzy sets* and *type-2 fuzzy sets*. For the former, the membership grade is a uniformly weighted interval of values, whereas for the latter the membership grade is a nonuniformly weighted interval of values. Obviously, interval-valued fuzzy sets are a special case of type-2 fuzzy sets and are therefore called by many (as we do in this book) *interval type-2 fuzzy sets*.

Why should using type-2 fuzzy sets be of interest to the fuzzy logic control community? This question is answered in great detail in this book, but two short answers are: (1) they are more robust to system uncertainties and can provide better control system performance than type-1 fuzzy sets; and (2) there is now more than a critical mass of papers that have been published that demonstrate these improvements for many real-world applications.

Because of the lack of basic calculation methods for type-2 fuzzy sets in their early days, type-2 fuzzy logic controllers (T2 FLCs) did not emerge until fairly recently. Things have changed a lot during the past decade, so that type-2 fuzzy logic control (which is still an emerging field) now has the attention of the fuzzy systems community, and, as a result of this, the number of publications on it is growing quickly.

Recall that the central themes of any control methodology, fuzzy or conventional, are (1) to analyze various aspects of a control system and (2) to design a control system to achieve given user specifications. This book focuses on both topics for T2 FLCs and type-2 fuzzy logic control systems. The analysis includes (1) the mathematical structure of some T2 FLCs, (2) stability of type-2 fuzzy logic control systems, and (3) robustness of the type-2 fuzzy logic control systems.

This book, the first one entirely on T2 FLC, shows how to design type-2 fuzzy logic control systems based on a variety of choices for the T2 FLC components and also demonstrates how to apply type-2 fuzzy logic control theory to applications. It has been written by five of the leading experts on type-2 fuzzy sets, systems, and control, with the help of six contributors. It will be useful to any technical person interested in learning type-2 fuzzy logic control theory and its applications, from students to practicing engineers.

This is an introductory book that provides theoretical, practical, and application coverage of type-2 fuzzy logic control, and uses a coherent structure and uniform mathematical notations to link chapters, which are closely related, reflecting the book's central themes—analysis and design of type-2 fuzzy logic control systems. It has been written with an educational focus rather than a pure research focus. Each chapter includes worked examples, and most refer to their computer codes (programs) accessible through the book's common website, and outline how to use them at some high level. It is a self-contained reference book suitable for engineers, researchers, and college graduate students who want to gain deep insights about type-2 fuzzy logic control.

The book begins with an easy-to-read chapter meant to whet the reader's appetite so that he or she will read on; it explains what the differences are between a type-1 fuzzy set and a type-2 fuzzy set, and a T2 FLC and a T1 FLC, and, it provides many real-world applications in which T2 FLCs have shown marked improvements in performance over T1 FLCs. Chapter 2 provides all of the background material that is needed about type-2 fuzzy sets so that you can read the rest of the book; its main emphasis is on interval type-2 fuzzy sets because at present they are the most widely used type-2 fuzzy sets in type-2 fuzzy logic control. Chapter 3 is about Mamdani and TSK interval T2 FLCs. Chapter 4 examines the analytical structure of various interval type-2 fuzzy PI and PD controllers. Chapter 5 is about ways to simplify interval type-2 fuzzy PI and PD controllers. Chapter 6 is about the rigorous design of interval type-2 TSK fuzzy controllers. Chapter 7 provides each of the five authors with an opportunity to look into the future of type-2 fuzzy logic control. The book's appendix describes Java-based software that will let the reader examine type-1, interval type-2, and even general type-2 FLCs. All references (which are very extensive) have been integrated into one list that is at the end of the book.

The book's software can be downloaded by means of the following procedure: Software for Examples 4.1 and 4.6 and the examples in Chapter 6 can be accessed at http://booksupport.wiley.com, and software for Appendix A, that supports T1, IT2 and GT2 FLCs, is available at http://juzzy.wagnerweb.net.

In addition to the five authors, six of their (former) graduate students contributed to this book, to whom the authors are greatly appreciative. Their names are listed in the Contributors List. More specifically, Christian Wagner contributed to Chapters 2, 3 and 7, and prepared the entire Appendix; Xinyu Du and Haibo Zhou contributed to Chapter 4; Maowen Nie and Dongrui Wu contributed to Chapter 5; and Mohammad Biglarbegian contributed to Chapter 6.

The authors gratefully acknowledge material quoted from books or journals published by Elsevier, IEEE, John Wiley & Sons, Mancy Publishing (www.maney.co.uk/journals/irs and www.ingentaconnect.com/content/maney/ias) and Pearsons Education, Inc. For a complete listing of quoted books or articles, please see the References.

JERRY M. MENDEL
Los Angeles, California

HANI HAGRAS
Colchester, UK

WOEI-WAN TAN
Singapore

WILLIAM W. MELEK
Waterloo, Canada

HAO YING
Detroit, Michigan

CONTRIBUTORS

Mohammad Biglarbegian, University of Guelph, Guelph Ontario, Canada

Xinyu Du, Wayne State University, Detroit, Michigan

Maowen Nie, A*Star Institute of Infocomm Research, Singapore

Christian Wagner, University of Nottingham, Nottingham, United Kingdom

Dongrui Wu, GE Global Research, New York

Haibo Zhou, Central South University, Changsha, China

CHAPTER 1

Introduction

1.1 EARLY HISTORY OF FUZZY CONTROL

Fuzzy control (also known as fuzzy logic control) is regarded as the most widely used application of fuzzy logic and is credited with being a well-accepted methodology for designing controllers that are able to deliver satisfactory performance in the face of uncertainty and imprecision (Lee, 1990; Sugeno, 1985; Feng, 2006). In addition, fuzzy logic theory provides a method for less skilled personnel to develop practical control algorithms in a user-friendly way that is close to human thinking and perception, and to do this in a short amount of time. Fuzzy logic controllers (FLCs) can sometimes outperform traditional control systems [like proportional–integral–derivative (PID) controllers] and have often performed either similarly or even better than human operators. This is partially because most FLCs are nonlinear controllers that are capable of controlling real-world systems (the vast majority of such systems are nonlinear) better than a linear controller can, and with minimal to no knowledge about the mathematical model of the plant or process being controlled.

Fuzzy logic controllers have been applied with great success to many real-world applications. The first FLC was developed by Mamdani and Assilian (1975), in the United Kingdom, for controlling a steam generator in a laboratory setting. In 1976, Blue Circle Cement and SIRA in Denmark developed a cement kiln controller (the first industrial application of fuzzy logic), which went into operation in 1982 (Holmblad and Ostergaard, 1982). In the 1980s, several important industrial applications of fuzzy logic control were launched successfully in Japan, including a water treatment system developed by Fuji Electric. In 1987, Hitachi put a fuzzy logic based automatic train operation control system into the Sendai city's subway system (Yasunobu and Miyamoto, 1985). These and other applications of FLCs motivated many Japanese engineers to investigate a wide range of novel applications for fuzzy logic. This led to a "fuzzy boom" in Japan, a result of close collaboration and technology transfer between universities and industry.

According to Yen and Langari (1999), in 1988, a large-scale national research initiative was established by the Japanese Ministry of International Trade and

Introduction to Type-2 Fuzzy Logic Control: Theory and Applications, First Edition.
Jerry M. Mendel, Hani Hagras, Woei-Wan Tan, William W. Melek, and Hao Ying.

Industry (MITI). The initiative established by MITI was a consortium called the Laboratory for International Fuzzy Engineering Research (LIFE). In late January 1990, Matsushita Electric Industrial (Panasonic) named their newly developed fuzzy-controlled automatic washing machine the fuzzy washing machine and launched a major commercial campaign of it as a *fuzzy* product. This campaign turned out to be a successful marketing effort not only for the product but also for fuzzy logic technology (Yen and Langari, 1999). Many other home electronics companies followed Panasonic's approach and introduced fuzzy vacuum cleaners, fuzzy rice cookers, fuzzy refrigerators, fuzzy camcorders (for stabilizing the image under hand jittering), fuzzy camera (for smart autofocus), and other applications. As a result, consumers in Japan recognized the now en-vogue Japanese word "fuzzy," which won the gold prize for a new word in 1990 (Hirota, 1995). Originating in Japan, the "fuzzy boom" triggered a broad and serious interest in this technology in Korea, Europe, the United States, and elsewhere. For example, Boeing, NASA, United Technologies, and other aerospace companies developed FLCs for space and aviation applications (Munakata and Jani, 1994).

Today FLCs are used in countless real-world applications that touch the lives of people all over the world, including white goods (e.g., washing machines, refrigerators, microwaves, rice cookers, televisions, etc.), digital video cameras, cars, elevators (lifts), heavy industries (e.g., cement, petroleum, steel), and the like.

While this book focuses on type-2 fuzzy logic control, it will also provide background material about type-1 fuzzy logic control. Indeed, before we can explain what type-2 fuzzy logic control is we must briefly explain what type-1 fuzzy sets, type-1 fuzzy logic control, and type-2 fuzzy sets are. In this chapter we do this from a high-level perspective without touching on the mathematical aspects in order to give a feel for the nature of fuzzy sets and their applications. Later chapters in this book provide rigorous treatments of mathematical underpinnings of the subjects just mentioned.

1.2 WHAT IS A TYPE-1 FUZZY SET?

Suppose that a group of people is asked about the temperature values they associate with the linguistic concepts Hot and Cold. If *crisp sets* are employed, as shown in Fig. 1.1a, then a threshold must be chosen above which temperature values are considered Hot and below which they are considered Cold. Reaching a consensus about such a threshold is difficult, and even if an agreement can be reached—for example, 18°C—, is it reasonable to conclude that 17.99999°C is Cold whereas 18.00001°C is Hot?

On the other hand, Hot and Cold can be represented as *type-1 fuzzy sets* (T1 FSs) whose membership functions (MFs) are shown in Fig. 1.1b. Note that, prior to the appearance of type-2 fuzzy sets, the phrase *fuzzy set* was used instead of the phrase *T1 fuzzy set*. Even today, in many publications that focus only on T1 FSs, such sets are called fuzzy sets. In this book we shall use the phrase *type-1 fuzzy set*. Returning to Fig. 1.1b, observe that no sharp boundaries exist between the two sets

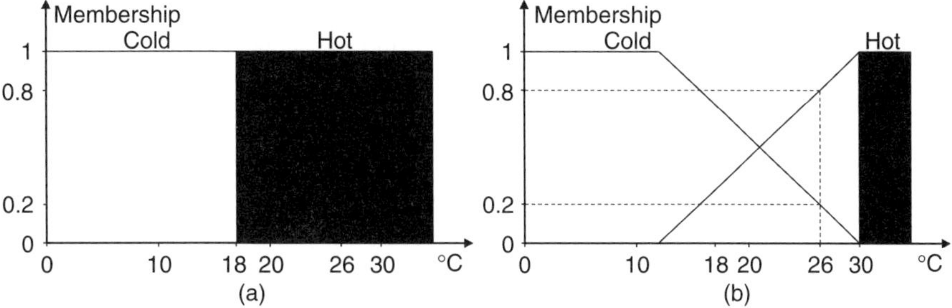

Figure 1.1 Representing Cold and Hot using (a) crisp sets, and (b) type-1 fuzzy sets.

and that each value on the horizontal axis may simultaneously belong to more than one T1 FS but with different degrees of membership. For example, 26°C, which is in the crisp Hot set with a membership value of 1.0 (Fig. 1.1a), is now in that set to degree 0.8, but is also in the Cold set to degree 0.2 (Fig. 1.1b).

Type-1 FSs provide a means for calculating intermediate values between the crisp values associated with being absolutely true (1) or absolutely false (0). Those values range between 0 and 1 (and can include them); thus, it can be said that a fuzzy set allows the calculation of shades of gray between white and black (or true and false). As will be seen in this book, the smooth transition that occurs between T1 FSs gives a good decision response for a type-1 fuzzy logic control system in the face of noise and other uncertainties.

1.3 WHAT IS A TYPE-1 FUZZY LOGIC CONTROLLER?

With the advent of type-2 fuzzy sets and type-2 fuzzy logic control, it has become necessary to distinguish between *type-2 fuzzy logic control* and all earlier fuzzy logic control that uses type-1 fuzzy sets (the distinctions between such fuzzy sets are explained in Section 1.4). We refer to fuzzy logic control that uses type-1 fuzzy sets as *type-1 fuzzy logic control*. When it does not matter whether the fuzzy sets are type-1 or type-2, we just use *fuzzy logic control* or *fuzzy control*.

Fuzzy logic control aims to mimic the process followed by the human mind when performing control actions. For example, when a person drives (controls) a car, he/she will not think:

> If the temperature is *10 degrees Celsius* and the rainfall is *70.5 mm* and the road is *40% slippery* and the distance between my car and the car in front of me is *3 meters*, then I will depress the acceleration pedal only *10%*.

Instead, it is much more likely that he/she thinks:

> If it is Cold and the rainfall is High and the road is Somewhat Slippery and the distance between my car and the car in front of me is Quite Close, then I will depress the acceleration pedal Slightly.

So, in systems controlled by humans, the control cycle starts by a person converting a physical quantity (e.g., a distance) from numbers into words or perceptions (e.g., Quite Close distance). The input words (or perceptions) then trigger a person's knowledge, accumulated through that person's experience, resulting in words representing actions (e.g., depress the acceleration pedal Slightly). The person then executes an action to actuate a given device that interfaces the person with the controlled system (e.g., depress the acceleration pedal only 10% might represent the person's implementation of "depress the accelerator pedal Slightly"). Because people think and reason by using imprecise linguistic information, FLCs try to mimic and convert linguistic control information into numerical control information that can be used in automatic control systems.

In its attempt to mimic human control actions, a type-1 FLC, whose structure is shown in Fig. 1.2, is composed of four main components: fuzzifier, rules, inference engine, and defuzzifier, where the operation of each component is summarized as follows:

- The fuzzifier maps each measured numerical input variable into a fuzzy set. One motivation for doing this is that measurements may be corrupted by noise and are somewhat uncertain (even after filtering). So, for example, a measured temperature of 26°C may be modeled as a triangular type-1 fuzzy set that is symmetrically centered around 26°C, where the base of the triangle is related to the uncertainty of this measurement. If, however, one believes that there is no measurement uncertainty, then the measurements can be modeled as crisp sets.
- Rules have an if–then structure, for example, *If Temperature is Low and Pressure is High, then Fan Speed is Low*. Each IF part of a rule is called its *antecedent*, and the THEN part of a rule is called its *consequent*. Rules relate input fuzzy sets to output fuzzy sets. All of the rules are collected into a rule base.

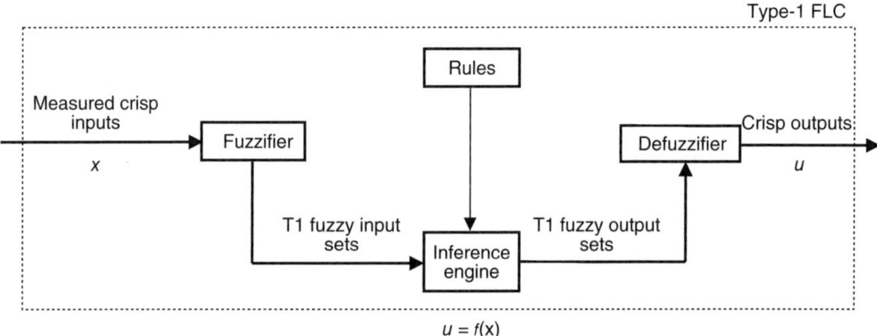

Figure 1.2 General structure of a type-1 FLC. The heavy lines with arrows indicate the path taken by signals during the actual operation of the FLC. Rules are used during the design of the FLC and are activated by the inference engine during the actual operation of the FLC (Mendel et al. (2006); © 2006, IEEE).

- The inference engine decides which rules from the rule base are fired and what their degrees of firing are, by using the fuzzy sets provided to it from the fuzzifier as well as some mathematics about fuzzy sets. The inference engine may also combine each rule's degree of firing with that rule's consequent fuzzy set to produce the rule's *output fuzzy set* (i.e., its fired-rule output set), and then combine all of those sets (across all of the fired rules) to produce an aggregated fuzzy output set using the mathematics of fuzzy sets; or it may send each rule's degree of firing directly to the defuzzifier where they are all aggregated in a different way.
- The defuzzifier receives either the aggregated fuzzy output sets from the inference engine or the degrees of firing for each rule plus some information about each consequent fuzzy set, and then processes this data to produce crisp outputs that are then passed to the physical actuators that control the actual plant.

In general, real-world control systems, such as fuzzy logic control systems, are affected by the following uncertainties:

- Uncertainties about the inputs to the FLC. For instance, sensor measurements can be affected by high noise levels and changing observation conditions such as changing environmental conditions, for example, wind, rain, humidity, and so forth. In addition to measurement noise, other possible inputs to the FLC, such as those estimated by an observer or computed using a process model, can also be imprecise and exhibit uncertainty.
- Uncertainties about control outputs that can occur because of changes in an actuator's characteristics due to wear and tear, environmental changes, and the like.
- Uncertainties about the change in operating conditions of the controller, such as changes in a plant's parameters.
- Uncertainties due to disturbances acting upon the system when those disturbances cannot be measured, for example, wind buffeting an airplane.

In a T1 FLC all of these uncertainties are handled by the T1 FSs in the antecedents and consequents of the rules, as well as through the chosen type of fuzzifier. Regarding the latter, one may choose to use: (1) a singleton fuzzifier in which a measured value is treated as perfect and is modeled as a crisp set; or (2) a type-1 fuzzifier in which a measured value is treated as signal plus stationary noise and is modeled as a normal, convex T1 FS (also called a *T1 fuzzy number*).

The type-1 FLC in Fig. 1.2 is a nonlinear controller that maps its inputs $\mathbf{x}$ into an output u, that is, $u = f(\mathbf{x})$, where f is a nonlinear function that is formed by fuzzy logic operations and the mathematics of fuzzy sets. Often, $f(\mathbf{x})$ is formed from linguistic rules that summarize human knowledge or experience (or may be constructed from data); thus, the type-1 FLC directly maps such knowledge or experience into a nonlinear control law whose explicit mathematical expression is unknown in most cases.

Many researchers (e.g., Wang, 1992; Wang and Mendel, 1992a; Castro, 1995; Kosko, 1994; Kreinovich et al. 1998) have shown that the type-1 FLC $f(\mathbf{x})$ can uniformly approximate any real continuous function on a compact domain to any degree of accuracy; hence, FLCs are known to be *universal approximators*. One way to interpret what this means is that the FLC $f(\mathbf{x})$ approximates a function by covering its graph with *fuzzy patches* (Kosko, 1994), where each rule in the FLC defines a fuzzy patch in system's input–output space, and it then averages overlapping patches. This approximation improves as the fuzzy patches grow in number and shrink in size; however, as more smaller patches are included, the complexity of the model increases (i.e., the number of fuzzy sets and rules increases).

Type-1 FLCs produce nonlinear control laws $f(\mathbf{x})$ that cannot be effectively generated by any other mathematical means because such $f(\mathbf{x})$ are derived from linguistic if–then rules. This has enabled fuzzy logic control to be used in complex ill-defined processes, especially those that can be controlled by a skilled human operator without the knowledge of their underlying dynamics (Mamdani and Assilian, 1975).

Recall that *variable structure control* (VSC) is a form of discontinuous nonlinear control that alters the dynamics of a nonlinear system through the application of high-frequency switching control. A T1 FLC can also be regarded as a variable structure controller by virtue of the mathematics of fuzzy sets and systems; that is, it partitions the state space *automatically* rather than by a planned design. This is because different rules are activated for different regions of the state space. Palm (1992) showed that an FLC can be regarded as an extension of a conventional variable structure controller with a boundary layer.

There are two widely used architectures for a type-1 FLC that mainly differ in their fuzzy rule consequents. Those architectures, both of which are examined in this book, are:

- Mamdani FLC, developed by Mamdani and Assilian (1975) in which the antecedents and consequents of the rules are linguistic terms, for example: If x_1 *is Low and* x_2 *is High, then u is Low*. The linguistic labels in a Mamdani FLC are represented by type-1 fuzzy sets.
- Takagi–Sugeno (TS) FLC or Takagi–Sugeno–Kang (TSK) FLC (Takagi and Sugeno, 1985) in which the antecedents of the rules are also linguistic terms (modeled as type-1 fuzzy sets), but each rule's consequent is modeled as a mathematical function of the input variables, for example: *If* x_1 *is Low and* x_2 *is High, then* $u = g(x_1, x_2)$, where $g(x_1, x_2)$ is a polynomial function of x_1 and x_2 (this can include a constant, a linear or affine function, a quadratic function, etc.). An example of a first-order TSK FLC rule, the most widely used order, is: *If* x_1 *is Low and* x_2 *is High, then* $u = c_0 + c_1x_1 + c_2x_2$, where c_0, c_1, and c_2 are the consequent parameters.

1.4 WHAT IS A TYPE-2 FUZZY SET?

Because T1 FSs (e.g., as in Fig. 1.1b) are themselves crisp and precise (i.e., their MFs are supposedly known perfectly), this does not allow for any uncertainties about membership values, which is a potential shortcoming when using such fuzzy sets. A *type-2 fuzzy set* (T2 FS) is characterized by a fuzzy MF, that is, the membership value for each element of this set is itself a fuzzy set in [0,1]. The MFs of T2 FSs are three dimensional (3D) and include a *footprint of uncertainty* (FOU) (which is shaded in gray in Fig. 1.3a). It is the new third dimension of T2 FSs (e.g., Fig. 1.4c) and its FOU that provide additional degrees of freedom that make it possible to directly model and handle MF uncertainties.

In Fig. 1.3a, observe that the 26°C membership value in Hot is no longer a crisp value of 0.8 (as was the case in Fig. 1.1b); instead, it is a function that takes values from 0.6 to 0.8 in the primary membership domain, and maps them into a triangular distribution in the third dimension (Fig. 1.3b), called a *secondary MF*. This triangular secondary MF weights the interval [0.6, 0.8] more strongly over its middle values and less strongly away from those middle values. Of course, other weightings are possible, including equal weightings, in which case the T2 FS is called an *interval type-2 FS* (IT2 FS). Being able to choose different kinds of secondary MFs demonstrates one of the flexibilities of T2 FSs.

Figure 1.4c depicts the 3D MF of a general T2 FS whose secondary MFs [$f_x(u)$] are triangles. By convention, such a T2 FS is called a *triangular T2 FS*. Its FOU is depicted in Fig. 1.4a and its secondary MF at x' [$f_{x'}(u)$] is depicted by the solid triangle in Fig. 1.4b. When the secondary membership values equal 1 for all the primary membership values (as in the dashed curve in Fig. 1.4b), this results in an interval-valued secondary membership function, and, as just mentioned, the resulting T2 FS is called an IT2 FS. In Fig. 1.4c, $\mu(x, u)$ denotes the MF value at (x, u).

Figure 1.5 depicts the FOU of an IT2 FS for Low. The three dashed functions that are embedded within that FOU are T1 FSs. Clearly, one can cover this FOU with a multitude of such T1 FSs. At this point it is not important whether there are a

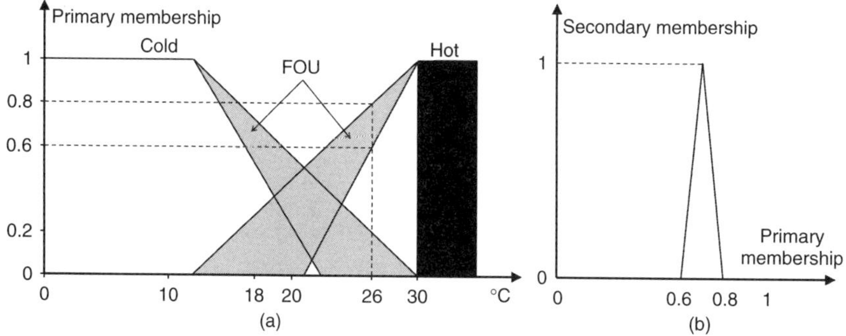

Figure 1.3 Type-2 fuzzy sets: (a) FOU and a primary membership and (b) a triangle secondary membership function.

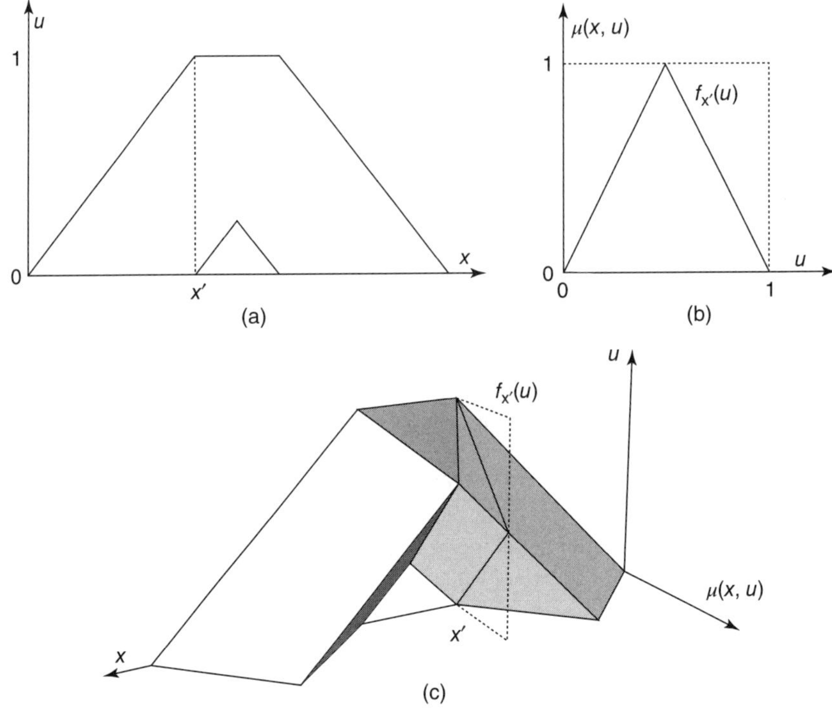

Figure 1.4 (a) FOU with primary membership (dashed) at x', (b) two possible secondary membership functions (triangle in solid line and interval in dashed line) associated with x', and, (c) the resulting 3D type-2 fuzzy set.

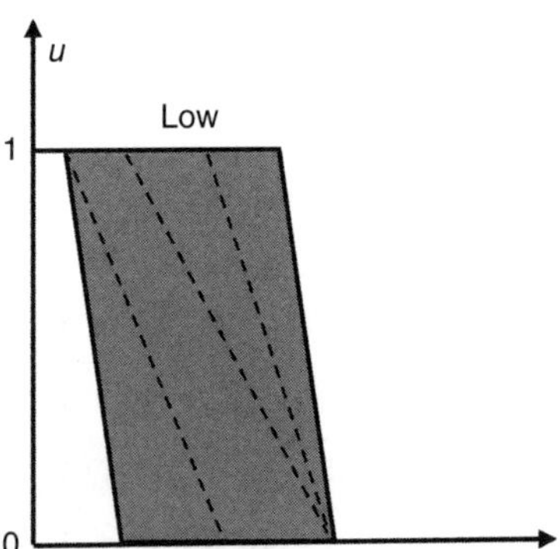

Figure 1.5 Three type-1 fuzzy sets that are embedded in the FOU of Low.

countable or uncountable number of such T1 FSs. What is important is interpreting an IT2 FS as the aggregation of a multitude of T1 FSs. This suggests that T1 FSs and everything that is already known about them can be used in derivations involving IT2 FSs, something that is exploited very heavily in this book. This interpretation also plays a very important role in understanding why an IT2 FLC may outperform a T1 FLC, something that we shall return to in the section below and in other chapters of this book.

1.5 WHAT IS A TYPE-2 FUZZY LOGIC CONTROLLER?

A type-2 FLC is depicted in Fig. 1.6. It contains five components: fuzzifier, rules, inference engine, type reducer, and defuzzifier. In a T2 FLC the inputs and/or outputs are represented by T2 FSs, and it operates as follows: crisp inputs, obtained from input sensors, are fuzzified into input T2 FSs, which then activate an inference engine that uses the same rules used in a T1 FLC to produce output T2 FSs. These are then processed by a type reducer that projects the T2 FSs into a T1 FS (this step is called *type reduction*) (Karnik et al., 1999; Liang and Mendel, 2000) after which that T1 FS is defuzzified to produce a crisp output that, for example, can be used as the command to an actuator in the control system. Type reduction followed by defuzzification is usually referred to as *output processing*.

In Section 1.3 we presented some sources of uncertainties that face real-world control systems in general. FLCs are also affected by:

- Linguistic uncertainties because the meaning of words that are used in the antecedents' and consequents' linguistic labels can be uncertain, that is, words mean different things to different FLC designers (Mendel, 2001).
- In addition, experts do not always agree and they often provide different consequents for the same antecedents. A survey of experts will usually lead to a histogram of possibilities for the consequent of a rule; this histogram represents the uncertainty about the consequent of a rule (Mendel, 2001).

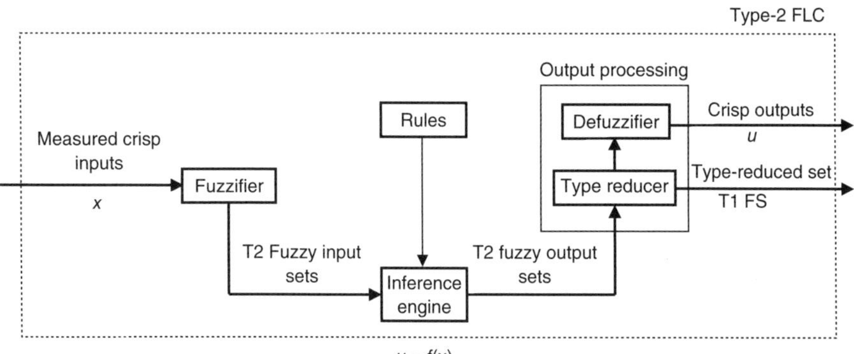

Figure 1.6 Overview of the architecture of a T2 FLC. The heavy lines with arrows indicate the path taken by signals during the actual operation of the FLC. Rules are used during the design of the FLC and are activated by the inference engine during the actual operation of the FLC (Mendel et al., 2006; © 2006, IEEE).

In a T2 FLC all of these uncertainties are modeled by the T2 FSs' MFs in the antecedents and/or consequents of the rules, as well as by the kind of fuzzifier. Regarding the latter, one may choose to use: (1) a singleton fuzzifier (as in a T1 FLC) in which a measured value is treated as perfect and is modeled as a crisp set; (2) a type-1 fuzzifier (as in a T1 FLC) in which a measured value is treated as signal plus stationary noise and is modeled as a normal, convex T1 FS (also called a *T1 fuzzy number*); or (3) a type-2 fuzzifier in which a measured value is treated as signal plus nonstationary noise and is modeled as a normal, convex T2 FS.

As we have explained in Section 1.4, a T2 FS can be thought of as a collection of many embedded T1 FSs (Mendel and John, 2002a). A T2 FLC may, therefore, be conceptually thought of as a collection of many (embedded) T1 FLCs whose crisp output is obtained by aggregating the outputs of all the embedded T1 FLCs (Karnik et al., 1999). Consequently, a T2 FLC has the potential to outperform a T1 FLC under certain conditions because it deals with uncertainties by aggregating a multitude of embedded T1 FLCs. The actual implementation of a T2 FLC does not actually require such an aggregation, but in this first chapter of this book, it is helpful to think of the output of a T2 FLC in this way.

Just as a T1 FLC is a variable structure controller so is a T2 FLC, and just as a T1 FLC has two architectures, Mamdani and TSK, a T2 FLC also has those two architectures. In a T2 Mamdani or TSK FLC, the fuzzy sets are type-2. Like their T1 FLC counterparts, T2 Mamdani and TSK FLCs are universal approximators (Ying, 2008, 2009). Both of these T2 FLC architectures will be covered in this book.

1.6 DISTINGUISHING AN FLC FROM OTHER NONLINEAR CONTROLLERS

Nonlinear control involves a nonlinear relationship between the controller's inputs and outputs and is more complicated than linear control; however, it is able to achieve better performance than linear control for many real-world control applications. Nonlinear control theory requires more challenging mathematical analysis and design than does linear control theory.

As mentioned in Section 1.3, an FLC is a nonlinear controller, that is, the function $f(\mathbf{x})$ is nonlinear. This will be demonstrated in later chapters of this book. What distinguishes an FLC, T1 or T2, from other nonlinear controllers is that it generates its nonlinear mapping function $f(\mathbf{x})$ through linguistic if–then rules and linguistic terms for the antecedents and consequents of the rules (e.g., Low Temperature, High Pressure). Such rules can be (easily) obtained from a human operator or can be postulated and learned from data. According to Kosko (1994), an FLC is unique in that it ties vague words like Low and High, and common sense rules, to state-space geometry.

According to Mamdani (1994), when tuned, the parameters of a PID controller affect the shape of the entire control surface. Because fuzzy logic control is a rule-based controller, the shape of the control surface can be individually manipulated for the different regions of the state space, thus limiting possible effects only to neighboring regions.

Fuzzy logic controllers have two important advantages over other classes of nonlinear controllers, namely (1) they are able to incorporate linguistic terms in the designs of the input–output membership functions, and (2) they are capable of handling uncertainties in inputs and state measurements more effectively. Moreover, similar to other classes of nonlinear controllers, they can be mathematically expressed, analyzed, and designed.

If the FLC rules are obtained from a group of experts, they may not all agree on the rule's consequents. By using T2 FSs, one is able to model the group's histogram of rule consequents, something that cannot be done by using a T1 FLC.

An FLC can be studied like any other nonlinear controller, for example, for the Mamdani FLC, stability and robustness studies can be performed by extensive simulations and by analyzing its control surface; see Fig.1.7, which depicts the mathematical function that maps robot controller inputs [e.g., right sensor front (RSF) and right sensor back (RSB)] into a control output (e.g., Steering). For a TSK FLC, it is possible to perform the same kinds of mathematical analyses that are applied to other nonlinear controllers, such as Lyapunov stability and robustness, and the like. Performance analyses of T2 Mamdani and TSK FLCs are given in later chapters of this book.

1.7 T2 FLCs VERSUS T1 FLCs

Type-1 FLCs use T1 FSs that have precise MFs, that is, there is nothing uncertain about such MFs. The following uncertainties that an FLC may encounter have been enumerated in Section 1.3: uncertainties about the inputs to the FLC, the control outputs, changing operating conditions of the controller, and disturbances acting upon the plant. Such uncertainties must somehow be mapped into MF uncertainties, and this is feasible to a greater extent in a T2 FLC than it is in a TI FLC because of the "noncrisp" nature of a T2 FS, the FOU for an IT2 FLC, or the combination of an FOU and secondary MFs for a general T2 FLC.

In addition to the above traditional kinds of uncertainties, which affect any kind of a controller, fuzzy or nonfuzzy, an FLC is also affected by the following additional uncertainties:

- Uncertainties about a rule's consequent, when rules are obtained from a group of experts, because, as we have mentioned above, experts do not generally all agree on the same consequent.
- Linguistic uncertainties about the meanings of the words used in a rule's antecedent and consequent linguistic terms, because *words mean different things to different people* (Mendel, 2001).
- Uncertainties associated with noisy training data that may be used to optimize (learn, tune) the MF parameters of an FLC.

It is difficult to directly model or minimize the effects of such uncertainties using T1 FSs. Consequently, using T1 FSs in an FLC may cause degradation in the performance of such a system.

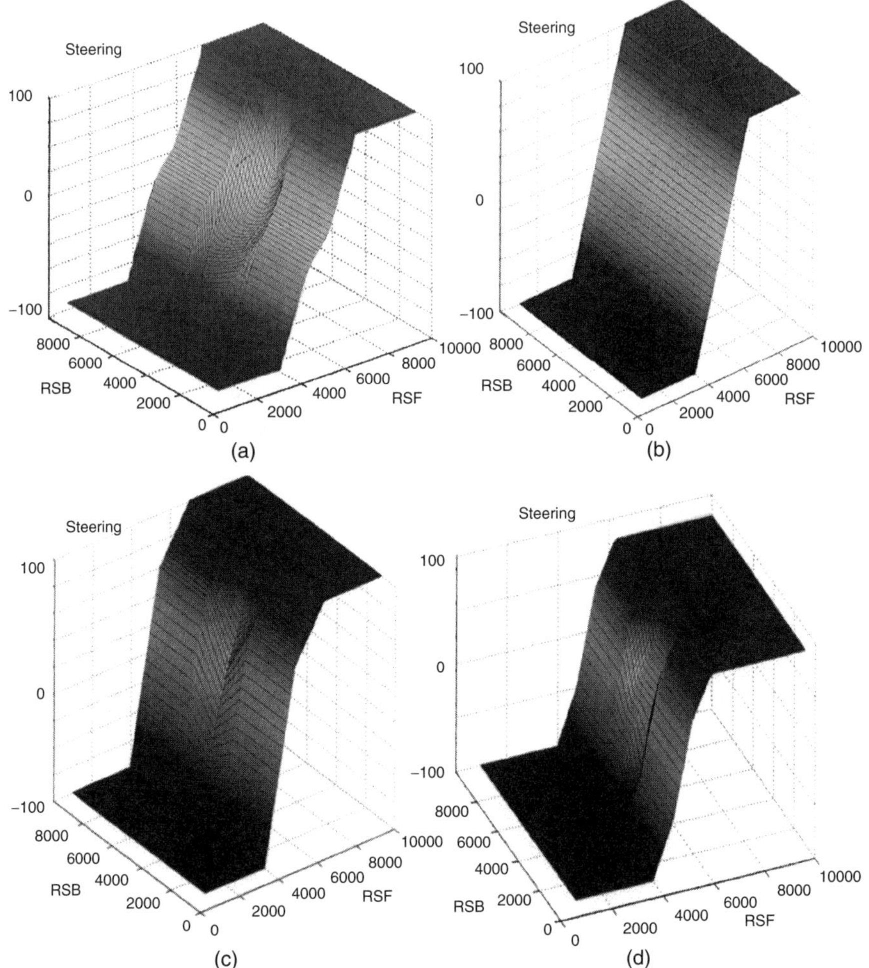

Figure 1.7 (a) Control surface of a robot T2 FLC with 4 rules, (b) control surface of a robot T1 FLC with 4 rules, (c) control surface of a robot T1 FLC with 9 rules, and (d) control surface of a robot T1 FLC with 25 rules (Hagras, 2004; © 2004, IEEE).

Because the MFs of a T2 FS are fuzzy, that is, have an FOU (and secondary MFs for a general T2 FS), they have more design degrees of freedom; hence, they have a greater *potential* to better model and handle all of the uncertainties just described in comparison to T1 FSs. Consequently, an FLC that is based on T2 FSs has the *potential* to produce better performance than a T1 FLC when dealing with such uncertainties. Observe that we have twice put emphasis on the word "potential." We have done this so as not to fool the reader into believing that a T2 fuzzy logic control system will always outperform a T1 fuzzy logic control system. The later chapters in this book will examine and compare the relative performances of both

T1 and T2 fuzzy logic control systems so that we may all better understand when or if a T2 fuzzy logic control system will outperform a T1 fuzzy logic control system.

As a preview to what will be demonstrated in those chapters, we note the following from Hagras (2004), Hagras (2007), and Wu (2012):

1. Using T2 FSs to represent the FLC inputs and outputs can lead to a smaller FLC rule base because MF uncertainties, represented by the FOUs of T2 FSs, let the T2 MFs cover the same range as T1 FSs, but with a smaller number of terms. This *rule reduction* (at the expense of more complicated MFs) increases as the number of FLC inputs increases.
2. A T2 FLC may give a *smoother control surface* than its T1 counterpart, especially in the region around the steady state [for a proportional–integral (PI) controller this means both the error and the change of error approach zero]. For example, Wu and Tan (2010) have shown that when a baseline T1 FLC implements a linear PI control law and the IT2 FSs of an IT2 FLC are obtained from symmetrical perturbations of the respective T1 FSs, the resulting IT2 FLC implements a variable gain PI controller around the steady state. These gains are smaller than the PI gains of the baseline T1 FLC, which results in a smoother control surface around the steady state. The PI gains of the IT2 FLC also change with the inputs, something that cannot be achieved by the baseline T1 FLC.
3. Type-2 FLCs *may realize more complex input–output relationships* than T1 FLCs. Karnik et al. (1999) pointed out that an IT2 fuzzy logic system can be thought of as a collection of many different embedded T1 fuzzy logic systems (as mentioned above). Additionally, Wu and Tan (2005) proposed a systematic method to identify the equivalent generalized T1 FSs that can be used to replace the FOU. They showed that the equivalent generalized T1 FSs are significantly different from traditional T1 FSs, and there are different equivalent generalized T1 FSs for different inputs. Du and Ying (2010) and Nie and Tan (2010) also showed that a symmetrical IT2 fuzzy PI [or the corresponding proportional–derivative (PD)] controller, obtained from a baseline T1 PI FLC, partitions the input domain into many small regions, and in each region the IT2 fuzzy PI controller is equivalent to a nonlinear PI controller with variable gains. The control law of the IT2 FLC in each small region is much more complex than that of the baseline T1 FLC, and hence it can realize more complex input–output relationships that cannot be achieved by a T1 FLC using the same rule base.
4. Type-2 FLCs have a *novelty* that does not exist in traditional T1 FLCs. Wu (2011) showed that in an IT2 FLC different membership grades from the same IT2 FS can be used in different rules (due to an IT2 FS being described by lower and upper MFs), whereas for a traditional T1 FLC the same membership grade from the same T1 FS is always used in the different rules. This further supports item 3, that an IT2 FLC can realize more complex input–output relationships than a T1 FLC, and that an IT2 FLC cannot be implemented by a T1 FLC using the same set of rules.

Figure 1.7, which shows control surfaces for an outdoor mobile robot, demonstrates how a T2 FLC with a rule base of only four rules (Fig. 1.7a) can produce a smoother control surface than its T1 counterparts that use a rule base of 4 (Fig. 1.7b), 9 (Fig. 1.7c), and 25 rules (Fig. 1.7d), respectively (Hagras, 2004). Observe, also, that as the T1 FLC rule base increases, its response approaches that of the T2 FLC because the latter includes a multitude of embedded type-1 FLCs.

1.8 REAL-WORLD APPLICATIONS OF IT2 MAMDANI FLCs

The last 10 years have witnessed a continuous increase in the deployment of IT2 Mamdani FLCs to real-world control problems. This trend promises to replicate the widespread use of type-1 FLCs to applications that touch the lives of people all over the world. The following subsections provide a brief *overview* of some of recent IT2 Mamdani FLCs for real-world control applications that are grouped into high-level application areas. We want to emphasize that all of the reported results are for specific systems and that we do not claim they apply universally. They are meant to whet the curiosity of the reader about potential performance improvements of IT2 FLC over T1 FLC, so as to encourage him or her to read the rest of this book.

1.8.1 Applications to Industrial Control

1.8.1.1 Speed Control of Marine Diesel Engines The first heavy-industry application of IT2 Mamdani FLCs was for the speed control of marine diesel engines (Lynch et al., 2005, 2006a, 2006b). These are huge engines classified according to their speeds, as slow-speed engines, medium-speed engines, or high-speed engines.

Due to their vast size and large power output, marine diesel engines require *accurate* and *robust* speed control/governing. Accurate speed control of such engines is of critical importance because significant deviations from the speed set point can be detrimental and damaging to the engine and its respective loads. Moreover, for applications such as power generation sets, the engine speed in revolutions per minute (rpm) must be stable in relation to multiples of the generated base frequency, that is, 50 Hz frequency requires the engine to operate at 1000 rpm, 1500 rpm, and so forth; hence, significant speed deviation can cause the generation of incorrect frequencies, resulting in loss of synchronization between the generator and its associated power grid, which is very problematic for any power generation system and its coupled loads.

Robustness in speed control is required for the marine diesel engine to overcome and recover quickly from the inherent instabilities and disturbances associated with the fast and dynamic changes of the environment, as well as load and operating conditions that marine diesel engines are exposed to on an everyday basis.

The ability to provide improved speed control response for marine diesel engines is not just desirable but is a requirement of the British Standard BS5514 "Reciprocating Internal Combustion Engines: Speed Governing," which details regulations concerning the speed controller's ability to recover from load changes and disturbances in terms of settling time, overshoot, and undershoot (British Standards).

Marine diesel engines operate in highly dynamic and uncertain environments and experience vast changes in ambient temperature, fuel, humidity, and load. There are many sources of uncertainty facing speed controllers of marine diesel engines, including:

- Uncertainties associated with the change in engine operation and load conditions due to varying loads, weather and sea conditions, wind strength, hull fouling (growth of algae, sea grass, and barnacles), and vessel displacement (which is dependent on cargo). For example, the resistance (the force working against the ship propulsion) as a result of weather and sea variations can, in general, increase by as much as 100% of the total ship resistance in calm weather. Also, experience shows that hull fouling may cause an increase of up to 40% in ship resistance. An increase in ship resistance can consequently cause a drastic reduction of the ship's speed and significant vibration that can affect the engine's sensors and actuators. These uncertainties are considered to be the most dynamic and severe uncertainties that can affect both the inputs and output of the FLC and can cause serious degradation in the performance of the marine diesel engine.
- Uncertainties affecting the inputs to the controller, because sensor measurements are affected by high noise levels from various sources, such as electromagnetic and radio frequency interference, and vibration-induced triboelectric cable charges.
- Uncertainties affecting the outputs of the controller, which can be due to the change of the actuator's characteristics because of wear and tear or environmental changes, for example, worn linkages between the actuator output and the fuel pump can result in excessive friction and/or backlash causing instability in the control loop.
- Linguistic uncertainties because the meanings of the words that are used in the antecedent's and consequent's linguistic labels are inherently uncertain, since words mean different things to different engineers, which causes uncertainties when designing the FLC for marine diesel engine control.

Due to the size and cost of marine diesel engines it is important to test and verify the engine speed controllers under different operating and load conditions before their deployment on a specific engine.

Speed controllers can be tested and verified by using the testing platform depicted in Fig. 1.8. This testing platform is designed to realistically reflect the characteristics and operating conditions of the marine diesel engines and has the ability to alter speed,[1] load, inertia, and torque. It uses the real-world noisy sensors that are used by a specific marine diesel engine and has the ability to introduce the same uncertainty levels faced by that engine.

[1]The speed of a marine diesel engine is associated with the rate of fuel delivery to its cylinders, which is a function of a hydraulic servoactuator that is controlled by an electronic embedded speed controller.

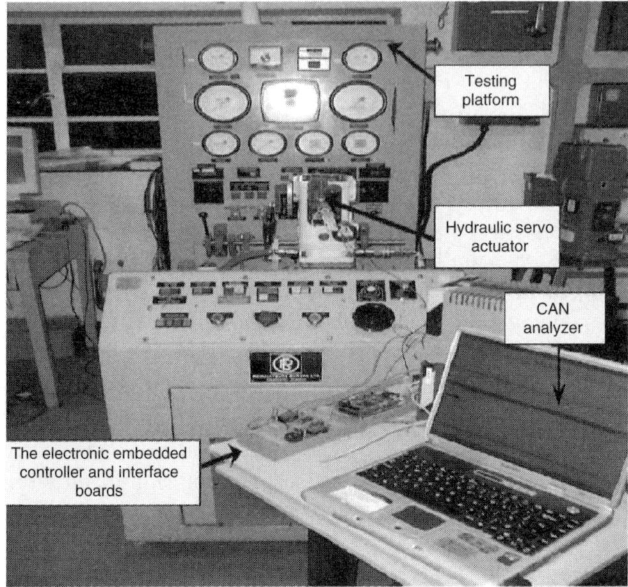

Figure 1.8 Marine diesel engine testing platform (Hagras, 2007; © 2007, IEEE).

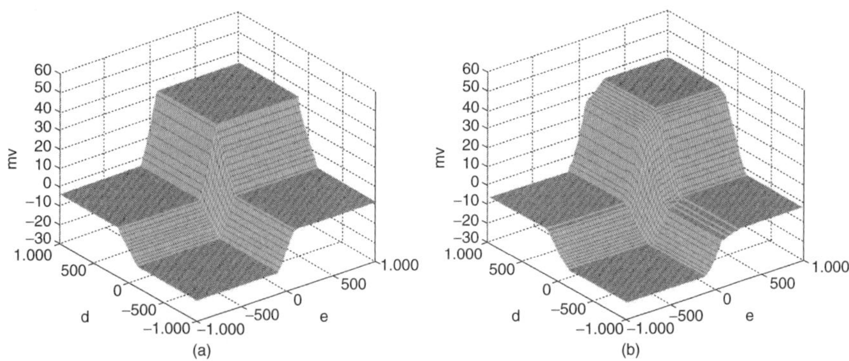

Figure 1.9 Control surfaces for (a) T1 Mamdani FLC and (b) IT2 Mamdani FLC (Hagras, 2007; © 2007, IEEE).

Figure 1.9a depicts the control surface for a T1 Mamdani FLC that was used in one of the marine diesel engine's speed controllers, and Fig. 1.9b depicts the control surface of an IT2 Mamdani FLC that was used for the same engine. Observe that the control surface for the T1 FLC is steep and nonsmooth, especially near the set point where the error (e) between the speed set point and the actual value, as well as the change of error (d), should both be equal to zero. Consequently, any small variations of e and d can cause considerable changes to the manipulated variable (mv) (i.e., the actuator controlling the fuel supply to the engine), which means that the T1 FLC is vulnerable to noise and uncertainties. Moreover, the larger the variations in e and d, due to the uncertainties, the larger are the disturbances to mv, which can cause instability and can potentially lead to the destruction of the engine.

The control surface that is depicted in Fig. 1.9b for the IT2 Mamdani FLC shows a very smooth and gradual response with no steep changes because it is (in theory) aggregating the outputs of a large number of embedded T1 FLCs. This smooth response gives very good control performance and can handle the uncertainties and disturbances that are near the set point where $e = 0$ and $d = 0$, that is, small variations in e and d do not cause significant changes to mv.

Many control experiments were performed in order to evaluate the performance of the IT2 and T1 Mamdani FLCs for handling uncertainties. The real operation of the diesel engines was mimicked where in each experiment the controllers were allowed to reach the set point and stabilize with no load, after which different loads were added suddenly to mimic the uncertainties associated with change of operation and load conditions. It is necessary for the diesel engine's speed controller to be able to deal quickly with the uncertainties associated with a change of load (for up to a 100% load addition) producing minimum overshoot/undershoot and settling times that must be in accordance with the British Standard BS5514 (British Standards).

In Lynch et al. (2006a), an IT2 Mamdani Real-Time Neuro-Fuzzy Controller[2] (RT2NFC) was developed. The performance of the RT2NFC was compared to the performances from a T1 FLC and a Viking 25 controller. The latter has been used in the past to control marine diesel engines and uses a PID algorithm with various nonlinear and gain-scheduling functions. Both the T1 and IT2 FLCs were coded in ANSI C and embedded in the industrial controller. For the engine testing platform, a set point of 905 rpm was chosen that corresponds with the requirements of medium-speed diesel engines.

All three controllers were tuned so that they could handle disturbances that were equivalent to 20% of the full load (which is a common disturbance for engines at a normal sea condition). It was noticed (not shown here) during the design process that the performances from all three controllers were very similar for the 20% load disturbance that they were designed to handle. However, as the uncertainty associated with the change of load increased to 100% load, the performance of both the Viking 25 and T1 FLC degraded significantly (see Fig. 1.10), producing large overshoots/undershoots as well as long settling times; hence, the performance of the Viking 25 and the T1 FLCs became unacceptable under these levels of uncertainties, which did not satisfy the desired standards.

A common practice in such situations is to retune the controller, which is a time-consuming process. The IT2 Mamdani FLC effectively handled the uncertainties associated with the change of the load and operating conditions to give a very good performance with small overshoots/undershoots as well as short settling times (see Fig. 1.10). The performance of the IT2 Mamdani FLC satisfied the required standards and required no further tuning. Therefore, the IT2 Mamdani

[2]A *neuro-fuzzy controller* is an FLC whose MF parameters are optimized using a tuning algorithm such as the back-propagation algorithm that is commonly used to tune the weights of a neural network.

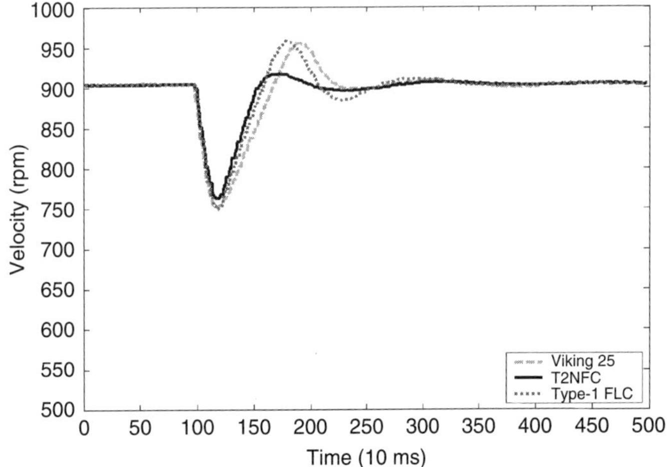

Figure 1.10 Comparison of the responses of the T1 FLC and Viking 25 against a T2NFC with 100% load addition (Lynch et al., 2006b; © 2006, IEEE).

FLC could be used effectively to produce accurate and robust speed controllers for marine diesel engines.

1.8.1.2 Liquid-Level Process Control In Wu and Tan (2004), a genetic algorithm[3] was used to design an IT2 Mamdani FLC to control a liquid-level process. The controlled process is the coupled tank apparatus depicted in Fig. 1.11a, which consists of two small tower-type tanks mounted above a reservoir that stores water that is pumped into the top of each tank by two independent pumps. The level of water in each tank is measured using a capacitive-type probe sensor, and each tank is outfitted with an outlet at the side near its base. Raising the baffle between the two tanks allows for water to flow between them. The amount of water that returns to the reservoir is approximately proportional to the square root of the height of the water column in the tank, and this is the main source of nonlinearity in this coupled-tank system. The volumetric flow rate of the pumps in the coupled-tank apparatus is nonlinear, and the system has nonzero transport delay.

It was observed (not shown here) that both the T1 and IT2 FLCs were able to attenuate oscillations when the modeling uncertainties were small. The liquid level in a tank eventually reached the desired set point, although the settling time was shorter when the IT2 FLC was used.

When, however, modeling uncertainties became larger, the T1 FLC gave rise to persistent oscillations (see Fig. 1.11b), whereas the IT2 FLC was able to eliminate these oscillations and the liquid level reached its desired height at steady state. Wu

[3]A *genetic algorithm* is a biologically inspired optimization algorithm that is used for tuning the MF parameters of the FLC as well as many other kinds of systems such as a neural network. See Section 3.6.2 for more details.

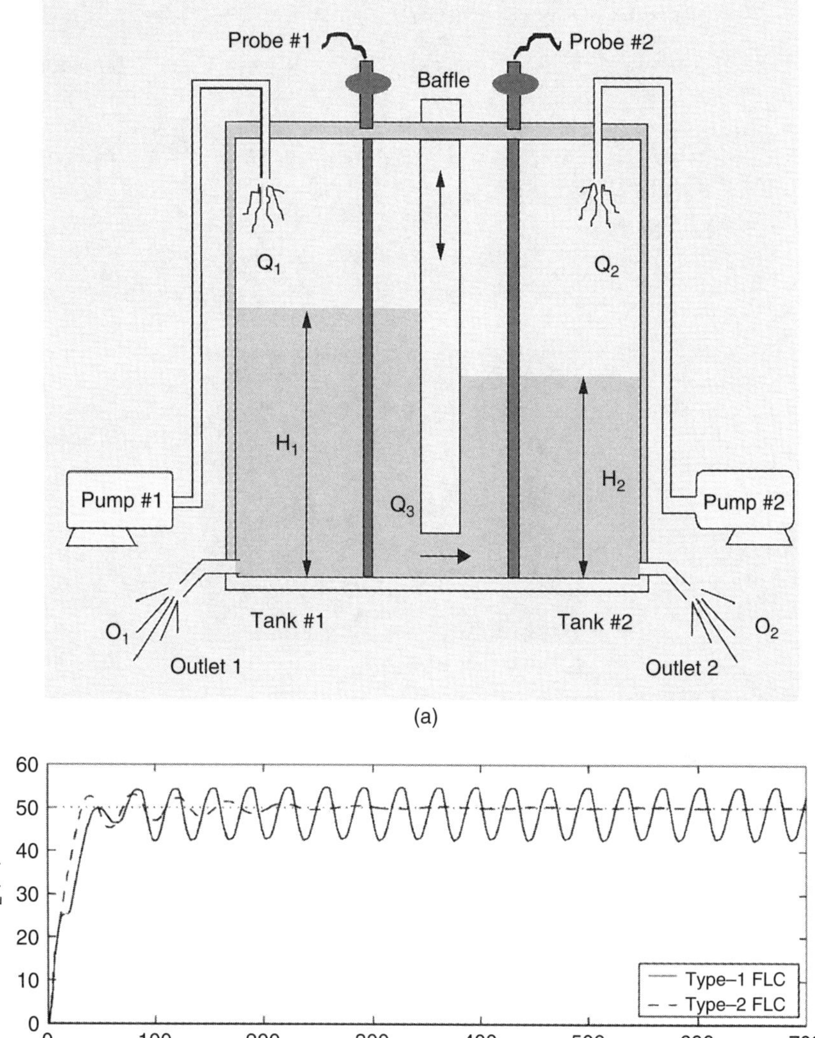

Figure 1.11 (a) Coupled-tank liquid-level control system and (b) T1 FLC (solid line) and IT2 FLC (dashed line) responses (Wu and Tan, 2004; © 2004, IEEE).

and Tan (2004) concluded that the IT2 FLC is more robust than the T1 FLC because the IT2 FLC outperformed its T1 FLC counterpart, especially when the uncertainty was large.

1.8.1.3 Control of Entry Temperature of a Steel Hot Strip Mill Mendez et al. (2010) applied a Mamdani IT2 FLC to control the coiling entry temperature of a steel hot strip mill (HSM). Figure 1.12a depicts an overview of an HSM from its

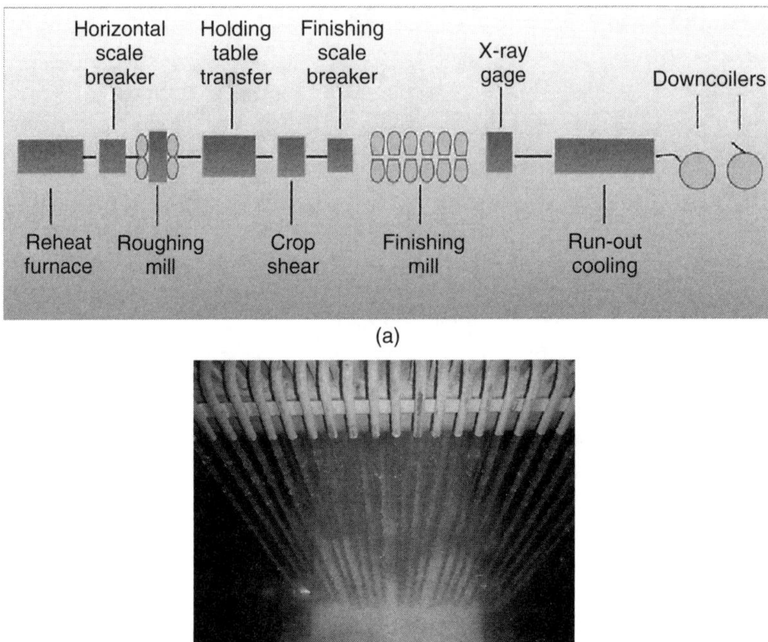

Figure 1.12 (a) Overview of a hot strip mill and (b) photo of a laminar cooling header at run-out table (Mendez et al., 2010).

initial stage at the reheat furnace entry to the final stage at the coiler side. In HSM there is a major need to satisfy quality requirements, for example, steel strip thickness, finishing temperature, and coiler temperature (the latter determines the final strip's mechanical properties). The most critical section of the coil is the head-end section due to the uncertainties involved at the head end of the incoming steel bar and the varying conditions from bar to bar.

As of 2010, in order to achieve head-end quality requirements, automation systems based on physical modeling were used, particularly for the reheat furnace, roughing mill (RM), finishing mill (FM), and the run-out cooling zone. As the market became more competitive, there was a need for flexible manufacturing capable of rolling a wider range of products in shorter periods of time. Such flexibility requirements yield higher time-varying conditions for the rolling process, thereby demanding automation systems that are better able to handle the encountered uncertainties. Most commercial systems employ proportional or proportional–integral controllers, which only compensate for the errors under current conditions; hence, the first batch in a given production cycle is usually below the given specifications.

A slab generally leaves the furnace at ~1200°C and is transported to the roughing mill by the transfer table. After several passes, the roughing stands adjust the slab thickness from ~200 to ~28 mm. The product from the roughing mill is called the *transfer bar*. The transfer bar is taken to the finishing mill where the finishing temperature and final width specifications have to be fulfilled. During the time the transfer bar travels from the roughing mill to the finishing mill scale forms on its surface. The scale breaker washes out the scale in order to allow proper rolling of the bar. Figure 1.12b shows a photograph of a top strip laminar cooling header. There are 34 top cooling headers divided into 6 sections of top header control. In addition, there are 27 bottom cooling headers divided into 3 sections of bottom spray control, giving 9 control sprays.

Strip resistance, and therefore force and gap setup, depend greatly on the strip temperature of the incoming bar, which is also essential for the speed setup, since strip temperature of the incoming bar depends on the entry bar thread speed, and the former is required to achieve both the specified finishing mill exit target head gauge and temperature. However, the bar surface temperature measurement at the scale breaker entry is not reliable due to scale formation and is therefore measured using a pyrometer located at the roughing mill exit side. Later, the head-end bar scale breaker entry temperature is estimated and used for the finishing mill and run-out cooling setup. The measurement at the roughing mill exit is affected by noise produced by transfer bar scale growth, environmental water steam, pyrometer location, calibration, resolution, and repeatability.

Experiments and results presented in Mendez et al. (2010) show that IT2 FLCs are able to model and control the cooling water flow to achieve the target coiler entry temperature in an HSM. They show that there is a substantial improvement in performance and stability of an IT2 Mamdani FLC over a T1 Mamdani FLC (e.g., Fig. 1.13). As can be seen from this figure, the IT2 FLC converged under real production conditions and had better performance in terms of the root-mean-square error (RMSE) than the T1 FLC. These results show the feasibility of the IT2 FLC for this particular industrial application.

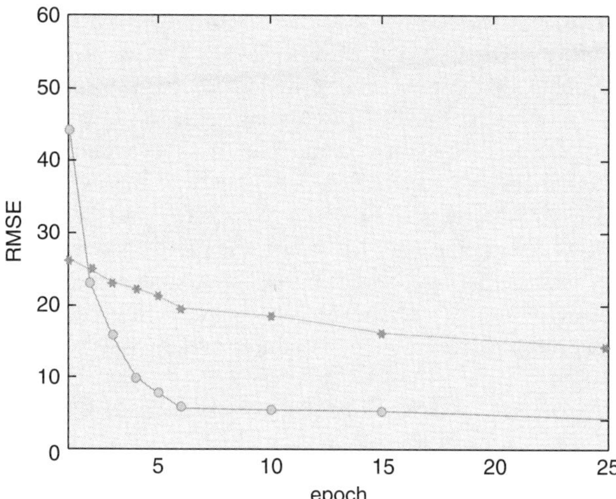

Figure 1.13 Root-mean-squared errors (RMSEs) for type-A cooling coil: (*) T1 FLC and (•) IT2 FLC models (Mendez et al., 2010).

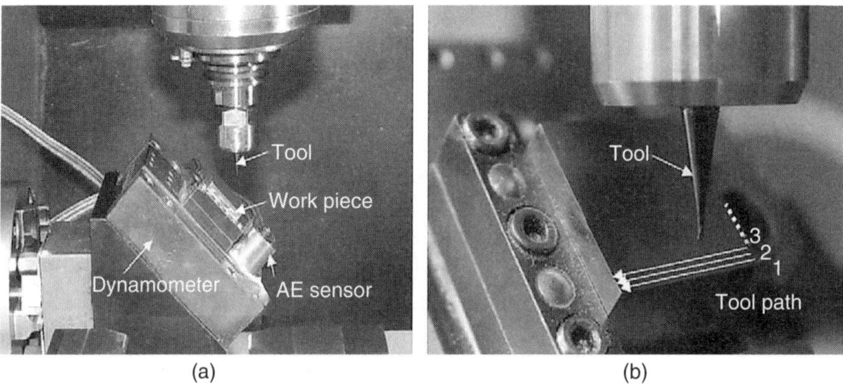

Figure 1.14 High-precision milling setup at Mondragón University (Spain). (a) Side view and (b) front view (Ren et al., 2010; © 2010, IEEE).

1.8.1.4 Modeling of Micromilling Cutting Forces Ren et al. (2010) designed an IT2 Mamdani FLC for the estimation of dynamic micromilling cutting forces. The resulting system was tested at the Micro-machining Laboratory at the Mondragón University in Spain. Figure 1.14 shows the actual setup. Researchers there noted that type-2 fuzzy estimation not only filters the noise and estimates the instantaneous cutting force in micromilling using observations acquired by sensors during cutting experiments but also assesses the uncertainties associated with the prediction caused by the manufacturing errors and signal processing. Moreover, the interval output of the type-2 fuzzy system gives very useful information to machine tool controllers in order to maximize material removal while controlling tool wear or tool failure to maintain part quality specifications.

1.8.1.5 Thyristor-Controlled Series Capacitor to Improve Power System Stability Tripathy and Mishra (2011) applied a Mamdani IT2 FLC to a thyristor-controlled series capacitor (TCSC) for improving power system stability. They report that the IT2 FLC along with the power system stabilizer (PSS) in the system satisfactorily damp out the speed and power oscillations following different critical faults. They show that the damping performance of the IT2 FLC is considerably better compared to its fixed gain bacteria-swarm-based tuned PSS and TCSC counterpart. Moreover, the performance of the IT2 FLC did not deteriorate even under uncertainty in the input signal to the controller, which shows the power of the IT2 Mamdani FLC in providing adequate performance even under conditions of increased uncertainty (in the inputs).

1.8.1.6 Control of Buck Direct-Current–Direct-Current (DC–DC) Convertors Lin et al. (2005) applied an IT2 Mamdani FLC to the control of buck DC–DC converters, which are nonlinear power electronic systems that convert one level of electrical voltage into another level by a switching action. They are used extensively in personal computers, computer peripherals, and adapters of consumer electronic devices.

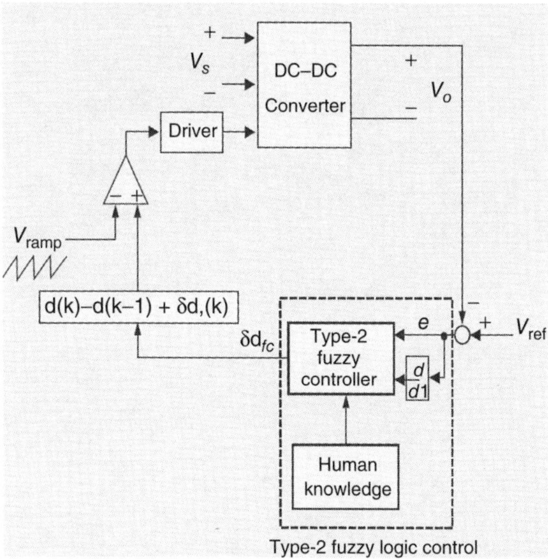

Figure 1.15 Block diagram of an IT2 FLC DC–DC converter system (Lin et al., 2005; © 2005, IEEE).

A control technique for DC–DC converters must cope with their wide input voltage and load variations to ensure stability in any operating condition while providing fast transient response. The control problem is to control the duty cycle so that the output voltage can supply a fixed voltage in the presence of input voltage uncertainty and load variations.

A block diagram of the IT2 Mamdani FLC DC–DC converter system is depicted in Fig. 1.15. Lin et al. (2005) have shown that the performance of an IT2 Mamdani FLC is better than its T1 counterpart, namely the rise time response of the IT2 Mamdani FLC is faster than that of T1 FLC and the former has no overshoot.

1.8.2 Airplane Altitude Control

Zaheer and Kim (2011) applied an IT2 Mamdani FLC to airplane altitude control for a propulsion-based airplane as shown in Fig. 1.16a. The throttle is used to regulate the speed of the airplane by varying the rotational speed of the propeller, the elevator is used to control the airplane's ascent and descent, the ailerons are used for airplane's lateral stabilization and midair turning, and the rudder is used for the on-ground taxiing of the airplane. They compared T1 and IT2 Mamdani FLCs for airplane control, and found that under high uncertainty levels, the IT2 Mamdani FLC outperformed the T1 FLC, namely that the T1 FLC showed oscillatory behavior around the reference altitude set points as shown in Figs. 1.16b and 1.16c.

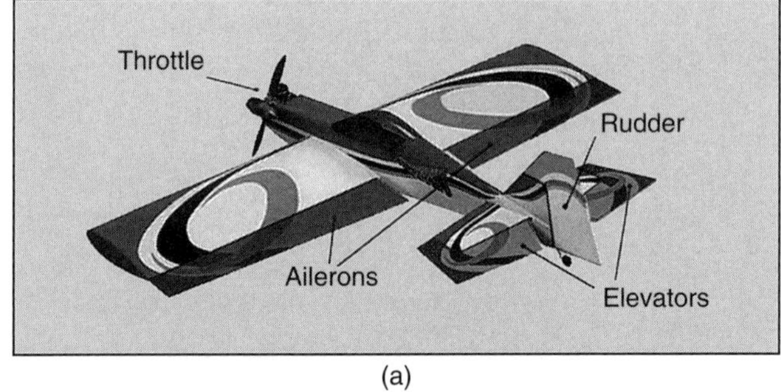

(a)

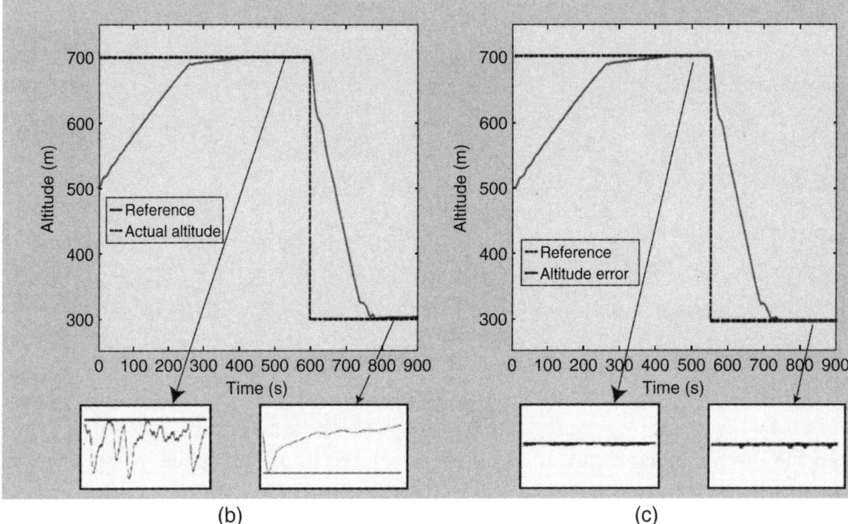

(b) (c)

Figure 1.16 (a) Basic airplane control; (b) results of the T1 FLC in the simulation setup with uncertainties [bottom blocks are the magnified steady-state responses (RMSE = 3.58 m)]; and (c) results of IT2 Mamdani FLC in the simulation setup with uncertainties [bottom blocks are the magnified steady-state responses (RMSE = 0.43 m)] (Zaheer and Kim, 2011; © 2011, IEEE).

1.8.3 Control of Mobile Robots

Autonomous mobile robots navigating in real-world unstructured environments must be able to operate under conditions of imprecision and uncertainties present in such environments, where the uncertainties can be in the form of numerical uncertainties[4] (that affect the inputs and/or outputs of the controller). The numerical uncertainties associated with changing unstructured environments cause

[4]Numerical uncertainties refer to noise and change of the sensor signal due to change of operating conditions, for example, an ultrasound sensor assumes that the speed of sound is constant, however, the speed of sound varies with wind, rain, humidity, and the like, so a sonar sensor at a distance of 1 m will read different readings in wind, rain, and so forth.

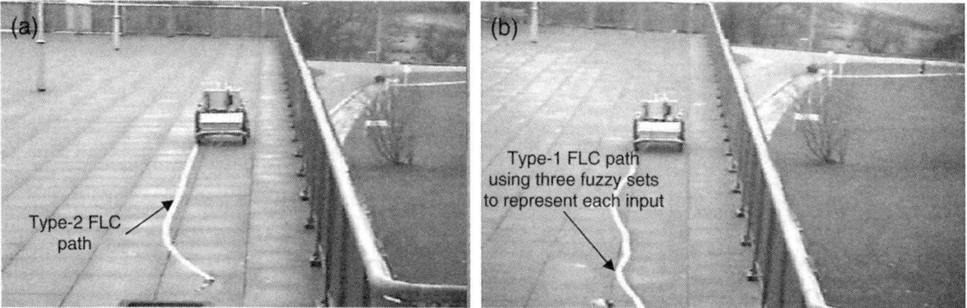

Figure 1.17 (a) Outdoor robot path using IT2 Mamdani FLC to implement the right edge following behavior to follow an irregular edge and (b) robot path using a T1 Mamdani FLC, which gave a poor response when the environment changed (windy weather) (Hagras, 2004; © 2004, IEEE).

problems in determining the exact and precise antecedents' and consequents' membership functions during the FLC design. The designed T1 fuzzy sets can be suboptimal for specific environment and robot conditions; however, as the robot operating conditions change from the design conditions, the T1 fuzzy sets will not be optimal any more, which can cause degradation in the mobile robot FLC performance. Hagras (2004) employed an IT2 Mamdani FLC for mobile robot control involving indoor and outdoor robots and found that the IT2 FLC always outperformed its T1 counterpart, and it also used a smaller number of rules. The former was demonstrated by examining robot paths and control surfaces (see Fig. 1.7). For the robot shown in Fig. 1.17a, the control surface of the IT2 Mamdani FLC has a smooth shape, which translated into a smooth control response that was able to deal effectively with uncertainty and imprecision. By means of control surface analyses, the more T1 fuzzy sets were used in the T1 FLC the more its response approached the smooth response of the IT2 Mamdani FLC (see Figs. 1.7b–1.7d). This is because the T2 fuzzy sets contain a large number of embedded T1 fuzzy sets, which allow for the detailed description of the control surface.

Hagras (2004) also performed experiments with robots in outdoor unstructured environments in order to evaluate the real-time performance of the robot IT2 FLC so as to see how they could handle large amounts of uncertainty and imprecision, as is present in such changing and dynamic environments. The robots were tested under different environmental conditions (e.g., rain, wind, sunshine), different ground conditions (e.g., slippery and dry ground), and at different times of the day. These experiments also involved using different challenging environmental features such as metallic and vegetation edges, which result in poor responses (i.e., echo) from the ultrasound sensor. They observed that the T1 FLC gave a good response under specific weather, ground, and robot conditions, but if any of these conditions changed, for example, when operating in windy weather conditions, then a nine-rule T1 FLC controlling the robot (see Fig. 1.17b) gave a poor oscillatory response because it could not handle the uncertainties associated with the outdoor environment conditions. On the other hand, they observed that the IT2 Mamdani FLC controlling the

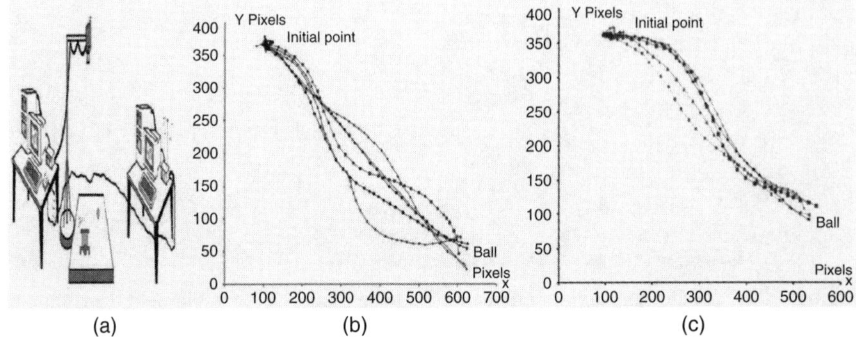

Figure 1.18 (a) Typical robot soccer platform, (b) player paths when a T1 FLC was used, and (c) player paths when an IT2 Mamdani FLC was used (Figueroa et al., 2005; © 2005, IEEE).

robot (see Fig. 1.17a) could handle such uncertainties and gave a better response while also using a smaller rule base.

Figueroa et al. (2005) described an IT2 Mamdani FLC for a robotic agent that tracks a mobile object in the context of robot soccer games, where the robotic agent has to track a ball accurately. In this application, the final goal of a player is to reach the position of the ball.

In robotic soccer games, positions of players and balls are captured through image processing because it is simple to do this. The basic configuration of a typical platform for robotic soccer games is shown in Fig. 1.18a; it comprises a football pitch (ground plane), a camera for image capture, one or two computers (server and client), and an radio frequency (RF) data transmitter.

Type-1 FLCs have been used in the past to control players; however, such FLCs face many sources of uncertainty, which include image processing algorithms (that cause uncertainties in the FLC inputs) as well as uncertainties in the actuators and networking resources. Hence, Figueroa et al. (2005) applied an IT2 Mamdani FLC to this problem and conducted two tests in order to evaluate the performance of the IT2 Mamdani FLC against its T1 counterpart.

The first test is called a *static ball test* and is one in which the way a "player" reaches the position of the ball is observed. During this test, the ball is positioned at a fixed point, for example, at one of the corners of the ground plane, and a player starts his movement from another point, usually the farthest corner. Figures 1.18b and 1.18c depict five static ball tests using the T1 and IT2 Mamdani FLCs, respectively. Observe that for both kinds of controllers the players' paths are always different (due to uncertainties); however, for the IT2 Mamdani FLC, the player only makes two corrections to reach the ball, whereas for the T1 FLC the player makes three corrections in order to reach the ball. Observe also that the paths followed by the T1 player have larger deviations than those of the T2 player, and that the shapes of those paths varied drastically. On the other hand, the paths followed by the T2 player were more regular. The control surface for the Mamdani IT2 FLC (not shown here) indicated that noisy sensors did not produce significant changes in a player's direction; however, for the T1 FLC, small variations in both the error and change of error produced a considerable change in direction, indicating that the T1 FLC was vulnerable to noise artefacts.

The second test is called a *mobile ball test* and is one in which the ball moves according to a defined trajectory and the player tries to track it. Figueroa et al. (2005) showed that, in all tests, the IT2 Mamdani FLC preserved a smaller average distance between the player and the moving ball. Additionally, they showed that the associated standard deviation was smaller for the IT2 Mamdani FLC than it was for the T1 FLC, which means that the paths followed by the IT2 player were closer to the ball's parabolic trajectory. They concluded, finally, that the IT2 Mamdani FLC was able to cope with uncertainties in a better way than the T1 FLC counterpart and also noted that the IT2 Mamdani FLC used a smaller rule base.

1.8.4 Control of Ambient Intelligent Environments

Ambient intelligence (AmI) provides basic criteria for the development of *ambient intelligent environments* (AIEs) in which intelligent computation that is enabled through simple and intuitive interactions with a user is invisibly embedded into the user's surrounding environments. The user is, therefore, empowered through a digital environment that is aware of her/his context and is sensitive, adaptive, and responsive to her/his needs in an unobtrusive manner.

Ambient intelligent environments rely on ubiquitous computing technologies that implement modular, low-powered devices and distributed high-bandwidth heterogeneous networks of sensors and actuators. They require distributed intelligence that uses modular units of intelligent behavior, such as intelligent agents, in order to create a pervasive distributed "layer of intelligence." Consequently, agents that are embedded in a user's environment (e.g., home, work, car, etc.) provide an intelligent "presence" by being able to recognize the user (or users) and autonomously program themselves to the users' needs by learning from their behaviors. The intelligence mechanisms employed within the agents must have low computational overheads, allowing them to be embedded into small hardware platforms or everyday consumer appliances. It is also important that these intelligent approaches provide their learned decisions in a form that is easily interpreted and analyzed by the end users.

One of the main underlying requirements for determining the kind of intelligent approach to use in the embedded agents is the ability to manage short-term and long-term uncertainties that arise due to changes in the environmental conditions along with changes in user behavior and activities over time. The AIEs face short-term uncertainties (within short-term time intervals) such as slight noise and imprecision associated with the inputs of the FLCs, as well as slight mood changes of the user. The AIEs also face long-term uncertainties because the environmental conditions and associated user activities change over longer durations of time due to:

- Seasonal variations in environmental conditions [e.g., external light level (the difference in the position of the sun can cause a difference between the late afternoon light levels in midsummer and the late afternoon light levels in midwinter), temperature, time of day (morning, afternoon, or evening)].

- People's behavior while occupying these environments because their behaviors, moods, and activities are dynamic, often nondeterministic and are subject to change with external factors such as time and season; there is also the fact that different words mean different things at different times of the year; for example, the values associated with *warm* temperature can vary from winter to summer.
- Changes in an actuator's characteristics as a result of wear and tear that occurs over time.

Hagras et al. (2007) describe an agent's architecture for the control of AIEs that uses an IT2 Mamdani FLC and a one-pass (noniterative) method to learn the user's particular behaviors and preferences in an online nonintrusive and seamless manner. The system learned the user's behavior by learning his/her particular rules and T2 membership functions. These rules and membership functions could then be adapted incrementally in a life-long learning mode to suit the changing environmental conditions and user's preferences. They developed a T2 agent architecture suitable for the embedded platforms used in AIEs, which have limited computational and memory capacities.

The agent based on IT2 Mamdani FLC was evaluated in the Essex Intelligent Dormitory (iDorm), depicted in Fig. 1.19a. The iDorm is a multiuser inhabited space that is fitted with a plethora of embedded sensors, actuators, processors, and heterogeneous networks that are cleverly concealed (buried in the walls and underneath furniture) so that the user is unaware of the hidden intelligent infrastructure of the room. It looks and feels like an ordinary study/bedroom environment, containing a mix of furnishings such as a bed, work desk, and wardrobe, which split the room into different areas of activity such as sleeping, working, and entertaining. Any networked embedded computer that can run a standard Java process can directly access and control the devices in the iDorm. The IT2 Mamdani FLC-based agent was embedded in an Internet Fridge (iFridge) computer.

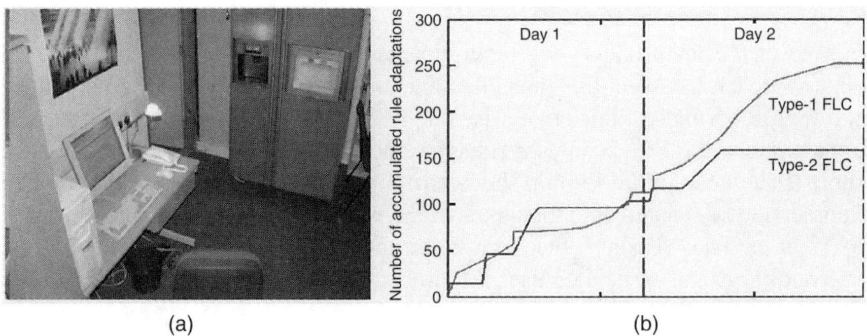

Figure 1.19 (a) iDorm and (b) number of accumulated online user adaptations (Hagras et al., 2007; © 2007, IEEE).

Experiments were conducted with various users during an extended period (spanning the course of a year) over which it was possible to evaluate and demonstrate how the agent could adapt in a life-long learning mode and could handle short- and long-term uncertainties. The agents based on IT2 Mamdani FLC were compared with T1-FLC-based agents regarding their ability to model a user's behavior while also handling long-term uncertainties. Results demonstrated that the IT2 FLC was better able to model a user's behavior and handle the short- and long-term uncertainties, and it used fewer rules than the T1 FLC.

Further experiments were conducted in the iDorm where *user satisfaction* was measured by monitoring how well the agents adjusted the iDorm environment to the user's preferences such that *user intervention* (which can be used as a measure of a user's satisfaction) was reduced over time. Figure 1.19b shows, for a two-day experiment, the number of rules that were adapted online every time the user had to override the agent's decision. Observe that agent based on the IT2 Mamdani FLC required significantly less user interaction than did the T1 agent. The curve for the T2 agent shows that user intervention initially was high but that it stabilized on the second day; therefore, the T2 agent only required the very short online tuning period of approximately one day. This is because the T2 agent better modeled user behavior and handled the short- and long-term uncertainties better than did the T1 agent. The curve for the T2 agent also shows it to be more stable (i.e., flat and not increasing with time) than the T1 agent in controlling the environment between the points when the user had to intervene in the agent's decisions to adapt the rules, that is, the curve for the T1 agent shows that user intervention continues to increase and does not properly stabilize by the end of the second day.

In conclusion, Hagras et al. (2007) show that T2 agents can adapt to user behaviors and that they generated fewer rules as compared with T1 agents. Fewer rules led to faster processing and more efficient memory usage. More specifically, the T2 agent was able to outperform the T1 agent achieving a 60% increase in processing speed as a result of a 50% reduction in the size of the rule base, thus reducing memory usage.

1.9 BOOK RATIONALE

Fuzzy control using familiar T1 FSs and logic has been extensively studied and applied to practical problems since 1974 and is considered a matured field. As mentioned above, fuzzy logic control relying on T2 FSs has now gained the attention of the fuzzy systems community, and the number of publications about it is growing rapidly.

Because of a lack of basic calculation methods in the early days of T2 FSs and logic, T2 FLCs have not emerged in popularity until recently. Now, T2 calculations can be done in real time.

As an emerging field, many different aspects of T2 fuzzy logic control need to be investigated in order to advance this new and powerful technology. This is the first book to bring together some of the latest developments on T2 fuzzy logic control in one place, so that interested researchers and practitioners can participate in this field. This book can be used to quickly understand the fundamentals of T2 fuzzy logic control and the latest theoretical developments about some important aspects of this new technology.

The central themes of any control methodology, fuzzy or conventional, are analysis and design. Analysis includes (1) describing the mathematical structure of T2 FLCs, (2) examining the stability of T2 fuzzy logic control systems, and (3) studying the robustness of T2 fuzzy logic control systems. Design means designing a T2 FLC (Mamdani or TSK) to control a given system to achieve user-desired performance, including stability. This book focuses on both topics for T2 FLCs and T2 fuzzy logic control systems, and also explains and demonstrates how to apply T2 fuzzy logic control to some important applications.

1.10 SOFTWARE AND HOW IT CAN BE ACCESSED

Software for Examples 4.1 and 4.6 and the examples in Chapter 6 can be accessed at http://booksupport.wiley.com/, and software for Appendix A, that supports T1, IT2 and GT2 FLCs, is available at http://juzzy.wagnerweb.net.

1.11 COVERAGE OF THE OTHER CHAPTERS

Chapter 2 provides background materials about IT2 FSs that are used in the rest of the book. To begin, T1 FSs are reviewed because T2 FSs build upon T1 FSs. Then a lot of information about interval T2 FSs is covered because this is needed in the rest of this book. Finally, general T2 FSs are introduced because such sets are the wave of the future.

Chapter 3 provides short reviews of T1 Mamdani and TSK FLCs so as to set the stage for the complete descriptions of IT2 Mamdani and TSK FLCs. These important IT2 FLCs are then developed in great detail, but using only T1 mathematics. The Wu–Mendel uncertainty bounds, which have let IT2 Mamdani FLCs run in real time, are stated; however, their derivations are included in Appendix 3A for completeness. Finally, some design methods for IT2 FLCs are described.

Chapter 4 describes techniques for rigorously deriving the precise mathematical relationships between the input and output of a variety of IT2 Mamdani and TSK FLCs. This is a relatively young area that started a few years ago. Some of the T2 FLCs are of the PI or PD type, and their derived relationships reveal them to be nonlinear variable PI or PD controllers that have variable proportional gain and integral gain (or derivative gain) plus variable control offset. Since many T1 fuzzy PI and PD controllers are already known to possess such structures, the structural characteristics of the T2 fuzzy PI controller can be (and are) compared to those of the corresponding T1 fuzzy PI controller. This chapter uses the derived relationships and structure characteristics analyses for insightfully understanding and studying the T2 FLCs and for developing their design guidelines.

Chapter 5 also focuses on the properties of IT2 proportional plus derivative (PD) and proportional plus integral (PI) FLCs. First, a class of IT2 PD/PI FLCs that has lower computational requirements, but still retains the properties previewed in Section 1.7, is introduced. The key idea is to only replace some critical T1 FSs by T2 FSs. Experimental results are presented that demonstrate the proposed simplified T2 FLC has the potential to be as robust as a conventional T2 FLC, while lowering the computational cost. Next, a methodology is presented, which is useful for theoretical studies, for deriving the analytical structure of IT2 PI/PD FLCs that have a symmetrical rule base. The methodology extends the analytical structure technique for T1 FLCs by leveraging a property of the Karnik–Mendel (KM) type reducer (which is derived and explained in Chapter 2) that constrains switch points to the locations of the consequent sets. Finally, examples are presented that illustrate how this framework may be applied to analyze IT2 FLCs.

Chapter 6 focuses on IT2 TSK FLCs. Its approach is based on rigorous mathematical analyses for both FLC analysis and design. It includes stability analysis and systematic methodologies for the design of adaptive and robust control, and introduces and provides some design approaches for practical control designs of such FLCs. Finally it includes several examples as well as an industrial application for modular and reconfigurable robotic systems.

Chapter 7 examines the future for T2 FLCs. Each of its sections has been written by one or more of the authors of this book and has a futuristic flavor.

CHAPTER 2

Introduction to Type-2 Fuzzy Sets

2.1 INTRODUCTION

This chapter provides background materials about type-2 fuzzy sets (T2 FSs) that are used in the rest of the book. To begin, a review of type-1 fuzzy sets (T1 FSs) is given in Section 2.2 because T2 FSs build upon T1 FSs. Then a lot of information about interval type-2 fuzzy sets (IT2 FSs) is given in Section 2.3 because this is needed in the rest of this book. Finally, an introduction to general type-2 fuzzy sets (GT2 FSs) is given in Section 2.4 because such sets are the wave of the future.

2.2 BRIEF REVIEW OF TYPE-1 FUZZY SETS[1]

Before discussions are given about T2 FSs, a short review is provided about T1 FSs. Doing this lets us establish common notations and definitions for T1 FSs.

2.2.1 Some Definitions

Definition 2.1 An *FS* (in this book called a *T1 FS*) A is comprised of a domain X of the real numbers (also called the *universe of discourse* of A) together with a *membership function* (MF) $\mu_A : X \rightarrow [0, 1]$. For each $x \in X$, the value of $\mu_A(x)$ is the *degree of membership*, or *membership grade*, of x in A. If $\mu_A(x) = 1$ or $\mu_A(x) = 0$ for $\forall x \in X$, then the FS A is said to be a *crisp set.*

Recall that a crisp set A can be described by listing all of its members, or by identifying the elements that are in A by specifying a condition or conditions that the elements must satisfy, or by using a zero–one MF (also called characteristic function, discrimination function, or indicator function) for A. On the other hand, a T1 FS can only be described by its MF; hence, the T1 FS A and its MF $\mu_A(x)$ are synonyms and are therefore used interchangeably, that is, $A \Longleftrightarrow \mu_A(x)$.

[1]Much of the material in this section is taken directly from Mendel and Wu (2010, Chapter 2; © 2010, IEEE).

Introduction to Type-2 Fuzzy Logic Control: Theory and Applications, First Edition.
Jerry M. Mendel, Hani Hagras, Woei-Wan Tan, William W. Melek, and Hao Ying.

When X is continuous (e.g., the real numbers), A is written as

$$A = \int_X \mu_A(x)/x \tag{2.1}$$

In this equation, the integral sign does not denote integration; it denotes the collection of all points $x \in X$ with associated MF $\mu_A(x)$. When X is discrete (e.g., the integers), it is denoted X_d, and A is written as

$$A = \sum_{X_d} \mu_A(x)/x \tag{2.2}$$

In this equation, the summation sign does not denote arithmetic addition; it denotes the collection of all points $x \in X_d$ with associated MF $\mu_A(x)$; hence, it denotes the set-theoretic operation of union. The slash in Eqs. (2.1) and (2.2) associates the elements in X or X_d with their membership grades, where $\mu_A(x) > 0$.

Sometimes a T1 FS may depend on more than a single variable in which case its MF is multivariate, for example, if the T1 FS B depends on two variables, x_1 and x_2, where $x_1 \in X_1$ and $x_2 \in X_2$, then, in general, its MF is $\mu_B(x_1, x_2)$ for $\forall (x_1, x_2) \in X_1 \times X_2$. This MF is three dimensional and can be quite complicated to establish. For more than two variables, it may be quite hopeless to establish a multivariate MF.

In this book *all multivariate MFs are assumed to be separable*, that is, $\mu_B(x_1, x_2)$ is expressed directly in terms of the univariate MFs $\mu_B(x_1)$ and $\mu_B(x_2)$, as

$$\mu_B(x_1, x_2) = \min\{\mu_B(x_1), \mu_B(x_2)\} \;\; \forall (x_1, x_2) \in X_1 \times X_2 \tag{2.3}$$

Definition 2.2 The *support* of a T1 FS A is the crisp set of all points $x \in X$ such that $\mu_A(x) > 0$. A T1 FS whose support is a single point in X with $\mu_A(x) = 1$ is called a (type-1) *fuzzy singleton.*

Definition 2.3 The *height* of a T1 FS is its maximum MF value. A *normal* T1 FS is one for which $\sup_{x \in X} \mu_A(x) = 1$, that is, its height equals 1.

Definition 2.4 A T1 FS A is *convex* if and only if $\mu_A(\lambda x_1 + (1 - \lambda)x_2) \geq \min[\mu_A(x_1), \mu_A(x_2)]$ where $x_1, x_2 \in X$ and $\lambda \in [0, 1]$ (Klir and Yuan, 1995).

This can be interpreted as (Lin and Lee, 1996): Take any two elements x_1 and x_2 in FS A; then the membership grade of all points between x_1 and x_2 must be greater than or equal to the minimum of $\mu_A(x_1)$ and $\mu_A(x_2)$. This will always occur when the MF of A is first monotonically nondecreasing and then monotonically nonincreasing.

The most commonly used shapes for MFs are triangular, trapezoidal, piecewise linear, Gaussian, and bell.

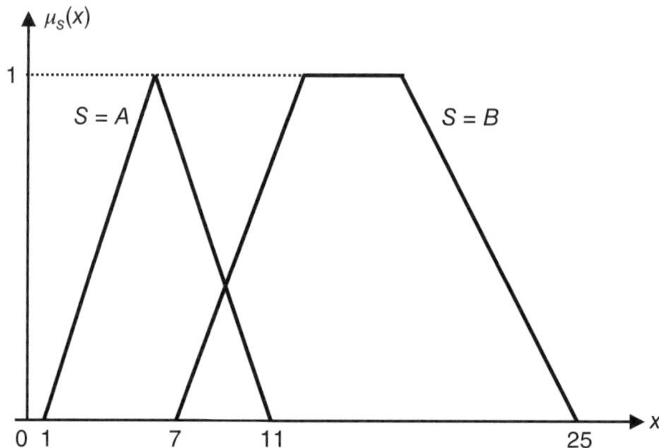

Figure 2.1 Examples of two T1 FSs, A and B (Mendel and Wu, 2010; © 2010, IEEE).

Example 2.1 Examples of triangle and trapezoidal MFs are depicted in Fig. 2.1. Observe that both of the T1 FSs are normal and convex, the support of T1 FS A is $(1, 11)$, the support of T1 FS B is $(7, 25)$, and for $x \in (7, 11)$ x resides simultaneously in both A and B but with different grades of membership.

In general, MFs can either be chosen arbitrarily, based on the preference of an individual (hence, the MFs for two individuals could be quite different depending upon their experiences, perspectives, cultures, etc.), or they can be designed using optimization procedures (e.g., Horikawa et al., 1992; Jang, 1992; Wang and Mendel, 1992a, 1992b). Much more will be said about how to choose MFs in other chapters of this book.

Definition 2.5 If a variable can take words in natural languages as its values, it is called a *linguistic variable*, where the words are characterized by FSs defined in the universe of discourse in which the variable is defined (Wang, 1997).

Let v denote the name of a *linguistic variable* (e.g., temperature, pressure, acceleration). Numerical (measured) values of a linguistic variable v are denoted x, where $x \in X$. A linguistic variable is usually decomposed into a set of terms, T, which cover its universe of discourse. This decomposition is based on syntactic rules (a grammar) for generating the terms. Examples of terms for temperature (pressure or acceleration) are very low, low, moderate, high, and very high. Each of the terms is treated as an FS and is modeled by an MF.

A more formal definition of a linguistic variable, due to Zadeh (1973, 1975), taken from Klir and Yuan (1995), follows.

Definition 2.5′ Each *linguistic variable* is fully characterized by a quintuple (v, T, X, g, m) in which v is the name of the variable, T is the set of linguistic terms of v that refer to a base variable whose values range over the universal set X, g is a syntactic rule for generating linguistic terms, and m is a semantic rule that assigns to each linguistic term $t \in T$ its meaning, $m(t)$, which is an FS on X, that is, $m: T \rightarrow F(X)$, where $F(X)$ denotes the set of all ordinary (i.e., T1) FSs of X, one such set for each $t \in T$.

2.2.2 Set-Theoretic Operations

Just as crisp sets can be combined using the union and intersection operations, so can T1 FSs; and, just as a crisp set can be complemented, so can a T1 FS.

Let T1 FSs A and B be two subsets of X that are described by their MFs $\mu_A(x)$ and $\mu_B(x)$. The *union* of A and B is described by the MF $\mu_{A \cup B}(x)$, where

$$\mu_{A \cup B}(x) = \max[\mu_A(x), \mu_B(x)] \quad \forall x \in X \tag{2.4}$$

The *intersection* of A and B is described by the MF $\mu_{A \cap B}(x)$, where

$$\mu_{A \cap B}(x) = \min[\mu_A(x), \mu_B(x)] \quad \forall x \in X \tag{2.5}$$

The *complement* of A is described by the MF $\mu_{A^c}(x)$ [also denoted $\mu_{A'}(x)$ or $\mu_{\overline{A}}(x)$], where

$$\mu_{A^c}(x) = 1 - \mu_A(x) \quad \forall x \in X \tag{2.6}$$

Although $\mu_{A \cup B}(x)$ and $\mu_{A \cap B}(x)$ can be described more generally by using t-conorms and t-norms (e.g., Klir and Yuan, 1995), in this book only the maximum t-conorm is used in Eq. (2.4), and either the minimum or product t-norm is used in Eq. (2.5) because those are the ones used in fuzzy logic control.

Example 2.2 The union and intersection of the two TI FSs A and B that are depicted in Fig. 2.1, as computed using Eqs. (2.4) and (2.5), are shown in Figs. 2.2a and 2.2b, respectively.

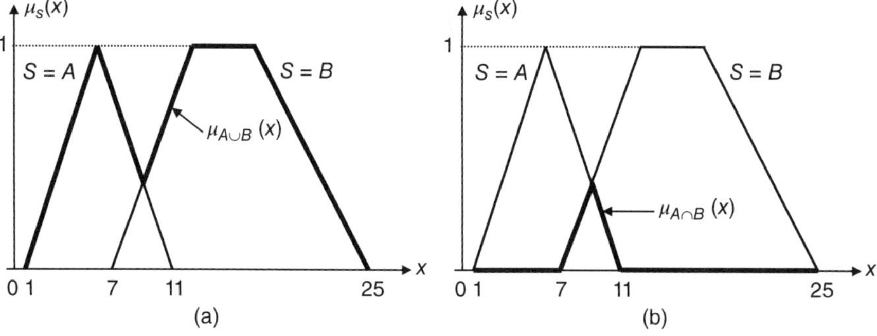

Figure 2.2 (a) Union and (b) intersection of the two T1 FSs A and B that are depicted in Fig. 2.1 (Mendel and Wu, 2010; © 2010, IEEE).

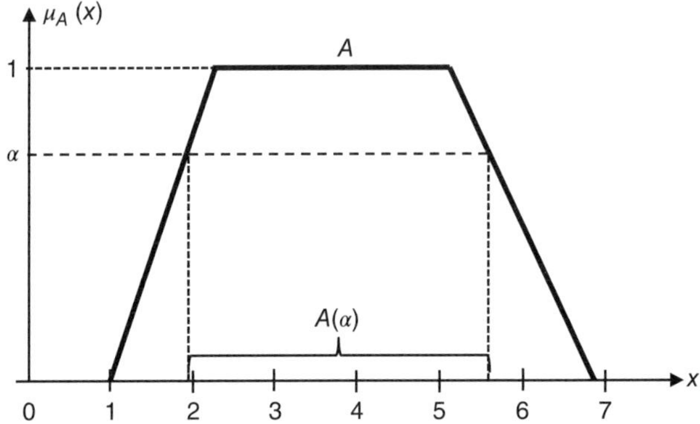

Figure 2.3 Trapezoidal T1 FS and an α-cut (Mendel and Wu, 2010; © 2010, IEEE).

2.2.3 Alpha cuts

Definition 2.6 The α-*cut* of T1 FS A, denoted $A(\alpha)$, is an interval of real numbers, defined as

$$A(\alpha) = \{x|\mu_A(x) \geq \alpha\} \tag{2.7}$$

where $\alpha \in [0, 1]$ (Klir and Yuan, 1995).

Example 2.3 An example of an α-cut is depicted in Fig. 2.3, and in this example, $A(\alpha) = [1.9, 5.5]$. Observe that the α-cut lies on the x axis.

Example 2.4 Given a specific A, it is easy to obtain formulas for the end points of the α-cut, for example, see Table 2.1. In order to obtain these formulas, such as the ones for the triangular distribution, solve the two equations $l(x) = \alpha$ for the left end point and $r(x) = \alpha$ for the right end point of $A(\alpha)$.

One of the major roles of α-cuts is their capability to represent a T1 FS. In order to do this, first the following *indicator function* is introduced:

$$I_{A(\alpha)}(x) = \begin{cases} 1 & \forall x \in A(\alpha) \\ 0 & \forall x \notin A(\alpha) \end{cases} \tag{2.8}$$

Associated with $I_{A(\alpha)}(x)$ is the following *square-well function*:

$$\mu_A(x|\alpha) \equiv \alpha I_{A(\alpha)}(x) = \alpha/A(\alpha) \tag{2.9}$$

This function, an example of which is depicted in Fig. 2.4, raises the α-cut $A(\alpha)$ off of the x axis to height α.

TABLE 2.1 Examples of T1 FSs and Their α-cut Formulas

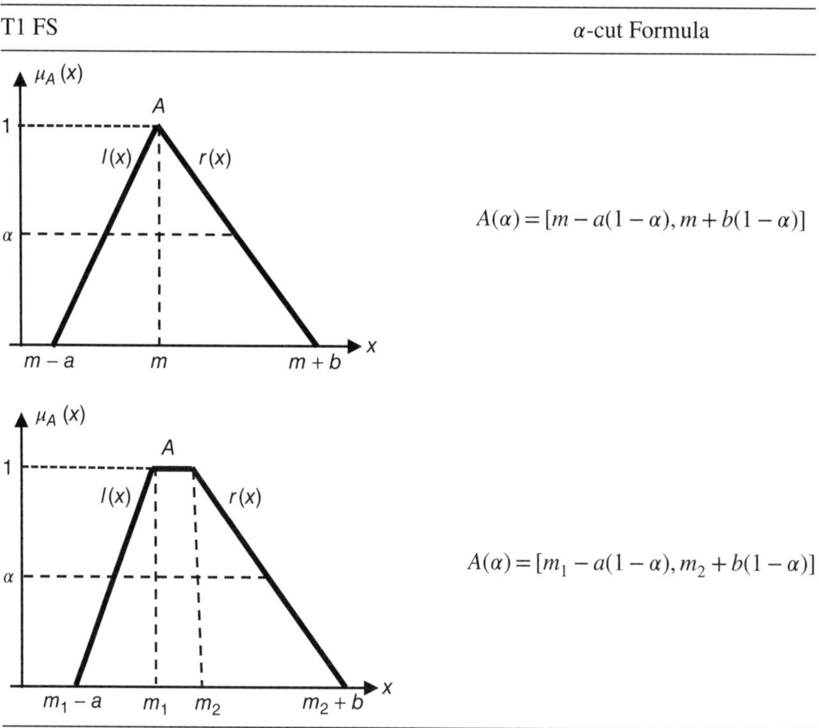

T1 FS	α-cut Formula
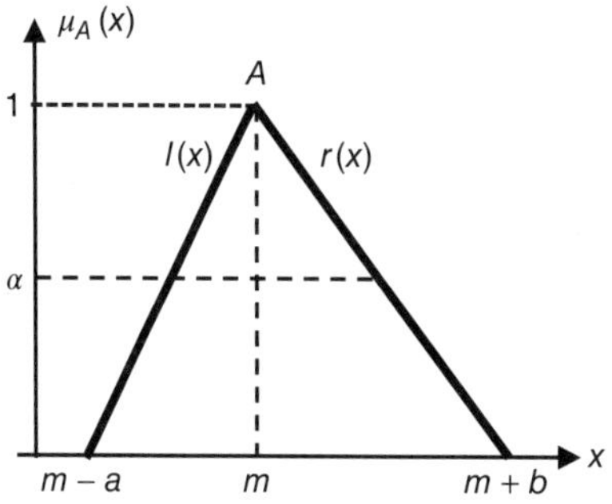	$A(\alpha) = [m - a(1-\alpha), m + b(1-\alpha)]$
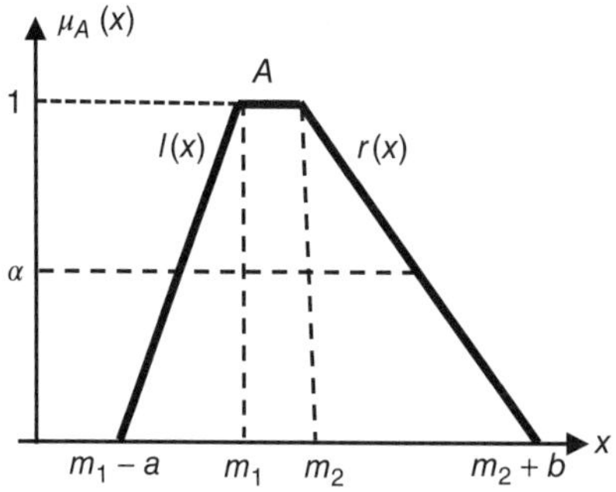	$A(\alpha) = [m_1 - a(1-\alpha), m_2 + b(1-\alpha)]$

Source: Mendel and Wu (2010; © 2010, IEEE).

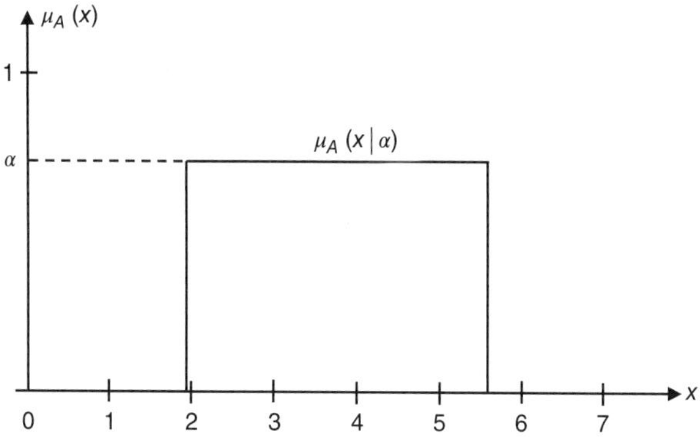

Figure 2.4 Square-well function $\mu_A(x|\alpha)$ (Mendel and Wu, 2010; © 2010, IEEE).

THEOREM 2.1 **(Decomposition Theorem)** A T1 FS A can be represented as

$$\mu_A(x) = \bigcup_{\alpha \in [0,1]} \mu_A(x|\alpha) = \sup_{\alpha \in [0,1]} \{\mu_A(x|\alpha)\} \quad \forall x \in X \tag{2.10}$$

where $\bigcup$ is the fuzzy union and $\mu_A(x|\alpha)$ is defined in Eq. (2.9) (Klir and Yuan, 1995).

This theorem is called a decomposition theorem because A is decomposed into a collection of square-well functions that are then aggregated using the union operation. Note that greater resolution is obtained by including more α-cuts, and the calculation of new α-cuts does not affect previously calculated α-cuts.

Example 2.5 Let $A = 0.2/x_1 + 0.4/x_2 + 0.6/x_3 + 0.8/x_4 + 1/x_5$. Some indicator functions for A are

$$\begin{aligned}
I_{A(0.2)}(x) &= 1/x_1 + 1/x_2 + 1/x_3 + 1/x_4 + 1/x_5 \\
I_{A(0.4)}(x) &= 0/x_1 + 1/x_2 + 1/x_3 + 1/x_4 + 1/x_5 \\
I_{A(0.6)}(x) &= 0/x_1 + 0/x_2 + 1/x_3 + 1/x_4 + 1/x_5 \\
I_{A(0.8)}(x) &= 0/x_1 + 0/x_2 + 0/x_3 + 1/x_4 + 1/x_5 \\
I_{A(1.0)}(x) &= 0/x_1 + 0/x_2 + 0/x_3 + 0/x_4 + 1/x_5
\end{aligned} \tag{2.11}$$

Their associated square-well functions are

$$\begin{aligned}
\mu_A(x|0.2) &= 0.2/x_1 + 0.2/x_2 + 0.2/x_3 + 0.2/x_4 + 0.2/x_5 \\
\mu_A(x|0.4) &= 0/x_1 + 0.4/x_2 + 0.4/x_3 + 0.4/x_4 + 0.4/x_5 \\
\mu_A(x|0.6) &= 0/x_1 + 0/x_2 + 0.6/x_3 + 0.6/x_4 + 0.6/x_5 \\
\mu_A(x|0.8) &= 0/x_1 + 0/x_2 + 0/x_3 + 0.8/x_4 + 0.8/x_5 \\
\mu_A(x|1.0) &= 0/x_1 + 0/x_2 + 0/x_3 + 0/x_4 + 1/x_5
\end{aligned} \tag{2.12}$$

Applying Eq. (2.10) to these functions, it follows that

$$A = \mu_A(x|0.2) \cup \mu_A(x|0.4) \cup \mu_A(x|0.6) \cup \mu_A(x|0.8) \cup \mu_A(x|1.0) \tag{2.13}$$

When performing these unions, we focus on a specific domain point, for example, $x = x_4$, for which

$$\mu_A(x_4) = \max(0, 0.2, 0.4, 0.6, 0.8)/x_4 = 0.8/x_4 \tag{2.14}$$

Performing these unions for the five domain points, whose MFs are nonzero, it is straightforward to recover $A = 0.2/x_1 + 0.4/x_2 + 0.6/x_3 + 0.8/x_4 + 1/x_5$.

The following properties hold for α-cuts:

$$(A \cup B)(\alpha) = A(\alpha) \cup B(\alpha)$$
$$(A \cap B)(\alpha) = A(\alpha) \cap B(\alpha) \qquad (2.15)$$
$$(A * B)(\alpha) = A(\alpha) * B(\alpha)$$

where * can be any of the four basic arithmetic operations—addition, subtraction, multiplication, and division. So the α-cut of the union equals the union of the α-cuts, and an equivalent property holds for the intersection as well as for arithmetic operations.

2.2.4 Compositions of T1 FSs

Consider the composition of fuzzy relations from different Cartesian product spaces that share a common set, namely $R(U, V)$ and $S(V, W)$, for example, pressure x is *lower* than pressure y, and pressure y is *close* to pressure z. Associated with fuzzy relation R is its membership function $\mu_R(x, y)$, where $\mu_R(x, y) \in [0, 1]$, and associated with fuzzy relation S is its membership function $\mu_S(y, z)$, where $\mu_S(y, z) \in [0, 1]$. The fuzzy composition of R and S, denoted $R \circ S$, whose MF is denoted $\mu_{R\circ S}(x, z)$, is given by the following *sup-star composition* of R and S (e.g., Wang, 1997):

$$\mu_{R\circ S}(x, z) = \sup_{y\in Y}[\mu_R(x, y) \star \mu_S(y, z)] \ \forall x \in X, z \in Z \qquad (2.16)$$

When X, Y, and Z are discrete universes of discourse, then the supremum operation is the *maximum*. The most commonly used sup-star compositions are the *sup-min* and *sup-product*.

Suppose the fuzzy relation R is just a fuzzy set, in which case $V = U$, so that $\mu_R(x, y)$ just becomes $\mu_R(x)$ [or, equivalently, $\mu_R(y)$], for example, "pressure y is *very high* and pressure y is *lower* than pressure z." In this case, the sup-star composition in Eq. (2.16) simplifies because $Y = X$, that is,

$$\sup_{y\in Y}[\mu_R(x, y) \star \mu_S(y, z)] = \sup_{x\in X}[\mu_R(x) \star \mu_S(x, z)] \ \forall z \in Z \qquad (2.17)$$

which is only a function of output variable z; hence, the notation $\mu_{R\circ S}(x, z)$ can be simplified to $\mu_{R\circ S}(z)$, so that *when R is just a fuzzy set*,

$$\mu_{R\circ S}(z) = \sup_{x\in X}[\mu_R(x) \star \mu_S(x, z)] \ \forall z \in Z \qquad (2.18)$$

Equation (2.18) plays the key role in obtaining the MF of a rule output.

2.2.5 Rules and Their MFs

A rule is the following if–then statement: "If x is A, then u is B," where $x \in X$ and $u \in U$. It has a membership function $\mu_{A \to B}(x, u)$ where $\mu_{A \to B}(x, u) \in [0, 1]$. Note that $\mu_{A \to B}(x, u)$ measures the degree of truth of the implication relation between x and u, and it resides in the Cartesian product space $X \times U$. An example of such a membership function is[2]

$$\mu_{A \to B}(x, u) = 1 - \min[\mu_A(x), 1 - \mu_B(u)] \tag{2.19}$$

In crisp logic, *modus ponens* is used as the inference rule, namely:

Premise: x is A.
Implication: If x is A, then u is B.
Consequent: u is B.

Modus ponens is associated with the implication "A implies B" ($A \to B$). In terms of propositions p and q, *modus ponens* is expressed as $(p \wedge (p \to q)) \to q$.

In fuzzy logic, *modus ponens* is extended to *generalized modus ponens*, namely:

Premise: x is A^*.
Implication: If x is A, then u is B.
Consequent: u is B^*.

In crisp logic a rule will be fired only if the premise is exactly the same as the antecedent of the rule, and the result of such rule firing is the rule's actual consequent. In FL, on the other hand, a rule is fired so long as there is a nonzero degree of similarity between the premise and the antecedent of the rule, and the result of such rule firing is a consequent that has a nonzero degree of similarity to the rule's consequent.

Generalized *modus ponens* is a fuzzy composition where the first fuzzy relation is merely the fuzzy set A^*. Consequently, using Eq. (2.18), $\mu_{B^*}(u)$ is obtained from the following sup-star composition:

$$\mu_{B^*}(u) = \sup_{x \in X} [\mu_{A^*}(x) \star \mu_{A \to B}(x, u)] \ \forall u \in U \tag{2.20}$$

This equation simplifies a lot when A^* is a fuzzy singleton, that is,

$$\mu_{A^*}(x) = \begin{cases} 1 & x = x' \\ 0 & x \neq x' \text{ and } \forall x \in X \end{cases} \tag{2.21}$$

This is called *singleton fuzzification*, and for it Eq. (2.20) becomes

[2] A proof that Eq. (2.19) gives correct results for material implication and crisp sets can be found in, for example, Mendel (1995) or Mendel (2001, Chapter 1).

$$\mu_{B^*}(u) = \sup_{x \in X}[\mu_{A^*}(x) \star \mu_{A \to B}(x, u)]$$
$$= \sup[\mu_{A \to B}(x', u), 0] = \mu_{A \to B}(x', u) \; \forall u \in U \quad (2.22)$$

regardless of whether one uses minimum or product for $\star$. Observe that for the singleton fuzzifier the supremum operation in Eq. (2.20) is very easy to evaluate because $\mu_{A^*}(x)$ is nonzero at only one point, x'.

Example 2.6 The two most popular implication MFs for fuzzy logic control are the following *minimum* and *product* implications, also known as Mamdani (1974) and Larsen (1980) implications, respectively, or many times just as Mamdani implications:

$$\mu_{A \to B}(x, u) \equiv \min[\mu_A(x), \mu_B(u)] \quad (2.23)$$

$$\mu_{A \to B}(x, u) \equiv \mu_A(x)\mu_B(u) \quad (2.24)$$

These implications have nothing to do with the material implication from propositional logic; hence, they are sometimes referred to as *engineering implications* (Mendel, 1995).

Figures 2.5 and 2.6 illustrate $\mu_{B^*}(u)$ in Eq. (2.22) for Eqs. (2.23) and (2.24), respectively, that is, when $\mu_{B^*}(u) = \min[\mu_A(x'), \mu_B(u)]$, $\forall u \in U$, and $\mu_{B^*}(u) = \mu_A(x')\mu_B(u)$, $\forall u \in U$. In these figures, the level shown for $\mu_A(x')$ was chosen arbitrarily. Observe, in Fig. 2.5, that given a specific antecedent $x = x'$ the result of firing a specific rule is a T1 FS whose support is finite and whose shape is a *clipped* version of $\mu_B(u)$. On the other hand, in Fig. 2.6, the result of firing a specific rule is a T1 FS whose support is finite but whose shape is a *scaled* (attenuated) version of $\mu_B(u)$.

So far, all of our discussions about rules have been for rules with single antecedents and consequents, for example, if x is A, then u is B. In Sections 2.4 and 2.5 and in later chapters, we characterize rules that have more than one antecedent, for example,

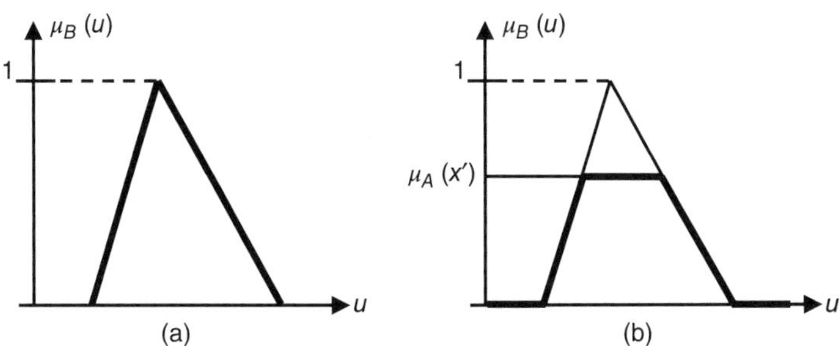

Figure 2.5 Construction of $\mu_{B^*}(u) = \min[\mu_A(x'), \mu_B(u)]$: (a) Consequent MF $\mu_B(u)$ and (b) construction of $\mu_{B^*}(u)$ (Mendel, 1995; © 1995, IEEE).

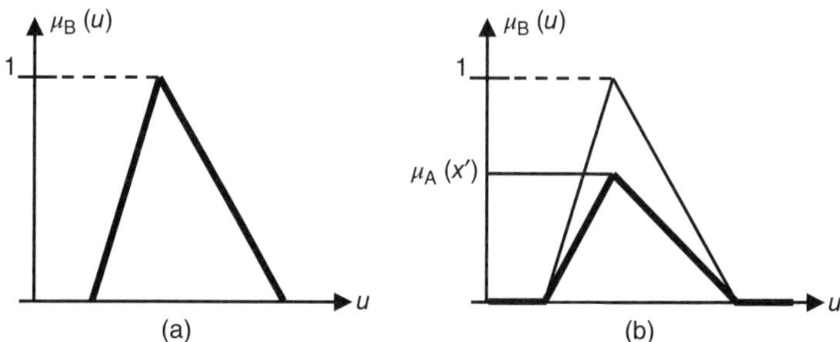

Figure 2.6 Construction of $\mu_{B^*}(u) = \mu_A(x')\mu_B(u)$: (a) Consequent MF $\mu_B(u)$ and (b) construction of $\mu_{B^*}(u)$ (Mendel, 1995; © 1995, IEEE).

$$\text{If } x_1 \text{ is } F_1 \text{ and } x_2 \text{ is } F_2 \text{ and } \dots \text{ and } x_p \text{ is } F_p, \text{ then } u \text{ is } G.$$

In such a multiple-antecedent rule, $x_1 \in X_1, \dots, x_p \in X_p$, $u \in U$, and $F_1, \dots, F_p$ and G are T1 FSs. If a rule with the same antecedents has L consequents, that is, a MIMO (multiple-input multiple-output) rule, then it can always be decomposed into L MISO (multiple-input single-output) rules, each having the same antecedent(s) but with only one consequent.

2.3 INTERVAL TYPE-2 FUZZY SETS

2.3.1 Introduction[3]

Imagine blurring[4] the T1 MF depicted in Fig. 2.7a by shifting the points on the triangle either to the left or to the right and not necessarily by the same amounts, as in Fig. 2.7b. Then, at a specific value of x, say x', there no longer is a single value

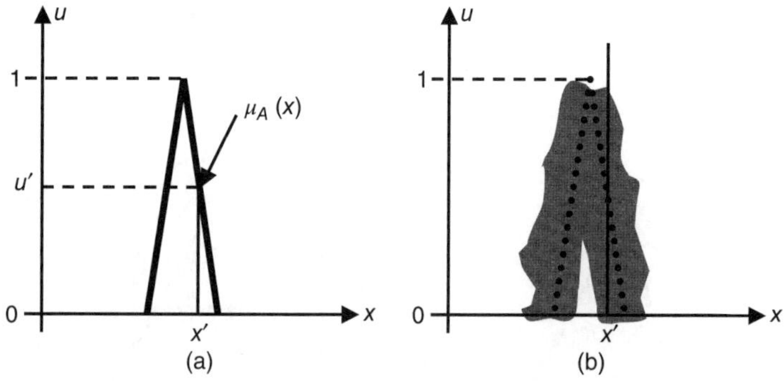

Figure 2.7 (a) T1 MF and (b) blurred T1 MF (Mendel et al., 2006; © 2006, IEEE).

[3]The material in this section is taken from Mendel et al. (2006; © 2006, IEEE).
[4]We are only using the idea of "blurring" as a pedagogical way to introduce a T2 FS. There are many other ways to create such an FS.

for the MF; instead, the MF takes on values wherever the vertical line intersects the blur. Those values need not all be weighted the same; hence, we can assign an amplitude distribution to all of those points. Doing this for all $x \in X$, we create a three-dimensional MF—a T2 MF—that characterizes a T2 FS.

2.3.2 Definitions[5]

Definition 2.7 A *T2 FS* [also called a *GT2 FS*], denoted $\tilde{A}$, is a bivariate function (Aisbett et al. 2010) on the Cartesian product $\mu : X \times [0, 1]$ into $[0, 1]$, where X is the universe for the *primary variable* of $\tilde{A}$, x. The 3D MF of $\tilde{A}$ is usually denoted $\mu_{\tilde{A}}(x, u)$, where $x \in X$ and $u \in U = [0, 1]$, that is,

$$\tilde{A} = \left\{ \left((x, u) , \mu_{\tilde{A}}(x, u) \right) | \forall x \in X, \ \forall u \in [0, 1] \right\} \tag{2.25}$$

in which $0 \le \mu_{\tilde{A}}(x, u) \le 1$. $\tilde{A}$ can also be expressed as

$$\tilde{A} = \int_{x \in X} \int_{u \in [0,1]} \mu_{\tilde{A}}(x, u) / (x, u) \tag{2.26}$$

where $\int\int$ denotes union[6] over all admissible x and u. For discrete universes of discourse $\int$ is replaced by $\sum$, and X and U by X_d and U_d.

In Definition 2.7, the first restriction that $\forall u \in [0, 1]$ is consistent with the T1 constraint that $0 \le \mu_A(x) \le 1$, that is, when uncertainties disappear, a T2 MF must reduce to a T1 MF, in which case the variable u equals[7] $\mu_A(x)$ and $0 \le \mu_A(x) \le 1$. The second restriction that $0 \le \mu_{\tilde{A}}(x, u) \le 1$ is consistent with the fact that the amplitudes of an MF should lie between or be equal to 0 and 1.

In Eqs. (2.25) and (2.26) u is called the *secondary variable* and has domain $U = [0, 1]$ at each $x \in X$.

Note: For a T2 FS, symbol u is widely used in the T2 literature as the variable label for the primary MF, so we are using this notation here as well. For control, symbol u, which denotes the control law, is always a function and will be shown as such, for example, as $u(t)$, where t is time; or $u(\mathbf{x})$ or $u(\mathbf{x}(t))$, where $\mathbf{x}$ is a state vector; or, $u(e(t), \dot{e}(t))$ or $u(\mathbf{e})$ or $u(\mathbf{e}(t))$, where e is error, and so forth.

Definition 2.8 When $\mu_{\tilde{A}}(x, u) = 1$ for $\forall x \in X$ and $\forall u \in U$, then $\tilde{A}$ is called an *interval T2 FS* (IT2 FS).

[5]Much of the material in this section is taken from Mendel et al. (2006; © 2006, IEEE).

[6]Recall that the union of two sets A and B is by definition another set that contains the elements in either A or B. When we view each element of a T2 FS as a subset, then the unions in Eq. (2.26) conform to the classical definition of union, since each element of that set is distinct. At a specific value of x and u only one term is activated in the union.

[7]In this case, the third dimension disappears.

Although the third dimension of the GT2 FS is no longer needed because it conveys no new information about the IT2 FS, the IT2 FS can still be expressed as a special case of the GT2 FS in Eq. (2.26) as

$$\tilde{A} = \int_{x \in X} \int_{u \in [0,1]} 1/(x, u) \tag{2.27}$$

In the rest of this section we will only be interested in IT2 FSs (we return to a GT2 FS in Section 2.7). Note, however, that in order to introduce the remaining widely used terminology of a GT2 FS we *temporarily* continue to retain the third dimension for an IT2 FS.

Definition 2.9 At each value of x, say $x = x'$, the 2D plane whose axes are u and $\mu_{\tilde{A}}(x', u)$ is called a *vertical slice* of $\mu_{\tilde{A}}(x, u)$. A *secondary MF* is a vertical slice of $\mu_{\tilde{A}}(x, u)$. It is $\mu_{\tilde{A}}(x = x', u)$ for $x' \in X$ and $\forall\, u \in [0, 1]$, that is,

$$\mu_{\tilde{A}}(x = x', u) \equiv \mu_{\tilde{A}}(x') = \int_{u \in J^u_{x'} \subseteq [0,1]} 1/u \tag{2.28}$$

where $J^u_{x'}$ is the subset of U that is the support of $\mu_{\tilde{A}}(x')$ and is called the *primary membership* of $\tilde{A}$. The amplitude of the secondary MF is called the *secondary grade*. The secondary grades of an IT2 FS are all equal to 1. Because $\forall\, x' \in X$, we drop the prime notation on $\mu_{\tilde{A}}(x')$ and refer to $\mu_{\tilde{A}}(x)$ as[8] a secondary MF; it is the MF of a T1 FS, which we also refer to as a *secondary set*, $\tilde{A}(x)$.

Example 2.7 The IT2 MF that is depicted in Fig. 2.8 has five vertical slices associated with it. The one at $x = 2$ is depicted in Fig. 2.9. The secondary MF at $x = 2$ is $\mu_{\tilde{A}}(2) = 1/0 + 1/0.2 + 1/0.4 + 1/0.6 + 1/0.8$, whereas the secondary MF at $x = 3$ (Fig. 2.8) is $\mu_{\tilde{A}}(3) = 1/0.6 + 1/0.8$.

Based on the concept of secondary sets, an IT2 FS can be reinterpreted as the union (see footnote 6) of all secondary sets, that is, using Eq. (2.28) in Eqs. (2.25) and (2.26), respectively, $\tilde{A}$ can be re-expressed in a *vertical-slice manner* as

$$\tilde{A} = \{(x, \mu_{\tilde{A}}(x)) | \forall x \in X\} \tag{2.29}$$

or, alternatively as

$$\tilde{A} = \int_{x \in X} \mu_{\tilde{A}}(x)/x = \int_{x \in X} \left[\int_{u \in J^u_x} 1/u \right] \Big/ x \qquad J^u_x \subseteq [0, 1] \tag{2.30}$$

[8] $\mu_{\tilde{A}}(x)$ is actually a function of secondary variable u; hence, a better notation for it is either $\mu_{\tilde{A}}(u|x)$ or $\mu_{\tilde{A}(x)}(u)$. Because the notation $\mu_{\tilde{A}}(x)$ is already widely used by the T2 FS community, it is not changed here.

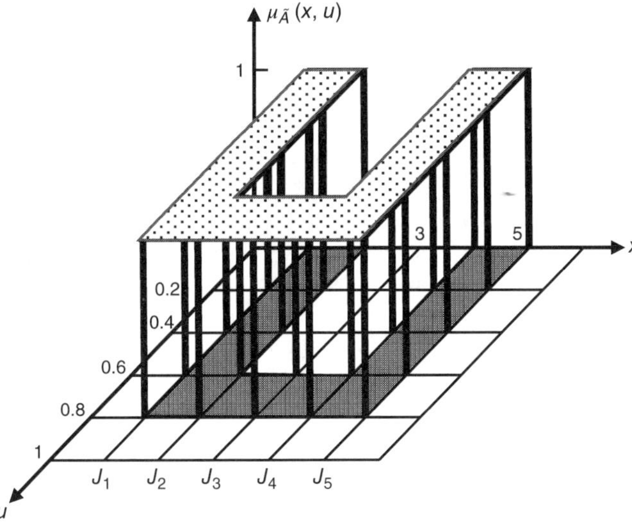

Figure 2.8 Example of an IT2 MF for discrete universes of discourse. The shaded area in the x–u plane is the FOU (Definition 2.10) (Mendel et al., 2006; © 2006, IEEE).

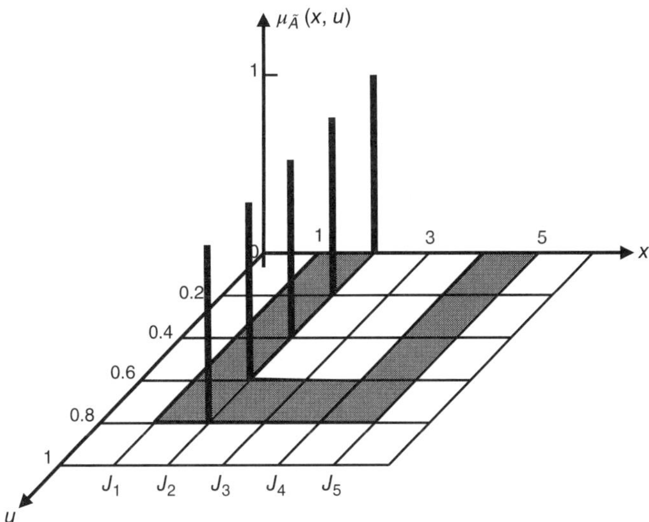

Figure 2.9 Example of a vertical slice for the IT2 MF depicted in Fig. 2.8 (Mendel et al., 2006; © 2006, IEEE).

If X and J_x^u are both discrete (either by problem formulation—as in Example 2.7—or by discretization of continuous universes of discourse), then the rightmost part of Eq. (2.30) can be expressed as

$$\tilde{A}=\sum_{x\in X}\left[\sum_{u\in J_x^u}1/u\right]\Big/x=\sum_{i=1}^{N}\left[\sum_{u\in J_{x_i}^u}1/u\right]\Big/x_i=\left[\sum_{k=1}^{M_1}1/u_{1k}\right]\Big/x_1+\cdots+\left[\sum_{k=1}^{M_N}1/u_{Nk}\right]\Big/x_N \tag{2.31}$$

In Eq. (2.31) + also denotes union. Observe that x has been discretized into N values and at each of these values u has been discretized into M_i values. The discretization along each u_{ik} does not have to be the same, which is why we have shown a different upper sum for each of the bracketed terms; however, if the discretization of each u_{ik} is the same, then $M_1=M_2=\cdots=M_N\equiv M$.

Example 2.8 In Fig. 2.8, the union of the five secondary MFs at $x=1,2,\ldots,5$ is $\mu_{\tilde{A}}(x,u)$. Observe that the primary memberships are

$$J_1=J_2=J_4=J_5=\{0,0.2,0.4,0.6,0.8\}\quad\text{and}\quad J_3=\{0.6,0.8\}$$

and we have only included values in J_3 for which $\mu_{\tilde{A}}(x,u)\neq 0$. Each of the spikes in Fig. 2.8 represents $\mu_{\tilde{A}}(x,u)$ at a specific (x,u) pair, and its amplitude of 1 is the secondary grade.

Definition 2.10 Uncertainty in the primary memberships of an IT2 FS, $\tilde{A}$, consists of a bounded region that is called the *footprint of uncertainty* (FOU). It is the two-dimensional support of $\tilde{A}$, that is (Aisbett et al., 2010),

$$\mathrm{FOU}(\tilde{A})=\{(x,u)\in X\times U\,|\,\mu_{\tilde{A}}(x,u)>0\} \tag{2.32}$$

$\mathrm{FOU}(\tilde{A})$ can also be expressed as the union of all primary memberships, that is,

$$\mathrm{FOU}(\tilde{A})=\bigcup_{x\in X}J_x^u \tag{2.33}$$

This is a *vertical-slice representation of the FOU*, because each of the primary memberships is the support of a vertical slice.

The shaded region on the x–u plane in Fig. 2.8 is an artistic rendition of the FOU.[9] Because the secondary grades of an IT2 FS convey no new information, *the FOU is a complete description of an IT2 FS*. The uniformly shaded FOU of an IT2 FS denotes that there is a uniform distribution that sits on top of it. The uniformly blurred T1 FS in Fig. 2.7b, where $x\in X$ and $u\in U$, is another example of the FOU of an IT2 FS.

[9]Strictly speaking, the FOU for discrete universes on $X_d\times U_d$ is just a collection of points [called a *domain of uncertainty* by Mendel and John (2002b)]. Artistic liberties have been taken in Fig 2.8 by calling the shaded region the FOU. If X_d was X and U_d was U, then the shaded region would indeed be the FOU.

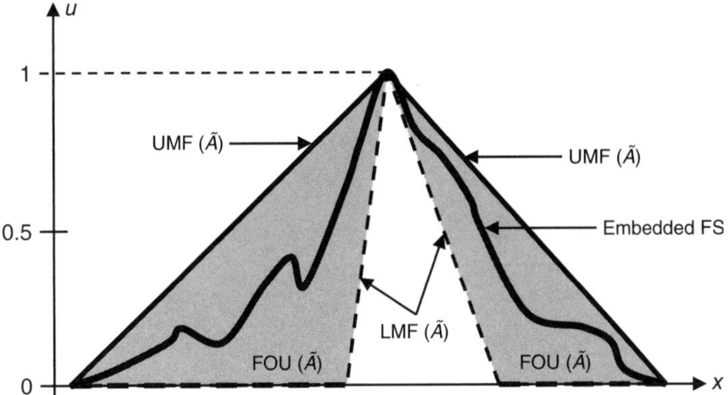

Figure 2.10 FOU (shaded), LMF (dashed), UMF (solid) and an embedded FS (wavy curve) for the IT2 FS $\tilde{A}$ (Mendel et al., 2006; © 2006, IEEE).

Definition 2.11 The *upper membership function* (UMF) and *lower membership function* (LMF) of $\tilde{A}$ are two T1 MFs that bound FOU($\tilde{A}$) (e.g., see Fig. 2.10). The UMF is associated with the upper bound of FOU($\tilde{A}$) and is denoted $\overline{\mu}_{\tilde{A}}(x)$ [or UMF($\tilde{A}$)], $\forall x \in X$, and the LMF is associated with the lower bound of FOU($\tilde{A}$) and is denoted $\underline{\mu}_{\tilde{A}}$ [or LMF($\tilde{A}$)],$\forall x \in X$, that is (Aisbett et al., 2010),

$$\overline{\mu}_{\tilde{A}}(x) = \text{UMF}(\tilde{A}) = \sup\{u | u \in [0,1], \mu_{\tilde{A}}(x,u) > 0\} \quad \forall x \in X \tag{2.34a}$$

$$\underline{\mu}_{\tilde{A}}(x) = \text{LMF}(\tilde{A}) = \inf\{u | u \in [0,1], \mu_{\tilde{A}}(x,u) > 0\} \quad \forall x \in X \tag{2.34b}$$

Note, also, that J_x^u can be expressed as

$$J_x^u = \{(x,u) : \forall u \in [\text{LMF}(\tilde{A}), \text{UMF}(\tilde{A})]\} \tag{2.35}$$

Definition 2.12 The *support* of LMF($\tilde{A}$) [UMF($\tilde{A}$)] is the crisp set of all points $x \in X$ such that LMF($\tilde{A}$) > 0 [UMF($\tilde{A}$) > 0]. The *support* of $\tilde{A}$ is the same as the support of UMF($\tilde{A}$).

Definition 2.13 IT2 FS $\tilde{A}$ is *convex* if both LMF($\tilde{A}$) and UMF($\tilde{A}$), which are T1 FSs, are convex over their respective supports (Definition 2.4).

In general, the supports of LMF($\tilde{A}$) and UMF($\tilde{A}$) are different, and the support of LMF($\tilde{A}$) is contained within the support of UMF($\tilde{A}$).

Definition 2.14 For discrete universes of discourse X_d and U_d, an *embedded IT2 FS* $\tilde{A}_e$ has N elements, where $\tilde{A}_e$ contains exactly one element from $J_{x_1}^u, J_{x_2}^u, \ldots$, and $J_{x_N}^u$, namely $u_1, u_2, \ldots, u_N$, each with a secondary grade equal to 1, that is,

$$\tilde{A}_e = \sum_{i=1}^{N} [1/u_i]/x_i \quad u_i \in J_{x_i}^u \subseteq U_d = \{0, \ldots, 1\} \tag{2.36}$$

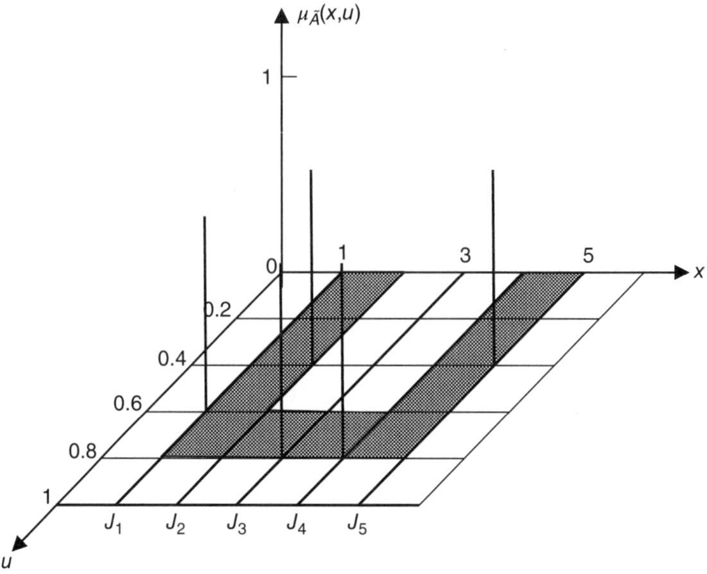

Figure 2.11 Example of an embedded IT2 FS associated with the IT2 MF depicted in Fig. 2.8 (Mendel et al., 2006; © 2006, IEEE).

Set $\tilde{A}_e$ is embedded in $\tilde{A}$, and, there are a total[10] of $\prod_{i=1}^{N} M_i\ \tilde{A}_e$.

Example 2.9 An example of an embedded IT2 FS is depicted in Fig. 2.10; it is the wavy curve for which its secondary grades (not shown) are all equal to 1. Other examples of $\tilde{A}_e$ are $1/\overline{\mu}_{\tilde{A}}(x)$ and $1/\underline{\mu}_{\tilde{A}}(x)$, $\forall x \in X$, where it is understood that in this notation the secondary grade equals 1 at all values of $\overline{\mu}_{\tilde{A}}(x)$ and $\underline{\mu}_{\tilde{A}}(x)$. Figure 2.11 depicts one of the possible $5^4 \times 2 = 1250$ embedded IT2 FSs for the T2 MF that is depicted in Fig. 2.8.

Definition 2.15 For discrete universes of discourse X_d and U_d, an *embedded T1 FS* A_e has N elements, one each from $J_{x_1}^u, J_{x_2}^u, \ldots$, and $J_{x_N}^u$, namely $u_1, u_2, \ldots, u_N$, that is,

$$A_e = \sum_{i=1}^{N} u_i/x_i \qquad u_i \in J_{x_i}^u \subseteq U_d = \{0, \ldots, 1\} \tag{2.37}$$

Set A_e is the union of all the primary memberships of set $\tilde{A}_e$ in Eq. (2.36), and, there are a total of $\prod_{i=1}^{N} M_i\ A_e$. Note that A_e acts as the domain for $\tilde{A}_e$.

[10]For a continuous IT2 FS, although there are an uncountable infinite number of embedded IT2 FSs, the concept of an embedded IT2 FS [as well as of an embedded T1 FS (Definition 2.15)] is still a theoretically useful one.

Example 2.10 An example of an embedded T1 FS is depicted in Fig. 2.10; it is the wavy curve. Note that embedded T1 FSs do not have to be convex or normal. For the FOU shown in Fig. 2.10, they will all be normal because they must all include the upper vertex of both the LMF and UMF, both of which occur at (1, 1). Other examples of A_e are $\overline{\mu}_{\tilde{A}}(x)$ and $\underline{\mu}_{\tilde{A}}(x)$, $\forall x \in X$. Observe, in Fig. 2.11, that the embedded T1 FS that is associated with the embedded IT2 FS is $A_e = 0.6/1 + 0.4/2 + 0.8/3 + 0.8/4 + 0.4/5$.

Comparing Eqs. (2.36) and (2.37), the embedded IT2 FS $\tilde{A}_e$ can be represented in terms of the embedded T1 FS A_e as

$$\tilde{A}_e = 1/A_e \tag{2.38}$$

with the understanding that this means putting a secondary grade of 1 at all points of A_e. We will make heavy use of this way to represent $\tilde{A}_e$ in the rest of this section.

So far we have emphasized the vertical-slice representation (decomposition) of an IT2 FS as given in Eq. (2.30). Next, we provide a different representation for such a fuzzy set that is in terms of so-called *wavy slices*. This representation, which makes very heavy use of embedded IT2 FSs (Definition 2.14), was first presented by Mendel and John (2002a) for an arbitrary T2 FS and is the bedrock for the rest of this chapter. We state this result for a discrete IT2 FS.

THEOREM 2.2 **(Wavy-Slice Representation Theorem)** For an IT2 FS, for which X and U are discrete, $\tilde{A}$ is the union of all of its embedded IT2 FSs, that is,

$$\tilde{A} = \sum_{j=1}^{n_A} \tilde{A}_e^j \tag{2.39}$$

where $(j = 1, \ldots, n_A)$

$$\tilde{A}_e^j = \sum_{i=1}^{N} [1/u_i^j]/x_i \quad u_i^j \in J_{x_i}^u \subseteq U_d = \{0, \ldots, 1\} \tag{2.40}$$

and

$$n_A = \prod_{i=1}^{N} M_i \tag{2.41}$$

in which M_i denotes the discretization levels of secondary variable u_i^j at each of the $N x_i$.

Comments

1. This theorem expresses $\tilde{A}$ as a union of simpler IT2 FSs, the $\tilde{A}_e^j$. They are simpler because their secondary MFs are singletons. Whereas Eq. (2.30) is a vertical-slice representation of $\tilde{A}$, Eq. (2.39) is a *wavy-slice* representation of $\tilde{A}$.
2. A detailed proof of this theorem appears in Mendel and John (2002a). Although it is important to have such a proof, we maintain that the results in Eq. (2.39) are obvious using the following simple geometric argument:
   - The MF of an IT2 FS is three dimensional (3D) (e.g., Fig. 2.8). Each of its embedded IT2 FSs is a 3D wavy slice (a foil). Create all of the possible wavy slices and take their union to reconstruct the original 3D MF. Some points, which occur in different wavy slices, only appear once in the set-theoretic union.

With reference to Fig. 2.10, Eq. (2.39) means collecting all of the embedded IT2 FSs into a *bundle* of such T2 fuzzy sets. Equivalently, because of Eq. (2.38), we can collect all of the embedded T1 FSs into a bundle of such T1 FSs.

COROLLARY 2.1 Because all of the secondary grades of an IT2 FS equal 1, Eqs. (2.39) and (2.40) can also be expressed as

$$\tilde{A} = 1/\text{FOU}(\tilde{A}) \tag{2.42}$$

where

$$\text{FOU}(\tilde{A}) = \begin{cases} \sum_{j=1}^{n_A} A_e^j = \left\{ \underline{\mu}_{\tilde{A}}(x), \ldots, \overline{\mu}_{\tilde{A}}(x) \right\} & \forall x \in X_d \\ \bigcup_{\forall A_e} A_e = [\underline{\mu}_{\tilde{A}}(x), \overline{\mu}_{\tilde{A}}(x)] & \forall x \in X \end{cases} \tag{2.43}$$

and [see Eq. (2.37)]

$$A_e^j = \sum_{i=1}^{N} u_i^j / x_i \quad u_i^j \in J_{x_i}^u \subseteq U_d = \{0, \ldots, 1\} \tag{2.44}$$

The top line of Eq. (2.43) is for a discrete universe of discourse, X_d, and contains n_A elements (functions), where n_A is given by Eq. (2.41), and the bottom line is for a continuous universe of discourse and is an interval set of functions, meaning that it contains an uncountable infinite number of functions that completely fills the space between $\overline{\mu}_{\tilde{A}}(x) - \underline{\mu}_{\tilde{A}}(x)$, for $\forall\, x \in X$.

Proof. From Eq. (2.38), each $\tilde{A}_e^j$ in Eq. (2.39) can be expressed as $1/A_e^j$; hence,

$$\tilde{A} = \sum_{j=1}^{n_A} \left(1/A_e^j\right) = 1/\sum\nolimits_{j=1}^{n_A} A_e^j \equiv 1/\mathrm{FOU}(\tilde{A}) \tag{2.45}$$

which is Eq. (2.42). Note that, as already mentioned, $\underline{\mu}_{\tilde{A}}(x)$ and $\overline{\mu}_{\tilde{A}}(x)$ are two legitimate elements of the n_A elements of A_e^j. In fact, they are the *lower* and *upper bounding functions*, respectively, for these n_A functions. For discrete universes of discourse, we can therefore express FOU($\tilde{A}$) as in the top line of Eq. (2.43), whereas for continuous universes of discourse we can express FOU($\tilde{A}$) as in the bottom line of Eq. (2.43) [using Eqs. (2.33) and (2.35)].

Equation (2.43) is called a *wavy-slice representation of* FOU($\tilde{A}$) because all A_e^j are functions, that is, they are wavy slices. We will see below that we do not need to know the explicit nature of any of the wavy slices in FOU($\tilde{A}$) other than $\underline{\mu}_{\tilde{A}}(x)$ and $\overline{\mu}_{\tilde{A}}(x)$.

2.3.3 Set-Theoretic Operations[11]

Our goal[11] in this section is to derive formulas for the union and intersection of two IT2 FSs and also the formula for the complement of an IT2 FS because these operations are widely used in an IT2 FLC. There are different approaches to doing this, for example, the extension principle (Zadeh, 1975), α cuts (Section 2.2.3), or interval arithmetic (e.g., Klir and Yuan, 1995). Our approach will be based entirely on the wavy-slice representation in Theorem 2.2, already well-known formulas for the union and intersection of two T1 FSs [Eqs. (2.4) and (2.5)], and the formula for the complement of a T1 FS [Eq. (2.6)]. By using Theorem 2.2, the set-theoretic operations that are needed for IT2 FSs will be derived using T1 FS mathematics.

THEOREM 2.3 (a) The *union* of two IT2 FSs, $\tilde{A}$ and $\tilde{B}$, is

$$\tilde{A} \cup \tilde{B} = 1/[\underline{\mu}_{\tilde{A}}(x) \vee \underline{\mu}_{\tilde{B}}(x), \overline{\mu}_{\tilde{A}}(x) \vee \overline{\mu}_{\tilde{B}}(x)] \quad \forall x \in X \tag{2.46}$$

(b) the *intersection* of two IT2 FSs, $\tilde{A}$ and $\tilde{B}$, is

$$\tilde{A} \cap \tilde{B} = 1/[\underline{\mu}_{\tilde{A}}(x) \wedge \underline{\mu}_{\tilde{B}}(x), \overline{\mu}_{\tilde{A}}(x) \wedge \overline{\mu}_{\tilde{B}}(x)] \quad \forall x \in X \tag{2.47}$$

and, (c) the *complement* of IT2 FS $\tilde{A}$, $\overline{\tilde{A}}$, is

$$\overline{\tilde{A}} = 1/[1 - \overline{\mu}_{\tilde{A}}(x), 1 - \underline{\mu}_{\tilde{A}}(x)] \quad \forall x \in X \tag{2.48}$$

[11]The material in this section is taken from Mendel et al. (2006).

In Eqs. (2.46) and (2.47) $\vee$ and $\wedge$ denote disjunction and conjunction operators, respectively, where in this book $\vee =$ maximum and $\wedge =$ minimum.

Proof. Because the proofs of parts (a) and (b) are so similar, we only provide the proofs for parts (a) and (c).

(a) Consider two IT2 FSs $\tilde{A}$ and $\tilde{B}$. From the wavy-slice representation in Theorem 2.2 and Corollary 2.1, it follows that[12]

$$\begin{aligned}\tilde{A} \cup \tilde{B} &= \sum_{j=1}^{n_A} \tilde{A}_e^j \cup \sum_{i=1}^{n_B} \tilde{B}_e^i = \sum_{j=1}^{n_A} \sum_{i=1}^{n_B} \tilde{A}_e^j \cup \tilde{B}_e^i \\ &= 1/\text{FOU}(\tilde{A} \cup \tilde{B}) \end{aligned} \tag{2.49}$$

where n_A and n_B denote the number of embedded IT2 FSs that are associated with $\tilde{A}$ and $\tilde{B}$, respectively, and [see the top part of Eq. (2.43)]

$$\text{FOU}(\tilde{A} \cup \tilde{B}) = \sum_{j=1}^{n_A} \sum_{i=1}^{n_B} A_e^j \cup B_e^i \tag{2.50}$$

What we must now do is compute the union of the $n_A \times n_B$ pairs of embedded T1 FSs A_e^j and B_e^i.

Using Eq. (2.4), it follows that[13]

$$A_e^j \cup B_e^i = \max\{\mu_{A_e^j}(x_k), \mu_{B_e^i}(x_k)\} \quad k = 1, 2, \ldots, N \tag{2.51}$$

Consequently, Eq. (2.50) is a collection of $n_A \times n_B$ functions that contain a lower-bounding function and an upper-bounding function since both $\mu_{A_e^j}(x_k)$ and $\mu_{B_e^i}(x_k)$ are bounded for all values of x_k.

In the case of IT2 FSs, for which the primary variable and primary memberships are defined over continuous domains, $n_A = \infty$ and $n_B = \infty$; however, Eq. (2.51) is still true, and the doubly infinite union of embedded T1 FSs in Eq. (2.50) still contains a lower-bounding function and an upper-bounding function because $\tilde{A}$ and $\tilde{B}$ each have a bounded FOU. We now obtain formulas for these bounding functions.

Recall (see Example 2.10) that the upper and lower (discrete, or, if continuous, sampled) MFs for an IT2 FS are also embedded T1 FSs. For $\tilde{A}$, $\overline{\mu}_{\tilde{A}}(x)$ and

[12]Equation (2.49) involves summation and union signs. As in the T1 case, where this mixed notation is used, the summation sign is simply shorthand for lots of + signs. The + indicates the union between members of a set, whereas the union sign represents the union of the sets themselves. Hence, by using both the summation and union signs, we are able to distinguish between the union of sets versus the union of members within a set.

[13]Although we present our derivation for maximum, it is also applicable for a general t-conorm.

$\underline{\mu}_{\tilde{A}}(x)$ denote its upper MF and lower MF, whereas for $\tilde{B}$, $\overline{\mu}_{\tilde{B}}(x)$ and $\underline{\mu}_{\tilde{B}}(x)$ denote its comparable quantities. It must therefore be true that

$$\begin{aligned}\sup_{\forall j,i} \max\{\mu_{A_e^j}(x_k), \mu_{B_e^i}(x_k)\} &= \max\{\overline{\mu}_{\tilde{A}}(x), \overline{\mu}_{\tilde{B}}(x)\} \quad \forall x \in X \\ &= \overline{\mu}_{\tilde{A}}(x) \vee \overline{\mu}_{\tilde{B}}(x) \quad \forall x \in X \end{aligned} \tag{2.52}$$

$$\begin{aligned}\inf_{\forall j,i} \max\{\mu_{A_e^j}(x_k), \mu_{B_e^i}(x_k)\} &= \max\{\underline{\mu}_{\tilde{A}}(x), \underline{\mu}_{\tilde{B}}(x)\} \quad \forall x \in X \\ &= \underline{\mu}_{\tilde{A}}(x) \vee \underline{\mu}_{\tilde{B}}(x) \quad \forall x \in X \end{aligned} \tag{2.53}$$

From Eqs. (2.49)–(2.53), we conclude that

$$\tilde{A} \cup \tilde{B} = 1/\sum_{j=1}^{n_A} \sum_{i=1}^{n_B} A_e^j \cup B_e^i = 1/[\underline{\mu}_{\tilde{A}}(x) \vee \underline{\mu}_{\tilde{B}}(x), \overline{\mu}_{\tilde{A}}(x) \vee \overline{\mu}_{\tilde{B}}(x)] \quad \forall x \in X \tag{2.54}$$

which is Eq. (2.46).

(c) Starting with Eq. (2.39), and Corollary 2.1, we see that

$$\overline{\tilde{A}} = \overline{\sum_{j=1}^{n_A} \tilde{A}_e^j} = \sum_{j=1}^{n_A} \overline{\tilde{A}}_e^j = 1/\text{FOU}(\overline{\tilde{A}}) \tag{2.55}$$

where [focusing on continuous universes of discourse; see also the second line of Eq. (2.43)]

$$\text{FOU}(\overline{\tilde{A}}) = \sum_{j=1}^{n_A} \overline{A}_e^j = [\underline{\mu}_{\overline{\tilde{A}}}(x), \overline{\mu}_{\overline{\tilde{A}}}(x)] \quad \forall x \in X \tag{2.56}$$

Using Eq. (2.6), it follows that

$$\mu_{\overline{A}_e^j}(x) = 1 - \mu_{A_e^j}(x) \tag{2.57}$$

Equation (2.56) is a bundle of functions that has a lower-bounding $[\underline{\mu}_{\overline{\tilde{A}}}(x)]$ and an upper-bounding $[\overline{\mu}_{\overline{\tilde{A}}}(x)]$ function; hence,

$$\overline{\mu}_{\overline{\tilde{A}}}(x) = \sup_{\forall j}[1 - \mu_{A_e^j}(x_k)] = 1 - \underline{\mu}_{\tilde{A}}(x) \quad \forall x \in X \tag{2.58}$$

$$\underline{\mu}_{\overline{\tilde{A}}}(x) = \inf_{\forall j}[1 - \mu_{A_e^j}(x_k)] = 1 - \overline{\mu}_{\tilde{A}}(x) \quad \forall x \in X \tag{2.59}$$

In obtaining the right-hand parts of Eqs. (2.58) and (2.59) we have used the facts that it is always true that $\overline{\mu}_{\tilde{A}}(x) \geq \underline{\mu}_{\tilde{A}}(x)$, consequently, it is always true that $1 - \underline{\mu}_{\tilde{A}}(x) \geq 1 - \overline{\mu}_{\tilde{A}}(x)$.

From Eqs. (2.55), (2.56), (2.58) and (2.59), we conclude that

$$\overline{\tilde{A}} = 1/\sum_{j=1}^{n_A} \overline{A}_e^j = 1/[1 - \overline{\mu}_{\tilde{A}}(x), 1 - \underline{\mu}_{\tilde{A}}(x)] \quad \forall x \in X \tag{2.60}$$

which is Eq. (2.48).

The generalizations of parts (a) and (b) of Theorem 2.3 to more than two IT2 FSs follows directly from Eqs. (2.46) and (2.47) and the associative property of T2 FSs, for example,

$$\tilde{A} \cup \tilde{B} \cup \tilde{C} = 1/[\underline{\mu}_{\tilde{A}}(x) \vee \underline{\mu}_{\tilde{B}}(x) \vee \underline{\mu}_{\tilde{C}}(x), \overline{\mu}_{\tilde{A}}(x) \vee \overline{\mu}_{\tilde{B}}(x) \vee \overline{\mu}_{\tilde{C}}(x)] \quad \forall x \in X$$

2.3.4 Centroid of an IT2 FS

The centroid of an IT2 FS, which is a very important computation in this book because it is used as a first step in the defuzzification of an IT2 FS, provides[14] a measure of the uncertainty of such an FS. This is explained more carefully at the end of this subsection. Using Eq. (2.43), the centroid of IT2 FS $\tilde{A}$, $C_{\tilde{A}}(x)$, is defined next.

Definition 2.16 Using the wavy-slice representation Theorem 2.2 for IT2 FS $\tilde{A}$, the *centroid* $C_{\tilde{A}}(x)$ of $\tilde{A}$ is the union of the centroids, $c(A_e)$, of all its embedded T1 FSs A_e. Associated with each of these numbers is a membership grade of 1 because the secondary grades of an IT2 FS are all equal to 1. This means (Karnik and Mendel, 2001a; Mendel, 2001)

$$\begin{aligned} C_{\tilde{A}}(x) &= 1/\bigcup_{\forall A_e} c_{\tilde{A}}(A_e) = 1/\bigcup_{\forall A_e} \frac{\sum_{i=1}^{N} x_i \mu_{A_e}(x_i)}{\sum_{i=1}^{N} \mu_{A_e}(x_i)} \\ &= 1/\{c_l(\tilde{A}), \ldots, c_r(\tilde{A})\} \equiv 1/[c_l(\tilde{A}), c_r(\tilde{A})] \end{aligned} \tag{2.61}$$

where

$$c_l(\tilde{A}) = \min_{\forall A_e} c_{\tilde{A}}(A_e) = \min_{\forall \theta_i \in [\underline{\mu}_{\tilde{A}}(x_i), \overline{\mu}_{\tilde{A}}(x_i)]} \frac{\sum_{i=1}^{N} x_i \theta_i}{\sum_{i=1}^{N} \theta_i} \tag{2.62}$$

$$c_r(\tilde{A}) = \max_{\forall A_e} c_{\tilde{A}}(A_e) = \max_{\forall \theta_i \in [\underline{\mu}_{\tilde{A}}(x_i), \overline{\mu}_{\tilde{A}}(x_i)]} \frac{\sum_{i=1}^{N} x_i \theta_i}{\sum_{i=1}^{N} \theta_i} \tag{2.63}$$

$C_{\tilde{A}}(x)$ is shown as an explicit function of x because the centroid of each embedded T1 FS falls on the x axis. $C_{\tilde{A}}(x)$ is a T1 interval fuzzy set (IFS). Note that it is customary in the IT2 FS literature to call $[c_l(\tilde{A}), c_r(\tilde{A})]$ the centroid of $\tilde{A}$, ignoring the uninformative MF grade of 1.

[14]Most of the material in this section is taken from Mendel and Wu (2010, Chapter 2; © 2010, IEEE).

Recall that there are n_A embedded T1 FSs that are contained within FOU($\tilde{A}$); hence, computing their centroids leads to a collection of n_A numbers, $\sum_{i=1}^{N} x_i\theta_i / \sum_{i=1}^{N} \theta_i$, that have both a smallest and largest element, $c_l(\tilde{A}) \equiv c_l$ and $c_r(\tilde{A}) \equiv c_r$, respectively.[15] That such numbers exist is because $\sum_{i=1}^{N} x_i\theta_i / \sum_{i=1}^{N} \theta_i$ is a bounded number. When the discretization of the primary variable and primary membership approach zero, $\{c_l(\tilde{A}), \ldots, c_r(\tilde{A})\} \to [c_l(\tilde{A}), c_r(\tilde{A})]$, an interval set.

Because x_i are sampled values of the primary variable, it is true that in Eqs. (2.62) and (2.63)

$$x_1 < x_2 < \cdots < x_N \tag{2.64}$$

in which x_1 denotes the smallest sampled value of x and x_N denotes the largest sampled value[16] of x.

Examining Eqs. (2.62) and (2.63) it seems that c_l and c_r could be computed by adding and then dividing interval sets. Klir and Yuan (1995) provide the following closed-form formula for the division of two interval sets:

$$\begin{aligned} [a,b]/[d,e] &= [a,b] \times \left[1/e, 1/d\right] \\ &= \left[\min(a/d, a/e, b/d, b/e), \max(a/d, a/e, b/d, b/e)\right] \end{aligned} \tag{2.65}$$

It would seem that this result could be applied to determine closed-form formulas for c_l and c_r. Unfortunately, this cannot be done because the derivation of this result assumes that a, b, d, and e are *independent*. Due to the appearance of θ_i in both the numerator and denominator of Eqs. (2.62) and (2.63), the required independence is not present; hence, this interesting closed-form result cannot be used to compute c_l and c_r.

Although $c_l(\tilde{A})$ and $c_r(\tilde{A})$ cannot be computed in closed-form, lower and upper bounds for them can be computed in closed form in terms of the geometry of an FOU (Mendel and Wu, 2006, 2007a).

Karnik and Mendel (2001a) have developed iterative algorithms—now known as *KM algorithms*—for computing c_l and c_r. These algorithms, which are very heavily used in many later chapters of this book, are derived and discussed in Section 2.3.6. The KM algorithms do not compute the exact values of c_l and c_r; instead, they lead to approximations of those values, $c_l(L)$ and $c_r(R)$, that have the following structures:

[15]When there is no ambiguity about the IT2 FS whose centroid is being computed, it is common to shorten $c_l(\tilde{A})$ and $c_r(\tilde{A})$ to c_l and c_r, respectively, something that is done in this book.

[16]If Gaussian MFs are used, then in theory $x_1 \to -\infty$ and $x_N \to \infty$; but, in practice when truncations are used x_1 and x_N are again finite numbers.

$$c_l(L) = \frac{\sum_{i=1}^{L} x_i \overline{\mu}_{\tilde{A}}(x_i) + \sum_{i=L+1}^{N} x_i \underline{\mu}_{\tilde{A}}(x_i)}{\sum_{i=1}^{L} \overline{\mu}_{\tilde{A}}(x_i) + \sum_{i=L+1}^{N} \underline{\mu}_{\tilde{A}}(x_i)} \approx c_l \tag{2.66}$$

$$c_r(R) = \frac{\sum_{i=1}^{R} x_i \underline{\mu}_{\tilde{A}}(x_i) + \sum_{i=R+1}^{N} x_i \overline{\mu}_{\tilde{A}}(x_i)}{\sum_{i=1}^{R} \underline{\mu}_{\tilde{A}}(x_i) + \sum_{i=R+1}^{N} \overline{\mu}_{\tilde{A}}(x_i)} \approx c_r \tag{2.67}$$

In these equations L and R are called *switch points*, and it is these switch points that are determined by the KM algorithms [as well as by many other algorithms, e.g., EKM (Wu and Mendel, 2009) and EIASC algorithms (Wu and Nie, 2011)]. Observe that in Eq. (2.66) when $i = L + 1$, θ_i switches from values on the UMF, $\overline{\mu}_{\tilde{A}}(x_i)$, to values on the LMF, $\underline{\mu}_{\tilde{A}}(x_i)$; and, in Eq. (2.67) when $i = R + 1$, θ_i switches from values on the LMF, $\underline{\mu}_{\tilde{A}}(x_i)$, to values on the UMF, $\overline{\mu}_{\tilde{A}}(x_i)$. An example that illustrates the two switch points is given in Fig. 2.12.

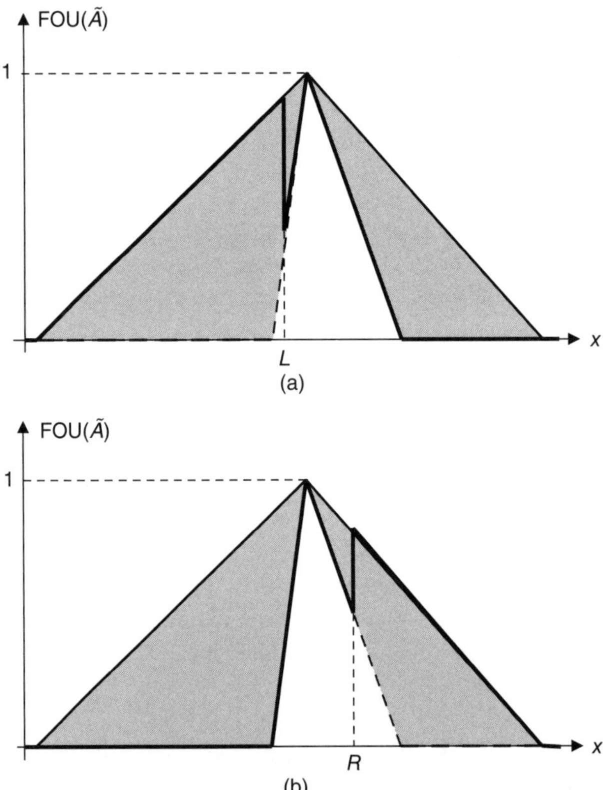

Figure 2.12 Embedded T1 FSs that are used to compute the switch points L and R are shown by the heavy lines in (a) and (b), respectively (Mendel, 2007; © 2007, IEEE).

When Eqs. (2.66) and (2.67) are used to approximate the centroid of IT2 FS $\tilde{A}$, then $c_l(L)$ $[c_r(R)]$ better approximates c_l $[c_r]$ as the sampling interval of primary variable x gets smaller and smaller. How close one actually needs the approximations to be to the actual values is application dependent. Many times in a "fuzzy" problem extremely high accuracy is not needed.

Wu (2011) has provided the following insights about the structure of Eq. (2.66) [(2.67)] as it relates to the solution of the optimization problem in Eq. (2.62) [(2.63)]: Examining the right-hand side of Eq. (2.62), in order to obtain the smallest value of $\sum_{i=1}^{N} x_i\theta_i / \sum_{i=1}^{N} \theta_i$, we want to associate the largest value of θ_i [namely, $\overline{\mu}_{\tilde{A}}(x_i)$] with the smallest values of x_i and then the smallest values of θ_i [namely, $\underline{\mu}_{\tilde{A}}(x_i)$] with the largest values of x_i. The opposite is true for obtaining the largest value of $\sum_{i=1}^{N} x_i\theta_i / \sum_{i=1}^{N} \theta_i$ in Eq. (2.63).

Note that Eqs. (2.66) and (2.67) can also be expressed as (e.g., X. Liu and Mendel, 2011)

$$c_l(k) = \frac{\sum_{i=1}^{k} x_i \overline{\mu}_{\tilde{A}}(x_i) + \sum_{i=k+1}^{N} x_i \underline{\mu}_{\tilde{A}}(x_i)}{\sum_{i=1}^{k} \overline{\mu}_{\tilde{A}}(x_i) + \sum_{i=k+1}^{N} \underline{\mu}_{\tilde{A}}(x_i)} \tag{2.68}$$

$$\begin{cases} L = \underset{k=1,\ldots,N-1}{\arg\min}\ c_l(k) \\ c_l(L) \approx c_l \end{cases} \tag{2.69}$$

$$c_r(k) = \frac{\sum_{i=1}^{k} x_i \underline{\mu}_{\tilde{A}}(x_i) + \sum_{i=k+1}^{N} x_i \overline{\mu}_{\tilde{A}}(x_i)}{\sum_{i=1}^{k} \underline{\mu}_{\tilde{A}}(x_i) + \sum_{i=k+1}^{N} \overline{\mu}_{\tilde{A}}(x_i)} \tag{2.70}$$

$$\begin{cases} R = \underset{k=1,\ldots,N-1}{\arg\max}\ c_r(k) \\ c_r(R) \approx c_r \end{cases} \tag{2.71}$$

As will be demonstrated in the next subsection, additional insight can be gained from the formulations in Eqs. (2.68)–(2.71).

It is well known from information theory that entropy provides a measure of the uncertainty of a random variable (Cover and Thomas, 1991). Recall that a one-dimensional random variable that is uniformly distributed over a region has entropy equal to the logarithm of the *length* of that region. Comparing the MF $\mu_{C_{\tilde{A}}}(x)$ of the IFS $C_{\tilde{A}}(x)$ [where $\mu_{C_{\tilde{A}}}(x) = 1$ when $x \in [c_l, c_r]$ and $\mu_{C_{\tilde{A}}}(x) = 0$ when $x \notin [c_l, c_r]$] with the probability density function $p_X(x)$ that is uniformly distributed over $[c_l, c_r]$ [where $p_X(x) = (c_r - c_l)^{-1}$ when $x \in [c_l, c_r]$ and $p_X(x) = 0$ when $x \notin [c_l, c_r]$], it is clear that they are similar to within a scale factor. It is therefore reasonable to consider the span of the IFS $C_{\tilde{A}}(x)$, $c_r - c_l$, as a measure of the extent of the uncertainty of an IT2 FS (Wu and Mendel, 2002).

TABLE 2.2 Properties[a] of $c_l(k)$ and $c_r(k)$

Name of Property	$c_l(k)$	$c_r(k)$
Location	$x_L \le c_l(L) < x_{L+1}$	$x_R \le c_r(R) < x_{R+1}$
Shape	$c_l(k) > x_k$ when $x_k < c_l$ $c_l(k) < x_k$ when $x_k > c_l$	$c_r(k) > x_k$ when $x_k < c_r$ $c_r(k) < x_k$ when $x_k > c_r$
Monotonicity	$c_l(k-1) \ge c_l(k)$ when $x_k < c_l$ $c_l(k+1) \ge c_l(k)$ when $x_k > c_l$	$c_r(k-1) \le c_r(k)$ when $x_k < c_r$ $c_r(k+1) \le c_r(k)$ when $x_k > c_r$

[a] Proofs of these properties are in Liu and Mendel (2008).

2.3.5 Properties of $c_l(k)$ and $c_r(k)$

Before obtaining the KM algorithms (as well as some others), it is useful to gain as much understanding about the two optimization problems in Eqs. (2.69) and (2.71) as possible. The three properties summarized in Table 2.2 provide a lot of insight into those problems (Liu and Mendel, 2008).

The *location* property locates $c_l(L)$ and $c_r(R)$ either between two specific adjacent values of x_i or at the left end point of these adjacent values [see[17] Fig. 2.13, where $x_4 < c_l(L) < x_5$].

The *shape* property explains the shapes of $c_l(k)$ and $c_r(k)$ both to the left and right of their respective minimum or maximum points, c_l and c_r (see Fig. 2.13). By this property we know that $c_l(k)$ [or $c_r(k)$] is above the line $y = x_k$ when x_k is to the left of c_l (c_r) and it is below that line when x_k is to the right of c_l (c_r).

The *monotonicity* property also helps us to understand the shapes of $c_l(k)$ and $c_r(k)$. When $c_l(k)$ is going in the downward direction (see Fig. 2.13), it cannot change that direction before $x_k = c_l$; and, after $x_k = c_l$, when it goes in the upward direction, it cannot change that direction. When $c_r(k)$ is going in the upward direction [we leave it to the reader to create a figure like Fig. 2.13 for $c_r(k)$], it cannot change that direction before $x_k = c_r$; and, after $x_k = c_r$, when it goes in the downward direction, it cannot change that direction.

From knowledge of the shapes[18] of $c_l(k)$ and $c_r(k)$, it should be clear to readers who are familiar with optimization theory that our two optimization problems are *easy*. Each problem has only one global extremum (Fig. 2.13) and there are no local extrema. Regardless of how one initializes any algorithm for finding the extremum, convergence will occur, that is, it is impossible to become trapped at a local extremum because of how each algorithm is initialized. It is also obvious, from the shapes of $c_l(k)$ and $c_r(k)$, that the algorithms that compute $c_l(L)$ and $c_r(R)$ will converge very quickly. In fact, the shapes of $c_l(k)$ and $c_r(k)$ suggest that quadratic convergence should be possible.

[17]For computing the centroid of an IT2 FS the x_i in Eq. (2.62) or (2.63) must be uniformly spaced, or else convergence of $c_l(L)$ to c_l [and $c_r(R)$ to c_r] will not occur; however, nonuniformly spaced x_i are important in other applications that use KM algorithms, such as center-of-sets type reduction, which is explained in Section 2.5.2.4.

[18]Apparently, such shape knowledge only became known in 2008 (Liu and Mendel, 2008); it was not known when the KM algorithms were invented.

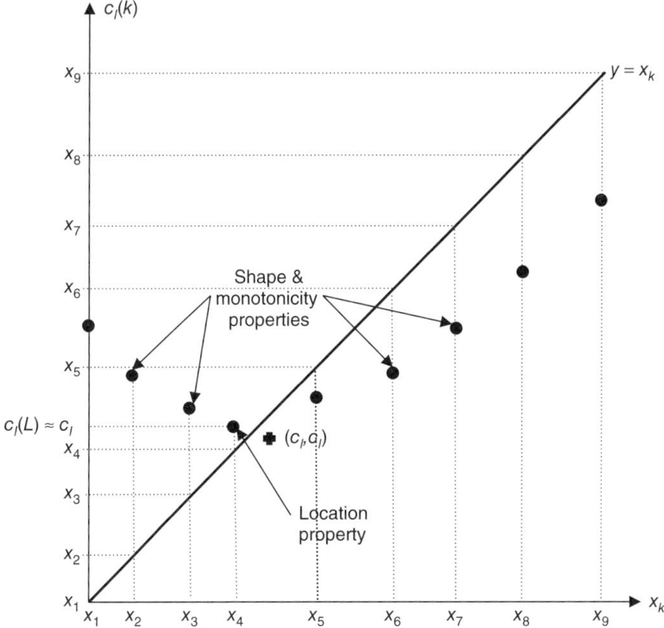

Figure 2.13 Illustration of the three properties associated with finding $c_l(L)$. The solid line shown for $y = x_k$ only has values at $x_1, \ldots, x_9$; the large dots are $c_l(k)$, $k = 1, \ldots, 9$; and (c_l, c_l) locates the theoretically optimal solution that can only be attained as sampling goes to zero.

2.3.6 KM Algorithms as Well as Some Others

Recall that any optimization algorithm *requires* a good way to (1) initialize it, (2) move from one step to the next, and (3) stop.

A *brute-force* way to find $c_l(L)$ $[c_r(R)]$ is to compute $\{c_l(k)\}_{k=1}^{N-1}$ $[\{c_r(k)\}_{k=1}^{N-1}]$ and then locate its smallest (largest) element. This is called "brute force" (or exhaustive) because it does not focus on a good way to initialize the process, move from one step to the next, or stop. All algorithms devised for finding $c_l(L)$ $[c_r(R)]$ focus on at least one of these three requirements of an optimization algorithm.

Karnik–Mendel algorithms do not provide a good way for their initialization, nor do they provide the best way for stopping them, but they do provide a good way to move from one step to the next. We derive the KM algorithms next.

Let[19]

$$y(\theta_1, \ldots, \theta_N) \equiv \frac{\sum_{i=1}^{N} x_i \theta_i}{\sum_{i=1}^{N} \theta_i} \tag{2.72}$$

If the usual calculus approach to optimizing $y(\theta_1, \ldots, \theta_N)$ is taken and it is differentiated with respect to any one of the N θ_i, say θ_k, it follows (after some algebra) that

[19]Our presentation follows that of Mendel and Wu (2007b).

$$\frac{\partial y(\theta_1, \dots, \theta_N)}{\partial \theta_k} = \frac{\partial}{\partial \theta_k}\left[\frac{\sum_{i=1}^{N} x_i\theta_i}{\sum_{i=1}^{N}\theta_i}\right] = \frac{x_k - y(\theta_1, \dots, \theta_N)}{\sum_{i=1}^{N}\theta_i} \tag{2.73}$$

Because $\sum_{i=1}^{N}\theta_i > 0$, it is easy to see from Eq. (2.73) that

$$\frac{\partial y(\theta_1, \dots, \theta_N)}{\partial \theta_k}\begin{cases}\geq 0 & \text{if } x_k \geq y\left(\theta_1, \dots, \theta_N\right)\\ < 0 & \text{if } x_k < y\left(\theta_1, \dots, \theta_N\right)\end{cases} \tag{2.74}$$

Unfortunately, equating $\partial y/\partial \theta_k$ to zero does not give us any information about the value of θ_k that optimizes $y(\theta_1, \dots, \theta_N)$, that is,

$$y(\theta_1, \dots, \theta_N) = x_k \Rightarrow \frac{\sum_{i=1}^{N} x_i\theta_i}{\sum_{i=1}^{N}\theta_i} = x_k \Rightarrow \frac{\sum_{i\neq k}^{N} x_i\theta_i}{\sum_{i\neq k}^{N}\theta_i} = x_k \tag{2.75}$$

The last part of Eq. (2.75) was obtained after cross multiplication and cancellation. Observe that θ_k no longer appears in the final expression in Eq. (2.75), so that the direct calculus approach does not work.

Equation (2.74) does give the direction in which θ_k should be changed in order to increase or decrease $y(\theta_1, \dots, \theta_N)$, that is,

$$\begin{cases}\text{If } x_k > y\left(\theta_1, \dots, \theta_N\right), \\ \qquad y(\theta_1, \dots, \theta_N) \text{ increases (decreases) as } \theta_k \text{ increases (decreases)} \\ \text{If } x_k < y(\theta_1, \dots, \theta_N), \\ \qquad y(\theta_1, \dots, \theta_N) \text{ increases (decreases) as } \theta_k \text{ decreases (increases)}\end{cases} \tag{2.76}$$

Recall [see Eqs. (2.62) and (2.63)] that the maximum value that θ_k can attain is $\overline{\mu}_{\tilde{A}}(x_k)$ and the minimum value that it can attain is $\underline{\mu}_{\tilde{A}}(x)$. Equation (2.76) therefore implies that $y(\theta_1, \dots, \theta_N)$ attains its *maximum value*, c_r, if

$$\theta_k = \begin{cases}\underline{\mu}_{\tilde{A}}\left(x_k\right) & \forall k \ni x_k < y(\theta_1, \dots, \theta_N)\\ \overline{\mu}_{\tilde{A}}(x_k) & \forall k \ni x_k > y(\theta_1, \dots, \theta_N)\end{cases} \tag{2.77}$$

Similarly, it can be deduced from (2.76) that $y(\theta_1, \dots, \theta_N)$ attains its *minimum value*, c_l, if

$$\theta_k = \begin{cases}\overline{\mu}_{\tilde{A}}\left(x_k\right) & \forall k \ni x_k < y(\theta_1, \dots, \theta_N)\\ \underline{\mu}_{\tilde{A}}(x_k) & \forall k \ni x_k > y(\theta_1, \dots, \theta_N)\end{cases} \tag{2.78}$$

Because there are only two possible choices for θ_k that are stated above, to compute $c_r(R)$ or $c_l(L)$, θ_k *switches only one time* between $\overline{\mu}_{\tilde{A}}(x_k)$ and $\underline{\mu}_{\tilde{A}}(x_k)$.

A KM algorithm (as well as all others) locates the switch point, and in general the switch point for $c_r(R)$, R, is different from the switch point for $c_l(L)$, L; hence, there are two KM algorithms, one for L and one for R. Regardless of what algorithms are used to compute the switch points, the final expression for $c_l(L)$ and $c_r(R)$ is Eq. (2.66) and Eq. (2.67), respectively.

Historically, the KM algorithms were the first, and are still the most widely used, algorithms for computing the switch points. They are summarized in Table 2.3. Observe that the first two steps of each KM algorithm are identical.

The enhanced KM (EKM) algorithms [Wu and Mendel, (2009)] start with the KM algorithms and modify them in three ways: (1) A better initialization is used to reduce the number of iterations; (2) the termination condition of the iterations is changed to remove an unnecessary iteration; and (3) a subtle computing technique is used to reduce the computational cost of each of the algorithm's iterations. The EKM algorithms are summarized in Table 2.4.

The better initializations are shown in step 1 of Table 2.4, and both were obtained from extensive simulations. A close examination of steps 2–5 in Table 2.3 reveals that the termination conditions can be moved one step earlier, something that is done in Table 2.4. The "subtle computing technique" uses the fact that very little changes from one iteration to the next, so instead of recomputing everything on the right-hand sides of $c_l(k)$ and $c_r(k)$, as is done in Table 2.3, only the portions of those right-hand sides that do change are recomputed, as is done in Table 2.4. For detailed explanations of how each of the three modifications are implemented, see Wu and Mendel (2009).

Extensive simulations have shown that on average the EKM algorithms can save about two iterations, which corresponds to a more than 39% reduction in computation time. Both the KM and EKM algorithms are quadratically convergent (Liu and Mendel, 2011) (see, also, the Comment at the end of this section).

Above, we discussed a *brute-force* algorithm, one that did not satisfy any of the three desired requirements of an optimization algorithm. Melgarejo (2007) and Duran et al. (2008) have beefed up the brute-force algorithm in their *iterative algorithm + stopping condition* (IASC), and D. Wu and Nie (2011) have made some improvements to it in their *enhanced IASC* (EIASC). Both the IASC and EIASC, which are iterative algorithms, are based on the shape and monotonicity properties (Table 2.2) so that $c_l(k)$ in Eq. (2.68) first monotonically decreases and then monotonically increases with the increase of k, and, $c_r(k)$ in Eq. (2.69) first monotonically increases and then monotonically decreases with the increase of k. The beautifully simple EIASC is given in Table 2.5. According to Wu and Nie (2011, p. 2135), "both the IASC and EIASC significantly outperformed the KM algorithms, especially when N is small ($N \leq 100$). And, EIASC outperformed IASC. The computational cost of both IASC and EIASC increase rapidly as N increases since they need to evaluate many possible switch points before finding the correct ones."

One could ask: Why wasn't each EIASC algorithm initialized as in the respective EKM algorithm (see step 1 in Table 2.4)? D. Wu actually tried this. His simulations (2011) showed that this further enhanced EIASC algorithm only outperformed the

TABLE 2.3 KM Algorithms for Computing the Centroid End Points of an IT2 FS $\tilde{A}$[a].

Step	KM Algorithm for $c_l(L)$	KM Algorithm for $c_r(R)$
	$c_l(L) = \min_{\forall \theta_i \in [\underline{\mu}_{\tilde{A}}(x_i), \overline{\mu}_{\tilde{A}}(x_i)]} \left(\sum_{i=1}^{N} x_i \theta_i \Big/ \sum_{i=1}^{N} \theta_i \right)$	$c_r(R) = \max_{\forall \theta_i \in [\underline{\mu}_{\tilde{A}}(x_i), \overline{\mu}_{\tilde{A}}(x_i)]} \left(\sum_{i=1}^{N} x_i \theta_i \Big/ \sum_{i=1}^{N} \theta_i \right)$
1	Initialize θ_i by setting $\theta_i = [\underline{\mu}_{\tilde{A}}(x_i) + \overline{\mu}_{\tilde{A}}(x_i)]/2,\ i = 1, \ldots, N$, and then compute $c' = c(\theta_1, \ldots, \theta_N) = \sum_{i=1}^{N} x_i \theta_i \Big/ \sum_{i=1}^{N} \theta_i$	
2	Find k $(1 \le k \le N-1)$ such that $x_k \le c' \le x_{k+1}$	
3	Set $\theta_i = \overline{\mu}_{\tilde{A}}(x_i)$ when $i \le k$, and $\theta_i = \underline{\mu}_{\tilde{A}}(x_i)$ when $i \ge k+1$, and then compute $c_l(k) \equiv \dfrac{\sum_{i=1}^{k} x_i \overline{\mu}_{\tilde{A}}(x_i) + \sum_{i=k+1}^{N} x_i \underline{\mu}_{\tilde{A}}(x_i)}{\sum_{i=1}^{k} \overline{\mu}_{\tilde{A}}(x_i) + \sum_{i=k+1}^{N} \underline{\mu}_{\tilde{A}}(x_i)}$	Set $\theta_i = \underline{\mu}_{\tilde{A}}(x_i)$ when $i \le k$, and $\theta_i = \overline{\mu}_{\tilde{A}}(x_i)$ when $i \ge k+1$, and then compute $c_r(k) = \dfrac{\sum_{i=1}^{k} x_i \underline{\mu}_{\tilde{A}}(x_i) + \sum_{i=k+1}^{N} x_i \overline{\mu}_{\tilde{A}}(x_i)}{\sum_{i=1}^{k} \underline{\mu}_{\tilde{A}}(x_i) + \sum_{i=k+1}^{N} \overline{\mu}_{\tilde{A}}(x_i)}$
4	Check if $c_l(k) = c'$. If yes, stop and set $c_l(k) = c_l(L)$ and call k L. If no, go to step 5.	Check if $c_r(k) = c'$. If yes, stop and set $c_r(k) = c_r(R)$ and call k R. If no, go to step 5.
5	Set $c' = c_l(k)$ and go to step 2.	Set $c' = c_r(k)$ and go to step 2.

[a] Note that $x_1 \le x_2 \le \ldots \le x_N$.

Source: Mendel and Wu (2010, Chapter 2; © 2010, IEEE).

TABLE 2.4 EKM Algorithms for Computing the Centroid End Points of an IT2 FS, $\tilde{A}$[a].

Step	EKM Algorithm for $c_l(L)$	EKM Algorithm for $c_r(R)$
	$c_l(L) = \min_{\forall \theta_i \in [\underline{\mu}_{\tilde{A}}(x_i), \overline{\mu}_{\tilde{A}}(x_i)]} \left(\sum_{i=1}^{N} x_i \theta_i \Big/ \sum_{i=1}^{N} \theta_i \right)$	$c_r(R) = \max_{\forall \theta_i \in [\underline{\mu}_{\tilde{A}}(x_i), \overline{\mu}_{\tilde{A}}(x_i)]} \left(\sum_{i=1}^{N} x_i \theta_i \Big/ \sum_{i=1}^{N} \theta_i \right)$
1	Set $k = [N/2.4]$ (the nearest integer to $N/2.4$) and compute: $a = \sum_{i=1}^{k} x_i \overline{\mu}_{\tilde{A}}(x_i) + \sum_{i=k+1}^{N} x_i \underline{\mu}_{\tilde{A}}(x_i)$ $b = \sum_{i=1}^{k} \overline{\mu}_{\tilde{A}}(x_i) + \sum_{i=k+1}^{N} \underline{\mu}_{\tilde{A}}(x_i)$	Set $k = [N/1.7]$ (the nearest integer to $N/1.7$) and compute $a = \sum_{i=1}^{k} x_i \underline{\mu}_{\tilde{A}}(x_i) + \sum_{i=k+1}^{N} x_i \overline{\mu}_{\tilde{A}}(x_i)$ $b = \sum_{i=1}^{k} \underline{\mu}_{\tilde{A}}(x_i) + \sum_{i=k+1}^{N} \overline{\mu}_{\tilde{A}}(x_i)$
	Compute $c' = a/b$	
2	Find $k' \in [1, N-1]$ such that $x_{k'} \le c' \le x_{k'+1}$	
3	Check if $k' = k$. If yes, stop and set $c' = c_l(L)$, and $k = L$. If no, go to step 4.	Check if $k' = k$. If yes, stop and set $c' = c_r(R)$, and $k = R$. If no, go to step 4.
4	Compute $s = \text{sign}(k' - k)$ and $a' = a + s \sum_{i=\min(k,k')+1}^{\max(k,k')} x_i [\overline{\mu}_{\tilde{A}}(x_i) - \underline{\mu}_{\tilde{A}}(x_i)]$ $b' = b + s \sum_{i=\min(k,k')+1}^{\max(k,k')} [\overline{\mu}_{\tilde{A}}(x_i) - \underline{\mu}_{\tilde{A}}(x_i)]$	Compute $s = \text{sign}(k' - k)$ and $a' = a - s \sum_{i=\min(k,k')+1}^{\max(k,k')} x_i [\overline{\mu}_{\tilde{A}}(x_i) - \underline{\mu}_{\tilde{A}}(x_i)]$ $b' = b - s \sum_{i=\min(k,k')+1}^{\max(k,k')} [\overline{\mu}_{\tilde{A}}(x_i) - \underline{\mu}_{\tilde{A}}(x_i)]$
	Compute $c''(k') = a'/b'$	
5	Set $c' = c''(k')$, $a = a'$, $b = b'$ and $k = k'$ and go to Step 2.	

[a] Note that $x_1 \le x_2 \le \dots \le x_N$.

Source: Mendel and Wu (2010, Chapter 2; © 2010, IEEE).

TABLE 2.5 EIASC

Step	EIASC for $c_l(L)$	EIASC for $c_r(R)$
1	Initialize $a = \sum_{i=1}^{N} x_i \underline{\mu}_{\tilde{A}}(x_i)$ $b = \sum_{i=1}^{N} \underline{\mu}_{\tilde{A}}(x_i)$ $L=0$	Initialize $a = \sum_{i=1}^{N} x_i \overline{\mu}_{\tilde{A}}(x_i)$ $b = \sum_{i=1}^{N} \overline{\mu}_{\tilde{A}}(x_i)$ $R=N$
2	Compute $L=L+1$ $a = a + x_L[\overline{\mu}_{\tilde{A}}(x_L) - \underline{\mu}_{\tilde{A}}(x_L)]$ $b = b + [\overline{\mu}_{\tilde{A}}(x_L) - \underline{\mu}_{\tilde{A}}(x_L)]$ $c_l(L)=a/b$	Compute $a = a + x_R[\overline{\mu}_{\tilde{A}}(x_R) - \underline{\mu}_{\tilde{A}}(x_R)]$ $b = b + [\overline{\mu}_{\tilde{A}}(x_R) - \underline{\mu}_{\tilde{A}}(x_R)]$ $c_r(R)=a/b$ $R=R-1$
3	If $c_l(L) \leq x_{L+1}$, stop, otherwise go to step 2	If $c_r(R) \geq x_R$, stop, otherwise go to step 2

Source: Wu and Nie (2011).

EIASC algorithm when $N > 1000$. For fuzzy logic controllers, usually $N \ll 1000$; hence, this further enhanced EIASC algorithm was not recommended.

The enhanced opposite direction searching (EODS) algorithm, developed by Hu et al. (2012) is based on the location property in Table 2.2. Although it is somewhat faster than the EIASC, it is much more complicated to understand; hence, we do not include it here.

Comments

1. Many studies about KM algorithms and improved KM algorithms provide separate simulation results for both the number of iterations required for their convergence and overall computation time. The former does not change as computers or hardware change or improve, but the latter does; hence, a more meaningful metric would be *computation time per iteration*. This number will, of course, become smaller and smaller as computers become faster and faster, something that always seems to occur. One may conjecture that, at some not-to-distant future time, computation time per iteration will be so small that it will not matter which improved KM algorithm is used because the differences in overall computation time will be imperceptible to a human. Although this is true for a human, it is very important to realize that when the KM algorithms are implemented in *hardware*, then the faster they can be performed frees up the hardware to perform other computations, something that can be very important for practical applications. Consequently, it is important to perform the KM calculations as quickly as possible, which is why there has been extensive work on improving the KM algorithms.

2. It is important to clear up some confusion that exists about the number of iterations that are required by the KM algorithms. Note that if parallel processing is available (as in a hardware implementation of the algorithms), then the two KM algorithms can be run concurrently because they are completely independent of one another.

The earliest KM study (Karnik and Mendel, 2001a) proved that each KM algorithm requires at most N iterations. This bound is extremely conservative but, unfortunately, is still being quoted by some researchers who have not followed the later literature about the KM algorithms. In Wu and Mendel (2002), it was stated (without proof) that the *average number* of iterations of each KM algorithm is $\leq (N+2)/4$. While this is a much lower bound than N, it is also still very conservative. In Liu and Mendel (2008, Appendix C), it is proved that the number of iterations of each KM algorithm is $\leq \lfloor (N+1)/2 \rfloor$, which is also much smaller than N. Superexponential (not linear) convergence for continuous versions of the KM (or EKM) algorithms was proven in Mendel and Liu (2007), and quadratic convergence for them was proven in Liu and Mendel (2011).

Many simulation studies have been performed in which it has been observed that for two significant figures (often this accuracy is adequate for an FLC) the KM algorithms achieve their final results in from two to six iterations, *regardless of N*. The size of N in Eqs. (2.66) and (2.67) can increase the ratio of computation time per iteration because the larger N is the more multiplications and additions have to be performed, and this can lead to a delay that may cause performance degradations in real-time applications. For non-real-time applications, this delay is of no consequence. Even with today's computers (as of the year 2009), computation time per iteration is between 10^{-5} and 10^{-3} sec (e.g., Wu and Mendel, 2009).

Example 2.11 Figure 2.14 depicts an FOU for an IT2 FS and four of its embedded T1 FSs (Mendel and Wu, 2010, pp. 54–55). The center of gravity for each of these embedded T1 FSs is: $c_{(a)} = 4.5603$, $c_{(b)} = 4.5961$, $c_{(c)} = 3.9432$, and $c_{(d)} = 5.2333$. By using the KM algorithms, it is established that $c_l(L) = c_{(c)}$ and $c_r(R) = c_{(d)}$; hence, this example should dispel any mistaken belief that the end points of the centroid of an IT2 FS are associated with the centroids of its lower- and upper-membership functions, $c_{(a)}$ and $c_{(b)}$. They are associated with embedded T1 FSs that involve segments from both the lower- and upper-membership functions.

Example 2.12 In this example we show details for the steps of the two KM algorithms given in Table 2.3, using the IT2 FS depicted in Fig. 2.15. This FOU is for the output $\tilde{A}$ = Left that is used in the Chapter 3 mobile robot example, Example 3.3. In order to compute the centroid of this IT2 FS, we use a very coarse discretization, namely $N = 4$, with $x_1 = -120$, $x_2 = -90$, $x_3 = -60$, and $x_4 = -30$. This is done only for illustrative purposes. When the centroid of this FOU is computed using a computer implementation of the KM algorithm, a much finer sampling would be used in order to ensure greater accuracy in the computations of c_l and c_r.

Numerical values for $\underline{\mu}_{\tilde{A}}(x_i)$, $\overline{\mu}_{\tilde{A}}(x_i)$, and $[\underline{\mu}_{\tilde{A}}(x_i) + \overline{\mu}_{\tilde{A}}(x_i)]/2$ $(i = 1, 2, 3, 4)$ are (see Fig. 2.15)

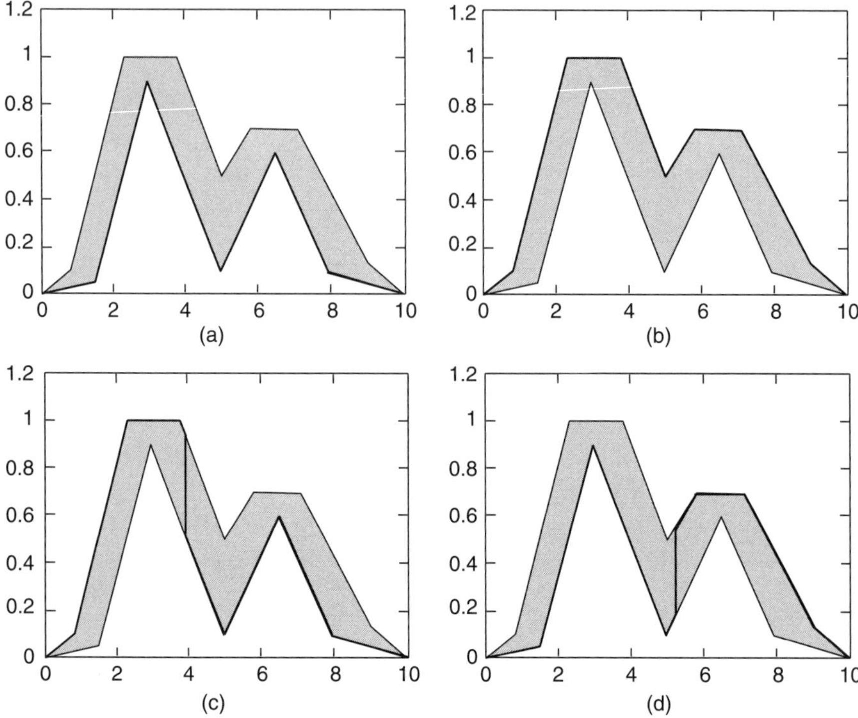

Figure 2.14 FOU and four embedded T1 FSs: (a) and (b) correspond to the lower and upper MFs, respectively; (c) and (d) (which are a result of using the KM algorithms) are associated with $c_l(L)$ and $c_r(R)$, respectively (Mendel and Wu, 2010, Chapter 2; © 2010, IEEE).

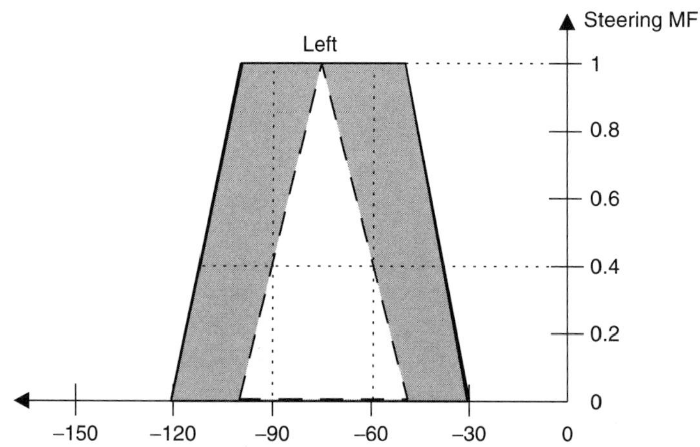

Figure 2.15 Very rough discretization of Left output IT2 FS.

$$\begin{cases} i=1 & x_1=-120 & \underline{\mu}_{\tilde{A}}(x_1)=0 & \overline{\mu}_{\tilde{A}}(x_1)=0 & \dfrac{\underline{\mu}_{\tilde{A}}(x_1)+\overline{\mu}_{\tilde{A}}(x_1)}{2}=0 \\ i=2 & x_2=-90 & \underline{\mu}_{\tilde{A}}(x_2)=0.4 & \overline{\mu}_{\tilde{A}}(x_2)=1 & \dfrac{\underline{\mu}_{\tilde{A}}(x_2)+\overline{\mu}_{\tilde{A}}(x_2)}{2}=0.7 \\ i=3 & x_3=-60 & \underline{\mu}_{\tilde{A}}(x_3)=0.4 & \overline{\mu}_{\tilde{A}}(x_3)=1 & \dfrac{\underline{\mu}_{\tilde{A}}(x_3)+\overline{\mu}_{\tilde{A}}(x_3)}{2}=0.7 \\ i=4 & x_4=-30 & \underline{\mu}_{\tilde{A}}(x_4)=0 & \overline{\mu}_{\tilde{A}}(x_4)=0 & \dfrac{\underline{\mu}_{\tilde{A}}(x_4)+\overline{\mu}_{\tilde{A}}(x_4)}{2}=0 \end{cases} \quad (2.79)$$

STEPS OF THE KM ALGORITHM FOR c_l (TABLE 2.3)

Iteration 1

Step 1: Initialize θ_i as $[\underline{\mu}_{\tilde{A}}(x_i)+\overline{\mu}_{\tilde{A}}(x_i)]/2$ using Eq. (2.79), and then compute

$$c' = \sum_{i=1}^{4} x_i\theta_i \Big/ \sum_{i=1}^{4} \theta_i$$

as:

$$c' = [(-120\times 0)+(-90\times 0.7)+(-60\times 0.7) + (-30\times 0)]/(0+0.7+0.7+0) = -75$$

Step 2: Find k $(1\le k\le 3)$ such that $x_k \le c' \le x_{k+1}$: Observe that -75 lies between $x_2=-90$ and $x_3=-60$; hence, $k=2$.

Step 3: Set $\theta_i=\overline{\mu}_{\tilde{A}}(x_i)$ when $i\le 2$, and $\theta_i=\underline{\mu}_{\tilde{A}}(x_i)$ when $i\ge 3$, and then compute $c_l(2)$ using Eq. (2.75) and the LMF and UMF values that are given in Eq. (2.79), as

$$c_l(2) = [(-120\times 0)+(-90\times 1)+(-60\times 0.4) + (-30\times 0)]/(0+1+0.4+0) = -81.43$$

Step 4: Check if $c_l(2)=c'$: Because $-81.43 \ne -75$, we go to step 5.

Step 5: Set $c'=-81.43$ and go to step 2 and begin iteration 2.

Iteration 2

Step 2: Find k $(1\le k\le 3)$ such that $x_k\le c'\le x_{k+1}$: Observe that $c'=-81.43$ lies between $x_2=-90$ and $x_3=-60$; hence, $k=2$, which is the same value as obtained in iteration 1.

Step 3: Set $\theta_i = \overline{\mu}_{\tilde{A}}(x_i)$ when $i \leq 2$, and $\theta_i = \underline{\mu}_{\tilde{A}}(x_i)$ when $i \geq 3$, and then compute c_l (2) using Eq. (2.75) and the LMF and UMF values that are given in Eq. (2.79). This will be the same value as obtained in iteration 1, namely $c_l(2) = -81.43$.

Step 4: Check if $c_l(2) = c'$: Because $-81.43 = -81.43$, we can stop and set $L = 2$.

STEPS OF THE KM ALGORITHM FOR $\boldsymbol{c_r}$ (TABLE 2.3)

Iteration 1

Steps 1 and 2 are exactly the same as in the KM algorithm for c_l; hence, $c' = -75$ and $k = 2$.

Step 3: Set $\theta_i = \underline{\mu}_{\tilde{A}}(x_i)$ when $i \leq 2$, and $\theta_i = \overline{\mu}_{\tilde{A}}(x_i)$ when $i \geq 3$, and then compute $c_r(2)$ using Eq. (2.75) and LMF and UMF values given in Eq. (2.79), as

$$\begin{aligned} c_r(2) =& [(-120 \times 0) + (-90 \times 0.4) + (-60 \times 1) \\ & + (-30 \times 0)]/(0 + 0.4 + 1 + 0) = -68.47 \end{aligned}$$

Step 4: Check if $c_r(2) = c'$: Because $-68.57 \neq -75$, we go to step 5.

Step 5: Set $c' = -68.57$ and go to step 2 and begin iteration 2.

Iteration 2

Step 2: Find k $(1 \leq k \leq 3)$ such that $x_k \leq c' \leq x_{k+1}$: Observe that $c' = -68.75$ lies between $x_2 = -90$ and $x_3 = -60$; hence, $k = 2$, which is the same value as obtained in iteration 1.

Step 3: Set $\theta_i = \underline{\mu}_{\tilde{A}}(x_i)$ when $i \leq 2$, and $\theta_i = \overline{\mu}_{\tilde{A}}(x_i)$ when $i \geq 3$, and then compute $c_l(2)$ using Eq. (2.75) and LMF and UMF values given in Eq. (2.79). This will be the same value as obtained in iteration 1, namely $c_l(2) = -68.57$.

Step 4: Check if $c_l(2) = c'$: Because $-68.57 = -68.57$, we can stop and set $R = 2$.

Note that when the EKM algorithm is used, stopping occurs in step 3 as soon as it is found that k found in step 2 of iteration 2 is the same as k found in iteration 1.

2.4 GENERAL TYPE-2 FUZZY SETS

2.4.1 α-Plane/zSlice Representation

A general T2 FS (GT2 FS) sits atop its FOU; hence, beginning with $\tilde{A}$ in Eq. (2.25), a GT2 FS can be expressed as [see, also Eqs. (2.30), (2.33), and (2.35)]

$$\begin{cases} \tilde{A} = \int_{\forall x \in X} \mu_{\tilde{A}}(x)/x = \int_{\forall x \in X} \left[\int_{\forall u \in J_x^u} f_x(u)/u \right] \Big/ x \\ J_x^u = \{(x,u) : \forall u \in [\mathrm{LMF}_{\mathrm{FOU}(\tilde{A})}(x), \mathrm{UMF}_{\mathrm{FOU}(\tilde{A})}(x)]\} \subseteq [0,1] \end{cases} \tag{2.80}$$

where $f_x(u)$ is the secondary grade of $\tilde{A}$. For a GT2 FS $f_x(u) \neq 1$ for $\forall x \in X$ and $\forall u \in J_x^u$. Equation (2.80) is the *vertical-slice* representation of a GT2 FS; it has found wide use in computations involving GT2 FSs.

Liu (2008) introduced the *horizontal-slice decomposition* (representation) of a GT2 FS. Because a horizontal slice is analogous to an α-cut raised to level α of a T1 FS (Section 2.2.3), it is called the *α-plane representation* of a GT2 FS, or, when x, y and z are used for the three coordinates of a GT2 FS, a *zSlice representation* (Wagner and Hagras, 2010).

Definition 2.17 An *α-plane* for the GT2 FS $\tilde{A}$, denoted $\tilde{A}_\alpha$, is the union of all primary memberships of $\tilde{A}$ whose secondary grades are greater than or equal to α $(0 \leq \alpha \leq 1)$, that is,

$$\begin{aligned} \tilde{A}_\alpha &= \{(x,u), \mu_{\tilde{A}}(x,u) \geq \alpha | \forall x \in X,\ \forall u \in [0,1]\} \\ &= \int_{\forall x \in X} \int_{\forall u \in [0,1]} \{(x,u) | f_x(u) \geq \alpha\} \end{aligned} \tag{2.81}$$

Note that

$$\mathrm{FOU}(\tilde{A}) = \tilde{A}_0 \tag{2.82}$$

and that, just as an α cut of a T1 FS resides on its 1D domain of support X, $\tilde{A}_\alpha$ resides on its 2D domain of support $X \times U$.

Each $\tilde{A}_\alpha$ can be converted to a *special IT2 FS* $R_{\tilde{A}_\alpha}$ (Mendel, 2010) where

$$R_{\tilde{A}_\alpha}(x,u) = \alpha / \tilde{A}_\alpha \ \forall x \in X, u \in [0,1] \tag{2.83}$$

Observe $R_{\tilde{A}_\alpha}(x,u)$ raises $\tilde{A}_\alpha$ to level-α so that $R_{\tilde{A}_\alpha}$ is an IT2 FS all of whose secondary MFs equal α (rather than 1 as would be the case for the usual IT2 FS). $R_{\tilde{A}_\alpha}$ is called an *α-level (zSlice) T2 FS*; it is also designated as $\tilde{A}(\alpha)$ (Liu, 2008).

An example that shows one $R_{\tilde{A}_\alpha}(x,u)$ is provided in Fig. 2.16 (Mendel, 2012). Observe that the GT2 FS $\tilde{A}$ can also be represented as the union of its vertical slices—the secondary MFs in Eq. (2.80)—and that each α-plane intersects a secondary MF at an α-cut of that MF raised to level-α.

THEOREM 2.4 The *α-plane (zSlice) representation* for a GT2 FS is

$$\tilde{A} = \bigcup_{\alpha \in [0,1]} R_{\tilde{A}_\alpha} = \bigcup_{\alpha \in [0,1]} \alpha / \tilde{A}_\alpha = \sup_{\alpha \in [0,1]} \left[\alpha / \tilde{A}_\alpha \right] \tag{2.84}$$

In Eq. (2.84), $\bigcup$ is the fuzzy union over α.

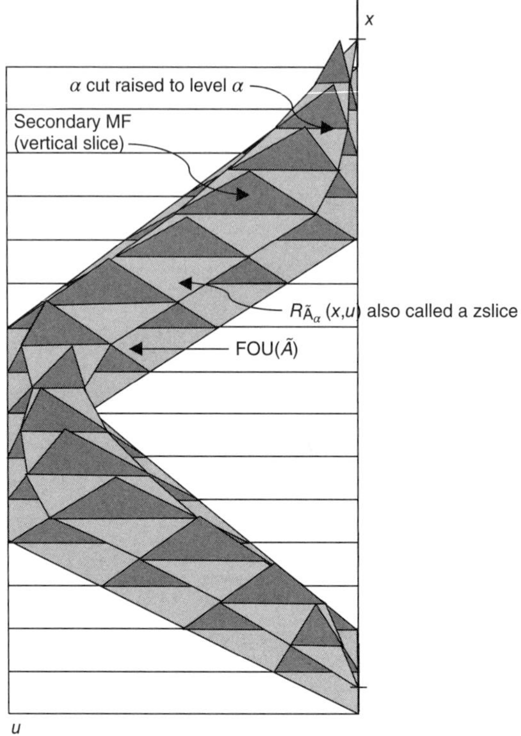

Figure 2.16 A 2–1/2D plot of α-plane representation of a GT2 FS (Mendel 2012; © 2012, IEEE). The $\mu(x, u)$ direction appears to come out of the page and is the new third dimension of a GT2 FS.

Proof. To begin, we express Eq. (2.80) as

$$\tilde{A} = \int_{\forall x \in X} \mu_{\tilde{A}}(x)/x = \int_{\forall x \in X} \left[\int_{\forall u \in [0,1]} f_x(u)/u \right] \Big/ x \tag{2.85}$$

Note that $\int_{\forall u \in [0,1]} f_x(u)/u$ is a T1 FS, called a *secondary MF* or a *vertical slice*, $\mu_{\tilde{A}}(x)$, that is,

$$\mu_{\tilde{A}}(x) = \int_{\forall u \in [0,1]} f_x(u)/u \tag{2.86}$$

This is a generalization of Eq. (2.28) from an IT2 FS to a GT2 FS. Because $\mu_{\tilde{A}}(x)$ is a T1 FS, it can also be expressed by means of the α-cut decomposition Theorem 2.1 as

$$\mu_{\tilde{A}}(x) = \bigcup_{\alpha \in [0,1]} \mu_{\tilde{A}}(x|\alpha) = \bigcup_{\alpha \in [0,1]} \alpha / \mu_{\tilde{A}}(\alpha) \tag{2.87}$$

where $\mu_{\tilde{A}}(\alpha)$ is the α-cut of the T1 FS $\mu_{\tilde{A}}(x)$, that is, [see Eq. (2.7)]

$$\mu_{\tilde{A}}(\alpha) = \{x | \mu_{\tilde{A}}(x) > \alpha\} \tag{2.88}$$

Substituting Eq. (2.87) into Eq. (2.85), it follows that

$$\begin{aligned}\tilde{A} &= \int_{\forall x \in X} \bigcup_{\alpha \in [0,1]} \alpha / \mu_{\tilde{A}}(\alpha) / x = \bigcup_{\alpha \in [0,1]} \int_{\forall x \in X} \alpha / \mu_{\tilde{A}}(\alpha) / x \\ &= \bigcup_{\alpha \in [0,1]} \alpha / \left[\int_{\forall x \in X} \mu_{\tilde{A}}(\alpha) / x \right]\end{aligned} \tag{2.89}$$

Using the concept of an α-plane (Definition 2.17), Eq. (2.89) can also be expressed as

$$\tilde{A} = \bigcup_{\alpha \in [0,1]} \alpha \bigg/ \left[\int_{\forall x \in X} \mu_{\tilde{A}}(\alpha) / x \right] = \bigcup_{\alpha \in [0,1]} \alpha / \tilde{A}_\alpha = \bigcup_{\alpha \in [0,1]} R_{\tilde{A}_\alpha}(x, u) \;\; \forall x \in X, u \in [0, 1] \tag{2.90}$$

which is Eq. (2.84).

Comment Chapter 7 uses the term "zSlice" instead of α-level T2FS and presents some of the material that is presented below directly in terms of those slices. We present this material in terms of α-planes so that the reader can easily see the direct connections between them and earlier T1 and IT2 results. Table 2.6 provides a comparison of α-plane and zSlice descriptions of some important items.

TABLE 2.6 Comparisons of α-Plane and zSlice Descriptions

Item	α-Plane Description	zSlice Description
Coordinates	(x, u, μ)	(x, y, z)
α-plane	$\tilde{A}_\alpha = \int_{\forall x \in X} \int_{\forall u \in [0,1]} \{(x, u) \vert f_x(u) \geq \alpha\}$	$\tilde{Z}_0$ projected onto $X \times Y$
zSlice	$\alpha / \tilde{A}_\alpha$	$\tilde{Z}_z = \int_{\forall x \in X} \int_{\forall y \in [0,1]} z/(x, y)$
FOU($\tilde{A}$)	$\tilde{A}_0$	$\tilde{Z}_0$
Vertical slice ($x_j \in X$)	$\int_{\forall \alpha \in [0,1]} \int_{\forall u \in [0,1]} \alpha / (x_j, u)$	$\int_{\forall z \in [0,1]} \int_{\forall y \in [0,1]} z/(x_j, y)$
Representation of $\tilde{A}$	$\tilde{A} = \bigcup_{\alpha \in [0,1]} \alpha / \tilde{A}_\alpha = \sup_{\alpha \in [0,1]} [\alpha / \tilde{A}_\alpha]$	$\tilde{A} = \bigcup_{z \in [0,1]} \tilde{Z}_z = \sup_{z \in [0,1]} [\tilde{Z}_z]$

2.4.2 Set-Theoretic Operations

All operations that have been developed for IT2 FSs can be applied to GT2 FSs, but for each of their α-level T2 FSs e.g., Liu (2008); Mendel et al. (2009); Wagner and Hagras (2010); Zhai and Mendel (2011b). This is a direct consequence of the α-plane (zSlice) representation for a GT2 FS.

Consider two GT2 FSs $\tilde{A}$ and $\tilde{B}$, where $\tilde{A}$ is defined in Eq. (2.80) and

$$\begin{cases} \tilde{B} = \displaystyle\int_{\forall x \in X} \mu_{\tilde{B}}(x)/x = \int_{\forall x \in X} \left[\int_{\forall w \in J_x^w} g_x(w)/w \right] \Big/ x \\ J_x^w = \{(x, w) : \forall w \in [\mathrm{LMF}_{\mathrm{FOU}(\tilde{B})}(x), \mathrm{UMF}_{\mathrm{FOU}(\tilde{B})}(x)]\} \subseteq [0, 1] \end{cases} \tag{2.91}$$

THEOREM 2.5 Let $(\tilde{A} \cup \tilde{B})_\alpha$ and $(\tilde{A} \cap \tilde{B})_\alpha$ be the α-planes of $\tilde{A} \cup \tilde{B}$ and $\tilde{A} \cap \tilde{B}$, respectively (Mendel et al., 2009); then

$$\tilde{A} \cup \tilde{B} = \bigcup_{\alpha \in [0,1]} \alpha/(\tilde{A} \cup \tilde{B})_\alpha = \bigcup_{\alpha \in [0,1]} \alpha/\tilde{A}_\alpha \cup \tilde{B}_\alpha \tag{2.92}$$

$$\tilde{A} \cap \tilde{B} = \bigcup_{\alpha \in [0,1]} \alpha/(\tilde{A} \cap \tilde{B})_\alpha = \bigcup_{\alpha \in [0,1]} \alpha/\tilde{A}_\alpha \cap \tilde{B}_\alpha \tag{2.93}$$

Proof. The proof of Eq. (2.92) can be found in Appendix A of Mendel et al. (2009). In that proof it is demonstrated that, because $\tilde{A}_\alpha$ and $\tilde{B}_\alpha$ are interval-valued sets, the calculations of $\tilde{A}_\alpha \cup \tilde{B}_\alpha$ [and $\tilde{A}_\alpha \cap \tilde{B}_\alpha$] only involve interval arithmetic and, in fact, reduce to the same computations that already exist for the union (and intersection) of IT2 FSs. The proof of Eq. (2.93) is similar to the proof of Eq. (2.92). In retrospect, these results are intuitively obvious.

To compute, for example, $\tilde{A}_\alpha \cup \tilde{B}_\alpha$, for each value of α:

1. Determine LMF($\tilde{A}_\alpha$), UMF($\tilde{A}_\alpha$), LMF($\tilde{B}_\alpha$), and UMF($\tilde{B}_\alpha$).
2. Compute $\tilde{A}_\alpha \cup \tilde{B}_\alpha = [\mathrm{LMF}(\tilde{A}_\alpha \cup \tilde{B}_\alpha), \mathrm{UMF}(\tilde{A}_\alpha \cup \tilde{B}_\alpha)]$, using Eq. (2.46).

Example 2.13 Figure 2.17a depicts $\tilde{A}_\alpha$ and $\tilde{B}_\alpha$ as well as their lower and upper MFs (Mendel et al. 2009). Once the two α-planes are drawn, it is a relatively simple matter to draw the α-plane of $\tilde{A}_\alpha \cup \tilde{B}_\alpha$. Just take the maximum value of UMF($\tilde{A}_\alpha$) and UMF($\tilde{B}_\alpha$) to find UMF($\tilde{A}_\alpha \cup \tilde{B}_\alpha$). Similarly, take the maximum value of LMF($\tilde{A}_\alpha$) and LMF($\tilde{B}_\alpha$) to find LMF($\tilde{A}_\alpha \cup \tilde{B}_\alpha$). These lower and upper MFs, as well as the α-plane $(\tilde{A} \cup \tilde{B})_\alpha = \tilde{A}_\alpha \cup \tilde{B}_\alpha$, are shown in Fig. 2.17b.

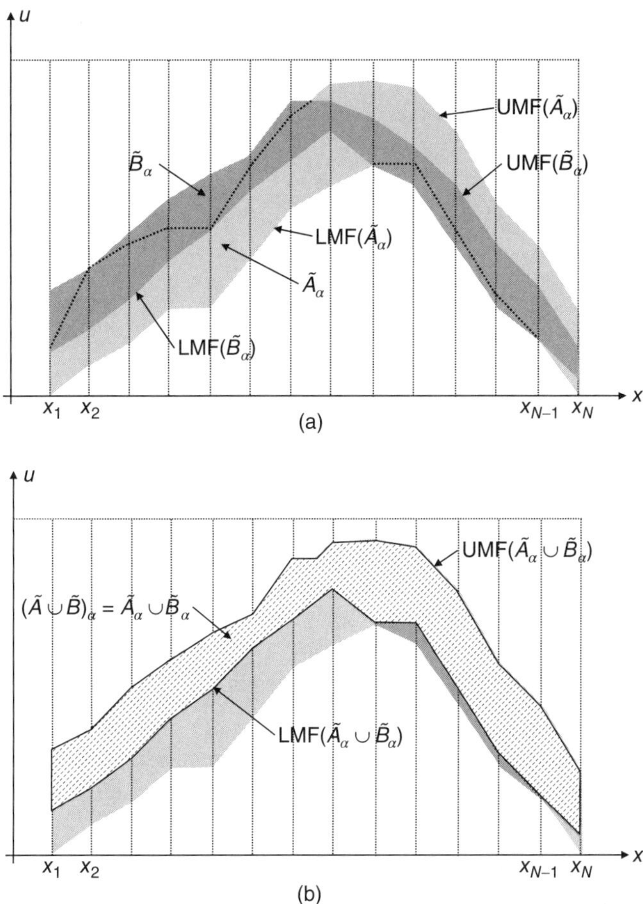

Figure 2.17 (a) α-planes $\tilde{A}_\alpha$ (lightly shaded) and $\tilde{B}_\alpha$(darkly shaded). (b) Their union (the outlined region) (Mendel et al., 2009; © 2009, IEEE). Note that in (a) the dotted curves indicate the portions of $\tilde{A}_\alpha$ that lie behind $\tilde{B}_\alpha$.

Instead of expressing $\tilde{A} \cup \tilde{B}$ and $\tilde{A} \cap \tilde{B}$ in terms of α-planes, they can also be expressed as the union of their vertical slices. The vertical slices of $\tilde{A} \cap \tilde{B}$ are called the *meet*, whereas the vertical slices of $\tilde{A} \cap \tilde{B}$ are called the *join*. Formulas for the meet and join can also be obtained directly by using the extension principle (Mendel, 2001; Karnik and Mendel, 2001b).

2.4.3 Centroid of a GT2 FS

Just as set-theoretic operations for GT2 FSs can be computed one α-plane at a time, and then aggregated by means of the fuzzy union over all values of α, so can the centroid of a GT2 FS be computed.

THEOREM 2.6 The centroid of GT2 FS $\tilde{A}$ (Liu, 2008), $C_{\tilde{A}}(x)$, is a T1 FS that can be computed using the α-plane representation of $\tilde{A}$, that is,

$$C_{\tilde{A}}(x) = \bigcup_{\alpha\in[0,1]} C_{R_{\tilde{A}_\alpha}}(x) = \bigcup_{\alpha\in[0,1]} \alpha/[c_l(R_{\tilde{A}_\alpha}), c_r(R_{\tilde{A}_\alpha})] \equiv \bigcup_{\alpha\in[0,1]} \alpha/[c_l(\alpha), c_r(\alpha)] \tag{2.94}$$

where $C_{R_{\tilde{A}_\alpha}}(x)$ is the centroid of the α-level T2 FS $R_{\tilde{A}_\alpha}$.

Proof. Obvious from the α-plane representation of $\tilde{A}$ that is given in Eq. (2.173).

A procedure for computing $C_{\tilde{A}}(x)$ is (Mendel et al. 2009):

1. Decide on how many α-planes will be used, where $\alpha \in [0, 1]$. Call that number k; its choice will depend on the accuracy that is required.
2. For each α, compute $\tilde{A}_\alpha$.
3. Compute $c_l(\alpha)$ and $c_r(\alpha)$ using two EKM algorithms (Table 2.4) or EIASC algorithms (Table 2.5). The accuracy of these results will depend upon the discretization of the primary variable.
4. Repeat steps 2 and 3 for the k values of α chosen in step 1.
5. Bring all of the $k\,C_{R_{\tilde{A}_\alpha}}(x)$ together using Eq. (2.94) to obtain $C_{\tilde{A}}(x)$.

If parallel processing is available, all of this can be performed using $2k$ processors; however this five-step procedure does not make use of the additional information that is available about the secondary MFs of a GT2 FS.

Zhai and Mendel (2011a) developed *centroid-flow algorithms* (*CFAs*). To begin, the centroid of the $\alpha = 0$ plane is computed using EKM algorithms, and then that centroid is propagated up to the next α-level T2 FS using formulas called *centroid-flow equations*. In order to reduce the accumulation of errors that can occur during this propagation procedure, they improved these algorithms in *enhanced centroid-flow algorithms* (Zhai and Mendel, 2012), which start at the $\alpha = 1/2$ level T2 FS and then move upward (downward) to the $\alpha = 1$ ($\alpha = 0$) level T2 FS. This reduces the error accumulation by 50%, which may be adequate for practical applications.

Yeh et al. (2011) compute the centroid of the $\alpha = 1$ level T2 FS using EKM algorithms that are initialized as indicated in Table 2.4 and then that centroid is used to *better initialize* the EKM algorithms for the computation of the centroid of the $\alpha = 1 - \delta$ level T2 FS, after which those results are used to *better initialize* the EKM algorithms for the computation of the centroid of the $\alpha = 1 - 2\delta$ level T2 FS, and so forth, until the last centroid is computed for the $\alpha = 0$ level T2 FS. A reason for beginning with the $\alpha = 1$ level T2 FS, instead of with the $\alpha = 0$ level T2 FS, is that when secondary MFs are all triangles, then the FOU of the $\alpha = 1$ level T2 FS is a T1 FS, and so one does not need to use EKM algorithms to compute the center of gravity (COG) of that function.

Both of these flow algorithms achieve a 70% reduction in computation time over using EKM algorithms for each α-level T2 FS.

Linda and Manic's (2012) *monotone centroid-flow algorithms* (*MCFAs*) do not require any EKM algorithms and are to date the fastest way to compute the centroid of a GT2 FS. The MCFAs start with the $\alpha = 1$ level T2 FS so as to use the exact COG calculation when all secondary MFs are triangles, or an approximate COG calculation if all secondary MFs are trapezoids. For the latter, they compute the COG of the average value of the lower and upper MFs (Nie and Tan, 2008). They then move down to the $\alpha = 1 - \delta$ level T2 FS but do not use the EKM algorithms at that plane. Instead, they return to *fundamentals* by focusing on the structure of the solutions in Eqs. (2.66) and (2.67) but for the $\alpha = 1 - \delta$ level T2 FS.

For triangle and trapezoidal secondary MFs [the MCFAs can also be applied to other secondary MFs; see Linda and Manic (2012) for details], the centroid is known to be an increasing function of α, as α goes from $\alpha = 1$ to $\alpha = 0$ (Mendel et al., 2009; Linda and Manic, 2012). This means that if the centroid at $\alpha = \alpha'$ is $[c_l(\alpha'), c_r(\alpha')]$, then the centroid at $\alpha = \alpha' - \delta$, $[c_l(\alpha' - \delta), c_r(\alpha' - \delta)]$, satisfies the following containment property:

$$[c_l(\alpha'), c_r(\alpha')] \subset [c_l(\alpha' - \delta), c_r(\alpha' - \delta)] \tag{2.95}$$

which means that

$$c_l(\alpha' - \delta) < c_l(\alpha') \quad \text{and} \quad c_r(\alpha' - \delta) > c_r(\alpha') \tag{2.96}$$

By using Eqs. (2.95), (2.96), (2.66), and (2.67), it is easy to find the switch points $L(\alpha' - \delta)$ and $R(\alpha' - \delta)$ from switch points $L(\alpha')$ and $R(\alpha')$ just by using arithmetic. By starting with $c_l(\alpha')$ in Eq (2.66) and changing $L(\alpha')$ to $L(\alpha') - k$ ($k = 1, \ldots$), one stops at the first value of k that violates $c_l(\alpha' - \delta) < c_l(\alpha')$, that is, the first value of k for which $c_l(\alpha' - \delta) > c_l(\alpha')$; and, by starting with $c_r(\alpha')$ in Eq. (2.67) and changing $R(\alpha')$ to $R(\alpha') + k$ ($k = 1, \ldots$), one stops at the first value of k that violates $c_r(\alpha' - \delta) > c_r(\alpha')$, that is, the first value of k for which $c_r(\alpha' - \delta) < c_r(\alpha')$. For each α-level T2 FS, Linda and Manic (2012) are in effect using the EIASC (Table 2.5).

Example 2.14 Shown in Fig. 2.18a is $\text{FOU}(\tilde{A})$ for which $\text{UMF}_{\text{FOU}(\tilde{A})}(x)$ and $\text{LMF}_{\text{FOU}(\tilde{A})}(x)$ are each the maximum of two piecewise linear functions (Mendel et al., 2009). This FOU is representative of one that might be obtained by computing the union of two fired-rule output sets in a GT2 FLC, and

$$\text{UMF}_{\text{FOU}(\tilde{A})}(x) = \max\left\{\begin{bmatrix} (x-1)/2 & 1 \le x \le 3 \\ (7-x)/4 & 3 \le x \le 7 \\ 0 & \text{otherwise} \end{bmatrix}, \begin{bmatrix} (x-2)/5 & 2 \le x \le 6 \\ (16-2x)/5 & 6 \le x \le 8 \\ 0 & \text{otherwise} \end{bmatrix}\right\} \tag{2.97}$$

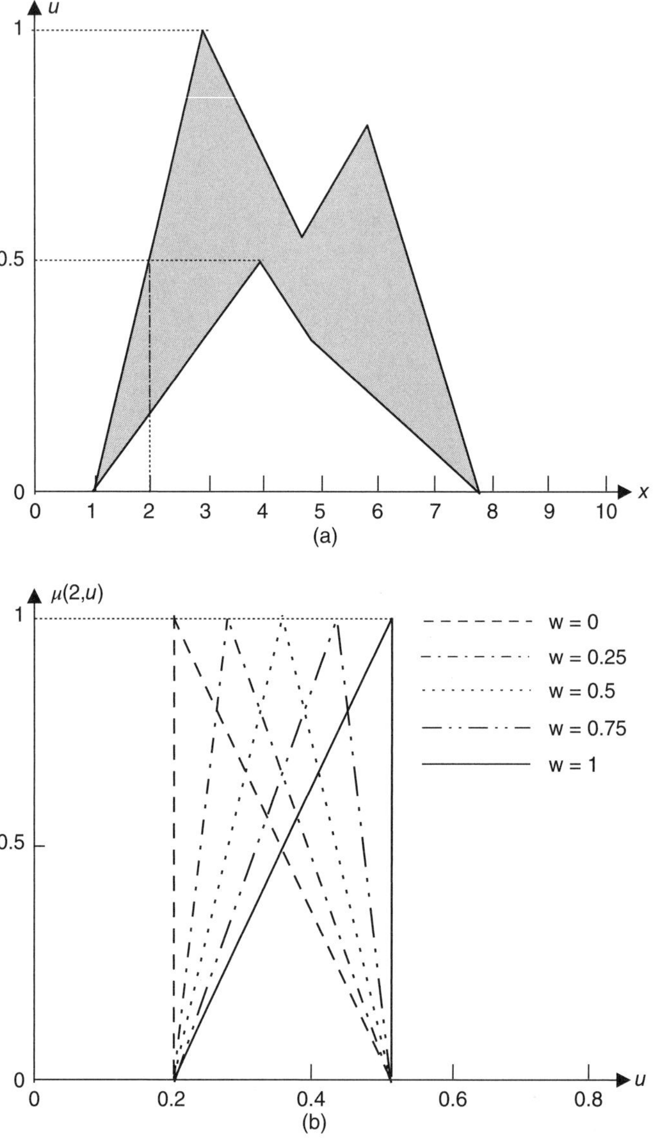

Figure 2.18 (a) FOU for Example 2.14. (b) Secondary MFs at $x = 2$ for five values of w (Mendel et al., 2009; © 2009, IEEE).

$$\text{LMF}_{\text{FOU}(\tilde{A})}(x) = \max\left\{\begin{bmatrix} (x-1)/6 & 1 \leq x \leq 4 \\ (7-x)/6 & 4 \leq x \leq 7 \\ 0 & \text{otherwise} \end{bmatrix}, \begin{bmatrix} (x-3)/6 & 3 \leq x \leq 5 \\ (8-x)/9 & 5 \leq x \leq 8 \\ 0 & \text{otherwise} \end{bmatrix}\right\} \tag{2.98}$$

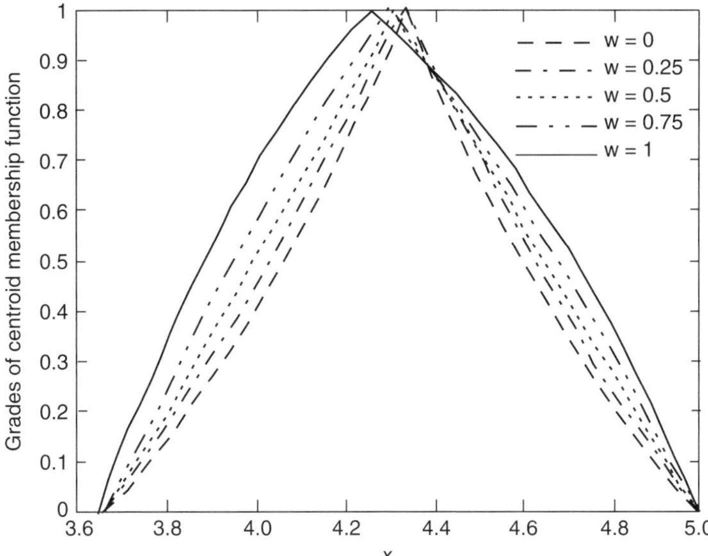

Figure 2.19 $C_{\tilde{A}}(x)$ for Example 2.14 when $w = 0, 0.25, 0.50, 0.75, 1$ (Mendel et al., 2009; © 2009, IEEE).

In this example, each secondary MF is chosen to be a triangle of height 1 whose base equals $\text{UMF}_{\text{FOU}(\tilde{A})}(x) - \text{LMF}_{\text{FOU}(\tilde{A})}(x)$ and whose apex location, Apex(x), is parameterized as

$$\text{Apex}(x) = \text{LMF}_{\text{FOU}(\tilde{A})}(x) + w[\text{UMF}_{\text{FOU}(\tilde{A})}(x) - \text{LMF}_{\text{FOU}(\tilde{A})}(x)] \qquad (2.99)$$

where $w = 0, 0.25, 0.5, 0.75, 1$. These secondary MFs are depicted in Fig. 2.18b when $x = 2$; so, $\tilde{A}$ is defined by Eqs. (2.97)–(2.99).

The centroids of $\tilde{A}$, which are T1 FSs, are depicted in Fig. 2.19 for the five values of w. Note that when $\alpha = 0$, $C_{R_{\tilde{A}_0}}(x) = [3.6605, 4.9917]$, and this is the support of all five centroids. Each of the centroids in Fig. 2.19 looks symmetrical; however, they are not exactly symmetrical.

2.5 WRAPUP

It is worthwhile to step back and consider the uncertainty models provided by the three kinds of fuzzy sets that have been discussed in this chapter. It is these uncertainty models that make the fundamental difference between the different kinds of FLCs, and it is these uncertainty models that will generally drive the selection of a specific kind of FLC for a given application.

In a T1 FLC, the degrees of its T1 FS memberships are specified as crisp numbers that belong to the interval [0, 1]. In a GT2 FLC, the degrees of its GT2 FS memberships are themselves fuzzy where each is specified as a T1 FS—a secondary MF.

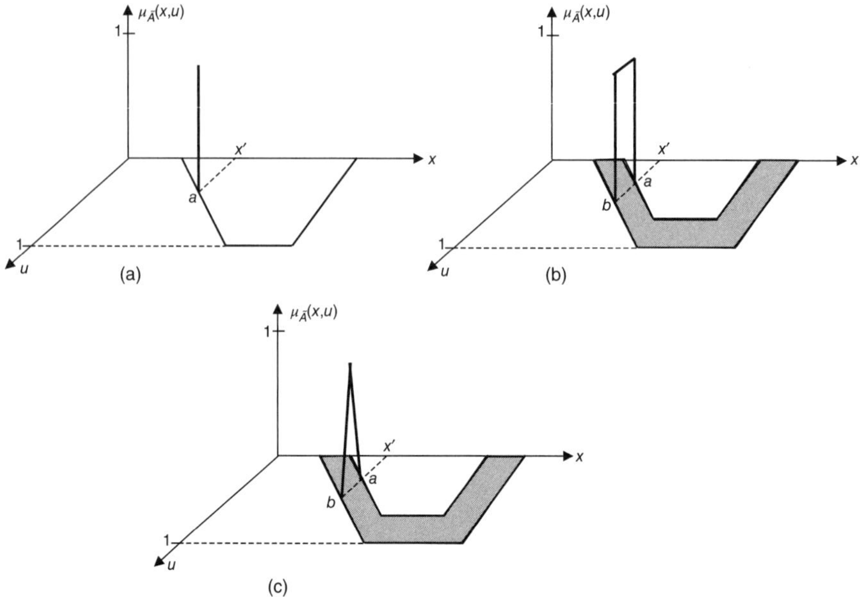

Figure 2.20 Secondary MFs at $x = x'$ for (a) T1 FS, (b) IT2 FS, and (c) GT2 FS.

When the secondary MF is always equal to 1, then each GT2 FS reduces to an IT2 FS, and the GT2 FLC reduces to an IT2 FLC.

Traditionally, a T1 FS is not visualized in three dimensions, whereas GT2 and IT2 FSs are (although only the FOU is needed for the latter). In order to level the playing field for these three kinds of FSs, it is also possible to visualize a T1 FS in three dimensions. This is easy to do by treating a T1 FS as an IT2 FS whose lower and upper MFs are the same, and whose secondary grades are all equal to 1. The three kinds of secondary membership functions for these three kinds of FSs are depicted in Fig. 2.20 at one specific value of the primary membership, $x = x'$.

Observe that for a T1 FS (Fig. 2.20a) the primary membership at $x = x'$ has only one value, a, for which the secondary grade equals 1; hence, in a T1 FS, for each x value there is no uncertainty associated with the primary membership value. For an IT2 FS (Fig. 2.20b), the primary membership at $x = x'$ has values within the interval $[a, b]$, where each point in this interval has a secondary membership equal to 1. As such, an IT2 FS encompasses a large amount of uncertainty that is spread evenly over the interval $[a, b]$. Finally, for a GT2 FS (Fig. 2.20c), the primary membership at $x = x'$ again has values within the interval $[a, b]$, but now each point in this interval can have a different secondary membership. This means the uncertainty that is associated with a GT2 FS, U_{GT2}, lies somewhere between the uncertainty of a T1 FS, U_{T1} and an IT2 FS, U_{IT2}, that is,

$$U_{\mathrm{T1}} \leq U_{\mathrm{GT2}} \leq U_{\mathrm{IT2}} \tag{2.100}$$

It appears, therefore, that using a GT2 FS will provide the most flexibility for modeling MF uncertainty.

While it is clear that there is a difference in the amount of uncertainty associated with the three kinds of FSs, it is still a very challenging problem to determine how to make the best use of the different FS's abilities to model real-world uncertainties. This challenge is particularly noteworthy in the context of designing an FLC, that is, should it be T1, IT2 or a GT2 FLC? And, is the hoped-for improvement in performance by either an IT2 or GT2 FLC worth their additional complexities? These questions can only be answered in the context of a specific application, and some of them are explored in the rest of this book.

2.6 MOVING ON

The background materials that you have just covered in this chapter will be used extensively in the rest of this book. In the next chapter they will all be used to obtain mathematical descriptions of Mamdani and TSK FLCs.

CHAPTER 3

Interval Type-2 Fuzzy Logic Controllers

3.1 INTRODUCTION

This chapter begins with short reviews of T1 Mamdani and TSK FLCs in Section 3.2 so as to set the stage for the complete descriptions of IT2 Mamdani and TSK FLCs that are given in Section 3.3. The Wu–Mendel uncertainty bounds, which have let IT2 Mamdani FLCs run in real time, are given in Section 3.4 (their derivations are included in Appendix 3A). Brief discussions about control performance for IT2 FLCs are given in Section 3.5. Two ways to determine FOU parameters of IT2 FLCs are described in Section 3.6.

3.2 TYPE-1 FUZZY LOGIC CONTROLLERS

3.2.1 Introduction[1]

Because our derivations of equations for an IT2 FLC in Section 3.3 use the equations for a T1 FLC, we provide a brief review of the latter here. As noted in Chapter 1, there are two major architectures for a T1 and T2 FLC, Mamdani and TSK. Both are reviewed in this section.

A generic T1 FLC is depicted in Fig. 3.1. In general, this FLC has p inputs $x_1 \in X_1, \ldots, x_p \in X_p$, $[\mathbf{x} \equiv \text{col}(x_1, \ldots, x_p)]$ and one output[2] $u(\mathbf{x}) \in U$, and is characterized by M rules. For a T1 Mamdani FLC, the sth rule has the form

$$R_M^s\text{: If } x_1 \text{ is } F_1^s \text{ and } \cdots \text{ and } x_p \text{ is } F_p^s, \text{then } u(\mathbf{x}) \text{ is } G^s \quad s = 1, \ldots, M \tag{3.1}$$

For a T1 TSK FLC, the sth rule has the form

$$R_{\text{TSK}}^s\text{: If } x_1 \text{ is } F_1^s \text{ and} \cdots \text{and } x_p \text{ is } F_p^s, \text{then } u^s(\mathbf{x}) \text{ is } g^s(x_1, \ldots, x_p) \quad s = 1, \ldots, M \tag{3.2}$$

The most common function used for $g^s(x_1, \ldots, x_p)$ is linear in $x_1, \ldots,$ and x_p, that is,

[1] Some of the material in this section has been taken from Mendel et al. (2006; © 2006, IEEE).

[2] In this chapter $u(\mathbf{x})$ and u are often used interchangeably, especially in derivations where it is not necessary to show the explicit dependence of u on $\mathbf{x}$.

Introduction to Type-2 Fuzzy Logic Control: Theory and Applications, First Edition.
Jerry M. Mendel, Hani Hagras, Woei-Wan Tan, William W. Melek, and Hao Ying.

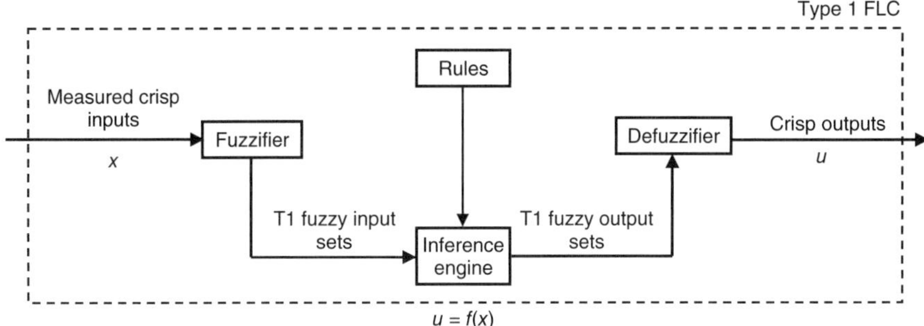

Figure 3.1 Type 1 FLC (Mendel et al., 2006; © 2006, IEEE).

$$u^s(\mathbf{x}) = c_0^s + c_1^s x_1 + \cdots + c_p^s x_p \tag{3.3}$$

In the rest of this section we explain what the computations are going from the input $\mathbf{x}$ to the crisp controller output $u(\mathbf{x})$ for both T1 Mamdani and TSK FLCs.

3.2.2 T1 Mamdani FLCs

The rule in Eq. (3.1) represents a T1 fuzzy relation between the input space $X_1 \times \cdots \times X_p$ and the output space, U, of the FLC. In the inference engine block of Fig. 3.1, fuzzy logic principles are used to combine fuzzy if–then rules from the fuzzy rule base into a mapping from T1 fuzzy input sets in $X_1 \times \cdots \times X_p$ to T1 fuzzy output sets in U. Each rule is interpreted as a fuzzy implication. With reference to Eq. (3.1), let $F_1^s \times \cdots \times F_p^s \equiv A^s$; then, Eq. (3.1) can be reexpressed as

$$R_M^s\colon\ F_1^s \times \cdots \times F_p^s \to G^s = A^s \to G^s \quad s = 1, \ldots, M \tag{3.4}$$

Rule R_M^s is described by the MF $\mu_{R_M^s}(\mathbf{x}, u)$, where

$$\mu_{R_M^s}(\mathbf{x}, u) = \mu_{A^s \to G^s}(\mathbf{x}, u) \tag{3.5}$$

and $\mathbf{x} = (x_1, \ldots, x_p)^T$. Consequently, $\mu_{R_M^s}(\mathbf{x}, u) = \mu_{R_M^s}(x_1, \ldots, x_p, u)$ and

$$\begin{aligned} \mu_{R_M^s}(\mathbf{x}, u) &= \mu_{A^s \to G^s}(\mathbf{x}, u) = \mu_{F_1^s \times \cdots \times F_p^s \to G^s}(\mathbf{x}, u) = \mu_{F_1^s \times \cdots \times F_p^s}(\mathbf{x}) \star \mu_{G^s}(u) \\ &= \mu_{F_1^s}(x_1) \star \cdots \star \mu_{F_p^s}(x_p) \star \mu_{G^s}(u) = [T_{m=1}^p \mu_{F_m^s}(x_m)] \star \mu_{G^s}(u) \end{aligned} \tag{3.6}$$

where it has been assumed that Mamdani implication (see Example 2.6) is used, multiple antecedents are connected by *and* (i.e., by t-norms) and T is short for a t-norm.

The p-dimensional input to R_M^s is given by the T1 fuzzy set $A_{\mathbf{x}}$ whose MF is that of a fuzzy Cartesian product, that is,

$$\mu_{A_{\mathbf{x}}}(\mathbf{x}) = \mu_{X_1}(x_1) \star \cdots \star \mu_{X_p}(x_p) = T_{m=1}^p \mu_{X_m}(x_m) \tag{3.7}$$

Each rule R_M^s determines a T1 fuzzy set $B^s = A_{\mathbf{x}} \circ R_M^s$ in U such that when one uses the sup-star composition in Eq. (2.20), one obtains ($s = 1, \ldots, M$):

$$\mu_{B^s}(u) = \mu_{A_{\mathbf{x}} \circ R_M^s}(u) = \sup_{\mathbf{x} \in \mathbf{X}} [\mu_{A_{\mathbf{x}}}(\mathbf{x}) \star \mu_{A^s \to G^s}(\mathbf{x}, u)] \quad u \in U \tag{3.8}$$

This equation is the input–output relationship in Fig. 3.1 between the T1 fuzzy sets that excite the sth rule and the T1 fuzzy set at the output of that rule.

Substituting Eqs. (3.6) and (3.7) into Eq. (3.8), we see that

$$\begin{aligned} \mu_{B^s}(u) &= \sup_{\mathbf{x} \in \mathbf{X}} [\mu_{A_{\mathbf{x}}}(\mathbf{x}) \star \mu_{A^s \to G^s}(\mathbf{x}, u)] \\ &= \sup_{\mathbf{x} \in \mathbf{X}} [T_{m=1}^p \mu_{X_m}(x_m) \star [T_{m=1}^p \mu_{F_m^s}(x_m)] \star \mu_{G^s}(u)] \\ &= \sup_{\mathbf{x} \in \mathbf{X}} \{[T_{m=1}^p \mu_{X_m}(x_m) \star \mu_{F_m^s}(x_m)] \star \mu_{G^s}(u)\} \\ &= \left\{ \left[\underbrace{\sup_{x_1 \in X_1} \mu_{X_1}(x_1) \star \mu_{F_1^s}(x_1)}_{\mu_{Q_1^s}(x_1)} \right] \star \cdots \star \left[\underbrace{\sup_{x_p \in X_p} \mu_{X_p}(x_p) \star \mu_{F_p^s}(x_p)}_{\mu_{Q_p^s}(x_p)} \right] \right\} \\ &\quad \star \mu_{G^s}(u) \quad u \in U \end{aligned} \tag{3.9}$$

The inputs to a T1 FLC can be a type-0 (i.e., crisp input) or a T1 FS, where, as mentioned below Eq. (2.21), the former is commonly referred to as a *singleton input*, with associated singleton fuzzification (SF), and the latter is commonly referred to as a *nonsingleton input*, with associated nonsingleton fuzzification (NSF). For a singleton input,

$$\mu_{X_i}(x_i) = \begin{cases} 1 & x_i = x_i' \\ 0 & x_i \neq x_i' \text{ and } \forall x_i \in X_i \end{cases} \tag{3.10}$$

Substituting Eq. (3.10) into Eq. (3.9) for SF, Eq. (3.9) can be expressed for *both* SF and NSF, as ($s = 1, \ldots, M$)

$$\mu_{B^s}(u) = \begin{cases} \left[T_{m=1}^p \mu_{F_m^s}(x_m') \right] \star \mu_{G^s}(u) = f^s(\mathbf{x}) \star \mu_{G^s}(u) & \text{SF} \\ \left[T_{m=1}^p \left(\underbrace{\sup_{x_m \in X_m} \mu_{X_m}(x_m) \star \mu_{F_m^s}(x_m)}_{\mu_{Q_m^s}(x_m)} \right) \right] \star \mu_{G^s}(u) = f^s(\mathbf{x}) \star \mu_{G^s}(u) & \text{NSF} \end{cases} \tag{3.11}$$

In Eq. (3.11) $f^s(\mathbf{x}')$ is called the *firing level* for the sth rule. What distinguishes NSF from SF is a more complicated way to compute its firing level. For NSF we must calculate $\sup_{x_m \in X_m} \mu_{X_m}(x_m) \star \mu_{F_m^s}(x_m)$, that is, we must first find $x^s_{m,\max}$, where[3]

$$x^s_{m,\max} = \arg \sup_{x_m \in X_m} \underbrace{\mu_{X_m}(x_m) \star \mu_{F_m^s}(x_m)}_{\mu_{Q_m^s}(x_m)} \tag{3.12}$$

and then determine $\mu_{Q_m^s}(x^s_{m,\max}) = \mu_{X_m}(x^s_{m,\max}) \star \mu_{F_m^s}(x^s_{m,\max})$. This can be done once MF formulas are specified for $\mu_{X_m}(x_m)$ and $\mu_{F_m^s}(x_m)$ (e.g., Mendel, 2001, Chapter 6). In this book (except for Section 7.2.3) we focus exclusively on singleton fuzzification because to date it is the only kind of fuzzification that is used in a real-world FLC.

As is well known, going from the fired rule output FSs in Eq. (3.11) to a number can be accomplished by means of defuzzification (Fig. 3.1) in many different ways, including[4]: (1) *centroid defuzzification*, where first the fired output FSs are unioned (using the maximum t-conorm) and then the centroid (center of gravity) of the union is computed; and (2) *center-of-sets defuzzification*, where the centroid of each of the fired rule output FSs are used in a different kind of centroid calculation.

When centroid defuzzification is used, then the output of the T1 Mamdani FLC can be expressed as

$$u_{M,1}(\mathbf{x}) = \frac{\sum_{i=1}^{N} u_i \mu_B(u_i)}{\sum_{i=1}^{N} \mu_B(u_i)} \tag{3.13}$$

where

$$B = \bigcup_{s=1}^{M} B^s \tag{3.14}$$

In Eq. (3.13) $\mu_B(u_i)$ denotes a sampled value of $\mu_B(u)$ and N denotes the number of sampled values that are used to compute the COG of B.

When center-of-sets defuzzification is used, then the output of the T1 Mamdani FLC can be expressed as

$$u_{M,1}(\mathbf{x}) = \frac{\sum_{s=1}^{M} f^s(\mathbf{x}) u^s}{\sum_{s=1}^{M} f^s(\mathbf{x})} \tag{3.15}$$

where u^s is the COG of the sth consequent fuzzy set G^s and M denotes the total number of rules.

Regardless of which defuzzification method is chosen, this now completes the chain of calculations for the T1 Mamdani FLC in Fig. 3.1.

[3]In Eq. (3.12) (as well as in other places in this book), "arg" means the value of x_m that is associated with performing the supremum operation. In Eq. (3.12) that value is called $x^s_{m,\max}$.

[4]Other defuzzification methods such as maximum and mean of maxima could also be used; however, in actual applications of an FLC, such defuzzification methods are rarely used.

Example 3.1 Here we consider pictorial descriptions of Eqs. (3.11), (3.14), and (3.13) for SF and the minimum t-norm (Mendel 2001). We do this because FLC designers are all familiar with such pictorial descriptions for a T1 FLC since they provide them with a good understanding of some of the operations of such a system. We also do this because we will provide comparable pictorial descriptions for T2 FLCs in Section 3.3, which can then be contrasted with the figures of this example to better understand the flow of uncertainties through a T2 FLC.

Figure 3.2 depicts input and antecedent operations [the terms in the bracket in the top equation (3.11)] for a two-antecedent–single-consequent rule. The firing level is a number equal to $f^s(\mathbf{x}') = \min[\mu_{F_1^s}(x_1'), \mu_{F_2^s}(x_2')]$. Observe, for example, that $\mu_{F_1^s}(x_1')$ occurs at the intersection of the vertical line at x_1' with $\mu_{F_1^s}(x_1)$. The firing level is then t-normed with the entire consequent set for the sth rule, that is, $\mu_{B^s}(u) = \min[f^s(\mathbf{x}'), \mu_{G^s}(u)]$, the result being the clipped triangle, which is a trapezoid, which is shown at the far right of this figure.

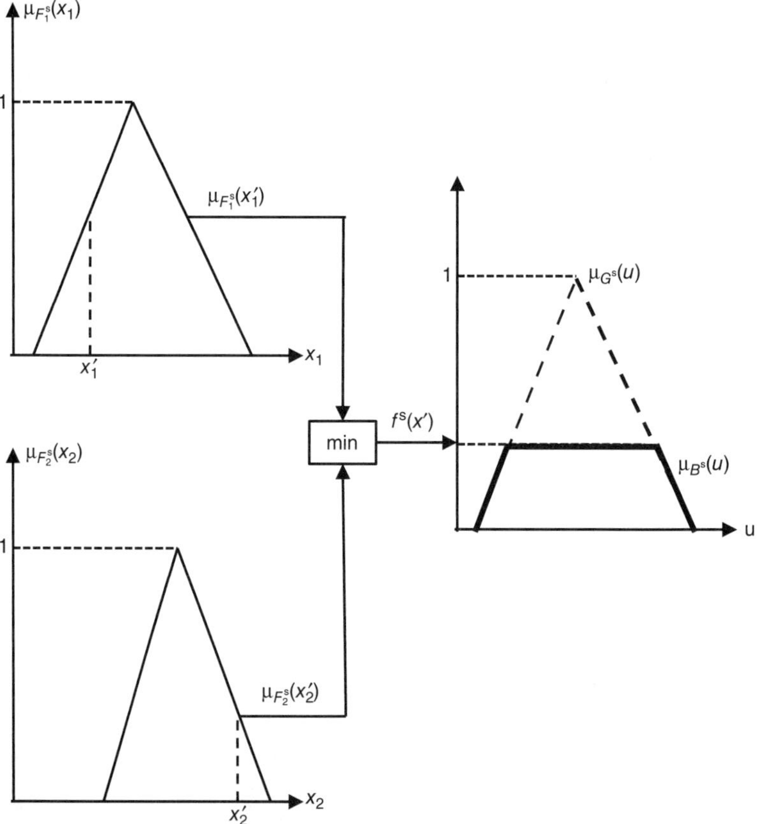

Figure 3.2 Pictorial description of input, antecedent, and consequent operations for one rule that has two antecedents, in a T1 Mamdani FLC with singleton fuzzification and minimum t-norm (Mendel, 2007; © 2007, IEEE).

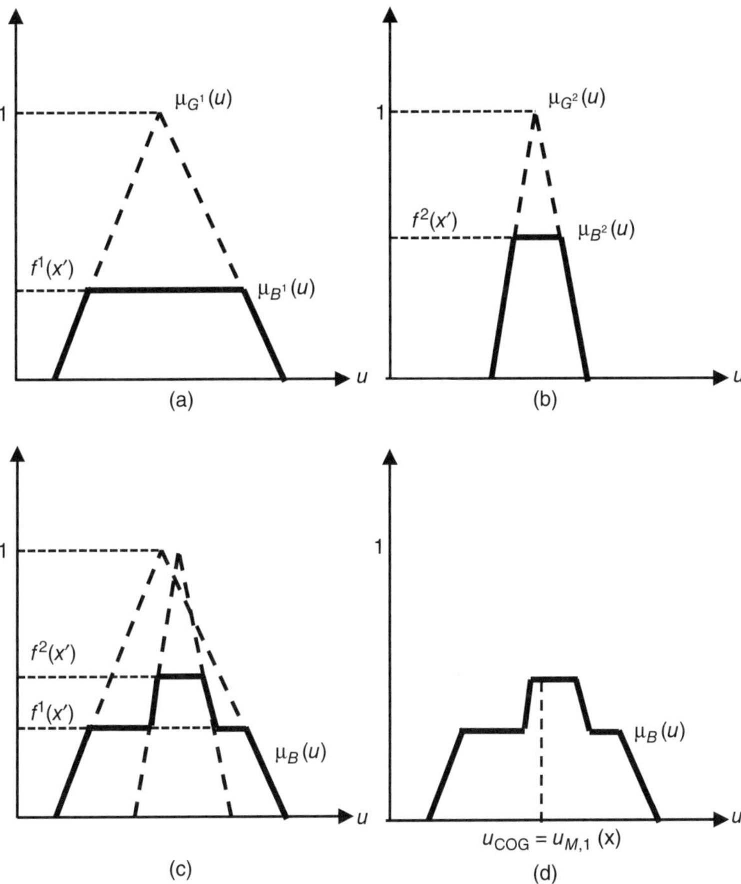

Figure 3.3 Pictorial description of (a), (b) consequent operations for two fired rules, (c) union of the two fired-rule output sets, and (d) defuzzified output.

Figures 3.3a and 3.3b depict $\mu_{B^s}(u)$ for two fired rules ($s = 1, 2$). They are each obtained as in Fig. 3.2. In Fig. 3.3c, the T1 FS $B = \cup_{s=1}^{2} B^s = \max[\mu_{B^1}(u), \mu_{B^2}(u)]$ is constructed for $\forall\, u \in U$. The result is the piecewise linear MF $\mu_B(u)$. The COG of $\mu_B(u)$, that is, Eq. (3.13), is shown in Fig. 3.3d.

We leave it to the reader to draw comparable figures for when the product t-norm is used in Eq. (3.11).

3.2.3 T1 TSK FLCs

Observe in Eq. (3.2) that membership functions are only associated with a TSK rule's antecedents; there is no consequent membership function; that is, the T1 TSK rule's consequent is an algebraic function of the p antecedent values. Hence, the rule also acts as the inference mechanism for a T1 TSK FLC. This means that it is not necessary to use the sup-star composition to obtain the output of a fired T1 TSK rule, which is quite different than what happens in a T1 singleton Mamdani FLC.

In this book we focus exclusively on a TSK rule whose consequent function is the linear one that is given in Eq. (3.3). The output, $u_{\mathrm{TSK},1}(\mathbf{x})$, of such a T1 TSK FLC is obtained by combining the outputs from the M rules in the following prescribed way:

$$u_{\mathrm{TSK},1}(\mathbf{x}) \equiv \frac{\sum_{s=1}^{M} f^s(\mathbf{x}) u^s(\mathbf{x})}{\sum_{s=1}^{M} f^s(\mathbf{x})} = \frac{\sum_{s=1}^{M} f^s(\mathbf{x})(c_0^s + c_1^s x_1 + c_2^s x_2 + \cdots + c_p^s x_p)}{\sum_{s=1}^{M} f^s(\mathbf{x})} \tag{3.16}$$

where the $f^s(\mathbf{x})$ are rule *firing levels* (strengths), defined as

$$f^s(\mathbf{x}) \equiv T_{k=1}^{p} \mu_{F_k^s}(x_k) \tag{3.17}$$

in which T again denotes a t-norm, usually minimum or product. In Eqs. (3.16) and (3.17), $\mathbf{x}$ denotes a specific input that is applied to the T1 TSK FLC.[5]

The T1 TSK FLC defined by Eqs. (3.16) and (3.17) is sometimes referred to as a *normalized* T1 TSK FLC because of the normalization of the weighted rule outputs in Eq. (3.16) by $\sum_{s=1}^{M} f^s(\mathbf{x})$. An *unnormalized* T1 TSK FLC (Tanaka et al., 1995; Tanaka and Sugeno, 1998) has for its output

$$u_{\mathrm{TSK},1}(\mathbf{x}) \equiv \sum_{s=1}^{M} f^s(\mathbf{x}) u^s(\mathbf{x}) \tag{3.18}$$

Note that in Eq. (3.16) when $u^s(\mathbf{x}) = c_0^s$ then $u_{\mathrm{TSK},1}(\mathbf{x})$ is exactly the same as $u_{M,1}(\mathbf{x})$ in Eq. (3.13). In this case it does not matter whether we call the T1 FLC TSK or Mamdani.

3.2.4 Design of T1 FLCs

In Section 3.6 we will describe how to complete the designs of IT2 FLCs. Since those FLCs are extensions of T1 FLCs from T1 FSs to IT2 FSs, we mention here that by "design" we mean the specification of such things as choice of the antecedents, number of terms/FSs used for each variable (antecedent in a rule), the shapes of the MFs for antecedents and consequents, the parameters that completely define each MF, number of rules, t-norm, and defuzzification method. Our emphasis will be on how to determine MF parameters.

3.3 INTERVAL TYPE-2 FUZZY LOGIC CONTROLLERS

3.3.1 Introduction

A general T2 FLC is depicted in Fig. 3.4. It is very similar to the T1 FLC in Fig. 3.1, the major structural difference being that the defuzzifier block of a T1

[5]Because no inference mechanism is associated with the statements of Eqs. (3.16) and (3.17), the issue of fuzzification does not appear. Hence, unlike Section 3.2.2, where we used $\mathbf{x}'$ to denote the measured value of input $\mathbf{x}$, no such distinction needs to be made between $\mathbf{x}$ and $\mathbf{x}'$ for a T1 TSK FLC.

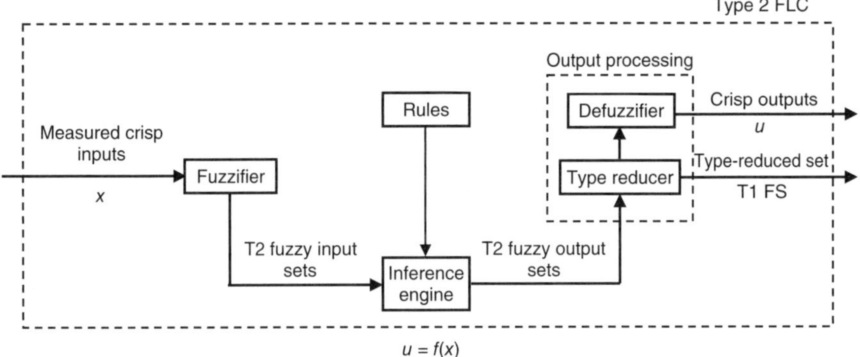

Figure 3.4 Type 2 FLC (Mendel et al., 2006; © 2006, IEEE).

FLS is replaced by the Output-processing block in a T2 FLC. That block consists of type reduction followed by defuzzification. Type reduction (TR) maps a T2 FS into a T1 FS, and then defuzzification, as usual, maps that T1 FS into a crisp number.

As for a T1 FLC, there are two major architectures for an IT2 FLC, Mamdani and TSK. Both are covered in this Section.

3.3.2 IT2 Mamdani FLCs

In the T1 case, we have Mamdani rules of the form stated in Eq. (3.1). The distinction between T1 and T2 is associated with the nature of the MFs, which is not important when forming the rules. The structure of the rules remains exactly the same in the T2 case, but now some or all of the FSs involved are T2. As for a T1 FLC, the T2 FLC has p inputs $x_1 \in X_1, \ldots, x_p \in X_p$, and one output and is characterized by M rules, where the sth rule now has the form

$$R^s\text{: IF } x_1 \text{ is } \tilde{F}_1^s \text{ and} \cdots \text{and } x_p \text{ is } \tilde{F}_p^s\text{, THEN } u \text{ is } \tilde{G}^s \quad s = 1, \ldots, M \tag{3.19}$$

Here we assume that *all* the antecedent and consequent fuzzy sets in Mamdani rules are T2; however, this need not necessarily be the case in practice. All results remain valid as long as just one FS is T2. This means that a FLC is T2 as long as any one of its antecedent or consequent (or input) FSs is T2.

When some or all of the antecedent and consequent T2 FSs are IT2 FSs, then we call the resulting T2 Mamdani FLC an *interval T2 FLC* (IT2 FLC). These are the FLCs that are focused on in this section.

Instead of using operations and associated mathematics that have been developed for general T2 FSs, we shall develop all of the results below using T1 FS mathematics, along the same lines as has already been done in the proof of Theorem 2.3.

3.3.2.1 Single-Antecedent Rule[6] In order to see the forest from the trees, so to speak, we shall focus initially on a single rule (i.e., $s = 1$) that has one antecedent

[6]Much of the materials in Sections 3.3.2.1–3.3.2.3 are taken from Mendel et al. 2006; © 2006, IEEE).

and that is activated by a crisp number (i.e., singleton fuzzification—SF), after which we shall show how those results can be extended first to multiple antecedents, and then to multiple rules.

In the rule[7]

$$\text{IF } x_1 \text{ is } \tilde{F}_1, \text{THEN } u \text{ is } \tilde{G} \tag{3.20}$$

let $\tilde{F}_1$ be an IT2 FS in the discrete universe of discourse X_{1d} for the antecedent, and $\tilde{G}$ be an IT2 FS in the discrete universe of discourse U_d for the consequent. Decompose $\tilde{F}_1$ into n_{F_1}-embedded IT2 FSs $\tilde{F}_{1e}^{j_1}$ ($j_1 = 1, \ldots, n_{F_1}$), whose domains are the embedded T1 FSs $F_{1e}^{j_1}$, and decompose $\tilde{G}$ into n_G-embedded IT2 FSs $\tilde{G}_e^j$ ($j = 1, \ldots, n_G$), whose domains are the embedded T1 FSs G_e^j. According to the wavy-slice representation Theorem 2.2 and Corollary 2.1, it follows that $\tilde{F}_1$ and $\tilde{G}$ can be expressed as

$$\tilde{F}_1 = \sum_{j_1=1}^{n_{F_1}} \tilde{F}_{1e}^{j_1} = 1/\text{FOU}(\tilde{F}_1) \tag{3.21}$$

where

$$\text{FOU}(\tilde{F}_1) = \sum_{j_1=1}^{n_{F_1}} F_{1e}^{j_1} = \sum_{j_1=1}^{n_{F_1}} \sum_{i=1}^{N_{x_1}} u_{1i}^{j_1}/x_{1i} \quad u_{1i}^{j_1} \in J_{x_{1i}}^u \subseteq U_d = \{0, \ldots, 1\} \tag{3.22}$$

and

$$\tilde{G} = \sum_{j=1}^{n_G} \tilde{G}_e^j = 1/\text{FOU}(\tilde{G}) \tag{3.23}$$

where

$$\text{FOU}(\tilde{G}) = \sum_{j=1}^{n_G} G_e^j = \sum_{j=1}^{n_G} \sum_{k=1}^{N_u} w_k^j/u_k \quad w_k^j \in J_{u_k}^w \subseteq U_d = \{0, \ldots, 1\} \tag{3.24}$$

Consequently, there are $n_{F_1} \times n_G$ possible combinations of embedded T1 antecedent and consequent FSs, so that the totality of fired output sets for all possible combinations of these embedded T1 antecedent and consequent FSs will be a *bundle of functions* FOU($\tilde{B}$) as depicted in Fig. 3.5, where

$$\text{FOU}(\tilde{B}) \triangleq \sum_{j_1=1}^{n_{F_1}} \sum_{j=1}^{n_G} \mu_{B(j_1,j)}(u) \quad \forall u \in U_d \tag{3.25}$$

in which the summations denote union.

[7]Although it is unnecessary to use the subscript 1 on x for a single-antecedent rule, doing so will make the multiple-antecedent case easier to understand.

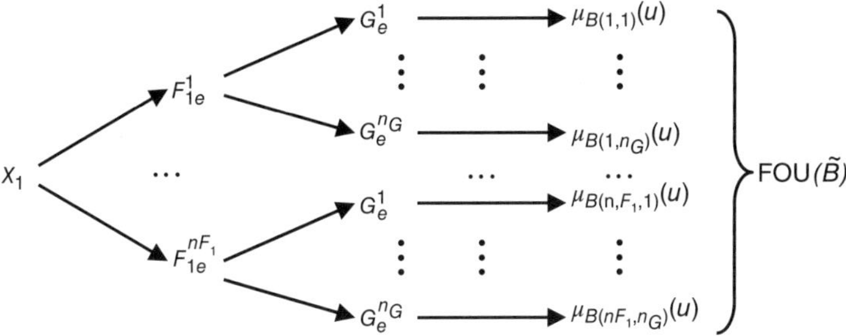

Figure 3.5 Fired output FSs for all possible $n_B = n_{F_1} \times n_G$ combinations of the embedded T1 antecedent and consequent FSs for a single-antecedent rule (Mendel et al., 2006; © 2006, IEEE).

The relationship between the bundle of functions FOU$(\tilde{B})$ in Eq. (3.25) and the FOU of the T2 fired output FS is summarized by the following theorem.

THEOREM 3.1 The bundle of functions FOU$(\tilde{B})$ in Eq. (3.25), computed using T1 FS mathematics, is the FOU of the T2 fired output FS, $\tilde{B} = 1/\text{FOU}(\tilde{B})$, and is given in Eqs. (3.30)–(3.32).

Proof. It follows from Fig. 3.5 that the fired output of the combination of the j_1th-embedded T1 antecedent FS and the jth-embedded T1 consequent FS can be computed for SF using Mamdani implication as in the top line of Eq. (3.11) with $p = 1$, that is,[8]

$$\mu_{B(j_1,j)}(u) = \mu_{F_{1e}^{j_1}}(x_1') \star \mu_{G_e^j}(u) \quad \forall u \in U_d \tag{3.26}$$

Since for any j_1 and j, $\mu_{B(j_1,j)}(u)$ in Eq. (3.26) is bounded in [0, 1], FOU$(\tilde{B})$ in Eq. (3.25) must also be a bounded function in [0, 1], which means that Eq. (3.25) can be expressed as

$$\text{FOU}(\tilde{B}) \equiv \{\underline{\mu}_{\tilde{B}}(u), \ldots, \overline{\mu}_{\tilde{B}}(u)\} \quad \forall u \in U_d \tag{3.27}$$

a set of $n_{F_1} \times n_G$ functions, where

$$\underline{\mu}_{\tilde{B}}(u) = \inf_{\forall j_1 j}(\mu_{B(j_1,j)}(u)) \quad \forall u \in U_d \tag{3.28}$$

$$\overline{\mu}_{\tilde{B}}(u) = \sup_{\forall j_1 j}(\mu_{B(j_1,j)}(u)) \quad \forall u \in U_d \tag{3.29}$$

denote the lower-bounding and upper-bounding functions of FOU$(\tilde{B})$, respectively.

[8]In Eq. (3.11), the superscript s denotes rule number. Since we are focusing on a single rule, we don't use this superscript here. Our superscripts are associated with specific embedded T1 FSs.

Recall that the upper and lower MFs for IT2 FSs are also embedded T1 FSs. Let $\overline{\mu}_{F_1}(x_1)$ and $\underline{\mu}_{F_1}(x_1)$ denote the embedded UMF and LMF T1 FSs associated with $\tilde{F}_1$, and $\overline{\mu}_G(u)$ and $\underline{\mu}_G(u)$ denote the corresponding embedded UMF and LMF T1 FSs of $\tilde{G}$. From Eq. (3.26), observe that to compute the *infimum* of $\mu_{B(j_1,j)}(u)$ one needs to choose the smallest embedded T1 FS of both the antecedent and consequent, namely $\underline{\mu}_{F_1}(x_1)$ and $\underline{\mu}_G(u)$, respectively. By doing this, one obtains the following equation for $\underline{\mu}_{\tilde{B}}(y)$:

$$\underline{\mu}_{\tilde{B}}(u) = \inf_{\forall j_1, j}(\mu_{B(j_1,j)}(u)) = \underline{\mu}_{F_1}(x'_1) \star \underline{\mu}_G(u) \quad \forall u \in U_d \tag{3.30}$$

Similarly, to compute the *supremum* of $\mu_{B(j_1,j)}(u)$, one needs to choose the largest embedded T1 FS of both the antecedent and consequent, namely $\overline{\mu}_{F_1}(x_1)$ and $\overline{\mu}_G(u)$, respectively. By doing this, one obtains the following equation for $\overline{\mu}_{\tilde{B}}(u)$:

$$\overline{\mu}_{\tilde{B}}(u) = \sup_{\forall j_1, j}(\mu_{B(j_1,j)}(u)) = \overline{\mu}_{F_1}(x'_1) \star \overline{\mu}_G(u) \quad \forall u \in U_d \tag{3.31}$$

Obviously, when the sample rate becomes infinite, the sampled universes of discourse X_{1d} and U_d can be considered as the continuous universes of discourse X_1 and U, respectively. In this case, FOU($\tilde{B}$) contains an uncountable infinite number of elements, which will still be bounded below and above by $\underline{\mu}_{\tilde{B}}(u)$ and $\overline{\mu}_{\tilde{B}}(u)$, respectively, where these functions are still given by the right-hand sides of Eqs. (3.30) and (3.31) (with $U_d \to U$), such that Eq. (3.27) can be expressed as

$$\text{FOU}(\tilde{B}) = [\underline{\mu}_{\tilde{B}}(u), \overline{\mu}_{\tilde{B}}(u)] \quad \forall u \in U \tag{3.32}$$

which completes the proof of this theorem.

Observe, from Eqs. (3.30) and (3.31), that FOU($\tilde{B}$) only uses the lower and upper MFs of the antecedent and consequent MFs, and both of these are T1 FSs.

3.3.2.2 Multiple-Antecedent Rule Next, we extend Theorem 3.1 from one antecedent to multiple antecedents. In the rule (3.19), let $\tilde{F}_1, \tilde{F}_2, \ldots, \tilde{F}_p$ be IT2 FSs in discrete universes of discourse $X_{1d}, X_{2d}, \ldots, X_{pd}$, respectively, and $\tilde{G}$ be an IT2 FS in the discrete universe of discourse U_d. Decompose each $\tilde{F}_i$ into its n_{F_i} $(i = 1, \ldots, p)$ embedded IT2 FSs $\tilde{F}^{j_i}_{ie}$, that is,

$$\tilde{F}_i = \sum_{j_i=1}^{n_{F_i}} \tilde{F}^{j_i}_{ie} = 1/\text{FOU}(\tilde{F}_i) \quad i = 1, \ldots, p \tag{3.33}$$

The domain of each $\tilde{F}^{j_i}_{ie}$ is the embedded T1 FS $F^{j_i}_{ie}$. As in the preceding subsection, $\tilde{G}$ is decomposed into n_G-embedded IT2 FSs $\tilde{G}^j_e$, whose domains are the embedded T1 FSs G^j_e, respectively; so Eqs. (3.23) and (3.24) remain unchanged for this case.

The Cartesian product of $\tilde{F}_1, \tilde{F}_2, \ldots, \tilde{F}_p$, $\tilde{F}_1 \times \tilde{F}_2 \times \cdots \times \tilde{F}_p$, has $\prod_{i=1}^{p} n_{F_i}$ combinations of the embedded T1 FSs, $F_{ie}^{j_i}$. Let F_e^n denote the nth combination of these embedded T1 FSs, that is,

$$F_e^n = F_{1e}^{j_1} \times \cdots \times F_{pe}^{j_p} \quad 1 \leq n \leq \prod_{i=1}^{p} n_{F_i} \quad \text{and} \quad 1 \leq j_i \leq n_{F_i} \tag{3.34}$$

This equation requires a combinatorial mapping from $(j_1, j_2, \ldots, j_p) \rightarrow n$; however, in the sequel we will not need to actually perform the specific mapping. All we need is to understand that it is theoretically possible to create such a mapping. To represent this mapping explicitly, we show $(j_1, j_2, \ldots, j_p) \rightarrow (j_1(n), j_2(n), \ldots, j_p(n))$, so that Eq. (3.34) can be expressed as

$$F_e^n = F_{1e}^{j_1(n)} \times \cdots \times F_{pe}^{j_p(n)} \quad 1 \leq n \leq \prod_{i=1}^{p} n_{F_i} \text{ and } 1 \leq j_i\,(n) \leq n_{F_i} \tag{3.35}$$

in which case

$$\mu_{F_e^n}(\mathbf{x}) = T_{m=1}^{p} \mu_{F_{me}^{j_m(n)}}(x_m) \quad 1 \leq n \leq \prod_{i=1}^{p} n_{F_i} \text{ and } 1 \leq j_m(n) \leq n_{F_m} \tag{3.36}$$

Additionally, let

$$n_F \equiv \prod_{m=1}^{p} n_{F_m} \tag{3.37}$$

With n_G-embedded T1 FSs for the consequent, we obtain $n_F \times n_G$ combinations of antecedent- and consequent-embedded T1 FSs, which generate the bundle of $n_F \times n_G$ fired output consequent T1 FS functions, that is,

$$\text{FOU}(\tilde{B}) = \sum_{n=1}^{n_F} \sum_{j=1}^{n_G} \mu_{B(n,j)}(u) \quad \forall u \in U_d \tag{3.38}$$

Observe how similar Eqs. (3.38) and (3.25) are to each other.

THEOREM 3.2 The bundle of functions $\text{FOU}(\tilde{B})$ in Eq. (3.38), computed using T1 FS mathematics, is the FOU of the T2 fired output FS, $\tilde{B} = 1/\text{FOU}(\tilde{B})$, and is given in Eqs. (3.32), (3.40), and (3.41).

Proof. The proof of this theorem is very similar to the proof of Theorem 3.1, but in the proof of that theorem the following changes must be made:

1. In Eq. (3.26), instead of computing $\mu_{B(j_1,j)}(u)$, we must now compute $\mu_{B(n,j)}(u)$, by using the top line of Eq. (3.11) in which $T_{m=1}^{p}\mu_{F_m^l}(x'_m)$ is replaced by Eq. (3.36), that is,

$$\mu_{B(n,j)}(u) = [T_{m=1}^{p}\mu_{F_{me}^{jm(n)}}(x'_m)] \star \mu_{G^j}(u) \quad \forall u \in U_d \tag{3.39}$$

2. Equation (3.27) is unchanged.
3. In Eqs. (3.28) and (3.29), replace the index j_1 by the index n.
4. Let $\overline{\mu}_{F_m}(x_m)$ and $\underline{\mu}_{F_m}(x_m)$ denote the embedded UMF and LMF T1 FSs associated with $\tilde{F}_m$. Note that $\overline{\mu}_{F_m}(x_m)$ and $\underline{\mu}_{F_m}(x_m)$ are two of the n_{F_m}-embedded T1 FSs that are associated with $\tilde{F}_m$. They will be the ones that are used in the next step.
5. Equations (3.30) and (3.31) are changed to

$$\underline{\mu}_{\tilde{B}}(u) = \inf_{\forall n,j}(\mu_{B(n,j)}(u)) = [T_{m=1}^{p}\underline{\mu}_{F_m}(x'_m)] \star \underline{\mu}_{G}(u) \quad \forall u \in U_d \tag{3.40}$$

$$\overline{\mu}_{\tilde{B}}(u) = \sup_{\forall n,j}(\mu_{B(n,j)}(u)) = [T_{m=1}^{p}\overline{\mu}_{F_m}(x'_m)] \star \overline{\mu}_{G}(u) \quad \forall u \in U_d \tag{3.41}$$

6. Equation (3.32) remains unchanged.

Once again, observe, now from Eqs. (3.40) and (3.41), that FOU($\tilde{B}$) only uses the lower and upper MFs of the p antecedent and consequent MFs, and all of these are T1 FSs.

COROLLARY 3.1 For an IT2 FLC the firing level becomes a *firing interval*, $F(\mathbf{x}')$, where

$$F(\mathbf{x}') \equiv [\underline{f}(\mathbf{x}'), \overline{f}(\mathbf{x}')] \tag{3.42}$$

$$\underline{f}(\mathbf{x}') \equiv T_{m=1}^{p}\underline{\mu}_{F_m}(x'_m) \tag{3.43}$$

$$\overline{f}(\mathbf{x}') \equiv T_{m=1}^{p}\overline{\mu}_{F_m}(x'_m) \tag{3.44}$$

Then, FOU($\tilde{B}$) can be expressed in terms of the firing interval, as

$$\text{FOU}(\tilde{B}) = [\underline{f}(\mathbf{x}') \star \underline{\mu}_{G}(u), \overline{f}(\mathbf{x}') \star \overline{\mu}_{G}(u)] = F(\mathbf{x}') \star [\underline{\mu}_{G}(u), \overline{\mu}_{G}(u)] \ \ \forall u \in U \tag{3.45}$$

Proof. Equations (3.42)–(3.44) are definitions that are motivated by the structures of the bracketed terms on the right-hand sides of Eqs. (3.40) and (3.41). Beginning with Eq. (3.32), and substituting Eqs. (3.40) and (3.41) into it, and then using Eqs. (3.43) and (3.44), one finds

$$\begin{aligned}\mathrm{FOU}(\tilde{B}) &= [\underline{\mu}_{\tilde{B}}(u), \overline{\mu}_{\tilde{B}}(u)] = [\underline{f}(\mathbf{x}') \star \underline{\mu}_G(u), \overline{f}(\mathbf{x}') \star \overline{\mu}_G(u)] \\ &= [\underline{f}(\mathbf{x}'), \overline{f}(\mathbf{x}')] \star [\underline{\mu}_G(u), \overline{\mu}_G(u)]\end{aligned} \tag{3.46}$$

Substituting Eq. (3.42) into this result, one obtains the right-hand side of Eq. (3.45).

3.3.2.3 Multiple Rules So far all of the derivations in Sections 3.3.2.1 and 3.3.2.2 have been for a single rule. In general, there are M rules that characterize an IT2 FLC, and frequently more than one rule fires when input x is applied to that system. What this means is that, as in the case of a T1 FLC, we need to include another index—s— in all of the IT2 FLC formulas. So, for example, for the sth rule, we would express Eqs. (3.32) and (3.40)–(3.45) as ($s = 1, \ldots, M$)

$$\mathrm{FOU}(\tilde{B}^s) = [\underline{\mu}_{\tilde{B}^s}(u), \overline{\mu}_{\tilde{B}^s}(u)] \quad \forall u \in U \tag{3.47}$$

$$\underline{\mu}_{\tilde{B}^s}(u) = \inf_{\forall n,j}(\mu_{B^s(n,j)}(u)) = [T_{m=1}^{p} \underline{\mu}_{F_m^s}(x_m')] \star \underline{\mu}_{G^s}(u) \quad \forall u \in U \tag{3.48}$$

$$\overline{\mu}_{\tilde{B}^s}(u) = \sup_{\forall n,j}(\mu_{B^s(n,j)}(u)) = [T_{m=1}^{p} \overline{\mu}_{F_m^s}(x_m')] \star \overline{\mu}_{G^s}(u) \quad \forall u \in U \tag{3.49}$$

$$F^s(\mathbf{x}') \equiv [\underline{f}^s(\mathbf{x}'), \overline{f}^s(\mathbf{x}')] \tag{3.50}$$

$$\underline{f}^s(\mathbf{x}') \equiv T_{m=1}^{p} \underline{\mu}_{F_m^s}(x_m') \tag{3.51}$$

$$\overline{f}^s(\mathbf{x}') \equiv T_{m=1}^{p} \overline{\mu}_{F_m^s}(x_m') \tag{3.52}$$

$$FOU(\tilde{B}^s) = [\underline{f}^s(\mathbf{x}') \star \underline{\mu}_{G^s}(u), \overline{f}^s(\mathbf{x}') \star \overline{\mu}_{G^s}(u)] = F^s(\mathbf{x}') \star [\underline{\mu}_{G^s}(u), \overline{\mu}_{G^s}(u)] \ \forall u \in U \tag{3.53}$$

As in the T1 case, fired rule sets are combined either before or as part of output processing. For illustrative purposes only,[9] let us assume that the s fired rule sets are combined using the union operation. In this case, we have the following theorem.

THEOREM 3.3 If the s fired rule sets are combined using the union operation, leading to a composite IT2 FS $\tilde{B}$, then

$$\tilde{B} = 1/\mathrm{FOU}(\tilde{B}) \tag{3.54}$$

$$\mathrm{FOU}(\tilde{B}) = [\underline{\mu}_{\tilde{B}}(u), \overline{\mu}_{\tilde{B}}(u)] \quad \forall u \in U \tag{3.55}$$

[9]We do not necessarily advocate combining IT2 FSs using the union operation, just as many people do not advocate combining fired T1 FSs in a T1 FLC using the union operation. This is explained in great detail in Mendel (2001) where more computationally tractable ways of blending the IT2 fired rule sets are described. Conceptually, one merely needs to think of some final (aggregated) IT2 FS, say $\tilde{B}(u)$ as having been obtained from the $\tilde{B}^s(u)$.

where

$$\underline{\mu}_{\tilde{B}}(u) = \underline{\mu}_{\tilde{B}^1}(u) \vee \underline{\mu}_{\tilde{B}^2}(u) \vee \cdots \vee \underline{\mu}_{\tilde{B}^M}(u) \tag{3.56}$$

$$\overline{\mu}_{\tilde{B}}(u) = \overline{\mu}_{\tilde{B}^1}(u) \vee \overline{\mu}_{\tilde{B}^2}(u) \vee \cdots \vee \overline{\mu}_{\tilde{B}^M}(u) \tag{3.57}$$

and specific formulas for $\underline{\mu}_{\tilde{B}^s}(u)$ and $\overline{\mu}_{\tilde{B}^s}(u)$ are given in Eqs. (3.48) and (3.49).

Proof. Equations (3.56) and (3.57) follow from $M - 1$ repeated applications of Eq. (2.46) to $\cup_{s=1}^{M} \tilde{B}^s$.

Example 3.2 This example parallels Example 3.1 and the construction of Figs. 3.2 and 3.3. First, we consider the pictorial description of Eq. (3.53) for the minimum t-norm. Figure 3.6 depicts input and antecedent operations for

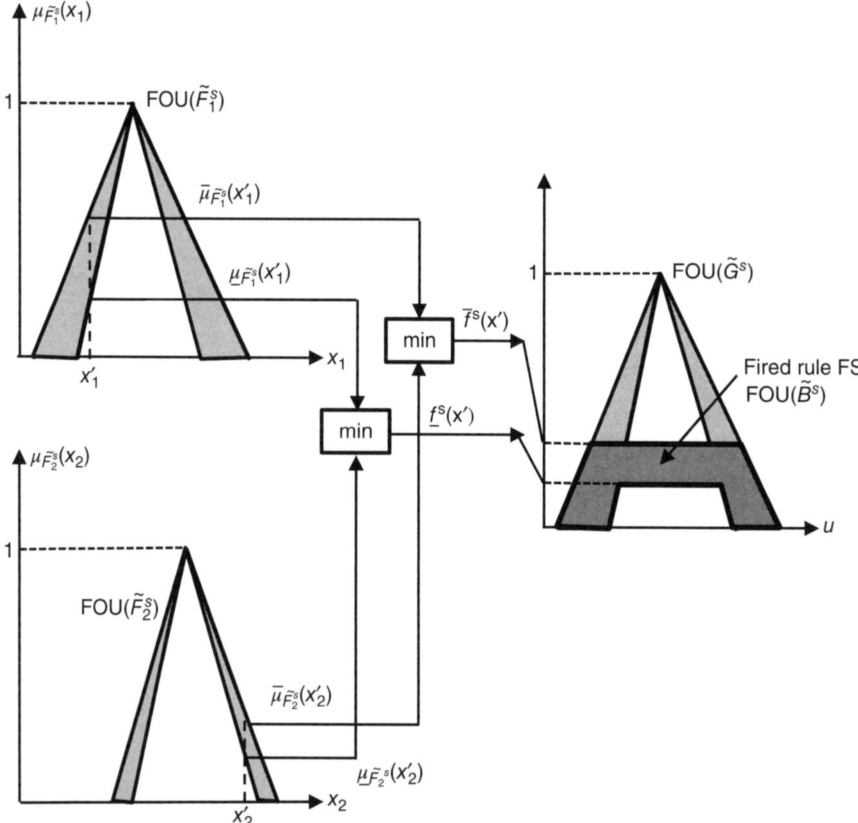

Figure 3.6 Pictorial description of input, antecedent, and consequent operations for rule s that has two antecedents, in an IT2 Mamdani FLC for singleton fuzzification and minimum t-norm (Mendel, 2007; © 2007, IEEE).

a two-antecedent single-consequent rule, SF, and minimum t-norm. The firing interval is an interval of real numbers equal to [using Eqs. (3.50)–(3.52)]

$$F^s(\mathbf{x}') \equiv [\min[\underline{\mu}_{F_1^s}(x_1'), \underline{\mu}_{F_2^s}(x_2')], \min[\overline{\mu}_{F_1^s}(x_1'), \overline{\mu}_{F_2^s}(x_2')]] \tag{3.58}$$

Observe, for example, that $\underline{\mu}_{\tilde{F}_1^s}(x_1')$ occurs at the intersection of the vertical line at x_1' with $\underline{\mu}_{\tilde{F}_1^s}(x_1)$, and $\overline{\mu}_{\tilde{F}_1^s}(x_1')$ occurs at the intersection of the vertical line at x_1' with $\overline{\mu}_{\tilde{F}_1^s}(x_1)$.

The firing interval is then t-normed with the entire consequent FOU for the sth rule [using Eq. (3.53)], that is,

$$\text{FOU}(\tilde{B}^s) = [\min(\underline{f}^s(\mathbf{x}'), \underline{\mu}_{G^s}(u)), \min(\overline{f}^s(\mathbf{x}'), \overline{\mu}_{G^s}(u))] \quad \forall u \in U \tag{3.59}$$

the result being the filled-in trapezoidal area that is shown at the far right of Fig. 3.6.

Comparing Figs. 3.6 and 3.2, observe how the uncertainties about the antecedent and consequent MFs have flowed through all of the computations, leading to FOU(B^s).

Next, we consider the pictorial description of Eqs. (3.55)–(3.57). Figures 3.7a and 3.7b depict FOU($\tilde{B}^s$) for two fired rules ($s = 1, 2$); they are each obtained as in Fig. 3.6. In Fig. 3.7c

$$\tilde{B} = \bigcup_{s=1}^{2} \tilde{B}^s = [\max(\underline{\mu}_{\tilde{B}^1}(u), \underline{\mu}_{\tilde{B}^2}(u)), \max(\overline{\mu}_{\tilde{B}^1}(u), \overline{\mu}_{\tilde{B}^2}(u))] \tag{3.60}$$

and this is constructed for $\forall y \in Y$. We leave it to the reader to draw comparable figures for when the product t-norm is used.

Comparing Figs. 3.7c and 3.3c, observe how the uncertainties about each fired rule's FOU have flowed further into the union of those FOUs.

3.3.2.4 Output Processing With reference to the T2 FLC depicted in Fig. 3.4, we now explain how to perform *output processing*. Type reduction is the first step of output processing and defuzzification is its second step.

There are as many type reduction methods as there are T1 defuzzification methods because each of the former is associated with one of the latter. Karnik and Mendel (2001a) [see, also, Mendel (2001, Chapter 9)] have developed centroid, center-of-sums, height, modified-height, and center-of-sets type reducers. Here we only explain centroid and center-of-sets type reduction because they are the most popular ones.

CENTROID TYPE REDUCTION In order to perform centroid type reduction one must begin with an FOU. This is obtained in an IT2 FLC by aggregating the fired rules using the union, as in Eqs. (3.55)–(3.57). *Centroid type reduction is equivalent to computing the centroid of an IT2 FS*. See Section 2.3.4 for how to do this. The most

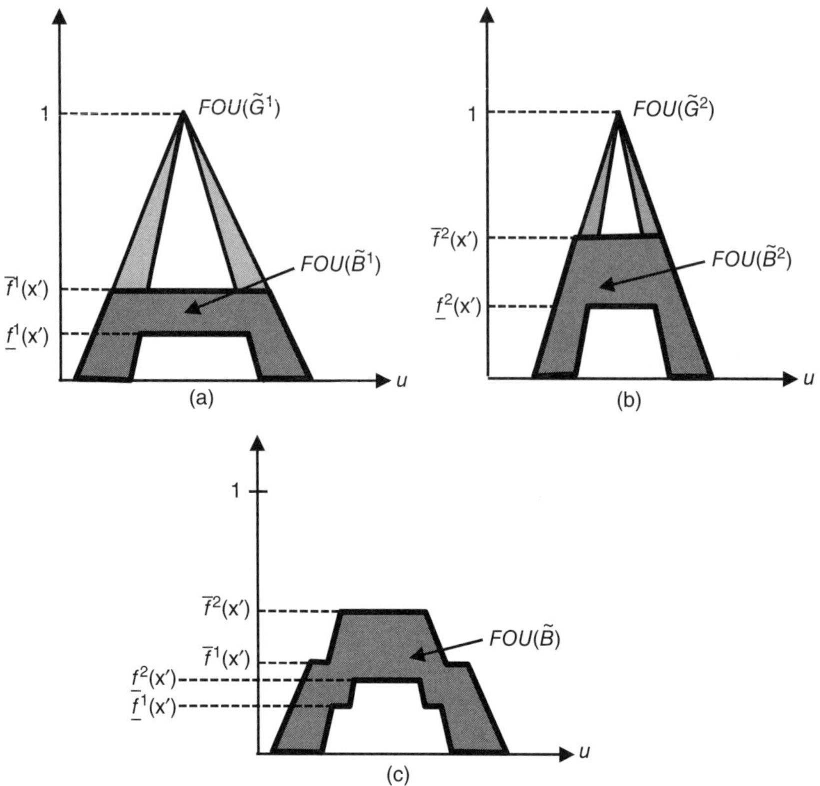

Figure 3.7 Pictorial description of (a), (b) consequent operations for two fired rules and (c) union of the two fired rule output sets.

widely used algorithms for doing this are the KM or EKM algorithms. The result from centroid type reduction is a T1 FS, $U_C(\mathbf{x})$, namely

$$U_C(\mathbf{x}) = 1/[c_l(\tilde{B}|\mathbf{x}), c_r(\tilde{B}|\mathbf{x})] \equiv 1/[u_l(\mathbf{x}), u_r(\mathbf{x})] \tag{3.61}$$

The explicit dependence of $U_C(\mathbf{x})$ on $\mathbf{x}$ in Eq. (3.61) emphasizes the fact that the type-reduced set depends on the inputs to the FLC. When $\mathbf{x}$ changes from one value to another, as it will in a fuzzy logic control system—from one time point to the next—then $U_C(\mathbf{x})$ also changes from one time point to the next.

Center-of-Sets (COS) Type Reduction In order to perform COS type reduction one begins with the firing intervals for the rules, $F^s(\mathbf{x}')$ in Eq. (3.50) (for $s = 1, \ldots, M$), and instead of combining them directly with their respective consequent FOU they are combined with the centroid of their respective consequent FOU during an averaging process.

Let the centroid of $\tilde{G}^s$ be expressed as

$$C_{\tilde{G}^s} = 1/[a(\tilde{G}^s), b(\tilde{G}^s)] \equiv 1/[a^s, b^s] \tag{3.62}$$

where a^s and b^s are computed using EKM algorithms. Consider the following average, which originally was called a *generalized centroid* (Karnik and Mendel, 2001a; Mendel, 2001) but is now known as in *interval weighted average* (Mendel and D. Wu, 2010, Chapter 5]:

$$U_{\text{IWA}}(\mathbf{x}) = 1 \Big/ \underset{\substack{u_s \in [a^s, b^s] \\ w_s \in [\underline{f}^s(\mathbf{x}), \bar{f}^s(\mathbf{x})]}}{\forall} \frac{\sum_{s=1}^{M} u_s w_s}{\sum_{s=1}^{M} w_s} \tag{3.63}$$

Clearly,

$$U_{IWA}(\mathbf{x}) = 1/[u_l(\mathbf{x}), u_r(\mathbf{x})] \tag{3.64}$$

where

$$u_l(\mathbf{x}) = \min_{\forall w_s \in [\underline{f}^s(\mathbf{x}), \bar{f}^s(\mathbf{x})]} \frac{\sum_{s=1}^{M} a^s w_s}{\sum_{s=1}^{M} w_s} \tag{3.65}$$

$$u_r(\mathbf{x}) = \max_{\forall w_s \in [\underline{f}^s(\mathbf{x}), \bar{f}^s(\mathbf{x})]} \frac{\sum_{s=1}^{M} b^s w_s}{\sum_{s=1}^{M} w_s} \tag{3.66}$$

Comparing Eqs. (3.65) and (2.62), observe that they are identical, when θ_i and x_i in Eq (2.62) are equated with w_s and a^s, respectively, in Eq. (3.65); hence, $u_l(\mathbf{x})$ can be computed by applying the EKM algorithm for the left end of a centroid to Eq. (3.65). To do this, a^s must first be reordered in increasing order and renumbered, and its associated w_s must then also be renumbered so that it remains associated with its corresponding value of a^s. Similarly, comparing Eqs. (3.66) and (2.63), observe that they are identical, when θ_i and x_i in Eq. (2.63) are equated with w_s and b^s, respectively, in Eq (3.66); hence, $u_r(\mathbf{x})$ can be computed by applying the EKM algorithm for the right end of a centroid to Eq. (3.66). To do this, b^s must first be reordered in increasing order and renumbered, and its associated w_s must then also be renumbered so that it remains associated with its corresponding value of b^s.

Center-of-sets type reduction is equivalent to computing $U_{IWA}(\mathbf{x})$ in Eq. (3.64): hence,

$$U_{\text{COS}}(\mathbf{x}) = U_{\text{IWA}}(\mathbf{x}) \tag{3.67}$$

DEFUZZIFICATION After type reduction, defuzzification is very simple, that is,

$$u(\mathbf{x}) = \tfrac{1}{2}[u_l(\mathbf{x}) + u_r(\mathbf{x})] \tag{3.68}$$

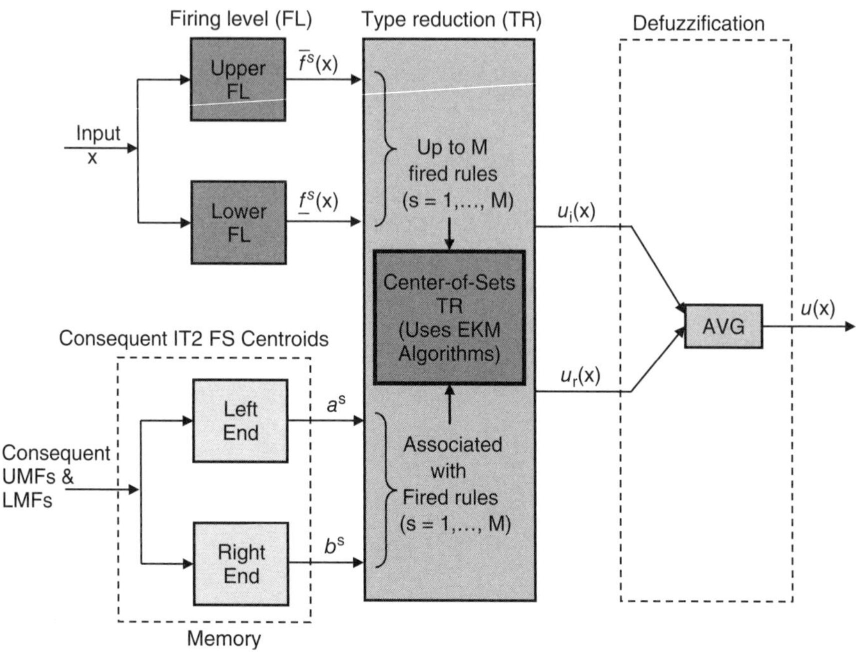

Figure 3.8 Block diagram summary of IT2 Mamdani FLC computations that use COS TR (Mendel 2007; © 2007, IEEE).

3.3.2.5 Summaries It is very helpful to summarize all of the computations for an IT2 Mamdani FLC. This can be done in different ways, two of which are given next.

BLOCK DIAGRAM SUMMARY The entire chain of computations is summarized in the block diagram of Fig. 3.8. Firing intervals are computed for all rules [using Eqs. (3.51) and (3.52)], and they depend explicitly on the input **x**. For COS TR, offline computations of the centroids are performed for each of the M consequent IT2 FSs using EKM algorithms [as in Eq. (3.62)] and are then stored in memory. COS TR combines the firing intervals and precomputed consequent centroids and uses EKM algorithms to perform the actual calculations [Eqs. (3.64)–(3.66), using the EKM algorithms given in Table 2.4]. Defuzzification is the simple averaging of the two end points of the type-reduced set [Eq. (3.68)].

NEUROFUZZY SUMMARY The entire chain of computations is also summarized in the layered architecture of Fig. 3.9, which resembles a neural network, although there is nothing "neural" about this FLC.[10] Layer 1 refers to the inputs, $x_1, \ldots, x_p$.

[10]Many variations of this figure appear in the literature (e.g., Aliev et al., 2011; Castro et al., 2009; Juang and Tsao, 2008; Lin and Chen, 2011), none of which look the same or are described in exactly the same way.

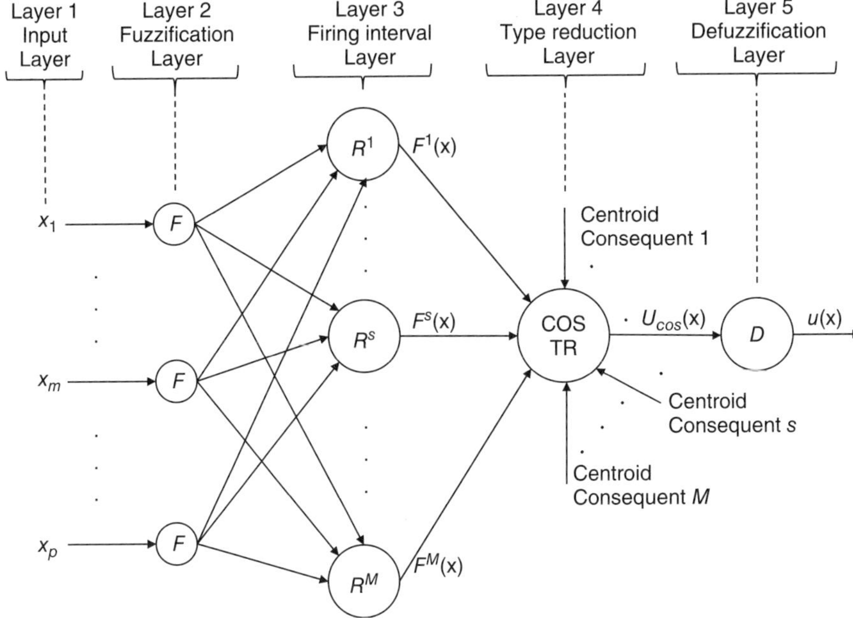

Figure 3.9 Neurofuzzy summary of IT2 Mamdani FLC computations that use COS TR.

Layer 2 is where the inputs are fuzzified (F) into IT2 FSs. When singleton fuzzification is used, then Eq. (3.10) applies. Layer 3 is where the firing interval is computed for each of the M rules [using Eqs. (3.51) and (3.52)]. Layer 4 is where TR is performed and is illustrated for COS TR,[11] for which centroid consequents, which have been stored in memory, also have to be used [Eqs. (3.64)–(3.66), using the EKM algorithms given in Table 2.4]. Layer 5 is where defuzzification is performed [Eq. (3.68)].

3.3.2.6 Comprehensive Example This example focuses on the full operation of the IT2 Mamdani FLC for a simple FLC that implements an edge-following behavior of an autonomous mobile robot (Fig. 1.16).

Example 3.3 The robot shown in Fig. 1.16a has two sonar sensors fixed to its right side. The sensor attached to the front right side of the robot is called the *right front sensor* (RFS) and the sensor attached to the front backside of the robot is called the *right back sensor* (RBS). Both sensors are modeled using two IT2 FSs, which are *Near* and *Far*, as depicted in Fig. 3.10a and 3.10b, respectively.

[11]If centroid TR is used, two additional layers have to be inserted between layers 3 and 4. In the first of these, fired rule output sets are computed for each of the M rules [using Eq. (3.53)], and in the second of these the fired rule output sets are aggregated by means of the union operation [using Eqs (3.55)–(3.57)].

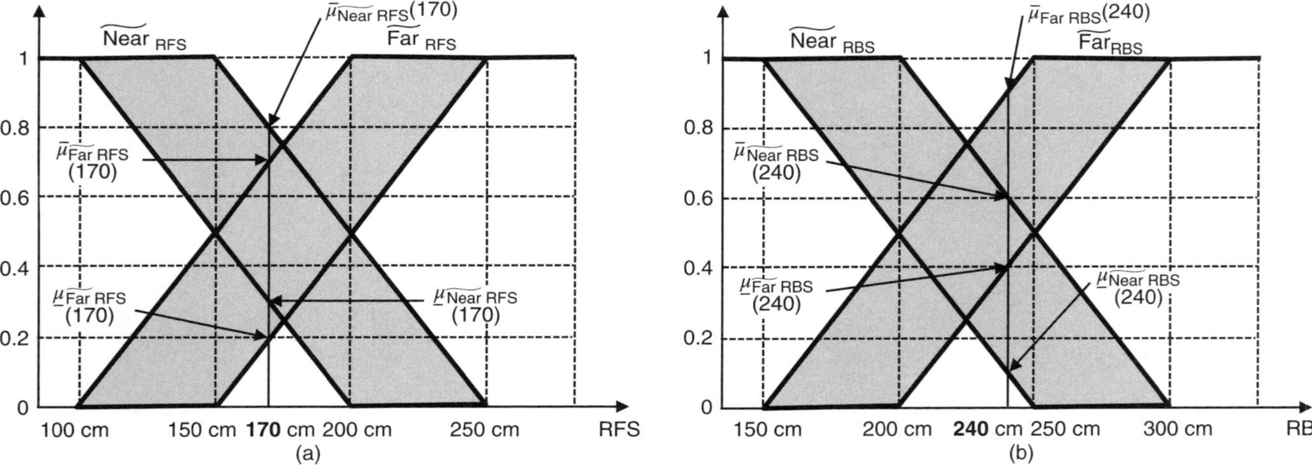

Figure 3.10 The IT2 FSs for: (a) RFS and (b) RBS.

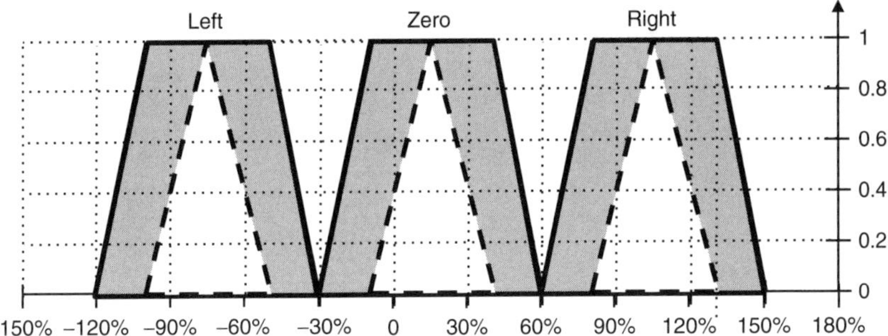

Figure 3.11 The IT2 FSs representing the steering output.

It is assumed that the robot moves at a constant speed; hence, the robot has one output, the steering that is represented by the three linguistic labels *Left*, *Zero*,[12] and *Right*, as depicted in Fig. 3.11.

The rule base consists of the following four rules:

$$\begin{aligned} &R^1\text{: IF RFS is } \mathit{Near} \text{ and RBS is } \mathit{Near}, \text{THEN steering is } \mathit{Left} \\ &R^2\text{: IF RFS is } \mathit{Near} \text{ and RBS is } \mathit{Far}, \text{THEN steering is } \mathit{Left} \\ &R^3\text{: IF RFS is } \mathit{Far} \text{ and RBS is } \mathit{Near}, \text{THEN steering is } \mathit{Zero} \\ &R^4\text{: IF RFS is } \mathit{Far} \text{ and RBS is } \mathit{Far}, \text{THEN steering is } \mathit{Right} \end{aligned} \tag{3.69}$$

The centroids of the corresponding rule consequents are given in Table 3.1; their numerical values were computed as in Example 2.12.

Here we consider an input vector $\mathbf{x}' = \text{col}(170, 240)$, whose components are shown in bold face in Fig. 3.10. From that figure[13] we obtain the following firing interval components for each of the IT2 FSs:

$$\begin{cases} [\underline{\mu}_{\widetilde{Near}_{\text{RFS}}}(170), \overline{\mu}_{\widetilde{Near}_{\text{RFS}}}(170)] = [0.3, 0.8] \\ [\underline{\mu}_{\widetilde{Far}_{\text{RFS}}}(170), \overline{\mu}_{\widetilde{Far}_{\text{RFS}}}(170)] = [0.2, 0.7] \\ [\underline{\mu}_{\widetilde{Near}_{\text{RBS}}}(240), \overline{\mu}_{\widetilde{Near}_{\text{RBS}}}(240)] = [0.1, 0.6] \\ [\underline{\mu}_{\widetilde{Far}_{\text{RBS}}}(240), \overline{\mu}_{\widetilde{Far}_{\text{RBS}}}(240)] = [0.4, 0.9] \end{cases} \tag{3.70}$$

[12]You may be wondering why the FOU for *Zero* is not centered about 0. This was actually caused by uncertainties in the steering output. Putting a 15% offset on the steering actuator gets the robot to move straight ahead, due to deterioration of the steering motor and mechanisms over time. The FOU depends on the surface on which the robot is traveling. On muddy surfaces there is a need to increase the 15% offset for the robot to proceed straight forward; on snowy surfaces there is a need to decrease the 15% offset for the robot to continue straight on; and on normal surfaces the 15% steering offset causes the robot to proceed straight ahead.

[13]These values can also be obtained by writing formulas for the LMFs and UMFs, as would be done in a computer simulation of the IT2 Mamdani FLC equations.

TABLE 3.1 Rule Base and Consequents of the IT2 FLS

	$x_2 = \text{RBS}$	
$x_1 = \text{RFS}$	*Near*	*Far*
Near	$C_{\tilde{G}^1} = [a^1, b^1] \equiv [-81.43, -68.57]$	$C_{\tilde{G}^2} = [a^2, b^2] \equiv [-81.43, -68.57]$
Far	$C_{\tilde{G}^3} = [a^3, b^3] \equiv [8.57, 21.42]$	$C_{\tilde{G}^4} = [a^4, b^4] \equiv [98.57, 111.42]$

TABLE 3.2 Firing Intervals

Rule Number	Firing Interval	Rule Consequent Centroid
R^1	$[\underline{f}^1, \bar{f}^1] = [\min(\underline{\mu}_{\widetilde{Near}_{\text{RFS}}}(170), \underline{\mu}_{\widetilde{Near}_{\text{RBS}}}(240)), \min(\bar{\mu}_{\widetilde{Near}_{\text{RFS}}}(170), \bar{\mu}_{\widetilde{Near}_{\text{RBS}}}(240))] = [\min(0.3, 0.1), \min(0.8, 0.6)] = [0.1, 0.6]$	$C_{\tilde{G}^1} = [a^1, b^1] \equiv [-81.43, -68.57]$
R^2	$[\underline{f}^2, \bar{f}^2] = [\min(\underline{\mu}_{\widetilde{Near}_{\text{RFS}}}(170), \underline{\mu}_{\widetilde{Far}_{\text{RBS}}}(240)), \min(\bar{\mu}_{\widetilde{Near}_{\text{RFS}}}(170), \bar{\mu}_{\widetilde{Far}_{\text{RBS}}}(240))] = [\min(0.3, 0.4), \min(0.8, 0.9)] = [0.3, 0.8]$	$C_{\tilde{G}^2} = [a^2, b^2] \equiv [-81.43, -68.57]$
R^3	$[\underline{f}^3, \bar{f}^3] = [\min(\underline{\mu}_{\widetilde{Far}_{\text{RFS}}}(170), \underline{\mu}_{\widetilde{Near}_{\text{RBS}}}(240)), \min(\bar{\mu}_{\widetilde{Far}_{\text{RFS}}}(170), \bar{\mu}_{\widetilde{Near}_{\text{RBS}}}(240))] = [\min(0.2, 0.1), \min(0.7, 0.6)] = [0.1, 0.6]$	$C_{\tilde{G}^3} = [a^3, b^3] \equiv [8.57, 21.42]$
R^4	$[\underline{f}^4, \bar{f}^4] = [\min(\underline{\mu}_{\widetilde{Far}_{\text{RFS}}}(170), \underline{\mu}_{\widetilde{Far}_{\text{RBS}}}(240)), \min(\bar{\mu}_{\widetilde{Far}_{\text{RFS}}}(170), \bar{\mu}_{\widetilde{Far}_{\text{RBS}}}(240))] = [\min(0.2, 0.4), \min(0.7, 0.9)] = [0.2, 0.7]$	$C_{\tilde{G}^4} = [a^4, b^4] \equiv [98.57, 111.42]$

The firing intervals for the four rules, computed using the minimum t-norm, are shown in Table 3.2.

Using the KM algorithms, it is found that (in two iterations) $L = 2$ and (in three iterations) $R = 3$; hence, $U_{\text{COS}}(\mathbf{x}') = [u_{l(\text{Steering})}(\mathbf{x}'), u_{r(\text{Steering})}(\mathbf{x}')]$, where [observe, in Table 3.2, that a^i and b^i are already ordered so that (coincidentally) $a^1 = a^2 < a^3 < a^4$ and $b^1 = b^2 < b^3 < b^4$; hence, these parameters do not have to be reordered for their use in the EKM algorithms]

$$
\begin{aligned}
u_{l(\text{Steering})}(\mathbf{x}') &= \frac{\bar{f}^1 a^1 + \bar{f}^2 a^2 + \underline{f}^3 a^3 + \underline{f}^4 a^4}{\bar{f}^1 + \bar{f}^2 + \underline{f}^3 + \underline{f}^4} \\
&= \frac{0.6 \times (-81.43) + 0.8 \times (-81.43) + 0.1 \times (8.57) + 0.2 \times (98.57)}{0.6 + 0.8 + 0.1 + 0.2} \\
&= -54.97
\end{aligned}
\tag{3.71}
$$

$$u_{r(\text{Steering})}(\mathbf{x}') = \frac{\underline{f}^1 b^1 + \underline{f}^2 b^2 + \underline{f}^3 b^3 + \bar{f}^4 b^4}{\underline{f}^1 + \underline{f}^2 + \underline{f}^3 + \bar{f}^4}$$
$$= \frac{0.1\times(-68.57)+0.3\times(-68.57) + 0.1 \times (21.42) + 0.7 \times (111.42)}{0.1 + 0.3 + 0.1 + 0.7}$$
$$= 43.92 \tag{3.72}$$

Finally, the defuzzified output is $u_{\text{Steering}}(\mathbf{x}') = (-54.97 + 43.92)/2 = -5.52$.

3.3.2.7 Novelty and Adaptiveness of an IT2 FLC In Section 1.7, it was mentioned that T2 FLCs have a *novelty* that does not exist in traditional T1 FLCs, namely (Wu, 2011) that in an IT2 FLC different membership grades from the same IT2 FS can be used in different rules (due to an IT2 FS being described by lower and upper MFs), whereas for traditional T1 FLC the same membership grade from the same T1 FS is always used in different rules. In Wu (2012, p. 838) this is stated in an equivalent way as: "*Novelty*, meaning that the UMF and LMF of the same IT2 FS may be used simultaneously in computing each bound of the type-reduced interval." This can be seen in Eqs. (3.71) and (3.72), where, for example, $\underline{f}^1$ is used to compute $u_{l(\text{Steering})}(\mathbf{x}')$ and $\bar{f}^1$ is used to compute $u_{r(\text{Steering})}(\mathbf{x}')$. From Table 3.2, observe that LMFs are used to compute $\underline{f}^1$, whereas UMFs are used to compute $\bar{f}^1$. So this example demonstrates the truth of *novelty*. Wu (2012, p. 838) concludes: "This novelty is impossible for a T1 FLC because it does not have embedded T1 FSs and the same MFs are always used in computing the firing levels of all rules."

Wu (2012, p. 837) also states that another fundamental difference between a T1 FLC and an IT2 FLC is *adaptiveness*, "meaning that the embedded T1 FSs used to compute the bounds of the type-reduced interval change as input changes." Consider $u_{r(\text{Steering})}(\mathbf{x}')$ in Eq. (3.72) for example. It uses the firing levels $\underline{f}^1, \underline{f}^2, \underline{f}^3$, and $\bar{f}^4$. At another time, when $\mathbf{x} = \mathbf{x}''$, $u_{r(\text{Steering})}(\mathbf{x}'')$ may depend on $\bar{f}^1, \bar{f}^2, \underline{f}^3$, and $\underline{f}^4$, or on $\bar{f}^1$, $\underline{f}^2$, $\underline{f}^3$, and $\underline{f}^4$, or on other combinations of the upper and lower values of the firing intervals. As Wu (2012) concludes: "This adaptiveness is impossible for a T1 FLC since it does not have embedded T1 FSs."

3.3.3 IT2 TSK FLCs

3.3.3.1 Introduction Consider an IT2 TSK FLC having p inputs $x_1 \in X_1, \ldots, x_p \in X_p$ and one output $u \in U$. An IT2 TSK FLC is also described by fuzzy if–then rules that represent input–output relations of a system. In a general first-order IT2 TSK FLC with a rule base of M rules, each having p antecedents, the sth rule can be expressed as (Liang and Mendel, 2001; Mendel, 2001, Chapter 13)

$$R^s_{\text{TSK}}\text{: IF } x_1 \text{ is } \tilde{F}^s_1 \text{ and} \cdots \text{and } x_p \text{ is } \tilde{F}^s_p, \text{ THEN}$$
$$U^s(\mathbf{x}) = C^s_0 + C^s_1 x_1 + C^s_2 x_2 + \cdots + C^s_p x_p \tag{3.73}$$

where $s = 1, \ldots, M$; C^s_j $(j = 0, 1, \ldots, p)$ are consequent T1 FSs; $U^s(\mathbf{x})$, the output of the sth rule, is also a T1 FS (because it is a linear combination of T1 FSs); and $\tilde{F}^s_k$ $(k = 1, \ldots, p)$ are IT2-antecedent fuzzy sets. These rules let one simultaneously account for uncertainty about antecedent MFs and consequent *parameter* values. Note that the latter is not the same as being able to account for uncertainty about a *linguistic* consequent, as can be done in an IT2 Mamdani FLC.

Because the antecedents and consequent in R^s_{TSK} are modeled using IT2 FSs and T1 FSs, respectively, this kind of TSK rule is referred to as the $A2-C1$ case (Liang and Mendel, 2001; Mendel, 2001, Chapter 13). It has not been used to date in an FLC because at present it is too complicated.

When the antecedents are T2 FSs and its consequents are crisp numbers (type-0 sets), then the TSK rule is referred to as the $A2-C0$ case (Liang and Mendel, 2001; Mendel, 2001, Chapter 13). This is the only case that is considered in the rest of this section because it is the one that is used in an FLC.

In the A2–C0 case, the rules in Eq. (3.73) simplify to

$$R^s_{\text{A2-C0}}\text{: IF } x_1 \text{ is } \tilde{F}^s_1 \text{ and} \cdots \text{and } x_p \text{ is } \tilde{F}^s_p, \text{ THEN}$$
$$u^s(\mathbf{x}) = c^s_0 + c^s_1 x_1 + c^s_2 x_2 + \cdots + c^s_p x_p \tag{3.74}$$

where $s = 1, \ldots, M$. In the sequel, $R^s_{\text{A2-C0}} \equiv R^s_{\text{TSK}}$.

3.3.3.2 *IT2 TSK FLC Computations* The output, $u_{\text{TSK},2}(\mathbf{x}')$, of an IT2 TSK FLC is obtained according to the following steps (Liang and Mendel, 2001; Mendel, 2001, Chapter 13):

1. Compute the value of each of the consequents of the A2–C0 rules, that is, compute the numbers $\{u^s(\mathbf{x})\}^M_{s=1}$. Then rank-order the $\{u^s(\mathbf{x})\}^M_{s=1}$ calling them $\{\gamma^k(\mathbf{x})\}^M_{k=1}$.
2. Compute the firing interval $F^s(\mathbf{x}) = [\underline{f}^s(\mathbf{x}), \overline{f}^s(\mathbf{x})]$ for each of the M rules using Eqs. (3.50)–(3.52). Relabel these firing intervals so that they conform to the $\{\gamma^k(\mathbf{x})\}^M_{k=1}$.
3. Compute the following interval weighted average:

$$U_{\text{TSK}}(\mathbf{x}) = \frac{\sum_{s=1}^M F^s(\mathbf{x})\gamma^s(\mathbf{x})}{\sum_{s=1}^M F^s(\mathbf{x})} = [u_l(\mathbf{x}), u_r(\mathbf{x})] \tag{3.75}$$

The actual computations of $u_l(\mathbf{x})$ and $u_r(\mathbf{x})$ are performed by using EKM algorithms that are applied to

$$u_l(\mathbf{x}) = \min_{\substack{f_j^s(\mathbf{x}) \in [\underline{f}^s(\mathbf{x}), \overline{f}^s(\mathbf{x})] \\ s=1,\ldots,M}} \frac{\sum_{s=1}^{M} f_j^s(\mathbf{x})\gamma^s(\mathbf{x})}{\sum_{s=1}^{M} f_j^s(\mathbf{x})} \tag{3.76}$$

$$u_r(\mathbf{x}) = \max_{\substack{f_j^s(\mathbf{x}) \in [\underline{f}^s(\mathbf{x}), \overline{f}^s(\mathbf{x})] \\ s=1,\ldots,M}} \frac{\sum_{s=1}^{M} f_j^s(\mathbf{x})\gamma^s(\mathbf{x})}{\sum_{s=1}^{M} f_j^s(\mathbf{x})} \tag{3.77}$$

Note that even though the same values of $\{\gamma^s(\mathbf{x})\}_{s=1}^{M}$ are used in both Eqs. (3.76) and (3.77), the EKM algorithms will give different values for the switch points L and R because Eq. (3.76) is a minimization problem and Eq. (3.77) is a maximization problem.

4. The interval $[u_l(\mathbf{x}), u_r(\mathbf{x})]$ is defuzzified to provide $y_{\text{TSK}}(\mathbf{x})$, where

$$u_{\text{TSK}}(\mathbf{x}) = 1/2[u_l(\mathbf{x}) + u_r(\mathbf{x})] \tag{3.78}$$

Although there is strong similarity between these computations and the ones for an IT2 Mamdani FLC that uses COS type reduction, there is also the following big difference: In the IT2 Mamdani FLC the centroids of its rule-consequent IT2 FSs are computed just once, and they are not a function of the input $\mathbf{x}$; but, in the IT2 TSK FLC, the interval-valued MF of each rule's consequent has to be recomputed for each value of $\mathbf{x}$.

3.3.4 Design of T2 FLCs

As mentioned in Section 3.2.4, Section 3.6 will describe how to complete the designs of IT2 FLCs. By "design" is meant the specification of such things as:

1. *For the IT2 Mamdani FLC*: choice of the antecedents, number of terms/FSs used for each variable (antecedent in a rule), the shapes of the FOUs for antecedents and consequents, the parameters that completely define each FOU, number of rules, t-norm, and type reduction method (if any).
2. *For the IT2 TSK FLC*: choice of the antecedents, number of terms/FSs used for each variable (antecedent in a rule), the shapes of the FOUs for antecedents, the parameters that completely define each FOU, the parameters that completely define each consequent, number of rules, and t-norm.

Our focus in Section 3.6 will be on how to determine the FOU parameters.

3.4 WU–MENDEL UNCERTAINTY BOUNDS

The iterative nature of the EKM algorithms introduces time delays that may lead to unpredictability, reduced performance, or even system instability in a fuzzy logic control system. This, of course, depends on the bandwidth of the plant and the system's sampling rate. Many approaches have been proposed for not using EKM algorithms. Some preserve the ability to either approximate or compute the type-reduced set, which may be a good thing because the type-reduced set provides a measure of the MF uncertainties as they flow through the IT2 FLC and plays a role that is analogous to a confidence interval. These approaches first compute a T1 FS from an IT2 FS and then defuzzify that T1 FS. The other approaches bypass type reduction completely and obtain the defuzzified output directly. They are unable to provide a measure of the MF uncertainties as they flow through the IT2 FLC, but they are faster than the methods that can do this. For a comprehensive survey of all of these different approaches, see Mendel (2013).

In this section we focus only on Wu and Mendel's uncertainty bounds (WM UBs) (Wu and Mendel, 2002) because they have already been used for real-time fuzzy logic control and will also be used in Chapter 6.

We already know that there are no closed-form formulas for type reduction. Wu and Mendel replace type reduction with lower and upper bounds—*uncertainty bounds*—for the end points of the type-reduced set, and those bounds, which are optimal in a mini–max sense, can be computed without having to perform type reduction.[14]

To begin, four centroids (also called *boundary T1 FLCs*) are defined, all of which can be computed once the left and right end points of the firing interval, $\underline{f}^s(\mathbf{x})$ and $\overline{f}^s(\mathbf{x})$ $(s = 1, \ldots, M)$, have been computed. In these centroids, a^s and b^s are the left and right end points of the centroid of the sth consequent IT2 FS, which have been reordered in ascending order [note that associated variables such as $\underline{f}^s(\mathbf{x})$ and $\overline{f}^s(\mathbf{x})$ need to be relabeled so they remain associated with the correct rule s]. These consequent centroids only have to be computed (and stored) one time after the IT2 FLC has been designed since they do not depend upon the input to the FLC. The boundary T1 FLC centroids are[15]

$$\{\text{LMFs, Left}\}\colon\ u_l^{(0)}(\mathbf{x}) = \frac{\sum_{s=1}^{M} \underline{f}^s a^s}{\sum_{s=1}^{M} \underline{f}^s} \tag{3.79}$$

$$\{\text{LMFs, right}\}\colon\ u_r^{(M)}(\mathbf{x}) = \frac{\sum_{s=1}^{M} \underline{f}^s b^s}{\sum_{s=1}^{M} \underline{f}^s} \tag{3.80}$$

[14]In Wu and Mendel (2002) there are detailed derivations of the uncertainty bounds for center-of-sets TR (because it handles nonsymmetrical shoulder MFs better than do other kinds of TR); however, these results are also applicable to other kinds of TR, as explained in their Table V.

[15]Note, for example, that in Eq. (3.79) {LMF, left} refers to the fact that this centroid only uses *lower* MFs of the firing interval and *left* end point values of the consequent set centroid. In addition, in Eqs. (3.79)–(3.82) and (3.85) and (3.86), $\underline{f}^s(\mathbf{x})$ and $\overline{f}^s(\mathbf{x})$ have been shortened to $\underline{f}^s$ and $\overline{f}^s$, respectively.

$$\{\text{UMFs, Left}\}:\ u_l^{(M)}(\mathbf{x}) = \frac{\sum_{s=1}^{M} \overline{f}^s a^s}{\sum_{s=1}^{M} \overline{f}^s} \tag{3.81}$$

$$\{\text{UMFs, right}\}:\ u_r^{(0)}(\mathbf{x}) = \frac{\sum_{s=1}^{M} \overline{f}^s b^s}{\sum_{s=1}^{M} \overline{f}^s} \tag{3.82}$$

THEOREM 3.4 (**WM UBs**) The end points $u_l(\mathbf{x})$ and $u_r(\mathbf{x})$ of the TR set of an IT2 FLC for the input $\mathbf{x}$, are bounded from below and above as $\underline{u}_l(\mathbf{x}) \le u_l(\mathbf{x}) \le \overline{u}_l(\mathbf{x})$ and $\underline{u}_r(\mathbf{x}) \le u_r(\mathbf{x}) \le \overline{u}_r(\mathbf{x})$, where:

$$\overline{u}_l(\mathbf{x}) = \min\{u_l^{(0)}(\mathbf{x}), u_l^{(M)}(\mathbf{x})\} \tag{3.83}$$

$$\underline{u}_r(\mathbf{x}) = \max\{u_r^{(0)}(\mathbf{x}), u_r^{(M)}(\mathbf{x})\} \tag{3.84}$$

$$\underline{u}_l(\mathbf{x}) = \overline{u}_l(\mathbf{x}) - \left[\frac{\sum_{s=1}^{M}\left(\overline{f}^s - \underline{f}^s\right)}{\sum_{s=1}^{M}\overline{f}^s \sum_{s=1}^{M}\underline{f}^s} \times \frac{\sum_{s=1}^{M}\underline{f}^s(a^s - a^1)\sum_{s=1}^{M}\overline{f}^s(a^M - a^s)}{\sum_{s=1}^{M}\underline{f}^s(a^s - a^1) + \sum_{s=1}^{M}\overline{f}^s(a^M - a^s)} \right] \tag{3.85}$$

$$\overline{u}_r(\mathbf{x}) = \underline{u}_r(\mathbf{x}) + \left[\frac{\sum_{s=1}^{M}\left(\overline{f}^s - \underline{f}^s\right)}{\sum_{s=1}^{M}\overline{f}^s \sum_{s=1}^{M}\underline{f}^s} \times \frac{\sum_{s=1}^{M}\overline{f}^s(b^s - b^1)\sum_{s=1}^{M}\underline{f}^s(b^M - b^s)}{\sum_{s=1}^{M}\overline{f}^s(b^s - b^1) + \sum_{s=1}^{M}\underline{f}^s(b^M - b^s)} \right] \tag{3.86}$$

Proof. See Appendix 3A.

Observe that the four bounds in Eqs. (3.83)–(3.86) can be computed without having to perform type reduction. Wu and Mendel (2002) then approximate the type-reduced set as

$$[u_l(\mathbf{x}), u_r(\mathbf{x})] \approx [\hat{u}_l(\mathbf{x}), \hat{u}_r(\mathbf{x})] = \left[\frac{\underline{u}_l(\mathbf{x}) + \overline{u}_l(\mathbf{x})}{2}, \frac{\underline{u}_r(\mathbf{x}) + \overline{u}_r(\mathbf{x})}{2} \right] \tag{3.87}$$

and compute the output of the IT2 FLC as

$$\hat{u}(\mathbf{x}) = \frac{1}{2}\left[\frac{\underline{u}_l(\mathbf{x}) + \overline{u}_l(\mathbf{x})}{2} + \frac{\underline{u}_r(\mathbf{x}) + \overline{u}_r(\mathbf{x})}{2} \right] \tag{3.88}$$

[instead of as $(u_l(\mathbf{x}) + u_r(\mathbf{x}))/2$]. So, *by using the WM UBs, they obtain both an approximate type-reduced set as well as a defuzzified output.*

Wu and Mendel (2002) also provide the following upper bound on the difference $\delta(\mathbf{x})$ between the defuzzified outputs of the actual type-reduced set and its approximation, namely:

$$\delta(\mathbf{x}) \equiv \left| \frac{u_l(\mathbf{x}) + u_r(\mathbf{x})}{2} - \frac{1}{2}\left[\frac{\underline{u}_l(\mathbf{x}) + \overline{u}_l(\mathbf{x})}{2} + \frac{\underline{u}_r(\mathbf{x}) + \overline{u}_r(\mathbf{x})}{2}\right] \right| \tag{3.89}$$

$$\delta(\mathbf{x}) \leq 1/4[(\overline{u}_l(\mathbf{x}) - \underline{u}_l(\mathbf{x})) + (\overline{u}_r(\mathbf{x}) - \underline{u}_r(\mathbf{x}))] \tag{3.90}$$

Lynch et al. (2006a, b) replace all of the Mamdani IT2 FLC computations with those in Eqs. (3.83)–(3.88), that is, Eqs. (3.83)–(3.88) are their IT2 FLC. This approach is summarized in Fig. 3.12.

A further approximation of Eqs. (3.85) and (3.86) is described in Chapter 6.

Two other ways to summarize the IT2 FLC in which type reduction has been replaced by the WM UBs are depicted in Figs. 3.13 and 3.14.

Block Diagram Summary The entire chain of computations is summarized in the block diagram of Fig. 3.13. Firing intervals are computed for all rules [using Eqs. (3.51) and (3.52)], and they depend explicitly on the input $\mathbf{x}$. Offline computations of the centroids are performed for each of the M consequent IT2 FSs using EKM algorithms [as in Eq. (3.62)] and are then stored in memory. Four boundary T1 FLC centroids are computed [using Eqs. (3.79)–(3.82)], after which the four uncertainty bounds are computed [using Eqs. (3.83)–(3.86)]. The bounds for the left (right) end of the approximated type-reduced set are then averaged [using the two equations on the right-hand side of Eq. (3.87)] after which those averages are again averaged [using Eq. (3.88)], the result being the mini–max approximation to the defuzzified value of the control signal, $\hat{u}(\mathbf{x})$.

Neurofuzzy Summary The entire chain of computations is also summarized in the layered architecture of Fig. 3.14, which resembles a neural network, although

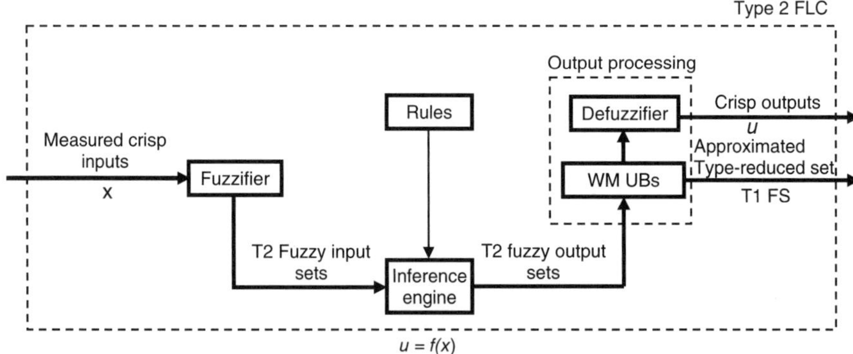

Figure 3.12 IT2 FLC in which type reduction has been replaced by the Wu–Mendel uncertainty bounds (WM UBs).

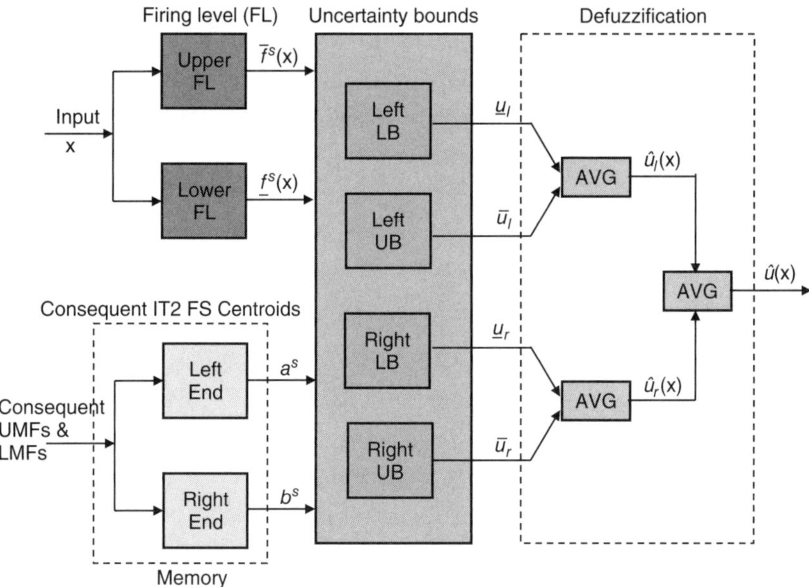

Figure 3.13 Block diagram summary of IT2 Mamdani FLC computations that use WM UBs (Mendel, 2007; © 2007, IEEE).

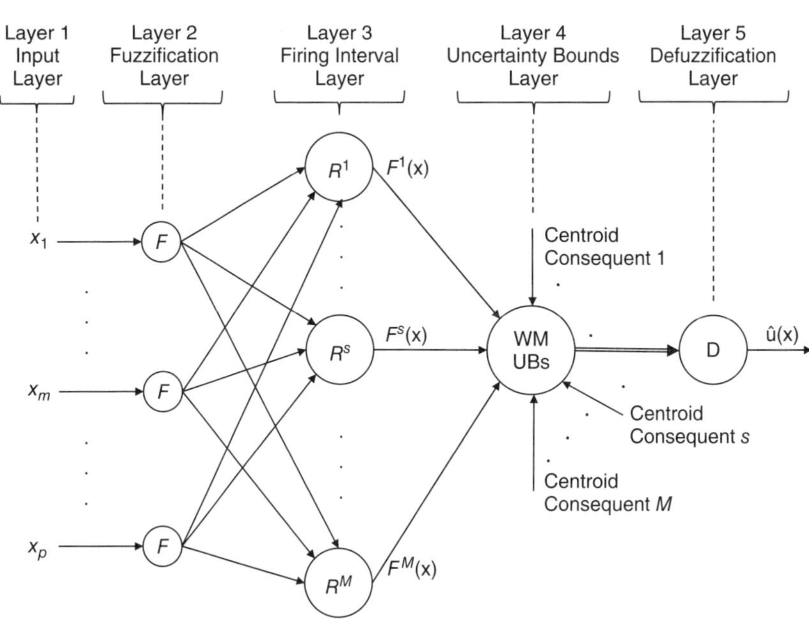

Figure 3.14 Neurofuzzy summary of IT2 Mamdani FLC computations that use WM UBs.

(again) there is nothing "neural" about this FLC. Layer 1 refers to the inputs, $x_1, \ldots, x_p$. Layer 2 is where the inputs are fuzzified (F) into IT2 FSs. When singleton fuzzification is used, then Eq (3.10) applies. Layer 3 is where the firing interval is computed for each of the M rules [using Eqs. (3.51) and (3.52)]. Layer 4 is where the WM UBs are computed [using Eqs. (3.79)–(3.86)]. Observe that the output from this layer is shown as a double-lined arrow, to indicate that four UBs are sent to Layer 5, where defuzzification is performed [Eq. (3.88], the result being the min–max approximation to the defuzzified value of the control signal, $\widehat{u}(\mathbf{x})$.

Example 3.4 This example is a continuation of Example 3.3. Beginning with the consequent centroids and firing intervals that are given, respectively, in Tables 3.1 and 3.2, for $\mathbf{x}' = \text{col}(170, 240)$, we compute $\overline{u}_{l(\text{Steering})}(\mathbf{x}')$, $\underline{u}_{l(\text{Steering})}(\mathbf{x}')$, $\underline{u}_{r(\text{Steering})}(\mathbf{x}')$, $\overline{u}_{r(\text{Steering})}(\mathbf{x}')$, $\widehat{u}_{l(\text{Steering})}(\mathbf{x})$, $\widehat{u}_{r(\text{Steering})}(\mathbf{x})$, and $\widehat{u}_{\text{Steering}}(\mathbf{x})$, using Eqs. (3.83)–(3.88). The results are

$$\begin{aligned}
\overline{u}_{l(\text{Steering})}(\mathbf{x}') &= \min\left\{\frac{\sum_{s=1}^{4}\underline{f}^s a^s}{\sum_{s=1}^{4}\underline{f}^s}, \frac{\sum_{s=1}^{4}\overline{f}^s a^s}{\sum_{s=1}^{4}\overline{f}^s}\right\} \\
&= \min\left\{\frac{0.1\times(-81.43)+0.3\times(-81.43)+0.1\times(8.57)+0.2\times(98.57)}{0.1+0.3+0.1+0.2},\right. \\
&\qquad \left.\frac{0.6\times(-81.43)+0.8\times(-81.43)+0.6\times(8.57)+0.7\times(98.57)}{0.6+0.8+0.6+0.7}\right\} \\
&= \min\{-17.17, -14.76\} = -17.17
\end{aligned} \tag{3.91}$$

$$\begin{aligned}
\underline{u}_{l(\text{Steering})}(\mathbf{x}') &= \overline{u}_{l(\text{Steering})}(\mathbf{x}') - \left[\frac{\sum_{s=1}^{4}\left(\overline{f}^s - \underline{f}^s\right)}{\sum_{s=1}^{4}\overline{f}^s\sum_{s=1}^{4}\underline{f}^s} \times \frac{\sum_{s=1}^{4}\underline{f}^s(a^s-a^1)\sum_{s=1}^{4}\overline{f}^s(a^4-a^s)}{\sum_{s=1}^{4}\underline{f}^s(a^s-a^1)+\sum_{s=1}^{4}\overline{f}^s(a^4-a^s)}\right] \\
&= -17.17 - \left[\frac{(0.6-0.1)+(0.8-0.3)+(0.6-0.1)+(0.7-0.2)}{(0.6+0.8+0.6+0.7)\times(0.1+0.3+0.1+0.2)} \times \frac{g}{h}\right] \\
&= -17.17 - \left[\frac{2}{1.89} \times \frac{14{,}230.08}{361.22}\right] = -58.67
\end{aligned} \tag{3.92}$$

where

$$\begin{aligned}
g &= [0.1\times(8.57+81.43)+0.2\times(98.57+81.43)] \\
&\quad \times [0.6\times(98.57+81.43)+0.8\times(98.57+81.43)+0.6\times(98.57+8.57)] \\
&= 14{,}230.08 \\
h &= [0.1\times(8.57+81.43)+0.2\times(98.57+81.43)] \\
&\quad + [0.6\times(98.57+81.43)+0.8\times(98.57+81.43)+0.6\times(98.57+8.57)] \\
&= 361.22
\end{aligned}$$

In a similar manner, it follows that

$$\underline{u}_{r(\text{Steering})}(\mathbf{x}') = -1.908 \tag{3.93}$$

$$\overline{u}_{r(\text{Steering})}(\mathbf{x}') = 57.208 \tag{3.94}$$

Consequently,

$$\widehat{u}_{l(\text{Steering})}(\mathbf{x}') = \tfrac{1}{2}(-17.17 - 58.67) = -37.92 \tag{3.95}$$

$$\widehat{u}_{r(\text{Steering})}(\mathbf{x}') = \tfrac{1}{2}(-1.908 + 57.208) = 27.65 \tag{3.96}$$

$$\widehat{u}_{\text{Steering}}(\mathbf{x}') = \tfrac{1}{2}(-37.92 + 27.65) = -5.14 \tag{3.97}$$

Comparing Eqs. (3.95)–(3.97) with Eqs. (3.71), (3.72), and $u_{\text{Steering}}(\mathbf{x}') = -5.52$, respectively, observe the following: (1) $\underline{u}_{l(\text{Steering})} \leq u_{l(\text{Steering})} \leq \overline{u}_{l(\text{Steering})}$, that is, $-58.67 \leq -54.97 \leq -17.17$; (2) $\underline{u}_{r(\text{Steering})} \leq u_{r(\text{Steering})} \leq \overline{u}_{r(\text{Steering})}$, that is, $-1.908 \leq 43.92 \leq 57.208$; (3) even though $\widehat{u}_{l(\text{Steering})}(\mathbf{x}') = -37.92$ is not a very good approximation to $u_{l(\text{Steering})}(\mathbf{x}') = -54.97$, and $\widehat{u}_{r(\text{Steering})}(\mathbf{x}') = 27.65$ is not a very good approximation to $u_{r(\text{Steering})}(\mathbf{x}') = 43.92$, $\widehat{u}_{\text{Steering}}(\mathbf{x}') = -5.14$ is a reasonably good approximation to $u_{\text{Steering}}(\mathbf{x}') = -5.52$.

Example 3.5[16] Section 1.8.1.1 compared a T1 FLC and an IT2 FLC for the speed control of a marine diesel engine. Although which kind of IT2 FLC was used was not stated in that section, we can now reveal that the results that are depicted in Figs. 1.9 and 1.10 were obtained for the IT2 FLC that used type reduction and KM algorithms.

Figure 3.15, which is analogous to Fig. 1.9, depicts the control surfaces for the IT2 FLC that used type reduction and KM algorithms and the IT2 FLC that used WM uncertainty bounds, and shows how similar they are. Figure 3.16 is analogous to Fig. 1.10 and shows that both the IT2 FLC that used type reduction and KM algorithms and the IT2 FLC that used WM uncertainty bounds give very similar responses to the 100% load addition.

Figures 3.15 and 3.16 strongly suggest that the WM UBs provide a very viable alternative to the IT2 FLC that uses type reduction and should be very useful for real-time applications.

3.5 CONTROL ANALYSES OF IT2 FLCs

Interval type-2 FLCs can be studied like any other nonlinear controller. For example, stability and robustness studies can be performed by means of extensive simulations

[16]The material in this example is taken from Lynch et al. (2006b; © 2006, IEEE).

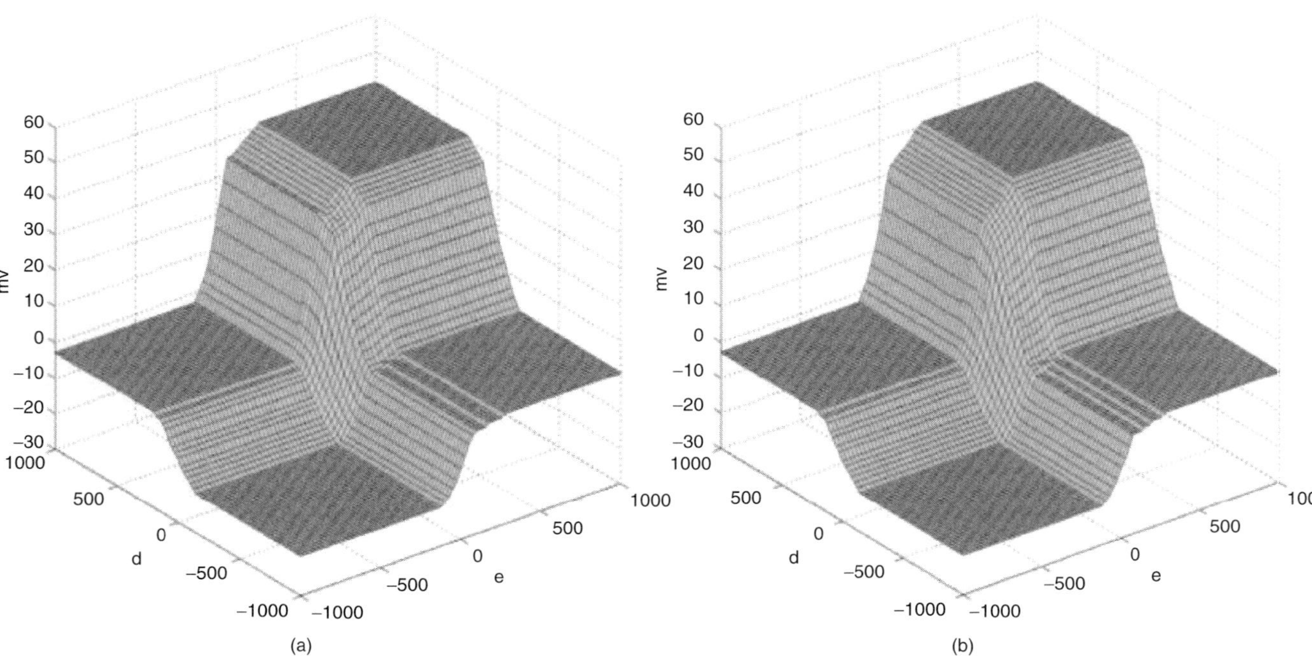

Figure 3.15 Control surfaces for IT2 FLCs that use (a) type reduction and (b) WM UBs (Hagras, 2007; © 2007, IEEE).

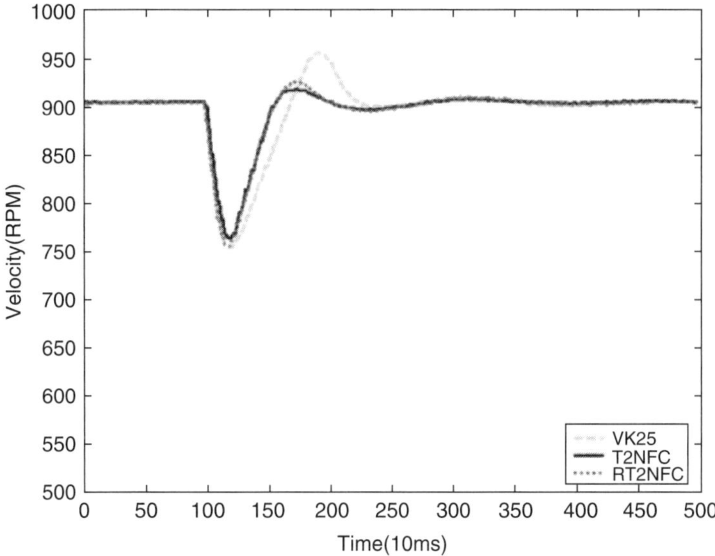

Figure 3.16 Comparisons of the responses of a Viking 25 against an IT2NFC (that uses type reduction) and a real-time IT2 FC (that uses WM UBs) with 100% load addition (Lynch et al., 2006b; © 2006, IEEE).

and by analyzing control surfaces (e.g., Fig. 1.9) showing the mathematical functions that map inputs into outputs.

For problems with low dimensionality (i.e., problems with one to two inputs), a control surface enables one to visualize the unknown input–output mapping function articulated by the IT2 FLC, which allows one to analyze the controller's response, for example, the smoother the shape of the control surface, generally, the better will be the controller's response. In addition, near the set point [e.g., in Fig. 1.9, when the error (e) and difference in error (d) are equal to zero], the control surfaces need to be smooth and must avoid steep changes so that uncertainties and disturbances in the inputs do not cause abrupt changes in the outputs, and so that overshoots, settling times, rise time, and steady-state errors will be as small as possible. Observe in Fig. 1.9a that the T1 FLC control surface is steep and nonsmooth, especially around the set point. Consequently, any small variations of e and d can cause considerable changes in the FLC output, which means that the T1 FLC in Fig. 3.1 is vulnerable to uncertainties. Moreover, the larger the variations in e and d due to uncertainties (associated with changes of operating and/or load conditions), the bigger will be the disturbances to the FLC output, which can cause instability.

Figures 1.9b, 3.10a, and 3.10b depict control surfaces for different kinds of IT2 FLCs. They all show a smooth response and result in very good control performance that can handle the uncertainties and disturbances, because near the set point where $e = 0$ and $d = 0$ small variations in e and d do not cause significant changes to the IT2 FLC output; the response proceeds gradually and smoothly with no steep changes.

Figure 3.17a depicts familiar step response plots. Similar plots can easily be obtained for all IT2 FLCs, as shown, for example, in Fig.3.17b for the speed control of the marine diesel engine that was described in Section 1.8.1.1.

Analyses of IT2 FLCs are not limited to examining control surfaces or step response plots. Chapters 4–6 present more detailed and mathematically oriented control analyses (Lyapunov stability analyses and robustness analyses) for different kinds of IT2 Mamdani and TSK FLCs.

3.6 DETERMINING THE FOU PARAMETERS OF IT2 FLCs

Section 3.3.4 explains what is meant by the design of either IT2 Mamdani or TSK FLCs. In this section we focus exclusively on determining the FOU parameters for such FLCs, and so we assume that decisions have already been made about the choice of the antecedents, number of MFs for each variable, number of rules, t-norm, and type reduction method (if one is used). Although there can be many methods for establishing the FOUs, in this section we focus on only two of them because they are the ones that are widely used for an FLC.

3.6.1 Blurring T1 MFs

In this method one begins with a T1 MF and then creates the FOU by blurring it (e.g., Mendel, 2001). Blurring can conceptually be thought of as moving the T1 MF to the left and right or only in one direction (e.g., Fig. 3.18). Exactly how much blurring should be done to create the FOU can be established either by trial and error or by any of the optimization procedures that are described in Section 3.6.2 in which the amount of blurring is treated as an FOU design parameter.

3.6.2 Optimizing FOU Parameters

Sometimes the parameters of an IT2 FLC are optimized (tuned, learned) during its design phase. The optimized parameters are then fixed during its operational phase, unless continued adaptation is required, in which case online changes to parameters take place.

In this approach one sets up a mathematical objective function, $J(\boldsymbol{\phi})$, that depends upon the design parameters, $\boldsymbol{\phi}$. The elements of $\boldsymbol{\phi}$ include all of the antecedent and consequent FOU parameters. For example, if antecedent and consequent FOUs are Gaussian with uncertain mean $m \in [m_1, m_2]$ and certain standard deviation, σ, then each FOU is described by three parameters. For M Mamdani rules (each with p antecedents and one consequent), $\boldsymbol{\phi}$ will contain $M(3p+3)$ elements; for M comparable TSK rules, $\boldsymbol{\phi}$ will contain $M[3p+(p+1)]$ elements since each rule consequent contains $p+1$ parameters.

The function $J(\boldsymbol{\phi})$ is a nonlinear function of $\boldsymbol{\phi}$ and so some sort of mathematical programming approach has to be used to optimize it. There are many different kinds of optimization algorithms that can be used to do this. For the rest of this section we assume that $\boldsymbol{\phi}$ has N_ϕ elements.

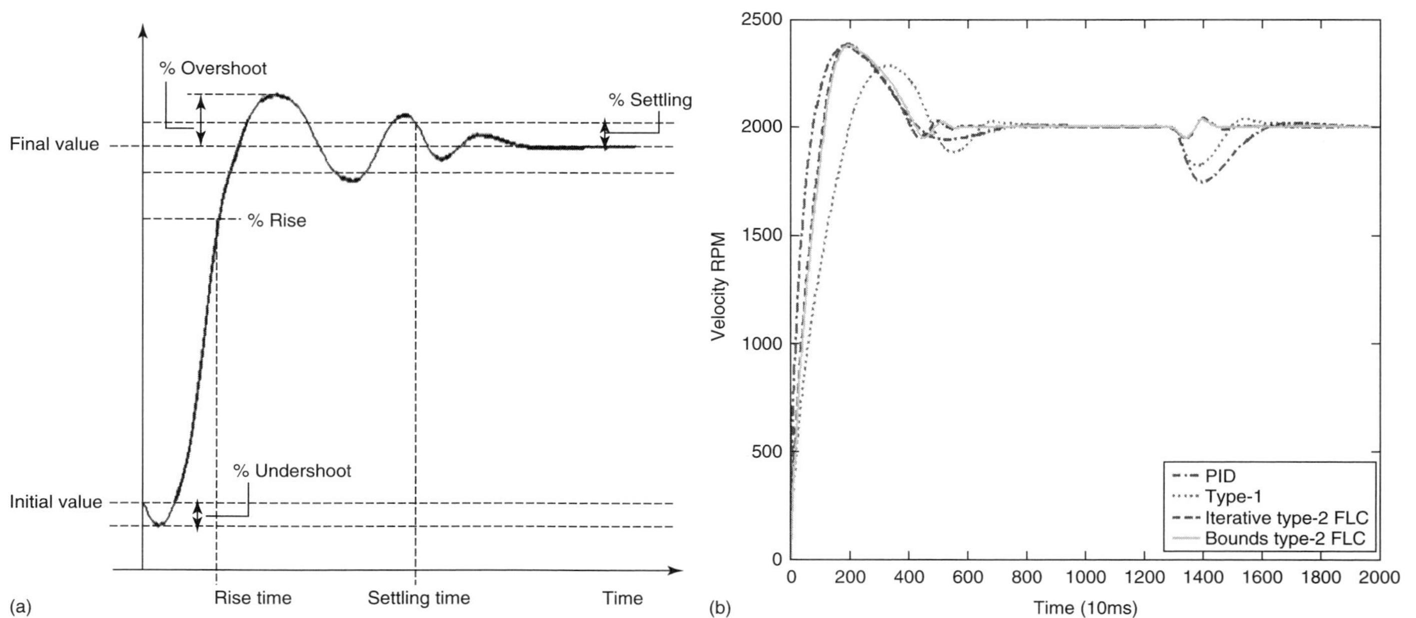

Figure 3.17 (a) Step response plot used in traditional control systems and (b) step response plot of an IT2 Mamdani FLC applied in the speed control of a marine diesel engine.

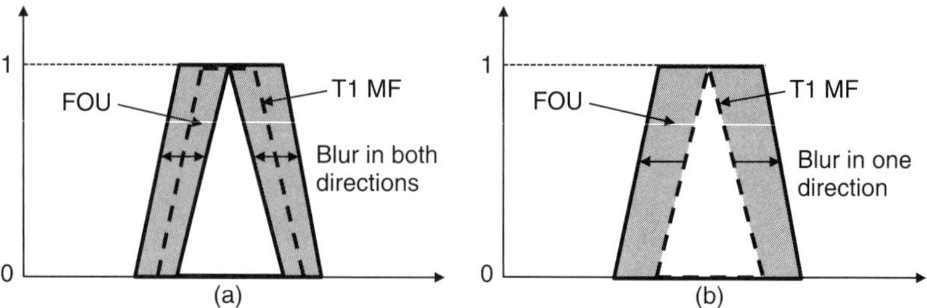

Figure 3.18 Blurring of T1 MFs to obtain an FOU: (a) A trapezoidal T1 MF is blurred by moving its legs both to the left and to the right, and (b) a triangle T1 MF is blurred by sliding its left leg to the left and its right leg to the right.

Until recently, most designs of IT2 FLSs used gradient-based optimization algorithms that require mathematical formulas for derivatives of $J(\boldsymbol{\phi})$ with respect to each of the design parameters. As stated in Mendel (2004 p. 85):

Generally, it is much more complicated to compute such derivatives for an IT2 FLC than it is for a T1 FLC, because:

- In an IT2 FLC the design parameters appear in *upper* and *lower* MFs, whereas in a T1 FLC they appear in a single MF.
- In an IT2 FLC, type reduction establishes the two parameters $L(\mathbf{x}')$ and $R(\mathbf{x}')$ (the switch points) which in turn establish the upper and lower firing interval MFs that are used to compute the left and right end points of the type-reduced set. There is no type reduction in a T1 FLC.

When EKM algorithms (or any of the other comparable algorithms) are used to compute the COS type-reduced set, they require (without exception) that the upper and lower firing intervals must first be reordered in an increasing order. Usually, the orderings are different for the upper and lower firing intervals. Derivative computations require that the upper and lower firing intervals must then be put back into their original rule ordering so that one knows exactly where the MFs are because the parameters in $\boldsymbol{\phi}$ occur in specific MFs. Keeping track of where the parameters are requires using permutation matrices, and detailed equations are provided in Mendel (2004).

It is very easy to make mistakes in computing the required derivatives. For example, Hagras (2006) points out incorrect derivative formulas that are in Wang et al. (2004). Not only that, but one must also include tests on the primary variables because lower and upper MFs can have different formulas for different subranges of the primary variables. Also, formulas for the derivatives depend upon the architecture chosen for the FLC and will be different for IT2 Mamdani, WM UB, and TSK architectures.

Additionally, gradient-based optimization algorithms (also, called back-propagation algorithms) are not globally convergent, so the solution one obtains by using them to optimize $J(\boldsymbol{\phi})$ will only be a local extremum. Of course, there are strategies for repeating the optimization by randomly rechoosing initial values for the parameters, but there is no guarantee that even doing this will provide the global extremum.

In conclusion, our recommendation is to not use gradient-based optimization algorithms for the designs of IT2 FLCs. For those readers who insist on using such optimization algorithms, see Mendel (2001, 2004) and Hagras (2006) for some derivative formulas, but only for IT2 Mamdani FLCs.

Fortunately, there now are globally convergent iterative search algorithms that do not require any derivatives, for example, simulated annealing, genetic algorithms (GA), particle swarm optimization (PSO), quantum particle swarm optimization (QPSO), ant colony optimization, and so forth. Many of these have already been applied to the designs of IT2 FLSs (e.g., Castillo and Melin, 2012; Hidalgo et al., 2012; Melin et al., 2013; Moldonado et al., 2013).

In the rest of this section we focus on two optimization algorithms (QPSO and GA) that do not require derivatives and that can be used for any of the FLC architectures in this book.

3.6.2.1 *Using QPSO to Optimize the Parameters in an IT2 FLC* QPSO is a globally convergent (Wei et al., 2010) iterative search algorithm that does not use derivatives, generally outperforms the original PSO (Kennedy and Eberhardt, 1995; Wang et al., 2011), and has fewer parameters to control. It is a population-based optimization technique, where a population is called a *swarm* that contains a set of different particles. Each *particle* represents a possible solution to an optimization problem. The position of each particle is updated (in each QPSO iteration) by using its most recent own best solution, mean of the personal best positions of all particles, and the global best solution found by all particles so far.

Quantum particle swarm optimization finds optimized $\boldsymbol{\phi}$ based on the following criterion: $\min_{\boldsymbol{\phi}_m} J(\boldsymbol{\phi}_m)$. The *current position* (vector) of the mth particle is defined as $(m = 1, \ldots, N_m)$

$$\boldsymbol{\phi}_m = \text{col}(\phi_{m,1}, \phi_{m,2}, \ldots, \phi_{m,N_\phi}) \tag{3.98}$$

A *particle best position* [i.e., the position that produces the minimal value of $J(\boldsymbol{\phi})$ over the entire history of that particle], $\mathbf{p}_m = \text{col}(p_{m,1}, p_{m,2}, \ldots, p_{m,N_\phi})$, is computed as $(t = 1, \ldots, N_\phi)$

$$p_{m,t}(g+1) = \eta \times p_{m,t}(g) + (1-\eta) \times p_{\text{gbest},t}(g) \tag{3.99}$$

where $g = 1, \ldots, G-1$ is the *index of a generation* (iteration), $p_{m,t}(1)$ is initialized by $\phi_{m,t}(1)$, η is a random variable uniformly distributed in (0,1], and $\mathbf{p}_{\text{gbest}}(g)$ [whose components are $p_{\text{gbest},t}(g)$] denotes the *global best* (*gbest*) *position* found in the history of the entire swarm, that is $(m = 1, \ldots, N_m)$,

$$\mathbf{p}_{\text{gbest}}(g) = \arg \min_{\mathbf{p}_m(g), \forall m=1 \ldots, N_m} J(\mathbf{p}_m(g)) \tag{3.100}$$

A global point, the *mean best position* of the population, is introduced into QPSO; it is denoted as $\mathbf{m}(g)$ and is defined as the sample mean of the $\mathbf{p}_m(g)$ positions of all N_m particles, that is,

$$\mathbf{m}(g) = \frac{1}{N_m} \sum_{m=1}^{N_m} \mathbf{p}_m(g) \tag{3.101}$$

At the end of each generation, a new position of a particle is obtained, as $(t = 1, \dots, N_\phi)$:

$$\phi_{m,t}(g+1) = p_{m,t}(g+1) \pm \beta |m_t(g) - \phi_{m,t}(g)| \ln \frac{1}{\rho} \tag{3.102}$$

where parameter β, called the *contraction–expansion coefficient*, can be tuned to control the convergence speed of the algorithm, and ρ is also a random variable uniformly distributed in (0,1]. In Eq. (3.102), the plus and minus signs are randomly selected to generate the new position of a particle.

Pseudocode for QPSO is given in Table 3.3.

Using IT2 FSs in an FLC has the potential to provide better control performance (in the strict sense of input-to-output mapping) for an FLC than using T1 FSs, which is why there has been so much interest in IT2 FLCs. In fact, as IT2 FSs generalize T1 FSs, control performance is certainly no worse for the IT2 FLC when doing a one-to-one comparison per control cycle (note that overall real-time control performance on standard computing equipment such as PCs may be affected by the faster execution and thus larger number of control cycles per timeframe of a T1 FLC in comparison to an equivalent IT2 FLC).

Using QPSO lets us do this by using the following *design procedure*.

1. Design a T1 FLC by optimizing its parameters using QPSO.
2. Design an IT2 FLC by optimizing its parameters using QPSO in which one particle is associated with the just designed T1 FLC.

Mendel (2013) has proven that by virtue of the QPSO algorithm, the performance of the optimized IT2 FLC cannot be worse than that of the optimized T1 FLC. This does not mean that the performance of the optimized IT2 FLC will be *significantly better* than that of the optimized T1 FLC. There is no analysis that is available to date that focuses on such relative performance improvements. Of course, relative improvements are very application dependent because objective function $J(\boldsymbol{\phi})$ is application dependent.

Example 3.6 In step 2 of the above design procedure, one particle is associated with a just designed T1 FLC. Suppose, for example, that the IT2 FLC is

TABLE 3.3 Pseudo Code for QPSO as Used in Optimal Designs of an IT2 FLC

```
Initialize ϕ_1(1) as a T1 FLC particle, and all other ϕ_m(1) randomly
(m = 2, ... , N_m)
Set p_m(1) = ϕ_m(1) (m = 1, ... , N_m)
For g = 1 to G-1
        Calculate m(g) = (1/N_m) Σ_{m=1}^{N_m} p_m(g)
        Calculate J(p_m(g)) (m = 1, ... , N_m)
        p_gbest(g) = arg min_{p_m(g), ∀m=1,...,N_m} J(p_m(g))
        for m = 1 to N_m (number of particles)
        Calculate J(ϕ_m(g))
                If J(ϕ_m(g)) < J(p_m(g))
                        p_m(g) = ϕ_m(g)
                end if
                for t = 1 to N_φ (number of components in each
                                  particle)
                        η = rand(0, 1)
                        p_{m,t}(g + 1) = η × p_{m,t}(g) + (1 − η) × p_{gbest,t}(g)
                        ρ = rand(0, 1)
                        if rand(0, 1) > 0.5 then
                                φ_{m,t}(g + 1) = p_{m,t}(g + 1) − β|m_t(g) − φ_{m,t}(g)| ln(1/ρ)
                        else
                                φ_{m,t}(g + 1) = p_{m,t}(g + 1) + β|m_t(g) − φ_{m,t}(g)| ln(1/ρ)
                        end if
                end for
        end for
end for
```

Mamdani + COS TR, and the antecedent and consequent FOUs are the Gaussian ones stated above. The structure of a particle for such an IT2 FLC is

$$\boldsymbol{\phi}_{\mathrm{IT2}} = \mathrm{col}(\underbrace{\overbrace{m_{11}^1, m_{12}^1, \sigma_1^1}^{\text{Antecedent 1}}, \ldots, \overbrace{m_{p1}^1, m_{p2}^1, \sigma_p^1}^{\text{Antecedent } p}, \overbrace{a^1, b^1}^{\text{Consequent}}}_{\text{Rule 1}};$$

$$\ldots; \underbrace{\overbrace{m_{11}^M, m_{12}^M, \sigma_1^M}^{\text{Antecedent 1}}, \ldots, \overbrace{m_{p1}^M, m_{p2}^M, \sigma_p^M}^{\text{Antecedent } p}, \overbrace{a^M, b^M}^{\text{Consequent}})}_{\text{Rule } M} \tag{3.103}$$

The T1 particle must be of the same length as this IT2 particle, begins with Eq. (3.103), and can be expressed as

$$\boldsymbol{\phi}_{\text{T1}} = \text{col}(\underbrace{\overbrace{m_1^1, m_1^1, \sigma_1^1}^{\text{Antecedent 1}}, \ldots, \overbrace{m_p^1, m_p^1, \sigma_p^1}^{\text{Antecedent } p}, \overbrace{y_{G^1}, y_{G^1}}^{\text{Consequent}}}_{\text{Rule 1}};$$

$$\ldots; \underbrace{\overbrace{m_1^M, m_1^M, \sigma_1^M}^{\text{Antecedent 1}}, \ldots, \overbrace{m_p^M, m_p^M, \sigma_p^M}^{\text{Antecedent } p}, \overbrace{y_{G^M}, y_{G^M}}^{\text{Consequent}})}_{\text{Rule } M} \tag{3.104}$$

Observe, in Eq. (3.104), that:

- All of the MF parameters are taken from the optimal T1 FLC design.
- By setting the values for the end points of the means to be the same, the uncertainty in each rule about the means for all p antecedents and its consequent has disappeared.
- In this way it is straightforward to embed a T1 particle into an IT2 particle.

3.6.2.2 *Using GA to Optimize the Parameters in an IT2 FLC* Genetic algorithms provide a stochastic optimization method that is based on Darwinian evolution. In a "survival of the fittest" approach, a large number of solutions is generated randomly, followed by selective "breeding" of the fittest solutions, through operations such as crossover and mutation, to obtain fitter (i.e., better at the task in question) solutions. Selection and breeding are repeated iteratively over time until a stopping criterion is reached, for example, completing a certain number of generations or discovering an individual with a certain desired fitness. Today, the founding work on modern GAs is attributed to Holland's seminal book (Holland, 1975) that laid the foundations for a vast research field in evolutionary algorithms, specifically GAs.

Generate algorithms have been widely applied to the configuration and/or optimization of IT2 FLCs [e.g., see Wu and Tan (2004), Wagner and Hagras (2007), Martinez et al. (2008), Martinez et al. (2009), and Shill et al. (2012).

In designing an IT2 FLC with GAs using a step-by-step approach, first an "optimal" T1 FLC (using a GA or other means) is designed, after which the T1 FSs are transformed to IT2 FSs using GA. In other words, the structure of the FLC (i.e., rule bases, number and type of fuzzy sets) is determined for the T1 case first, then the T1 sets are blurred (see Section 3.6.1), after which the parameters of the "blurred" IT2 sets are optimized by a GA. This approach has the advantage of a smaller set of parameters for the GA to optimize (because the overall task is split into two smaller subtasks, resulting in fewer possible permutations of parameters), thus offering faster convergence. Further, in many applications, the T1 FLC developed in the first stage of the process is employed as a baseline reference for comparison and evaluation of the utility of the IT2 FLC. An example of this approach can be found in Wu and Tan (2004) and is also described in Section 5.2.5.3.

A second approach to designing IT2 FLCs using GAs relies on the direct configuration and optimization of all the IT2 FLC parameters (i.e., number and kind of IT2 FSs, parameters for all IT2 FSs, and the rule base). This approach has the advantage that it avoids a significant pitfall of the step-by-step approach, namely, it does not rely on the assumption that the basic design of the optimal IT2 FLC is necessarily the same as that of an optimal T1 FLC. As such, the second approach is not dependent on any previous optimization process and as such results in an optimal IT2 FLC where all its design parameters are considered directly. An obvious downside to this approach is the larger number of parameters that need to be optimized by the GA in comparison to the two-step approach. An example of this approach can be found in Wagner and Hagras (2007).

Because the direct approach enables a direct and overall optimization of an IT2 FLC using a GA, we focus on it in the rest of this section. Applying a GA to optimize an IT2 FLC involves a number of steps that are enumerated below. For each of the steps a variety of options are possible. We focus on illustrating a common approach based loosely on the GA/IT2 FLC design presented in Wagner and Hagras (2007).

Step 0: Configuration – A GA is used to evolve the parameters of the IT2 FLC FOUs and its rules. As such, the GA chromosome includes the IT2 FOU parameters for both the inputs and outputs of the T2 FLC as well as a representation for the rules.

Real-value encoding can be used to encode each MF parameter of the FLC as a gene in a chromosome, whereas symbolic encoding can be used to encode the rule base of the FLC (e.g., using a 0 to represent the logical OR and a 1 to represent the logical AND). In other words, the chromosome becomes a list of all the real- and symbol-valued parameters of the FLC. Furthermore, a population size P_{Size} for the GA needs to be defined, where P_{Size} refers to the number of individuals in each generation. A selection strategy is chosen that defines how the best individuals are chosen for reproduction through crossover and mutation in each generation. Common strategies are *elitist* or *roulette wheel* (or a combination of both). Finally, the crossover and mutation operations are defined where the crossover operation is specified, for example, as single- or multipoint crossover, and the mutation rate is fixed (commonly to a very low level of about 5%). It is crucial to ensure that the crossover and mutation operations result in valid FLCs (e.g., that symbolic and real-valued representations in the chromosomes are not mixed).

Step 1: Initialization – P_{Size} chromosomes are generated randomly taking into account the grammatical correctness of the chromosome (e.g., the standard deviation of an LMF must be less than that of the UMF for any given IT2 FS). The “chromosome counter” is set to 1 (the first chromosome), and the “generation counter” is set to 1 (the first generation).

Step 2: Fitness Evaluation – An IT2 FLC is constructed using the chromosome identified by the chromosome counter and the resulting FLC is executed [in simulation or in the real world (such as in a mobile robot)]. A *fitness function* is used to determine the specific FLC's performance/fitness (e.g., how accurately the controller follows a given set point). If chromosome counter $<P_{\text{Size}}$, then the chromosome counter is incremented by 1 and step 2 is repeated, otherwise one proceeds to step 3.

Step 3: Selection – A number of FLC chromosomes are selected for the reproduction step based on the chosen selection strategy, for example, using a roulette wheel strategy, the fitter an individual is, the higher the probability is for its selection.

Step 4: Reproduction – Crossover and mutation are applied, as defined in step 0 to the chromosomes selected in step 3 and a new population of size P_{Size} is created.

Step 5: Iteration – The generation counter is incremented. If a stopping criterion is not achieved, for example, if generation counter < the maximum number of generations and/or if the desired performance is not achieved, the chromosome counter is reset to 1 and step 2 is returned to; otherwise, step 6 is reached.

Step 6: Termination – The chromosome/FLC that results in the best fitness is kept and a final design has been achieved.

Pseudocode for GA is given in Table 3.4.

3.7 MOVING ON

The materials you have just covered in this chapter will be used Chapters 4–6. In the next chapter some of it will be used to rigorously derive precise mathematical relationships between the input and output of a variety of IT2 Mamdani and TSK FLCs. Some of the T2 FLCs are of the PI or PD type, and their derived relationships reveal them to be nonlinear variable PI or PD controllers that have variable proportional-gain and integral-gain (or derivative-gain) plus variable control offset.

TABLE 3.4 Pseudocode for GA as Used in Optimal Designs of an IT2 FLC[a]

Set GA Input Variables:

$P =$ Population size
GC = Generation counter
CC = Chromosome counter
ϕ_m = Chromosome ($m = 1, \ldots, P$)
F_m = Chromosome fitness ($m = 1, \ldots, P$)
S_m = Fitness-based selection probability ($m = 1, \ldots, P$)
C_{Rate} = Crossover rate
M_{Rate} = Mutation rate
$G =$ Maximum generation counter ($g = 1, \ldots, G$)

TABLE 3.4 *(Continued)*

```
Initialize Φ_P (Step 1): initialize all φ_m chromosomes randomly (but
so that they result in valid IT2 FLCs)
WHILE g ≤ G
{
 Evaluate Fitness (Step 2):
    for m=1 to m=P
    F_m =  fitness of FLC based on chromosome φ_m
    end for
 Selection (Step 3):
    for m=1 to m=P
        S_m = F_m / Σ_{i=1}^{P} F_i
    end for
    select P chromosomes based on S_m and store as Φ'
 Reproduction (Step 4):
    Apply Crossover to chromosomes in Φ' based on C_Rate
    Apply Mutation to chromosomes in Φ' based on M_Rate
 Iteration (Step 5):
    g=g+1
    IF stopping criterion (e.g., g>G) is reached
       break WHILE (Step 6)
}
```

[a] Many variations of GA designs are possible (e.g., including elitism).

APPENDIX 3A. PROOF OF THEOREM 3.4

This theorem, whose proof is taken from Appendix IV of Wu and Mendel (2002; © 2002, IEEE), is proved in two steps. First, it is shown that $[\overline{u}_l(\mathbf{x}), \underline{u}_r(\mathbf{x})]$ [where $\overline{u}_l(\mathbf{x})$ and $\underline{u}_r(\mathbf{x})$ are given by Eqs. (3.83) and (3.84), respectively] is an inner bound for the type-reduced set. Then the outer-bound set $[\underline{u}_l(\mathbf{x}), \overline{u}_r(\mathbf{x})]$ [where $\underline{u}_l(\mathbf{x})$ and $\overline{u}_r(\mathbf{x})$ are given by Eqs. (3.85) and (3.86), respectively] is derived, based on the distance between the type-reduced set and its inner-bound set.

3A.1 Inner-Bound Set $[\overline{u}_l(\mathbf{x}), \underline{u}_r(\mathbf{x})]$

For COS type reduction [see Eqs. (3.65) and (3.66)],

$$u_l(\mathbf{x}) = \min_{0 \le L \le M} \frac{\sum_{s=1}^{L} \overline{f}^s a^s + \sum_{s=L+1}^{M} \underline{f}^s a^s}{\sum_{s=1}^{L} \overline{f}^s + \sum_{s=L+1}^{M} \underline{f}^s} \tag{3A.1}$$

$$u_r(\mathbf{x}) = \max_{0 \le R \le M} \frac{\sum_{s=1}^{R} \underline{f}^s b^s + \sum_{s=R+1}^{M} \overline{f}^s b^s}{\sum_{s=1}^{R} \underline{f}^s + \sum_{s=R+1}^{M} \overline{f}^s} \tag{3A.2}$$

Considering the two extreme end points for L (namely, $L=0$ and $L=M$) for both Eqs. (3A.1) and (3A.2), it must be true that

$$u_l(\mathbf{x}) \le u_l^{(0)}(\mathbf{x}) \quad \text{and} \quad u_l(\mathbf{x}) \le u_l^{(M)}(\mathbf{x}) \tag{3A.3}$$

and

$$u_r(\mathbf{x}) \ge u_r^{(0)}(\mathbf{x}) \quad \text{and} \quad u_r(\mathbf{x}) \ge u_r^{(M)}(\mathbf{x}) \tag{3A.4}$$

In these equations, $u_l^{(0)}(\mathbf{x})$, $u_r^{(M)}(\mathbf{x})$, $u_l^{(M)}(\mathbf{x})$, and $u_r^{(0)}(\mathbf{x})$ are defined in Eqs. (3.79)–(3.82), respectively, and the $\le$ in Eq. (3A.3) is due to the minimum operation in Eq (3A.1), whereas the $\ge$ in Eq. (3A.4) is due to the maximum operation in Eq. (3A.2). From the inequalities in Eqs. (3A.3) and (3A.4), it follows that

$$u_l(\mathbf{x}) \le \min\{u_l^{(0)}(\mathbf{x}), u_l^{(M)}(\mathbf{x})\} \equiv \overline{u}_l(\mathbf{x}) \tag{3A.5}$$

$$u_r(\mathbf{x}) \ge \max\{u_r^{(0)}(\mathbf{x}), u_r^{(M)}(\mathbf{x})\} \equiv \underline{u}_r(\mathbf{x}) \tag{3A.6}$$

The right-hand sides of Eqs. (3A.5) and (3A.6) are the same as Eqs. (3.83) and (3.84), respectively. We leave it to the reader to show that $\overline{u}_l(\mathbf{x}) \le \underline{u}_r(\mathbf{x})$ so that $[\overline{u}_l(\mathbf{x}), \underline{u}_r(\mathbf{x})]$ is a valid set.

3A.2 Outer-Bound Set $[\underline{u}_l(\mathbf{x}), \overline{u}_r(\mathbf{x})]$

We provide the details for $\underline{u}_l(\mathbf{x})$, and leave the details of $\overline{u}_r(\mathbf{x})$ to the reader because they are so similar to the ones for $\underline{u}_l(\mathbf{x})$.

In Eq. (3A.5) we have shown that $\overline{u}_l(\mathbf{x}) - u_l(\mathbf{x}) \ge 0$. We next show that $\overline{u}_l(\mathbf{x}) - u_l(\mathbf{x})$ is bounded from above, that is, $\overline{u}_l(\mathbf{x}) - u_l(\mathbf{x}) \le c$, from which it follows that $u_l(\mathbf{x}) \ge \overline{u}_l(\mathbf{x}) - c \equiv \underline{u}_l(\mathbf{x})$. To determine c, we begin by using the following inequality:

$$\min(A, B) \le \zeta A + (1-\zeta)B \tag{3A.7}$$

for $0 \le \zeta \le 1$. To understand this inequality we can, without loss of generality, assume $A \ge B$, in which case $\min(A, B) = B$. Thus, we need only show that $B \le \zeta A + (1-\zeta)B = \zeta(A-B) + B$; but this inequality is valid since $A \ge B$ and $\zeta \ge 0$. Although Eq. (3A.7) is in terms of a free parameter ζ, in the following we determine an optimal value for ζ, ζ^*.

From Eq. (3.83), observe that

$$\overline{u}_l(\mathbf{x}) - u_l(\mathbf{x}) = \min\{u_l^{(0)}(\mathbf{x}) - u_l(\mathbf{x}), u_l^{(M)}(\mathbf{x}) - u_l(\mathbf{x})\} \tag{3A.8}$$

Applying Eq. (3A.7) to this equation, we find

$$\overline{u}_l(\mathbf{x}) - u_l(\mathbf{x}) \le \zeta[u_l^{(0)}(\mathbf{x}) - u_l(\mathbf{x})] + (1-\zeta)[u_l^{(M)}(\mathbf{x}) - u_l(\mathbf{x})] \tag{3A.9}$$

Before finding ζ^*, we obtain upper bounds for $u_l^{(0)}(\mathbf{x}) - u_l(\mathbf{x})$ and $u_l^{(M)}(\mathbf{x}) - u_l(\mathbf{x})$.

Expressing $u_l^{(0)}(\mathbf{x})$ using Eq. (3.79), and using Eq (3A.1) for $u_l(\mathbf{x})$, with L^* as the winning value of L, we find

$$u_l^{(0)}(\mathbf{x}) - u_l(\mathbf{x}) = \frac{\sum_{s=1}^{M} \underline{f}^s u_l^s}{\sum_{s=1}^{M} \underline{f}^s} - \frac{\sum_{s=1}^{L^*} \bar{f}^s a^s + \sum_{s=L^*+1}^{M} \underline{f}^s a^s}{\sum_{s=1}^{L^*} \bar{f}^s + \sum_{s=L^*+1}^{M} \underline{f}^s} \geq 0 \tag{3A.10}$$

where the latter inequality follows from Eq. (3A.3) when applied to the left-hand side of Eq. (3A.10); hence,

$$\begin{aligned} u_l^{(0)}(\mathbf{x}) - u_l(\mathbf{x}) &= \frac{\overbrace{\sum_{s=1}^{M} \underline{f}^s a^s}^{a}}{\underbrace{\sum_{s=1}^{M} \underline{f}^s}_{c}} - \frac{\overbrace{\sum_{s=1}^{M} \underline{f}^s a^s}^{a} + \overbrace{\sum_{s=1}^{L^*} (\bar{f}^s - \underline{f}^s) a^s}^{b}}{\underbrace{\sum_{s=1}^{M} \underline{f}^s}_{c} + \underbrace{\sum_{s=1}^{L^*} (\bar{f}^s - \underline{f}^s)}_{d}} \\ &= \frac{ad - bc}{c(c+d)} = \frac{a}{c} \cdot \frac{d}{c+d} - \frac{b}{c+d} \end{aligned} \tag{3A.11}$$

Upon further simplification of this equation, it can be expressed as

$$u_l^{(0)}(\mathbf{x}) - u_l(\mathbf{x}) = u_l^{(0)}(\mathbf{x}) \cdot \frac{\sum_{s=1}^{L^*} (\bar{f}^s - \underline{f}^s)}{\sum_{s=1}^{L^*} \bar{f}^s + \sum_{s=L^*+1}^{M} \underline{f}^s} - \frac{\sum_{s=1}^{L^*} (\bar{f}^s - \underline{f}^s) a^s}{\sum_{s=1}^{L^*} \bar{f}^s + \sum_{s=L^*+1}^{M} \underline{f}^s} \tag{3A.12}$$

Because $a^1 \leq a^2 \leq \cdots \leq a^{L^*}$, it follows that

$$u_l^{(0)}(\mathbf{x}) - u_l(\mathbf{x}) \leq u_l^{(0)}(\mathbf{x}) \cdot \frac{\sum_{s=1}^{L^*} (\bar{f}^s - \underline{f}^s)}{\sum_{s=1}^{L^*} \bar{f}^s + \sum_{s=L^*+1}^{M} \underline{f}^s} - a^1 \frac{\sum_{s=1}^{L^*} (\bar{f}^s - \underline{f}^s)}{\sum_{s=1}^{L^*} \bar{f}^s + \sum_{s=L^*+1}^{M} \underline{f}^s} \tag{3A.13}$$

$$u_l^{(0)}(\mathbf{x}) - u_l(\mathbf{x}) \leq \frac{[u_l^{(0)}(\mathbf{x}) - a^1] \sum_{s=1}^{L^*} (\bar{f}^s - \underline{f}^s)}{\sum_{s=1}^{L^*} \bar{f}^s + \sum_{s=L^*+1}^{M} \underline{f}^s} \tag{3A.14}$$

In a similar manner, we find

$$u_l^{(M)}(\mathbf{x}) - u_l(\mathbf{x}) = \frac{\sum_{s=1}^{M} \bar{f}^s a^s}{\sum_{s=1}^{M} \bar{f}^s} - \frac{\sum_{s=1}^{L^*} \bar{f}^s a^s + \sum_{s=L^*+1}^{M} \underline{f}^s a^s}{\sum_{s=1}^{L^*} \bar{f}^s + \sum_{s=L^*+1}^{M} \underline{f}^s} \geq 0 \tag{3A.15}$$

where the latter inequality also follows from Eq. (3.83); hence,

$$u_l^{(M)}(\mathbf{x}) - u_l(\mathbf{x}) = \frac{\sum_{s=L^*+1}^{M}(\overline{f}^s - \underline{f}^s)a^s}{\sum_{s=1}^{L^*}\overline{f}^s + \sum_{s=L^*+1}^{M}\underline{f}^s} - u_l^{(M)}(\mathbf{x})\frac{\sum_{s=L^*+1}^{M}(\overline{f}^s - \underline{f}^s)}{\sum_{s=1}^{L^*}\overline{f}^s + \sum_{s=L^*+1}^{M}\underline{f}^s} \tag{3A.16}$$

$$u_l^{(M)}(\mathbf{x}) - u_l(\mathbf{x}) \le a^M \frac{\sum_{s=L^*+1}^{M}(\overline{f}^s - \underline{f}^s)}{\sum_{s=1}^{L^*}\overline{f}^s + \sum_{s=L^*+1}^{M}\underline{f}^s} - u_l^{(M)}(\mathbf{x})\frac{\sum_{s=L^*+1}^{M}(\overline{f}^s - \underline{f}^s)}{\sum_{s=1}^{L^*}\overline{f}^s + \sum_{s=L^*+1}^{M}\underline{f}^s} \tag{3A.17}$$

$$u_l^{(M)}(\mathbf{x}) - u_l(\mathbf{x}) \le \frac{[a^M - u_l^{(M)}(\mathbf{x})]\sum_{s=L^*+1}^{M}(\overline{f}^s - \underline{f}^s)}{\sum_{s=1}^{L^*}\overline{f}^s + \sum_{s=L^*+1}^{M}\underline{f}^s} \tag{3A.18}$$

Substituting Eqs. (3A.14) and (3A.18) into Eq. (3A.9), we find that

$$\overline{u}_l(\mathbf{x}) - u_l(\mathbf{x}) \le \zeta \frac{[u_l^{(0)}(\mathbf{x}) - a^1]\sum_{s=1}^{L^*}(\overline{f}^s - \underline{f}^s)}{\sum_{s=1}^{L^*}\overline{f}^s + \sum_{s=L^*+1}^{M}\underline{f}^s} + (1-\zeta)\frac{[a^M - u_l^{(M)}(\mathbf{x})]\sum_{s=L^*+1}^{M}(\overline{f}^s - \underline{f}^s)}{\sum_{s=1}^{L^*}\overline{f}^s + \sum_{s=L^*+1}^{M}\underline{f}^s} \tag{3A.19}$$

Note that it is always possible to express $\sum_{s=1}^{L^*}(\overline{f}^s - \underline{f}^s)$ and $\sum_{s=L^*+1}^{M}(\overline{f}^s - \underline{f}^s)$ in terms of $\sum_{s=1}^{M}(\overline{f}^s - \underline{f}^s)$ as follows:

$$\sum_{s=1}^{L^*}(\overline{f}^s - \underline{f}^s) = t\sum_{s=1}^{M}(\overline{f}^s - \underline{f}^s) \tag{3A.20}$$

and

$$\sum_{s=L^*+1}^{M}(\overline{f}^s - \underline{f}^s) = (1-t)\sum_{s=1}^{M}(\overline{f}^s - \underline{f}^s) \tag{3A.21}$$

where $t \equiv t(\mathbf{x}) \in [0, 1]$ is determined by $L^*(\mathbf{x})$ [just solve Eq. (3A.20) for t].

Substituting Eqs. (3A.20) and (3A.21) into Eq. (3A.19), we then find that

$$\overline{u}_l(\mathbf{x}) - u_l(\mathbf{x}) \le \frac{\zeta t[u_l^{(0)}(\mathbf{x}) - a^1] + (1-\zeta)(1-t)[a^M - u_l^{(M)}(\mathbf{x})]}{t\sum_{s=1}^{M}\overline{f}^s + (1-t)\sum_{s=1}^{M}\underline{f}^s} \times \sum_{s=1}^{M}(\overline{f}^s - \underline{f}^s) \tag{3A.22}$$

which can be expressed as

$$\overline{u}_l(\mathbf{x}) - u_l(\mathbf{x}) \le g(\zeta, t)\sum_{s=1}^{M}(\overline{f}^s - \underline{f}^s) \le \max_{t\in[0,1]} g(\zeta, t)\sum_{s=1}^{M}(\overline{f}^s - \underline{f}^s) \tag{3A.23}$$

where

$$g(\zeta, t) \equiv \frac{\zeta t[u_l^{(0)}(\mathbf{x}) - a^1] + (1-\zeta)(1-t)[a^M - u_l^{(M)}(\mathbf{x})]}{t\sum_{s=1}^{M} \bar{f}^s + (1-t)\sum_{s=1}^{M} \underline{f}^s} \tag{3A.24}$$

Next, we treat ζ and t as independent variables and use the min–max method to find ζ^*, that is,

$$\zeta^* = \arg\min_{\zeta\in[0,1]} \left[\max_{t\in[0,1]} g(\zeta, t)\right] = \arg\min_{\zeta\in[0,1]} g(\zeta, t^*) \tag{3A.25}$$

Notice that $g(\zeta, t^*)$ is the maximum of $g(\zeta, t)$ with respect to t and $g(\zeta^*, t^*)$ is the minimum of $g(\zeta, t^*)$ with respect to ζ. Since $t\in[0, 1]$ is indirectly determined by the input $\mathbf{x}$ [see the line after Eq. (3A.21)], we cannot arbitrarily choose its value; instead, we consider the worst case (i.e., the maximum value) of $g(\zeta, t)$ with respect to t, to find the upper bound for $\bar{u}_l(\mathbf{x}) - u_l(\mathbf{x})$. On the other hand, since $\zeta \in [0, 1]$ is a free parameter, we can choose its value arbitrarily to find a tight upper bound for $\bar{u}_l(\mathbf{x}) - u_l(\mathbf{x})$.

To find t^* and $g(\zeta, t^*)$, we calculate the partial derivative of $g(\zeta, t)$ in Eq. (3A.24) with respect to t as follows:

$$\frac{\partial g(\zeta, t)}{\partial t} = \frac{\zeta[u_l^{(0)}(\mathbf{x}) - a^1]\sum_{s=1}^{M} \underline{f}^s - (1-\zeta)[a^M - u_l^{(M)}(\mathbf{x})]\sum_{s=1}^{M} \bar{f}^s}{\left[t\sum_{s=1}^{M} \bar{f}^s + (1-t)\sum_{s=1}^{M} \underline{f}^s\right]^2} \tag{3A.26}$$

Because the numerator of Eq. (3A.26) is not a function of t, t^* cannot be determined by setting $\partial g(\zeta, t)/\partial t = 0$. Instead, Eq. (3A.26) must be analyzed in order to determine t^*. To begin, note that $\partial g(\zeta, t)/\partial t$ is a linear function of ζ, that is, its slope does not depend on ζ:

$$\frac{\partial}{\partial \zeta}\left[\frac{\partial g(\zeta, t)}{\partial t}\right] = \frac{[u_l^{(0)}(\mathbf{x}) - a^1]\sum_{s=1}^{M} \underline{f}^s + [a^M - u_l^{(M)}(\mathbf{x})]\sum_{s=1}^{M} \bar{f}^s}{\left[t\sum_{s=1}^{M} \bar{f}^s + (1-t)\sum_{s=1}^{M} \underline{f}^s\right]^2} \tag{3A.27}$$

Note, also, that (remember that the a^s are rank ordered in increasing order, and the lower and upper values of the firing intervals are positive)

$$u_l^{(0)}(\mathbf{x}) - a^1 = \frac{\sum_{s=1}^{M} \underline{f}^s a^s}{\sum_{s=1}^{M} \underline{f}^s} - a^1 = \frac{\sum_{s=2}^{M} \underline{f}^s (a^s - a^1)}{\sum_{s=1}^{M} \underline{f}^s} > 0 \tag{3A.28}$$

and

$$a^M - u_l^{(M)}(\mathbf{x}) = a^M - \frac{\sum_{s=1}^{M} \bar{f}^s a^s}{\sum_{s=1}^{M} \bar{f}^s} = \frac{\sum_{s=1}^{M} \bar{f}^s (a^M - a^s)}{\sum_{s=1}^{M} \bar{f}^s} > 0 \tag{3A.29}$$

Substituting Eqs. (3A.28) and (3A.29) into Eq. (3A.27) we see that the slope of $\partial g(\zeta, t)/\partial t$ is positive.

Next, set $\partial g(\zeta, t)/\partial t = 0$ in Eq. (3A.26) and solve for ζ, calling that value ζ^*, where

$$\zeta^* \equiv \frac{[a^M - u_l^{(M)}(\mathbf{x})]\sum_{s=1}^{M} \bar{f}^s}{[u_l^{(0)}(\mathbf{x}) - a^1]\sum_{s=1}^{M} \underline{f}^s + [a^M - u_l^{(M)}(\mathbf{x})]\sum_{s=1}^{M} \bar{f}^s} \tag{3A.30}$$

The situation now is as summarized in Fig. 3.19. Observe that:

1. When ζ is chosen so that

$$0 \le \zeta < \zeta^* \tag{3A.31}$$

then $\partial g(\zeta, t)/\partial t < 0$, which means that $g(\zeta, t)$ is a monotonically decreasing function with respect to t and, therefore, its maximum value with respect to $t \in [0, 1]$ occurs at $t^* = 0$, that is [see Eq. (3A.24)],

$$g(\zeta, t^*) = \max_{t\in[0,1]} g(\zeta, t) = g(\zeta, 0) = \frac{(1-\zeta)[a^M - u_l^{(M)}(\mathbf{x})]}{\sum_{s=1}^{M} \underline{f}^s} \tag{3A.32}$$

In this case, $g(\zeta, t^*)$ is a monotonically decreasing function with respect to ζ and, therefore, substituting Eq. (3A.30) into (3A.32), we find

$$\begin{aligned} \inf_{0\le\zeta<\zeta^*} g(\zeta, t^*) &= \lim_{\zeta\to\zeta^*} g(\zeta, t^*) \\ &= \frac{[u_l^{(0)}(\mathbf{x}) - a^1][a^M - u_l^{(M)}(\mathbf{x})]}{[u_l^{(0)}(\mathbf{x}) - a^1]\sum_{s=1}^{M} \underline{f}^s + [a^M - u_l^{(M)}(\mathbf{x})]\sum_{s=1}^{M} \bar{f}^s} \end{aligned} \tag{3A.33}$$

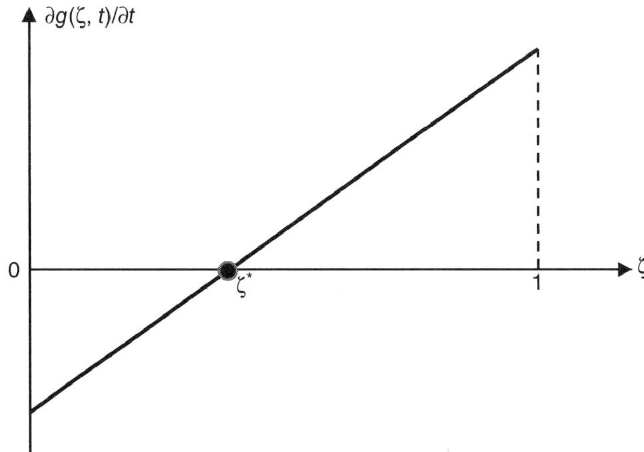

Figure 3.19 $\partial g(\zeta, t)/\partial t$ versus ζ.

2. When ζ is chosen so that

$$\zeta = \zeta^* \tag{3A.34}$$

then $\partial g(\zeta, t)/\partial t = 0$, which means that $g(\zeta, t)$ is independent of t. Without loss of generality, we may set $t = 1$ (or 0, or any value from 0 to 1); hence,

$$g(\zeta^*, t^*) = \max_{t\in[0,1]} g(\zeta^*, t) = g(\zeta^*)$$
$$= \frac{[u_l^{(0)}(\mathbf{x}) - a^1][a^M - u_l^{(M)}(\mathbf{x})]}{[u_l^{(0)}(\mathbf{x}) - a^1]\sum_{s=1}^{M} \underline{f}^s + [a^M - u_l^{(M)}(\mathbf{x})]\sum_{s=1}^{M} \overline{f}^s} \tag{3A.35}$$

3. When ζ is chosen so that

$$\zeta^* < \zeta \leq 1 \tag{3A.36}$$

then $\partial g(\zeta, t)/\partial t > 0$, which means that $g(\zeta, t)$ is a monotonically increasing function with respect to t, and, therefore, its maximum value with respect to $t \in [0, 1]$ occurs at $t^* = 1$, that is [see Eq. (3A.24)],

$$g(\zeta, t^*) = \max_{t\in[0,1]} g(\zeta, t) = g(\zeta, 1) = \frac{\zeta[u_l^{(0)}(\mathbf{x}) - a^1]}{\sum_{s=1}^{M} \overline{f}^s} \tag{3A.37}$$

In this case, $g(\zeta, t^*)$ is a monotonically increasing function with respect to ζ and, therefore, substituting Eq. (3A.30) into Eq. (3A.37), we find

$$\inf_{\zeta^*<\zeta\leq 1} g(\zeta, t^*) = \lim_{\zeta\to\zeta^*} g(\zeta, t^*) = \frac{[u_l^{(0)}(\mathbf{x}) - a^1][a^M - u_l^{(M)}(\mathbf{x})]}{[u_l^{(0)}(\mathbf{x}) - a^1]\sum\limits_{s=1}^{M} \underline{f}^s + [a^M - u_l^{(M)}(\mathbf{x})]\sum_{s=1}^{M} \overline{f}^s} \tag{3A.38}$$

Equations (3A.33), (3A.35), and (3A.38) show that in all cases the min-max of $g(\zeta, t)$ is achieved when $\zeta = \zeta^*$ and this is independent of t, that is,

$$\min_{\zeta\in[0,1]} [\max_{t\in[0,1]} g(\zeta, t)] = g(\zeta^*) = \frac{[u_l^{(0)}(\mathbf{x}) - a^1][a^M - u_l^{(M)}(\mathbf{x})]}{[u_l^{(0)}(\mathbf{x}) - a^1]\sum\limits_{s=1}^{M} \underline{f}^s + [a^M - u_l^{(M)}(\mathbf{x})]\sum_{s=1}^{M} \overline{f}^s} \tag{3A.39}$$

Substituting $g(\zeta^*)$ into Eq. (3A.23), we find

$$\overline{u}_l(\mathbf{x}) - u_l(\mathbf{x}) \leq g(\zeta^*)\sum_{s=1}^{M}(\overline{f}^s - \underline{f}^s) = \frac{[u_l^{(0)}(\mathbf{x}) - a^1][a^M - u_l^{(M)}(\mathbf{x})]\sum_{s=1}^{M}(\overline{f}^s - \underline{f}^s)}{[u_l^{(0)}(\mathbf{x}) - a^1]\sum_{s=1}^{M} \underline{f}^s + [a^M - u_l^{(M)}(\mathbf{x})]\sum_{s=1}^{M} \overline{f}^s} \tag{3A.40}$$

Using the definitions of $u_l^{(0)}(\mathbf{x})$ and $u_l^{(M)}(\mathbf{x})$, which are given in Eqs. (3.79) and (3.81), respectively, it follows that Eq. (3A.40) simplifies to

$$\overline{u}_l(\mathbf{x}) - u_l(\mathbf{x}) \le \frac{\sum_{s=1}^{M} \underline{f}^s(a^s - a^1)\sum_{s=1}^{M} \overline{f}^s(a^M - a^s)}{\sum_{s=1}^{M} \underline{f}^s(a^s - a^1) + \sum_{s=1}^{M} \overline{f}^s(a^M - a^s)} \times \frac{\sum_{s=1}^{M}(\overline{f}^s - \underline{f}^s)}{\sum_{s=1}^{M} \overline{f}^s \sum_{s=1}^{M} \underline{f}^s} \tag{3A.41}$$

Hence

$$u_l(\mathbf{x}) \ge \overline{u}_l(\mathbf{x}) - \left[\frac{\sum_{s=1}^{M}\left(\overline{f}^s - \underline{f}^s\right)}{\sum_{s=1}^{M} \overline{f}^s \sum_{s=1}^{M} \underline{f}^s} \times \frac{\sum_{s=1}^{M} \underline{f}^s(a^s - a^1)\sum_{s=1}^{M} \overline{f}^s(a^M - a^s)}{\sum_{s=1}^{M} \underline{f}^s(a^s - a^1) + \sum_{s=1}^{M} \overline{f}^s(a^M - a^s)} \right] \tag{3A.42}$$

The right-hand side of Eq. (3A.42) is defined as $\underline{u}_l(\mathbf{x})$, that is, $\underline{u}_l(\mathbf{x})$ is as stated in Eq. (3.85).

CHAPTER 4

Analytical Structure of Various Interval Type-2 Fuzzy PI and PD Controllers

4.1 INTRODUCTION

A T2 fuzzy controller, like its T1 counterpart, is presently viewed and treated by most fuzzy control practitioners and researchers as a black box that is a function generator that produces a desired nonlinear mapping between input and output of the controller (we call the mapping analytical structure). The analytical structure's implicit mathematical representation, $u=f(x)$ as shown in Fig. 1.2, is the nonlinear control solution being sought. In this chapter, innovative techniques capable of deriving the explicit mathematical representation of $f(x)$ for some common classes of interval T2 fuzzy controllers of both the Mamdani type and the TSK type will be presented. Connections between the resulting analytical structures and the conventional nonlinear controllers will be shown and insightful analyses will be conducted. The $f(x)$ of the T2 fuzzy controllers will be compared with those of the comparable T1 fuzzy controllers, and their relative advantages and disadvantages will be exposed. Based on the exposed analytical structure of the T2 fuzzy controllers, some design guidelines are developed to assist in determining the controller parameters (more than 10).

Numerous techniques have been developed in the literature for analyzing and designing a wide variety of fuzzy control systems of both the Mamdani and TSK type. They are mostly for the T1 fuzzy controllers at this point (Feng, 2006), but a growing number of techniques are for T2 controllers. The literature can be classified into two groups of methodology: (1) the model-based approach and (2) the knowledge-based approach, which is a model-free approach. When the model-based approach is used, the precise mathematical model of the system to be controlled must be assumed explicitly available, whereas the knowledge-based approach does not make such an assumption. The models of interest should be nonlinear because a practical system is always nonlinear. Strictly speaking, a linear system does not exist—a linear model is an approximate model of the nonlinear system valid for a region around one of the system operation points. There is little point to applying fuzzy control, T2 or T1, to a linear model owing

Introduction to Type-2 Fuzzy Logic Control: Theory and Applications, First Edition.
Jerry M. Mendel, Hani Hagras, Woei-Wan Tan, William W. Melek, and Hao Ying.

to the existence of the much more effective linear control theory. While the model availability assumption makes theoretical development more mathematically tractable and convenient for the model-based approach, it does not realistically reflect practical constraints. The fact of the matter is this—it is always challenging to attain a credible nonlinear mathematical model for most systems in the real world. This is partially because assuming a plausible mathematical structure for a given physical system in practice is a tough task to begin with, and validating it is another demanding task. The pitfall of the model availability assumption holds not only for fuzzy control but also equally for conventional control. Emerging in the 1990s, this approach provides mathematical convenience at the cost of practicality. It has produced a large volume of publications; nevertheless, its usefulness in practice has yet to be confirmed. In short, without knowing the nonlinear model, many of the model-based analysis and design methods developed are simply inapplicable.

The first fuzzy controller invented in 1974 was a knowledge-based controller, which opened a new era of knowledge-based control approach. The underlying idea was to construct a nonlinear controller without the need of the mathematical model of the nonlinear system. This approach employed fuzzy sets, fuzzy logic, fuzzy rules, and fuzzy reasoning to capture, represent, and process human operator's control expertise to construct the controller. The knowledge-based approach works because the behavior and dynamics of the system are reflected in the human control knowledge and hence are implicitly built into the controller. This approach has achieved a huge success in a vast variety of real-world applications and commercial products largely because it does not require the system model, making it possible to develop a product with reduced development time and cost.

After a fuzzy controller, T2 or T1, is constructed or designed, its analytical structure is implicit [e.g., Eq. (3.68)] because $f(x)$ does not spell out the explicit relationship between the input variables $\mathbf{x}$ and the output variable u. In other words, it shows there is a relationship but does not reveal exactly what it is. When the model-based approach utilizes the implicit $f(x)$ to develop a design or analysis method, it treats the fuzzy controller as a black box function generator capable of producing the nonlinear mapping between the input and output of the controller being sought. On the other hand, the knowledge-based approach does not start with $f(x)$. Rather, it relies on a systematic procedure consisting of a number of steps for practically constructing $f(x)$ through manipulating, often in a trial-and-error fashion, fuzzy sets, fuzzy rules, fuzzy inference, among other components. For each component, the developer will face choices. For instance, for the input fuzzy sets (i.e., the fuzzy sets for fuzzifying the input variables), the developer has to decide how many of them should be used, what type should be used (e.g., triangular vs. Gaussian), and whether a mixture of different types should be used. This is just one of the several components that the developer has to specify (the other components include the output fuzzy sets and fuzzy rules). Coupled with computer simulation, this approach often suffices for the practitioner to build a satisfactory fuzzy control system as a solution to the real-world problem at hand. Importantly, this tactic usually works even when the mathematical model of the system is not available. Apart from the approaches (model-based or knowledge-based), once built, the fuzzy controller remains a black box in that the explicit expression of $f(x)$ is unknown. The components work together to generate a value for $f(x)$ for any given value of **x**. Obviously, the explicit expression of $f(x)$ depends on how the components are selected. Nevertheless, the nature of the implicit $f(x)$ changes little no matter how the components are composited.

Either approach is in sharp contrast to the traditional control theories. In conventional control, once a controller is chosen by the developer according to the system to be controlled, the controller's analytical structure, linear or nonlinear, is always explicitly ready for analysis and design of the control system. The linear and nonlinear control theories have matured with many powerful time-tested analysis and design schemes. The primary technical difficulty for controller design lies in how to first select or design $f(x)$ and then determine its parameter values based on the given system model so that the designed control system performance will meet the control system user's performance specifications. The $f(x)$ value is explicitly known after the control system design is completed. Control system analysis, stability, control performance, and other system characteristics are analyzed and determined based on the explicitly given $f(x)$ and the system model. To bring fuzzy control to the same level of sophistication and acceptance as the conventional control theories, fuzzy control needs to overcome two hurdles pertinent only to fuzzy control and irrelevant to conventional control. The first hurdle is the unavailability of $f(x)$ in an explicit form after it is designed/constructed, and the second relates to the fundamental question of whether $f(x)$ can be an arbitrary nonlinear function (this issue, referred to as fuzzy systems as universal approximators in the literature, has been extensively addressed for the T1 fuzzy controllers (Ying, 1994a, 1998a) and somewhat investigated for the T2 controllers (Ying, 2008, 2009). To a large extent, analytically and rigorously studying fuzzy control, T2 or T1, is inherently more challenging than studying typical nonlinear control problems. Not explicitly knowing $f(x)$ puts both the model-based and model-free fuzzy control approaches in a disadvantageous position because knowing the analytical structures of both the controller and system can make it possible for the system analysis and design more precise and effective and less conservative.

No matter if a T2 or T1 fuzzy controller is theoretically designed using a model-based scheme or is empirically constructed via a model-free method, revealing controllers' analytical structure can be significantly beneficial because we can:

1. Understand how and why the fuzzy controller works in the same sense as we understand how a conventional controller functions.
2. Find a possible connection between the fuzzy controller and a conventional controller.
3. Explore more rigorously the differences between the T2 and T1 fuzzy controllers and their relative merits and pitfalls (e.g., control performance and structural complexity).

4. Take advantage of the nonlinear control theory to develop more effective analysis and design methods for the T2 control system as the fuzzy control problem has transformed into a nonlinear control problem.
5. Make T2 fuzzy control more acceptable to safety-critical fields such as clinical medicine and nuclear industry where people are reluctant to employ a black box as a controller.

We stress that the analytical structure of a fuzzy controller should be investigated in such a way that the structure is sensible in the context of control theory. This is to say that deriving the explicit structure is only a first step, after which the structure should be represented in a form clearly understandable from a control theory standpoint to gain the full potential benefits in system analysis and design.

In the next section, we will provide a brief review of the linear PID, PI, and PD controllers as most of the T2 fuzzy controllers studied in this chapter are related to the PI and PD controllers. In Section 4.3, the common components of the T2 fuzzy controllers are defined. In Sections 4.4–4.7, we show various techniques for deriving the explicit analytical structures of four different types of T2 Mamdani fuzzy controllers with two input variables and link the resulting structures to the PI and PD controllers. Section 4.8 provides yet another derivation technique for a class of T2 TSK fuzzy controllers with two input variables and shows how the analytical structure obtained ties to the PI and PD controllers. We conduct analysis on part of the derived analytical structures of the T2 Mamdani fuzzy controllers in Section 4.9, including their connection to the comparable T1 fuzzy controllers. Based on the materials in the prior sections, Section 4.10 establishes some design guidelines for the T2 Mamdani fuzzy controllers. The work in this chapter realizes the first four benefits listed above to varying extents.

4.2 PID, PI, AND PD CONTROLLERS AND THEIR RELATIONSHIPS

All the fuzzy controllers studied in this chapter have two input variables and are related to the PI and PD controllers. To facilitate a better understanding of the rest of this chapter, we provide a brief review of the PID control first.

4.2.1 Two Forms of PID Controller—Position Form and Incremental Form

The continuous-time linear PID controller in position form (i.e., using full controller output) is described by the following expression:

$$u(t) = K\left(e(t) + \frac{1}{T_i}\int_0^t e(\tau)\,d\tau + T_d\frac{de(t)}{dt}\right)$$

Here, $e(t)$ is the error signal defined as $e(t) = \mathrm{SP}(t) - y(t)$ where $\mathrm{SP}(t)$ is the system output reference signal, and $y(t)$ the output of the system under control; K is gain,

T_i is integration time, and T_d is derivative time. The corresponding discrete-time position form is

$$u(k) = K\left[e(k) + \frac{T}{T_i}\sum_{i=0}^{k} e(i) + \frac{T_d}{T}r(k)\right] = Ke(k) + \frac{KT}{T_i}\sum_{i=0}^{k} e(i) + \frac{KT_d}{T}r(k)$$

$$= K_p e(k) + K_i \sum_{i=0}^{k} e(i) + K_d r(k) \tag{4.1}$$

where $r(k) = e(k) - e(k-1)$ is the change of the error, and T is sampling period; and K_p, K_i, and K_d are proportional-gain, integral-gain, and derivative-gain of the PID controller, respectively.

The three gains are constants for the linear PID controller. If the value of at least one of the gains varies with system state, the PID controller becomes nonlinear. There are various forms of nonlinear PID controllers. For instance, a PID controller with an antiwindup mechanism is a nonlinear PID controller.

The above PID control algorithms are in position form because they directly compute the controller output itself. The PID controller is often used in the incremental form, which calculates change of the controller output. Note that at time $k-1$,

$$u(k-1) = K_p e(k-1) + K_i \sum_{i=0}^{k-1} e(i) + K_d r(k-1)$$

Hence, the incremental form of the PID controller (i.e., using change of controller output) corresponding to Eq. (4.1) is

$$\Delta u(k) = u(k) - u(k-1) = K_p r(k) + K_i e(k) + K_d d(k) \tag{4.2}$$

where

$$d(k) = r(k) - r(k-1) \tag{4.3}$$

4.2.2 PI and PD Controllers and Their Relationship

In practice, full PID control is not always desired. Instead, partial PID control in the form of PI or PD control is more effective and appropriate. This is because the derivative term tends to amplify noise and hence should be avoided if the system output is noisy. On the other hand, the integral term can cause slower system response and larger system overshoot. It should not be included in certain applications of the PID control. For these reasons, PI control and PD control should not be merely considered as incomplete PID control. Rather, they are controllers on their own with distinctive merits in comparison with the full PID control, and they may be viewed as separate classes of controllers. Indeed, many studies treat PI, PD, and PID controllers separately and differently.

When K_d is set to zero in Eq. (4.2), the PID controller becomes a PI controller in incremental form:

$$\Delta u(k) = K_i e(k) + K_p r(k) \tag{4.4}$$

whereas when $K_i = 0$ in Eq. (4.2), the PID controller reduces to a PD controller in incremental form:

$$\Delta u(k) = K_p r(k) + K_d d(k) \tag{4.5}$$

A PI controller in incremental form is related to a PD controller in position form. Letting $K_i = 0$ in Eq. (4.1), we obtain a PD controller in position form:

$$u(k) = K_p e(k) + K_d r(k) \tag{4.6}$$

Now, comparing Eq. (4.6) with (4.4), one sees that the PD controller in position form becomes the PI controller in incremental form, if (1) $e(k)$ and $r(k)$ exchange, (2) K_d is replaced by K_i, and (3) $\Delta u(k)$ and $u(k)$ exchange positions. Furthermore, comparing Eq. (4.4) with (4.5), the PI controller in incremental form becomes the PD controller in incremental form, if (1) $e(k)$ is replaced by $d(k)$, and (2) K_i is replaced by K_d.

These two structural relationships between the PI and PD controllers are important for the analytical structure derivation and analyses of the fuzzy PI and PD controllers in this chapter. Analytical structure derived for the fuzzy PI controller can directly be extended to the corresponding fuzzy PD controller and vice versa. All one needs to do is to use proper input and output variables for the fuzzy controllers (we will show how in Section 4.4). Consequently, we will focus on the fuzzy PI controllers only when presenting a variety of structure derivation techniques.

4.3 COMPONENTS OF THE INTERVAL T2 FUZZY PI AND PD CONTROLLERS

The interval T2 fuzzy PI controllers are a subset of the T2 FLCs. They employ two input variables, the error $e(k)$ and the change of the error $r(k)$, and one output variable $u(k)$ (see Fig. 4.1). Two scaling factors, k_e and k_r, are used to scale $e(k)$ and $r(k)$, respectively:

$$E(k) = k_e e(k) = k_e[\mathrm{SP}(k) - y(k)] \tag{4.7}$$

$$R(k) = k_r r(k) = k_r[e(k) - e(k-1)] \tag{4.8}$$

The output variable $u(k)$ is related to the incremental output $\Delta u(k)$ and the previous output $u(k-1)$:

$$u(k) = u(k-1) + \Delta u(k) \tag{4.9}$$

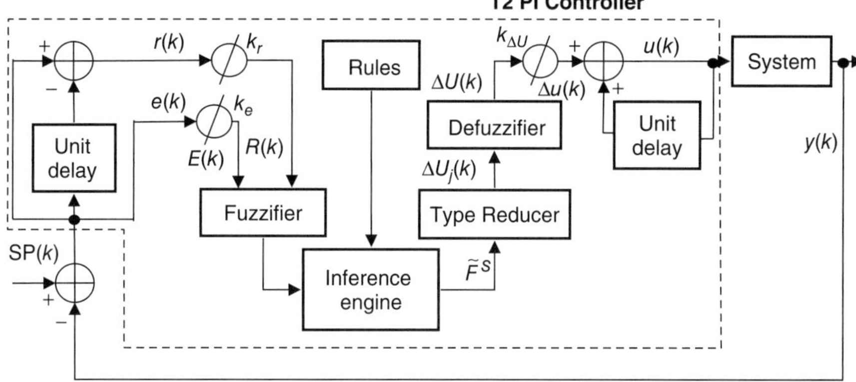

Figure 4.1 Structure of the T2 fuzzy PI controllers.

An intermediate output variable $\Delta U(k)$ and its scaling factor $k_{\Delta U}$ will be needed, and they are related to $\Delta u(k)$ through

$$\Delta u(k) = k_{\Delta U} \Delta U(k) \tag{4.10}$$

The variables $E(k)$ and $R(k)$ are defined over $[L_1, R_1]$ and $[L_2, R_2]$, respectively. There are N_1 interval T2 fuzzy sets, $\widetilde{A}_{1i}$, $i = 1, \ldots, N_1$, defined for $E(k)$ and N_2 interval T2 fuzzy sets, $\widetilde{A}_{2j}$, $j = 1, \ldots, N_2$, defined for $R(k)$. They act as a fuzzifier in that they translate the crisp values of $E(k)$ and $R(k)$ to membership values of the interval T2 fuzzy sets. Their UMFs and LMFs are designated as $\overline{\mu}_{\widetilde{A}_{1i}}$ and $\overline{\mu}_{\widetilde{A}_{2j}}$ and $\underline{\mu}_{\widetilde{A}_{1i}}$ and $\underline{\mu}_{\widetilde{A}_{2j}}$, respectively. They meet the requirements set in Definitions 2.11–2.13. Depending on the configuration in the sections below, $\Delta U(k)$ uses either interval T2 fuzzy sets $\widetilde{B}_m$ or T1 fuzzy sets B_m, $m = 1, \ldots, N_3$, but not both at the same time. They can be T2 or T1 singleton fuzzy sets. The support of such fuzzy set's UMF and LMF is a single point in the universe of discourse of the output variable $C_0^s + C_1^s E(k) + C_2^s R(k)$.

The interval T2 fuzzy PI controllers can use either Mamdani fuzzy rules or TSK fuzzy rules, but not both simultaneously. Based on the notations in Eqs. (3.19), (3.73), and (3.74), the sth fuzzy rule in the form of Mamdani type, $s = 1, \ldots, M$, is

$$R^s\text{: If } E(k) \text{ is } \widetilde{F}_1^s \text{ and } R(k) \text{ is } \widetilde{F}_2^s, \text{ then } \Delta U(k) \text{ is } \widetilde{G}^s \tag{4.11}$$

where $\widetilde{F}_1^s \in \{\widetilde{A}_{1i} | i = 1, \ldots, N_1\}$, $\widetilde{F}_2^s \in \{\widetilde{A}_{2j} | j = 1, \ldots, N_2\}$, and $\widetilde{G}^s \in \{\widetilde{B}_m | m = 1, \ldots, N_3\}$ (or $\widetilde{G}^s \in \{B_m | m = 1, \ldots, N_3\}$). A TSK fuzzy rule is

$$R^s\text{: If } E(k) \text{ is } \widetilde{F}_1^s \text{ and } R(k) \text{ is } \widetilde{F}_2^s, \text{ then } \Delta U(k) = C_0^s + C_1^s E(k) + C_2^s R(k) \tag{4.12}$$

where C_0^s, C_1^s, and C_2^s are either all constants or all T1 fuzzy sets.

Applying the interval T2 fuzzy inference to these rules, the firing interval for rule R^s of either the Mamdani type or the TSK type is described in Eqs. (3.50) and (3.51). In this chapter, only the Zadeh AND operator (i.e., the minimum t-norm) is in use. Therefore, the firing interval for R^s is

$$\begin{aligned}\widetilde{F}^s(E(k),\ R(k)) &= [\underline{f}^s(E(k),\ R(k)),\ \overline{f}^s(E(k),\ R(k))]\\ &= [\min(\underline{\mu}_{\widetilde{F}_1^s}(E(k)),\ \underline{\mu}_{\widetilde{F}_2^s}(R(k)),\ \min(\overline{\mu}_{\widetilde{F}_1^s}(E(k)), \overline{\mu}_{\widetilde{F}_2^s}(R(k))] \end{aligned} \quad (4.13)$$

The Zadeh AND operator and the product AND operator are the only two AND operators widely used in the theory and applications of fuzzy control. As the reader will see in Sections 4.4–4.8, much of the analytical structure derivation difficulties can be attributed to the use of the Zadeh AND operator. Deriving the analytical structure of a fuzzy controller, T1 or T2, with the product AND operator is, relatively speaking, a lot easier. Importantly, the two operators lead to substantially different (analytical) structures. Consequently, one operator cannot be said to be better than the other; thus one cannot replace the other.

The center-of-sets type of reducer and the centroid type of reducer will be used by the T2 controllers in this chapter because they are the most popular type reducers. For a T2 fuzzy controller with the center-of-sets type of reducer, the interval weighted-average method represented by Eq. (3.63) will be employed to realize the type reduction. This method will use the results produced by Eq. (4.13) to reduce the composite T2 output fuzzy set, $\widetilde{G}$, which is formed by the M fuzzy rules, to an interval T1 fuzzy set. See Section 3.3.2.4 for the details. If a fuzzy controller uses the centroid type of reducer instead, one will first need to use Eqs. (3.48), (3.49), (3.56), and (3.57) and min( ) to compute the lower and upper memberships of $\widetilde{G}$ as follows:

$$\begin{aligned}\underline{\mu}_{\widetilde{G}}(\Delta U) &= \max_{s=1}^{M}\left[\min\left(\underline{f}^s(E(k),\ R(k)), \underline{\mu}_{\widetilde{G}^s}(\Delta U(k))\right)\right]\\ &= \max_{s=1}^{M}\left[\min\left(\underline{\mu}_{\widetilde{F}_1^s}(E(k)),\ \underline{\mu}_{\widetilde{F}_2^s}(R(k)), \underline{\mu}_{\widetilde{G}^s}(\Delta U(k))\right)\right]\end{aligned} \quad (4.14)$$

$$\begin{aligned}\overline{\mu}_{\widetilde{G}}(\Delta U) &= \max_{s=1}^{M}\left[\min\left(\overline{f}^s(E(k),\ R(k)), \overline{\mu}_{\widetilde{G}^s}(\Delta U(k))\right)\right]\\ &= \max_{s=1}^{M}\left[\min\left(\overline{\mu}_{\widetilde{F}_1^s}(E(k)),\ \overline{\mu}_{\widetilde{F}_2^s}(R(k)), \overline{\mu}_{\widetilde{G}^s}(\Delta U(k))\right)\right]\end{aligned} \quad (4.15)$$

The resulting memberships will then be used to determine the interval T1 fuzzy set (Section 2.3.4 provides the details).

The outcome of either type reduction method is the left and right end points of the interval of the interval T1 fuzzy set for $\Delta U(k)$. The interval will then be converted to a crisp value for $\Delta U(k)$ by a defuzzifier. The most popular defuzzifier is the centroid defuzzifier, which uses the average of the left and right end points of the interval as its output.

As an alternative to the interval weighted-average method plus the centroid defuzzifier, a modified interval weighted-average method and an average defuzzifier were recently developed for the T2 controllers that use T1 or T2 singleton fuzzy sets as output fuzzy sets (Du and Ying, 2008, 2010). They work together to implement both the center-of-sets type of reducer and the defuzzification. The modified interval weighted-average method computes all the M terms in the numerator of Eq. (3.63), which are $u_s w_s$ where u_s and w_s take the terminal values of the respective intervals in Eq. (3.63). The average defuzzifier then calculates the average of these M values as its output. One merit of this approach is avoiding the iterative calculations in finding the left and right end points of the interval T1 fuzzy set required by KM algorithms and the like. The new approach provides an approximate solution to what a KM algorithm (or other applicable algorithm) plus the centroid defuzzifier will generate. One advantage associated with this approach over other existing methods is that it is less challenging to derive the explicit analytical structure of the T2 fuzzy controller that uses this approach.

Let's see in more detail how the modified interval weighted-average method works. When the output fuzzy sets are singleton, either T1 or T2, each firing interval is an interval with the left and right end points being, respectively, the lower and upper memberships of the singleton fuzzy set $\widetilde{F}_1^s$ [i.e., $\underline{f}^s$ and $\overline{f}^s$ in Eq. (4.13)] that has nonzero value only at G^s of the universe of discourse. Referring to Eq. (3.63), in our case $a^s = b^s = G^s$ and $w_s = [\underline{f}^s, \overline{f}^s]$. Instead of using the interval weighted-average method for the center-of-sets type of reducer, the modified interval weighted-average method computes the following:

$$\Delta U_p(k) = \frac{\sum_{s=1}^{M} G^s w^s}{\sum_{s=1}^{M} w^s} \qquad p = 1, 2, \ldots, 2^M \tag{4.16}$$

where w^s is either $\underline{f}^s$ or $\overline{f}^s$ and hence for a given p, the numerator of Eq. (4.16) represents one of the total 2^M different combinations of $\underline{f}^s$ or $\overline{f}^s$. The 2^M values of $\Delta U_p(k)$ form a special interval T1 fuzzy set whose membership function is 1 at these 2^M locations of the universe and 0 elsewhere. The interval's left and right end points are, respectively, the smallest and the largest values of $\Delta U_p(k)$. After the type reduction, the average defuzzifier simply calculates the average value of the 2^M points to produce the defuzzification result:

$$\Delta U(k) = \frac{1}{2^M} \sum_{p=1}^{2^M} \Delta U_p(k) \tag{4.17}$$

The modified interval weighted-average method works in a similar fashion when the centroid type of reducer is involved instead. One just needs to replace $\underline{f}^s$ by $\underline{\mu}_{\widetilde{G}}(\Delta U)$ and $\overline{f}^s$ by $\overline{\mu}_{\widetilde{G}}(\Delta U)$ in Eq. (4.16).

The fuzzy PI and PD controllers in Sections 4.7 and 4.8 employ the center-of-sets type of reducer, which is implemented by the iterative KM algorithm besides the interval weighted-average method. That is, the centroid of G^s, $[a^s, b^s]$, of the sth rule consequents in Eq. (3.62) are computed by the iterative KM algorithm. The firing interval $[f^s, \overline{f}^s]$ in Eq. (3.42) are precomputed first. Then the iterative procedure in Table 2.3 is adopted to carry out the type reduction. It's worth noting that the symbols x_i, θ_i, N, i, k, $c_l(L)$, $c_r(R)$, and c' in Table 2.3 are replaced with a^s (or b^s), f^s, M, s, L (or R), u_l, u_r, and u_l' (or u_r') in Sections 4.7 and 4.8, respectively.

We now show different techniques needed for deriving the explicit analytical structure of various interval T2 fuzzy PI and PD controllers (Table 4.1), all of which use Zadeh AND operator. The fuzzy controllers studied become more and more complex in terms of (1) type of input fuzzy sets (from T2 triangular in Section 4.4 to any T2 nonlinear in Section 4.7), (2) number of input fuzzy sets (from 2 in Section 4.4 to any number in Section 4.7), and (3) type of output fuzzy sets (from T1 singleton in Section 4.4 to T2 singleton in Section 4.6 to any T1 or T2 fuzzy sets in Section 4.7). Section 4.8 deals with TSK T2 fuzzy controllers, whereas Sections 4.4–4.7 cover Mamdani T2 fuzzy controllers.

The reader may wonder given a control problem how to choose a T2 fuzzy controller. Here, we assume that the control problem is a nonlinear control problem (a linear control problem can be easily addressed by the linear control theory) and the reader has tried conventional control schemes (e.g., the PID control) as well as the T1 fuzzy control techniques and failed to achieve satisfactory control performance. The T2 fuzzy controllers are all nonlinear controllers as we will show below. Because one cannot claim that one nonlinear controller is better or worse than the other based on their mathematical representations, one T2 fuzzy controller cannot be deemed to be superior or inferior to the other for the same reason. Nevertheless, in engineering applications, the simplest solution is usually considered the best solution. In this spirit, a simpler T2 fuzzy controller should be preferred. A more complex fuzzy controller is employed only when its use can be justified. Finally, we stress that it is meaningless to discuss selection of a fuzzy controller, T1 or T2, without knowing the system to be controlled by the fuzzy controller.

We will use the fuzzy PI controllers only to present the derivation techniques. The analytical structure of the corresponding fuzzy PD controllers can easily be obtained according to the relationship between the PI and PD controllers, and we will show how to do this.

4.4 MAMDANI FUZZY PI AND PD CONTROLLERS—CONFIGURATION 1[1]

4.4.1 Fuzzy PI Controller Configuration

This configuration requires T2 triangular input fuzzy sets, T1 singleton output fuzzy sets, Mamdani fuzzy rules, the center-of-sets type of reducer implemented by the interval weighted-average method, and the centroid defuzzifier (Table 4.1). More

[1] Part of the material in this section is adapted from Du and Ying (2010; © 2010, IEEE).

TABLE 4.1 Configurations of Various Interval T2 Fuzzy PI and PD Controllers Whose Analytical Structure are Explicitly Derived in This Chapter

Configuration	Type	Type / Number of Input Fuzzy Sets	Type of Output Fuzzy Sets	Type Reducer	Type-Reducer Implementation	Defuzzifier	Section
1	Mamdani	T2 triangular/2	T1 singleton	Center of sets	Interval weighted-average algorithm	Centroid	4.4
2	Mamdani	T2 triangular/2	T1 singleton	Center of sets	Modified interval weighted-average algorithm	Average	4.5
3	Mamdani	T2 triangular/2	T2 singleton	Centroid	Modified interval weighted-average algorithm	Average	4.6
4	Mamdani	T2 any nonlinear/any	Any T1 or T2	Center of sets	Interval weighted-average algorithm and KM algorithm	Centroid	4.7
5	TSK	T2 any nonlinear/any	T2 singleton	Center of sets	Interval weighted-average algorithm and KM algorithm	Centroid	4.8

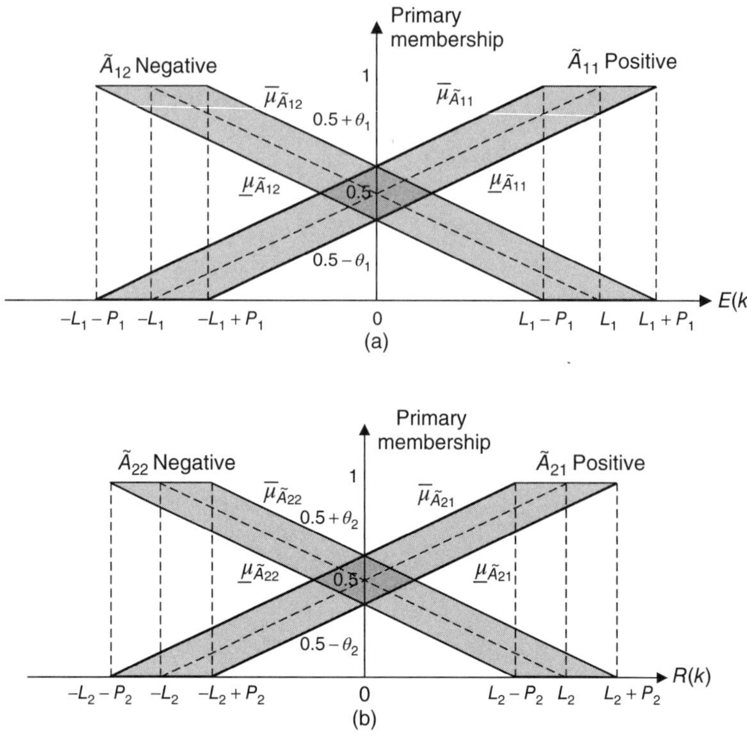

Figure 4.2 Primary membership functions of the interval T2 fuzzy sets (a) for $E(k)$ and (b) for $R(k)$. On the primary membership axes, θ_1 and θ_2 are the distance between 0.5 and intersection of the upper (or lower) membership functions.

specifically, two interval T2 fuzzy sets, namely $\tilde{A}_{11}$ and $\tilde{A}_{12}$ (Fig. 4.2a), are defined for $E(k)$ and another two, $\tilde{A}_{21}$ and $\tilde{A}_{22}$ (Fig. 4.2b), are defined for $R(k)$ where $\tilde{A}_{11}$ and $\tilde{A}_{21}$ are also linguistically named "positive" and $\tilde{A}_{12}$ and $\tilde{A}_{22}$ "negative." These fuzzy sets can be considered as parts of trapezoidal T2 fuzzy sets. The primary memberships of $E(k)$ and $R(k)$ are bounded by the upper and lower membership functions that are trapezoidal T1 fuzzy sets— $\bar{\mu}_{\tilde{A}_{11}}$, $\underline{\mu}_{\tilde{A}_{11}}$, $\bar{\mu}_{\tilde{A}_{12}}$, $\underline{\mu}_{\tilde{A}_{12}}$, $\bar{\mu}_{\tilde{A}_{21}}$, $\underline{\mu}_{\tilde{A}_{21}}$, $\bar{\mu}_{\tilde{A}_{22}}$, and $\underline{\mu}_{\tilde{A}_{22}}$. Design parameters θ_1 and θ_2 are used to specify different degrees of uncertainties represented by the T2 fuzzy sets. In Fig. 4.2, because there are two similar right triangles, $0.5/L_1 = \theta_1/P_1$, and hence $P_1 = 2L_1\theta_1$. Similarly, $P_2 = 2L_2\theta_2$. Being of the interval type, the secondary membership functions of these T2 fuzzy sets are all equal to 1 for the entire universes of the discourses.

Four T1 singleton fuzzy sets are defined for $\Delta U(k)$. They have nonzero membership value only at $\Delta U(k) = b_1$, b_2, b_3, and b_4, respectively, as shown in Fig. 4.3. The order of the four values in the figure represents one of the many possibilities and is illustrative only. Their actual relative positions depend on the values assigned by the controller designer.

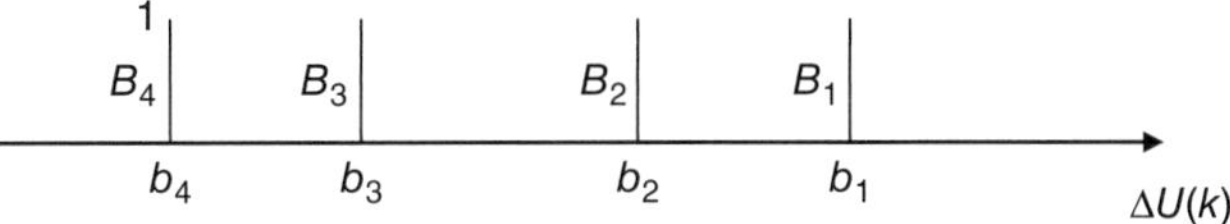

Figure 4.3 Four T1 singleton fuzzy sets for $\Delta U(k)$. The nonzero memberships are 1.

Because there are two fuzzy sets for $E(k)$ and two fuzzy sets for $R(k)$, the following four fuzzy rules are employed to cover all the four different combinations of B_m and $R(k)$:

$$R^1: \text{If } E(k) \text{ is } \widetilde{A}_{11} \text{ and } R(k) \text{ is } \widetilde{A}_{21}, \text{ then } \Delta U(k) \text{ is } B_1.$$

$$R^2: \text{If } E(k) \text{ is } \widetilde{A}_{11} \text{ and } R(k) \text{ is } \widetilde{A}_{22}, \text{ then } \Delta U(k) \text{ is } B_2.$$

$$R^3: \text{If } E(k) \text{ is } \widetilde{A}_{12} \text{ and } R(k) \text{ is } \widetilde{A}_{21}, \text{ then } \Delta U(k) \text{ is } B_3.$$

$$R^4: \text{If } E(k) \text{ is } \widetilde{A}_{12} \text{ and } R(k) \text{ is } \widetilde{A}_{22}, \text{ then } \Delta U(k) \text{ is } B_4.$$

For set-point control problems [i.e., SP(k) is either a constant or changes in a stair step fashion], which is one of the most commonly encountered control problems in practice, the four rules should be adequate in many cases as there are only four different scenarios, each of which is taken care of by one of the rules if the values of b_1, b_2, b_3, and b_4 are proper (Ying, 2000; Ying et al., 1990) (Fig. 4.4). Let's see a simple example where $b_1 > 0$, $b_2 = b_3 = 0$, and $b_4 < 0$. Rule R^1 covers the situation in which system output is below the target output and is still decreasing. Controller

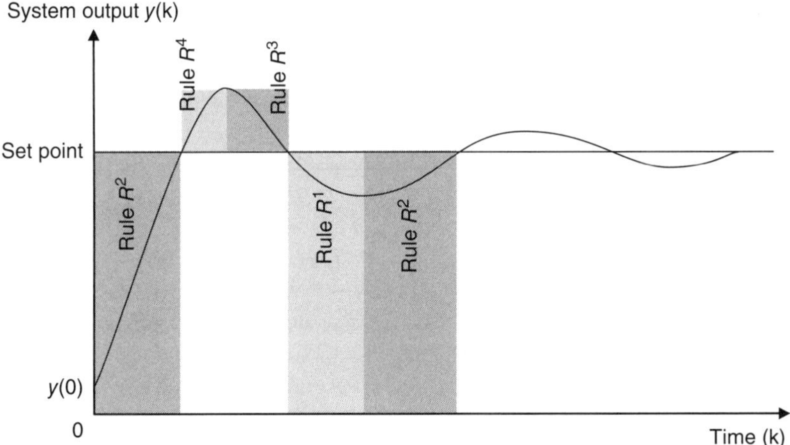

Figure 4.4 Illustration of how the four fuzzy rules can cover all possible control needs of a set-point control problem.

output should be increased. Rule R^4 deals with the opposite circumstance—system output is larger than the target level and is still increasing. Logically, controller output should be reduced. The two remaining scenarios are: (1) system output is below the target level but is increasing, and (2) system output is above the target output but is decreasing. In either case, it is sensible to let controller output stay at the same level, hoping system output will land on the target level smoothly on its own. This is what rules R^2 and R^3 can achieve.

When applying the interval T2 fuzzy inference to these rules, each of the T1 fuzzy sets in the rule consequents needs to be treated as a special interval T2 fuzzy set. Specifically, it can be regarded as a special interval T2 fuzzy set whose upper and lower primary membership values are both 1. The firing intervals as a result of Zadeh AND operation are

$$\widetilde{F}^1 = \left[\underline{f}^1, \overline{f}^1\right] = \left[\min\left(\underline{\mu}_{\widetilde{A}_{11}}, \underline{\mu}_{\widetilde{A}_{21}}\right), \min\left(\overline{\mu}_{\widetilde{A}_{11}}, \overline{\mu}_{\widetilde{A}_{21}}\right)\right] \tag{4.18}$$

$$\widetilde{F}^2 = \left[\underline{f}^2, \overline{f}^2\right] = \left[\min\left(\underline{\mu}_{\widetilde{A}_{11}}, \underline{\mu}_{\widetilde{A}_{22}}\right), \min\left(\overline{\mu}_{\widetilde{A}_{11}}, \overline{\mu}_{\widetilde{A}_{22}}\right)\right] \tag{4.19}$$

$$\widetilde{F}^3 = \left[\underline{f}^3, \overline{f}^3\right] = \left[\min\left(\underline{\mu}_{\widetilde{A}_{12}}, \underline{\mu}_{\widetilde{A}_{21}}\right), \min\left(\overline{\mu}_{\widetilde{A}_{12}}, \overline{\mu}_{\widetilde{A}_{21}}\right)\right] \tag{4.20}$$

$$\widetilde{F}^4 = \left[\underline{f}^4, \overline{f}^4\right] = \left[\min\left(\underline{\mu}_{\widetilde{A}_{12}}, \underline{\mu}_{\widetilde{A}_{22}}\right), \min\left(\overline{\mu}_{\widetilde{A}_{12}}, \overline{\mu}_{\widetilde{A}_{22}}\right)\right] \tag{4.21}$$

The center-of-sets type of reducer is implemented by the interval weighted-average method. The centroid defuzzifier is used.

4.4.2 Method for Deriving the Analytical Structure

The key to the mathematical structure derivation of the fuzzy controller is to determine the outcomes of the min( ) operations in the four rules. The nature of this issue is the same as what one encounters when deriving the analytical structure of the T1 fuzzy controllers whose fuzzy rules involve the Zadeh AND operator (Ying, 2000; Ying, et al., 1990). The principle of the structure-deriving technique in Ying et al. (1990) is readily applicable to the T2 fuzzy controller and indeed will be utilized to develop the new derivation method for the T2 controller.

Without loss of generality, we consider the case when $E(k)$ is inside $[-L_1-P_1,\ L_1+P_1]$ and $R(k)$ is inside $[-L_2-P_2,\ L_2+P_2]$ (Fig. 4.2). The structure expressions for the remaining cases [i.e., when $E(k) \in (-\infty, -L_1-P_1)$ or (L_1+P_1, ∞), $R(k) \in (-\infty,\ -L_2-P_2)$ or $(L_2+P_2,\ \infty)$] can be derived similarly. To derive the explicit expressions, we first need to determine the left and right membership end points of the firing intervals in Eqs. (4.18)–(4.21). In each of the min( ) expressions, the membership values vary with $E(k)$ and $R(k)$. Therefore, to decide which membership value is smaller [i.e., the outcome of the min( )], one must divide the two-dimensional input space $[-L_1-P_1,\ L_1+P_1] \times [-L_2-P_2,\ L_2+P_2]$ into a number of regions so that for each region, either the membership value of $E(k)$ is always greater than that of $R(k)$ or it is the other way around, but not both.

Such a region is called an IC (input combination). Figure 4.5 shows the input space division results for R^1 to R^4. Note that the axes of $E(k)$ and $R(k)$ in the figure are utilized to just indicate some key points of $E(k)$ and $R(k)$. The axes do not form a coordinate system. This is also the case for some of the other figures in this chapter.

As an example, Fig. 4.5a shows the region divisions for $\overline{f}^1 = \min(\overline{\mu}_{\tilde{A}_{11}}, \overline{\mu}_{\tilde{A}_{21}})$ of R^1. When $E(k)$ and $R(k)$ are in the IC-labeled $\overline{f}^1$_IC1, $\overline{\mu}_{\tilde{A}_{11}} = 1$ and $\overline{\mu}_{\tilde{A}_{21}} = 1$. Hence, $\overline{f}^1 = 1$, which is marked in IC1 in the figure. In $\overline{f}^1$_IC2, $\overline{\mu}_{\tilde{A}_{11}} < \overline{\mu}_{\tilde{A}_{21}}$, hence $\overline{f}^1 = \overline{\mu}_{\tilde{A}_{11}}$. Finally, in $\overline{f}^1$_IC3, $\overline{\mu}_{\tilde{A}_{11}} > \overline{\mu}_{\tilde{A}_{21}}$, leading to $\overline{f}^1 = \overline{\mu}_{\tilde{A}_{21}}$. On the boundaries of the two adjacent regions, either result can be used as they are equal.

Figure 4.5 only covers what happens to the min( ) operation when each rule is evaluated one by one. Nevertheless, at any time k, all four rules are actually executed at the same time. Thus, they should be considered simultaneously. This amounts to superimposing all eight figures in Fig. 4.5 (Ying 2006). The number and shape of the final region divisions after the superimposing depend on the parameter values of the input fuzzy sets. For example when $L_1 > L_2$ and $L_1 \times \theta_1 = L_2 \times \theta_2$ (i.e., $P_1 = P_2$), the superimposing result is a total of 25 ICs as shown in Fig. 4.6a. The results of the left and right end points of the firing intervals in Eqs. (4.18)–(4.21) for these 25 ICs are listed in Table 4.2. If we suppose that $L_1 > L_2$ but $\theta_1 = \theta_2$ (i.e., $P_1 \neq P_2$), the superimposing will create 16 ICs instead (Fig. 4.6b). The corresponding left and right endpoints of the intervals are the same as those of the first 16 of the 25 ICs in Table 4.2. Obviously, the superimposing result depends on the fuzzy controller's configuration and hence varies from controller to controller.

For each IC, put the eight membership functions (i.e., the eight entries in a row in Table 4.2) into Eq. (4.16), and after some mathematical manipulations, the explicit expressions of $\Delta U_p(k)$, $p = 1, \ldots, 16$, for that IC will be obtained. Table 4.3 illustrates mathematically $\Delta U_p(k)$ for IC1 in Fig. 4.6b. It is easy to see that the 16 expressions in Table 4.3 share the same mathematical structure below [note that $E(k) = k_e e(k)$ and $R(k) = k_r r(k)$]:

$$\Delta U_p(k) = \frac{\varphi_1^p k_e e(k) + \varphi_2^p k_r r(k) + \varphi_3^p}{\xi_1^p k_e e(k) + \xi_2^p k_r r(k) + \xi_3^p} \tag{4.22}$$

where φ_k^p and ψ_k^p, $k = 1, 2, 3$, are constants in the IC. Their values not only depend on the design parameters of the fuzzy controller (e.g., L_1, L_2, $k_{\Delta U}$, and b_m) but also on IC. That means their values may be different in different ICs. $\Delta U_p(k)$ for the rest of the 15 ICs confirm to this pattern because both the denominator and the numerator of Eq. (4.16) are always linear functions of $e(k)$ and/or $r(k)$.

Recall that the centroid defuzzifier coupled with the interval weighted-average method produces

$$\Delta U(k) = \frac{\Delta U_p^{\min}(k) + \Delta U_p^{\max}(k)}{2} \tag{4.23}$$

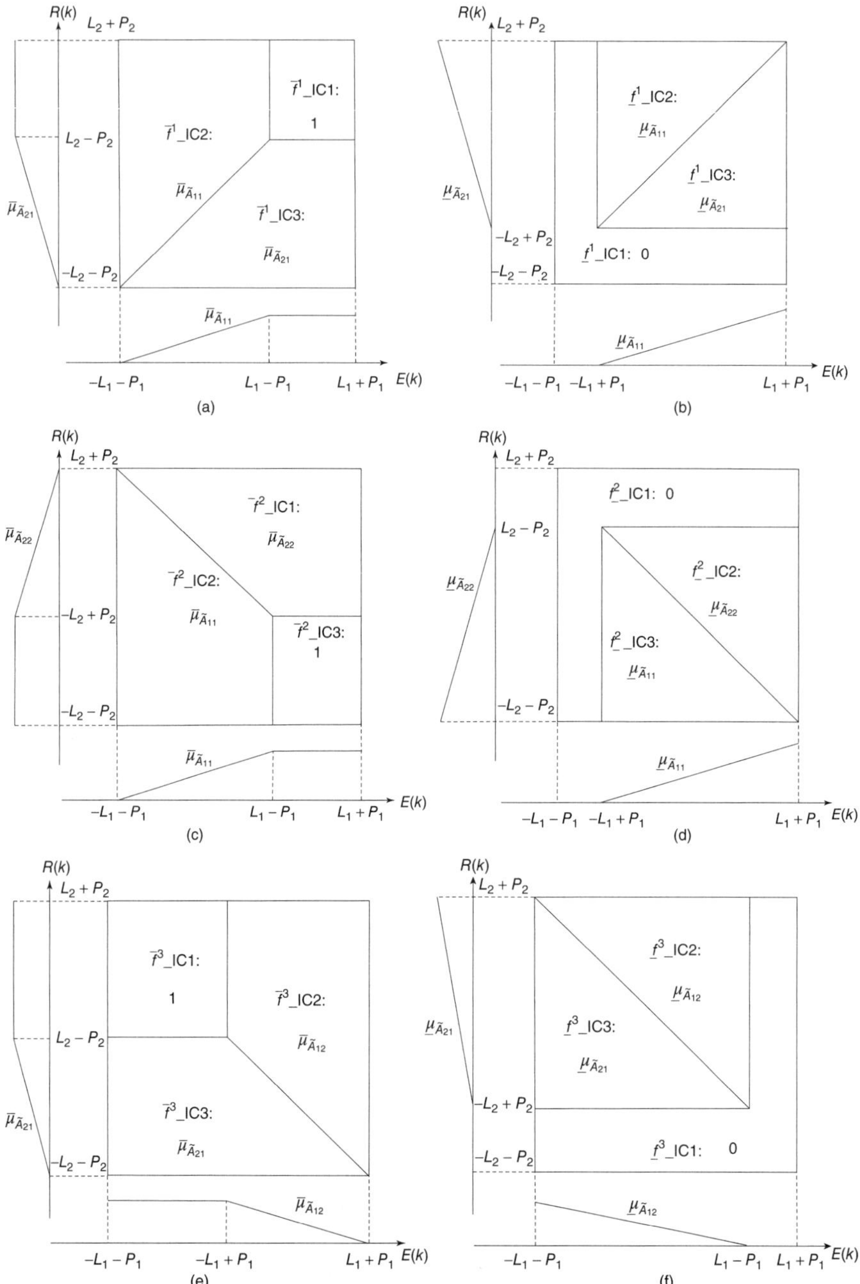

Figure 4.5 ICs for the upper and lower bounds of each of the four fuzzy rules. Figures (a)–(h) correspond to $\bar{f}^1$, $\underline{f}^1$, $\bar{f}^2$, $\underline{f}^2$, $\bar{f}^3$, $\underline{f}^3$, $\bar{f}^4$, and $\underline{f}^4$ in Eqs. (4.18)–(4.21), respectively.

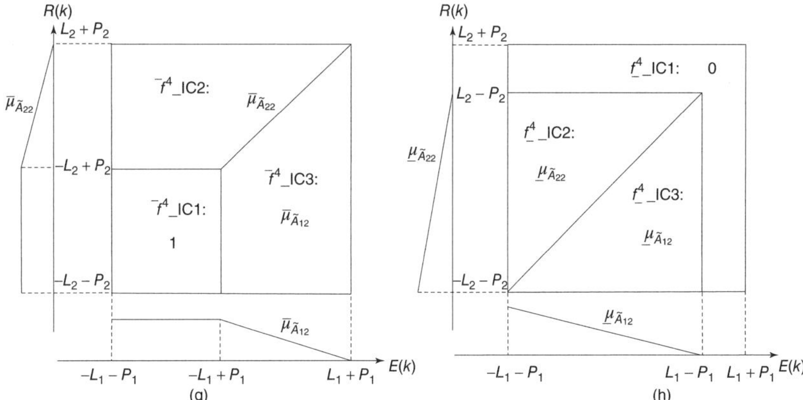

Figure 4.5 *(Continued)*

where $\Delta U_p^{\min}(k)$ and $\Delta U_p^{\max}(k)$ are, respectively, the minimum value and maximum value of the 16 $\Delta U_p(k)$ at time k. It is important to emphasize that unlike most studies in the literature, $\Delta U_p(k)$, $\Delta U_p^{\min}(k)$, and $\Delta U_p^{\max}(k)$ in Eq. (4.23) are all mathematical expressions, as opposed to numerical values. This also holds true for the other four controller configurations in this chapter, and the reader needs to keep it in mind. Because $\Delta U_p^{\min}(k)$ and $\Delta U_p^{\max}(k)$ depend on the values of the two input variables, $\Delta U(k)$ varies with k because the input variables change with time. For any given input values, we can always determine which of the 16 $\Delta U_p(k)$ expressions are $\Delta U_p^{\min}(k)$ and $\Delta U_p^{\max}(k)$ by plugging the input values into the 16 expressions. The expressions producing the minimum and maximum are $\Delta U_p^{\min}(k)$ and $\Delta U_p^{\max}(k)$, respectively. This issue will be discussed more after Theorem 4.1 is established below. At the current stage, knowing the fact that one of the 16 $\Delta U_p(k)$ is $\Delta U_p^{\min}(k)$ and another $\Delta U_p^{\max}(k)$ is sufficient to continue our derivation. Putting Eq. (4.22) into (4.23) will lead to the result in the form of

$$\Delta u(k) = \frac{1}{2}\sum_{j=1}^{2} K_{\Delta U}\frac{\varphi_1^j k_e e(k) + \varphi_2^j k_r r(k) + \varphi_3^j}{\psi_1^j k_e e(k) + \psi_2^j k_r r(k) + \psi_3^j}$$

where $j = 1$ and $j = 2$ are for $\Delta U_p^{\min}(k)$ and $\Delta U_p^{\max}(k)$, respectively (the two p values corresponding to $j = 1$ and $j = 2$ can always be found). Designating

$$K_p(e(k),\ r(k)) = \frac{1}{2}\sum_{j=1}^{2}\frac{k_{\Delta U} k_r \varphi_2^j}{\psi_1^j k_e e(k) + \psi_2^j k_r r(k) + \psi_3^j}$$

$$K_i(e(k),\ r(k)) = \frac{1}{2}\sum_{j=1}^{2}\frac{k_{\Delta U} k_e \varphi_1^j}{\psi_1^j k_e e(k) + \psi_2^j k_r r(k) + \psi_3^j}$$

TABLE 4.2 Left and Right End Points of Four Firing Intervals in Eqs. (4.18)–(4.21) for IC1–IC25 Shown in Fig. 4.6

	Rule 1		Rule 2		Rule 3		Rule 4	
IC No.	$\overline{f}^1$	$\underline{f}^1$	$\overline{f}^2$	$\underline{f}^2$	$\overline{f}^3$	$\underline{f}^3$	$\overline{f}^4$	$\underline{f}^4$
1	$\overline{\mu}_{\tilde{A}_{21}}$	$\underline{\mu}_{\tilde{A}_{21}}$	$\overline{\mu}_{\tilde{A}_{22}}$	$\underline{\mu}_{\tilde{A}_{22}}$	$\overline{\mu}_{\tilde{A}_{12}}$	$\underline{\mu}_{\tilde{A}_{12}}$	$\overline{\mu}_{\tilde{A}_{12}}$	$\underline{\mu}_{\tilde{A}_{12}}$
2	$\overline{\mu}_{\tilde{A}_{11}}$	$\underline{\mu}_{\tilde{A}_{11}}$	$\overline{\mu}_{\tilde{A}_{22}}$	$\underline{\mu}_{\tilde{A}_{22}}$	$\overline{\mu}_{\tilde{A}_{12}}$	$\underline{\mu}_{\tilde{A}_{12}}$	$\overline{\mu}_{\tilde{A}_{22}}$	$\underline{\mu}_{\tilde{A}_{22}}$
3	$\overline{\mu}_{\tilde{A}_{11}}$	$\underline{\mu}_{\tilde{A}_{11}}$	$\overline{\mu}_{\tilde{A}_{11}}$	$\underline{\mu}_{\tilde{A}_{11}}$	$\overline{\mu}_{\tilde{A}_{21}}$	$\underline{\mu}_{\tilde{A}_{21}}$	$\overline{\mu}_{\tilde{A}_{22}}$	$\underline{\mu}_{\tilde{A}_{22}}$
4	$\overline{\mu}_{\tilde{A}_{21}}$	$\underline{\mu}_{\tilde{A}_{21}}$	$\overline{\mu}_{\tilde{A}_{11}}$	$\underline{\mu}_{\tilde{A}_{11}}$	$\overline{\mu}_{\tilde{A}_{21}}$	$\underline{\mu}_{\tilde{A}_{21}}$	$\overline{\mu}_{\tilde{A}_{12}}$	$\underline{\mu}_{\tilde{A}_{12}}$
5	1	$\underline{\mu}_{\tilde{A}_{21}}$	$\overline{\mu}_{\tilde{A}_{22}}$	0	$\overline{\mu}_{\tilde{A}_{12}}$	0	$\overline{\mu}_{\tilde{A}_{12}}$	0
6	1	$\underline{\mu}_{\tilde{A}_{11}}$	$\overline{\mu}_{\tilde{A}_{22}}$	0	$\overline{\mu}_{\tilde{A}_{12}}$	0	$\overline{\mu}_{\tilde{A}_{22}}$	0
7	$\overline{\mu}_{\tilde{A}_{11}}$	0	$\overline{\mu}_{\tilde{A}_{22}}$	0	1	$\underline{\mu}_{\tilde{A}_{12}}$	$\overline{\mu}_{\tilde{A}_{22}}$	0
8	$\overline{\mu}_{\tilde{A}_{11}}$	0	$\overline{\mu}_{\tilde{A}_{11}}$	0	1	$\underline{\mu}_{\tilde{A}_{21}}$	$\overline{\mu}_{\tilde{A}_{22}}$	0
9	$\overline{\mu}_{\tilde{A}_{11}}$	0	$\overline{\mu}_{\tilde{A}_{11}}$	0	$\overline{\mu}_{\tilde{A}_{21}}$	0	1	$\underline{\mu}_{\tilde{A}_{22}}$
10	$\overline{\mu}_{\tilde{A}_{11}}$	0	$\overline{\mu}_{\tilde{A}_{11}}$	0	$\overline{\mu}_{\tilde{A}_{21}}$	0	1	$\underline{\mu}_{\tilde{A}_{12}}$
11	$\overline{\mu}_{\tilde{A}_{21}}$	0	1	$\underline{\mu}_{\tilde{A}_{11}}$	$\overline{\mu}_{\tilde{A}_{21}}$	0	$\overline{\mu}_{\tilde{A}_{12}}$	0
12	$\overline{\mu}_{\tilde{A}_{21}}$	0	1	$\underline{\mu}_{\tilde{A}_{22}}$	$\overline{\mu}_{\tilde{A}_{12}}$	0	$\overline{\mu}_{\tilde{A}_{12}}$	0
13	$\overline{\mu}_{\tilde{A}_{21}}$	$\underline{\mu}_{\tilde{A}_{21}}$	$\underline{\mu}_{\tilde{A}_{21}}$	$\underline{\mu}_{\tilde{A}_{22}}$	$\overline{\mu}_{\tilde{A}_{12}}$	0	$\overline{\mu}_{\tilde{A}_{12}}$	0
14	$\overline{\mu}_{\tilde{A}_{11}}$	$\underline{\mu}_{\tilde{A}_{11}}$	$\underline{\mu}_{\tilde{A}_{21}}$	0	$\overline{\mu}_{\tilde{A}_{12}}$	$\underline{\mu}_{\tilde{A}_{12}}$	$\underline{\mu}_{\tilde{A}_{21}}$	0
15	$\overline{\mu}_{\tilde{A}_{11}}$	0	$\overline{\mu}_{\tilde{A}_{11}}$	0	$\overline{\mu}_{\tilde{A}_{21}}$	$\underline{\mu}_{\tilde{A}_{21}}$	$\underline{\mu}_{\tilde{A}_{21}}$	$\underline{\mu}_{\tilde{A}_{22}}$
16	$\overline{\mu}_{\tilde{A}_{21}}$	0	$\overline{\mu}_{\tilde{A}_{11}}$	$\underline{\mu}_{\tilde{A}_{11}}$	$\overline{\mu}_{\tilde{A}_{21}}$	0	$\overline{\mu}_{\tilde{A}_{12}}$	$\underline{\mu}_{\tilde{A}_{12}}$
17	$\overline{\mu}_{\tilde{A}_{11}}$	$\underline{\mu}_{\tilde{A}_{21}}$	$\underline{\mu}_{\tilde{A}_{21}}$	0	$\overline{\mu}_{\tilde{A}_{12}}$	$\underline{\mu}_{\tilde{A}_{12}}$	$\overline{\mu}_{\tilde{A}_{12}}$	0
18	$\overline{\mu}_{\tilde{A}_{11}}$	$\underline{\mu}_{\tilde{A}_{11}}$	$\overline{\mu}_{\tilde{A}_{11}}$	0	$\overline{\mu}_{\tilde{A}_{12}}$	$\underline{\mu}_{\tilde{A}_{21}}$	$\underline{\mu}_{\tilde{A}_{21}}$	0
19	$\overline{\mu}_{\tilde{A}_{11}}$	0	$\overline{\mu}_{\tilde{A}_{11}}$	$\underline{\mu}_{\tilde{A}_{11}}$	$\overline{\mu}_{\tilde{A}_{21}}$	0	$\overline{\mu}_{\tilde{A}_{12}}$	$\underline{\mu}_{\tilde{A}_{22}}$
20	$\overline{\mu}_{\tilde{A}_{21}}$	0	$\overline{\mu}_{\tilde{A}_{11}}$	$\underline{\mu}_{\tilde{A}_{22}}$	$\overline{\mu}_{\tilde{A}_{12}}$	0	$\overline{\mu}_{\tilde{A}_{12}}$	$\underline{\mu}_{\tilde{A}_{12}}$
21	$\overline{\mu}_{\tilde{A}_{11}}$	$\underline{\mu}_{\tilde{A}_{21}}$	$\underline{\mu}_{\tilde{A}_{21}}$	$\underline{\mu}_{\tilde{A}_{22}}$	$\overline{\mu}_{\tilde{A}_{12}}$	$\underline{\mu}_{\tilde{A}_{12}}$	$\overline{\mu}_{\tilde{A}_{12}}$	$\underline{\mu}_{\tilde{A}_{22}}$
22	$\overline{\mu}_{\tilde{A}_{11}}$	$\underline{\mu}_{\tilde{A}_{11}}$	$\overline{\mu}_{\tilde{A}_{11}}$	$\underline{\mu}_{\tilde{A}_{22}}$	$\overline{\mu}_{\tilde{A}_{12}}$	$\underline{\mu}_{\tilde{A}_{21}}$	$\underline{\mu}_{\tilde{A}_{21}}$	$\underline{\mu}_{\tilde{A}_{22}}$
23	$\overline{\mu}_{\tilde{A}_{11}}$	$\underline{\mu}_{\tilde{A}_{21}}$	$\overline{\mu}_{\tilde{A}_{11}}$	$\underline{\mu}_{\tilde{A}_{11}}$	$\overline{\mu}_{\tilde{A}_{21}}$	$\underline{\mu}_{\tilde{A}_{21}}$	$\overline{\mu}_{\tilde{A}_{12}}$	$\underline{\mu}_{\tilde{A}_{22}}$
24	$\overline{\mu}_{\tilde{A}_{21}}$	$\underline{\mu}_{\tilde{A}_{21}}$	$\overline{\mu}_{\tilde{A}_{11}}$	$\overline{\mu}_{\tilde{A}_{22}}$	$\overline{\mu}_{\tilde{A}_{12}}$	$\underline{\mu}_{\tilde{A}_{21}}$	$\overline{\mu}_{\tilde{A}_{12}}$	$\underline{\mu}_{\tilde{A}_{12}}$
25	$\overline{\mu}_{\tilde{A}_{11}}$	$\underline{\mu}_{\tilde{A}_{21}}$	$\overline{\mu}_{\tilde{A}_{11}}$	$\overline{\mu}_{\tilde{A}_{22}}$	$\overline{\mu}_{\tilde{A}_{12}}$	$\underline{\mu}_{\tilde{A}_{21}}$	$\overline{\mu}_{\tilde{A}_{12}}$	$\overline{\mu}_{\tilde{A}_{22}}$

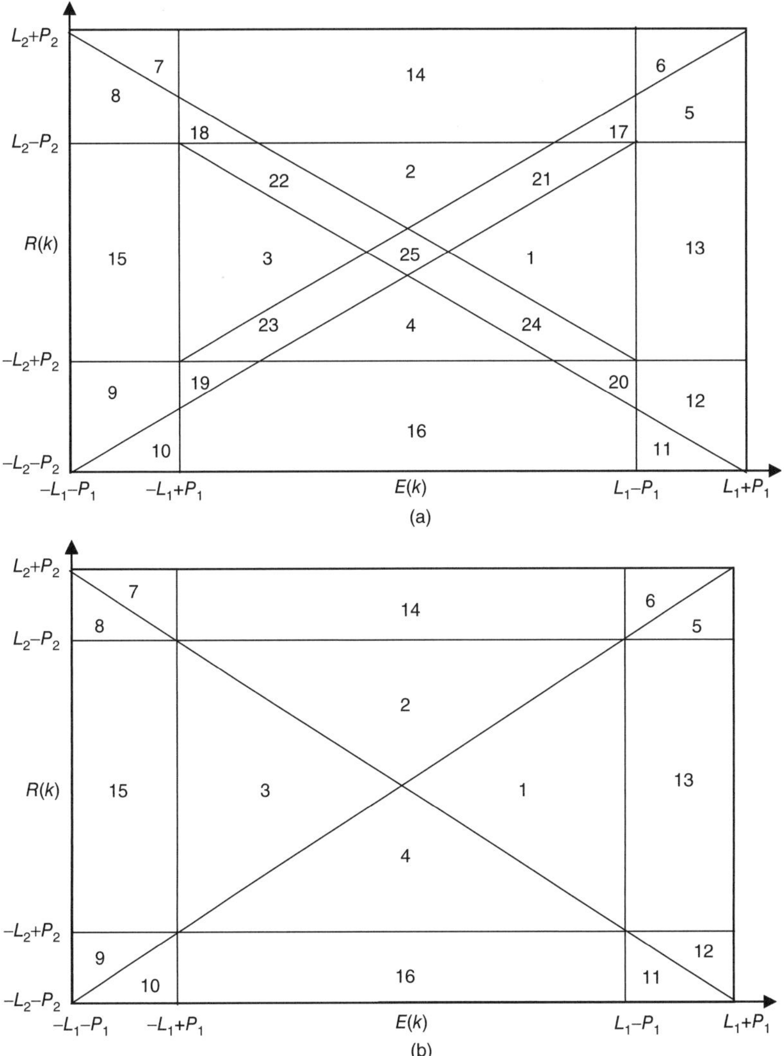

Figure 4.6 Division of the input space when the four rules are considered simultaneously (the numbers represent the IC numbers): (a) when $L_1 > L_2$ and $P_1 = P_2$ and (b) when $L_1 > L_2$ and $P_1 > P_2$.

$$\delta(e(k),\ r(k)) = \frac{1}{2}\sum_{j=1}^{2} \frac{k_{\Delta U}\varphi_3^j}{\psi_1^j k_e e(k) + \psi_2^j k_r r(k) + \psi_3^j}$$

then

$$\Delta u(k) = K_p(e(k), r(k)) \cdot r(k) + K_i(e(k), r(k)) \cdot e(k) + \delta(e(k), r(k)) \tag{4.24}$$

This expression represents a nonlinear PI controller in incremental form with variable proportional gain and integral gain plus a variable control offset term

TABLE 4.3 Analytical Expressions of $\Delta U_p(k)$ for IC1 in Fig. 4.6b

p	$\Delta U_p(k)$
1	$\dfrac{(b_3+b_4)L_2k_ee(k)-(b_1-b_2)L_1k_rr(k)-2L_1L_2[(0.5+\theta_2)(b_1+b_2)+(0.5+\theta_1)(b_3+b_4)]}{2L_2k_ee(k)-4L_1L_2(1+\theta_1+\theta_2)}$
2	$\dfrac{(b_3+b_4)L_2k_ee(k)-(b_1-b_2)L_1k_rr(k)-2L_1L_2[(0.5+\theta_2)(b_1+b_2)+(0.5+\theta_1)b_3+(0.5-\theta_1)b_4]}{2L_2k_ee(k)-4L_1L_2(1+\theta_2)}$
3	$\dfrac{(b_3+b_4)L_2k_ee(k)-(b_1-b_2)L_1k_rr(k)-2L_1L_2[(0.5+\theta_2)(b_1+b_2)+(0.5-\theta_1)b_3+(0.5+\theta_1)b_4]}{2L_2k_ee(k)-4L_1L_2(1+\theta_2)}$
4	$\dfrac{(b_3+b_4)L_2k_ee(k)-(b_1-b_2)L_1k_rr(k)-2L_1L_2[(0.5+\theta_2)(b_1+b_2)+(0.5-\theta_1)(b_3+b_4)]}{2L_2k_ee(k)-4L_1L_2(1+\theta_2-\theta_1)}$
5	$\dfrac{(b_3+b_4)L_2k_ee(k)-(b_1-b_2)L_1k_rr(k)-2L_1L_2[(0.5+\theta_2)b_1+(0.5-\theta_2)b_2+(0.5+\theta_1)(b_3+b_4)]}{2L_2k_ee(k)-4L_1L_2(1+\theta_1)}$
6	$\dfrac{(b_3+b_4)L_2k_ee(k)-(b_1-b_2)L_1k_rr(k)-2L_1L_2[(0.5+\theta_2)b_1+(0.5-\theta_2)b_2+(0.5+\theta_1)b_3+(0.5-\theta_1)b_4]}{2L_2k_ee(k)-4L_1L_2}$
7	$\dfrac{(b_3+b_4)L_2k_ee(k)-(b_1-b_2)L_1k_rr(k)-2L_1L_2[(0.5+\theta_2)b_1+(0.5-\theta_2)b_2+(0.5-\theta_1)b_3+(0.5+\theta_1)b_4]}{2L_2k_ee(k)-4L_1L_2}$
8	$\dfrac{(b_3+b_4)L_2k_ee(k)-(b_1-b_2)L_1k_rr(k)-2L_1L_2[(0.5+\theta_2)b_1+(0.5-\theta_2)b_2+(0.5-\theta_1)(b_3+b_4)]}{2L_2k_ee(k)-4L_1L_2(1-\theta_1)}$

9	$\frac{(b_3+b_4) L_2 k_e e(k) - (b_1-b_2) L_1 k_r r(k) - 2L_1L_2 \left[(0.5-\theta_2) b_1 + (0.5+\theta_2) b_2 + (0.5+\theta_1)(b_3+b_4)\right]}{2L_2 k_e e(k) - 4L_1L_2 (1+\theta_1)}$
10	$\frac{(b_3+b_4) L_2 k_e e(k) - (b_1-b_2) L_1 k_r r(k) - 2L_1L_2 \left[(0.5-\theta_2) b_1 + (0.5+\theta_2) b_2 + (0.5+\theta_1) b_3 + (0.5-\theta_1) b_4\right]}{2L_2 k_e e(k) - 4L_1L_2}$
11	$\frac{(b_3+b_4) L_2 k_e e(k) - (b_1-b_2) L_1 k_r r(k) - 2L_1L_2 \left[(0.5-\theta_2) b_1 + (0.5+\theta_2) b_2 + (0.5-\theta_1) b_3 + (0.5+\theta_1) b_4\right]}{2L_2 k_e e(k) - 4L_1L_2}$
12	$\frac{(b_3+b_4) L_2 k_e e(k) - (b_1-b_2) L_1 k_r r(k) - 2L_1L_2 \left[(0.5-\theta_2) b_1 + (0.5+\theta_2) H_2 + (0.5-\theta_1)(b_3+b_4)\right]}{2L_2 k_e e(k) - 4L_1L_2 (1-\theta_1)}$
13	$\frac{(b_3+b_4) L_2 k_e e(k) - (b_1-b_2) L_1 k_r r(k) - 2L_1L_2 \left[(0.5-\theta_2)(b_1+b_2) + (0.5+\theta_1)(b_3+b_4)\right]}{2L_2 k_e e(k) - 4L_1L_2 (1-\theta_2+\theta_1)}$
14	$\frac{(b_3+b_4) L_2 k_e e(k) - (b_1-b_2) L_1 k_r r(k) - 2L_1L_2 \left[(0.5-\theta_2)(b_1+b_2) + (0.5+\theta_1) b_3 + (0.5-\theta_1) b_4\right]}{2L_2 k_e e(k) - 4L_1L_2 (1-\theta_2)}$
15	$\frac{(b_3+b_4) L_2 k_e e(k) - (b_1-b_2) L_1 k_r r(k) - 2L_1L_2 \left[(0.5-\theta_2)(b_1+b_2) + (0.5-\theta_1) b_3 + (0.5+\theta_1) b_4\right]}{2L_2 k_e e(k) - 4L_1L_2 (1-\theta_2)}$
16	$\frac{(b_3+b_4) L_2 k_e e(k) - (b_1-b_2) L_1 k_r r(k) - 2L_1L_2 \left[(0.5-\theta_2)(b_1+b_2) + (0.5-\theta_1)(b_3+b_4)\right]}{2L_2 k_e e(k) - 4L_1L_2 (1-\theta_2-\theta_1)}$

$\delta(e(k), r(k))$. Notice that the denominator of $\delta(e(k), r(k))$ can have $e(k)$ or $r(k)$ or both. But the numerator cannot have either variable. This is an important requirement, or Eq. (4.24) may not be unique as one may move part of the first or second term in Eq. (4.24), if it has more than one term in the numerator, into $\delta(e(k), r(k))$.

Recall that the structure of the linear PD controller in position form is the same as that of the linear PI controller in incremental form. Therefore, if $U(k)$ is employed in the four fuzzy rules instead of $\Delta U(k)$, the result will be a nonlinear PD controller with variable proportional gain and derivative gain plus a variable offset term:

$$u(k) = K_p(e(k), r(k)) \cdot e(k) + K_d(e(k), r(k)) \cdot r(k) + \delta(e(k), r(k))$$

where the mathematical expressions of $K_p(e(k), r(k))$ and $K_d(e(k), r(k))$ are exactly the same as $K_i(e(k), r(k))$ and $K_p(e(k), r(k))$ in Eq. (4.24), respectively. These results can be summarized formally as in Theorem 4.1.

THEOREM 4.1 The T2 fuzzy PI (or the corresponding PD) controller in this section is structurally equivalent to a nonlinear PI (or PD) controller with variable gains and a variable control offset.

Usually, after a theorem is stated, one would provide a proof. In this chapter, we do not wish to use this format. Rather, we choose to first give sufficient explicit information that supports a theorem and then formally describe the theorem. Our approach improves readability at the expense of rigor. (Based on the material presented above, it is obvious that a rigorous proof of can be carried out. But the process will be long and tedious as we will have to prove the analytical structure for each and every IC.)

In light of the theorem, it would be interesting for the reader to know that many T1 fuzzy PI (or PD) controllers are also equivalent to nonlinear PI (or PD) controllers with variable proportional gain and integral gain (or derivative gain) (Ying, 1993a, 1998b, 1998c). More discussion in this regard will be provided in Section 4.9.

Even though the forms of the explicit structures of the fuzzy PI and PD controllers have been revealed [e.g., Eq. (4.24)], there are two issues remaining: (1) For any point in the input space, we need to determine which of the 16 $\Delta U_p(k)$ is $\Delta U_p^{\min}(k)$ or $\Delta U_p^{\max}(k)$ at any time k, and (2) different parts of some of the ICs found based on the min( ) operations may have different $\Delta U_p^{\min}(k)$ [and/or $\Delta U_p^{\max}(k)$]. The second issue rises due to the nature of the IC formation—an IC is formed by comparing the two membership functions of the input fuzzy sets in a fuzzy rule. The process does not involve $\Delta U(k)$ and consequently does not guarantee the uniqueness of $\Delta U_p^{\min}(k)$ [and/or $\Delta U_p^{\max}(k)$] in the IC. Therefore, whenever the second issue becomes relevant to an IC, the IC will need to be divided into smaller ICs in such a way that each of the smaller ICs has the same $\Delta U_p^{\min}(k)$ [and/or $\Delta U_p^{\max}(k)$]. Resolving either issue analytically seems to be difficult and remains a research topic. Alternatively, one can attain numerical solutions by writing a (simple) computer program, preferably using a symbolic software package, such as Mathematica or

MATLAB Symbolic Math Toolbox, to determine, at each point of the input space, which $\Delta U_p(k)$ is $\Delta U_p^{\min}(k)$ or $\Delta U_p^{\max}(k)$. The connected adjacent points with the same expression of $[\Delta U_p^{\min}(k) + \Delta U_p^{\max}(k)]/2$ collectively form an additional new IC. Initially, the numbers and shapes of these additional ICs' boundaries found may vary with the number of points involved. This is because at some points, more than one $\Delta U_p(k)$ may have the same $\Delta U_p^{\min}(k)$ or $\Delta U_p^{\max}(k)$ value. To differentiate between such points, one needs to check what $\Delta U_p^{\min}(k)$ [or $\Delta U_p^{\max}(k)$] is for its immediately surrounding points. If they employ the same $\Delta U_p^{\min}(k)$ [or $\Delta U_p^{\max}(k)$] as the point does, then assign that $\Delta U_p^{\min}(k)$ [or $\Delta U_p^{\max}(k)$] to that point. This procedure continues until the numbers and shapes of the new ICs no longer vary with the number of points used (obviously, the more points are used, the smoother the boundaries of the ICs will be). At that time, the new IC boundaries found can be regarded as the underlying ones and the total number of ICs (the original ICs plus the additional ICs) will be the smallest. In an IC, $\Delta U_p^{\min}(k)$ and $\Delta U_p^{\max}(k)$ can then be substituted into Eq. (4.23) to derive the fuzzy controller's analytical structure for that IC.

We should point out that the derivation technique as well as the derived analytical structure and its connection to the PI and PD controllers are equally valuable. This is not only true for the present section but also for the related sections below. The derived structure may seem to be mathematically complex. This, however, is a genuine description of the T2 fuzzy controller. The derivation technique itself is rigorous—there is no omission or approximation in the derivation process.

In this section, the T2 fuzzy controller is shown to be a nonlinear PI (or PD) controller with variable gains and a variable control offset term. This is a step forward—one can now understand the nature of the T2 fuzzy controller from a conventional control theory standpoint. Before the analytical structure has been revealed, the fuzzy controller functioned as a number-crunching blackbox—it took values of input variables and computed an output value. The entire process was numerical, not analytical (i.e., symbolic). Another value of the analytical structure derivation is that it tells us how restrictive a T2 fuzzy controller is. That is, it reveals what kinds of nonlinear control laws it can and cannot produce. This is not even an issue for conventional nonlinear control theory as a controller can have any mathematical representation. Any type of fuzzy controller does not nearly have this kind of flexibility. Yet another benefit of having the analytical structure is that it enables one to rigorously compare it with the analytical structure of the comparable T1 fuzzy controller. We will show how to do so in Section 4.9.1.

It is possible to use the derived analytical structure for further analysis (e.g., stability) and system design, which should be viewed as a bonus. This will certainly require nonlinear control theory and is open research issues.

Example 4.1 A fuzzy controller is constructed using the configuration given in Section 4.4.1. More specifically, it uses the following parameters: $\theta_1 = \theta_2 = 0.1$, $L_1 = L_2 = 1$, $b_1 = 1$, $b_2 = b_3 = 0$, and $b_4 = -1$. Also $E(k) \in [-1.2, 1.2]$ and $R(k) \in [-1.2, 1.2]$. Determine the analytical structure of this controller.

The variable $\Delta U_p(k)$ was first derived for this fuzzy controller. Table 4.3 showing the 16 $\Delta U_p(k)$ for IC1 when the parameter setting is general is used to illustrate this particular example. Putting the parameter values specified for this example into Table 4.3 will generate the 16 $\Delta U_p(k)$ for IC1 for this example.

The intervals of the input variables, $[-1.2, 1.2]$, were divided in the same way to create the following points in the intervals: $-1.2, -1.18, -1.16, \ldots, 1.18, 1.2$ (there were 121 such points in total for each variable). In the $E(k)$–$R(k)$ coordinate system, draw a line parallel to the $R(k)$ axis that goes through each of the 121 points of $E(k)$. Also, draw a line that is parallel to the $E(k)$ axis and passes through each of the 121 points of $R(k)$. These 242 lines formed a 2D grid of 121×121 (=14,641) points. A MATLAB program was developed to determine at each of the grid points (1) the ICs and (2) $\Delta U_p^{\min}(k)$ and $\Delta U_p^{\max}(k)$. Twenty-four ICs shown in Fig. 4.7 were found (points with the same gray level form an IC). For each IC, $\Delta U_p^{\min}(k)$ and $\Delta U_p^{\max}(k)$ are listed in Table 4.4. The MATLAB programs producing Fig. 4.7a and Table 4.4 are available for the reader to download. As a result, the analytical structure is readily derivable, which is $[\Delta U_p^{\min}(k) + \Delta U_p^{\max}(k)]/2$. One may notice that the ICs as well as their analytical structure expressions are symmetrical with respect to the origin, leading to similar symmetry of the controller output.

The reader may wonder how he/she can know for sure whether the analytical structure of a fuzzy controller derived in this chapter is correct? A more general and practical question is how can the reader validate the analytical structure of a fuzzy controller of his/her own configuration that he/she has derived using the method presented in this chapter? The easiest way for validation is to write a computer program using a programming language such as MATLAB, Mathematica, Maple, C, C++, Java, and so on. Part A of the program will implement the fuzzy controller as the blackbox controller by constructing it through the controller components (e.g., input and output fuzzy sets and fuzzy rules). Part B of the program will implement the derived input–output mathematical expressions of the same fuzzy controller. Then, compare the controller output values computed by parts A and B of the program for arbitrarily selected points in the input space (the quantity is chosen by the reader). If the two output values are always identical for the same input point, the derived input–output expressions are correct. Otherwise, there is something wrong with the derivation. This approach works for any T1 as well as T2 fuzzy controllers.

4.5 MAMDANI FUZZY PI AND PD CONTROLLERS—CONFIGURATION 2[2]

This configuration consists of T2 triangular input fuzzy sets, T1 singleton output fuzzy sets, Mamdani fuzzy rules, the center-of-sets type of reducer implemented via the modified interval weighted-average method, and the average defuzzifier (Table 4.1). The differences between this configuration and the configuration in Section 4.4 are the type of reducer implementation method and defuzzifier.

[2]Part of the material in this section is adapted from Du and Ying (2010; © 2010, IEEE).

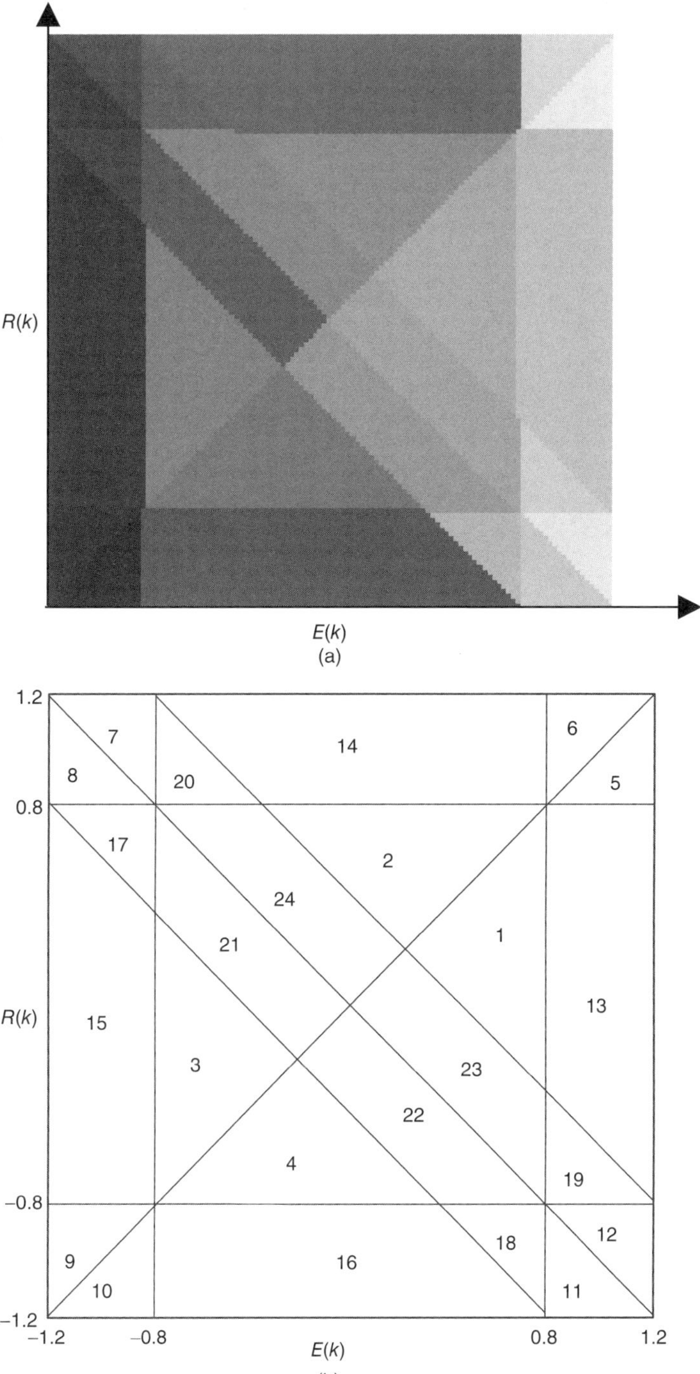

Figure 4.7 Division of the input space (a) ICs found by the MATLAB program and (b) numbering the ICs in (a).

TABLE 4.4 $\Delta U_p^{\min}(k)$ and $\Delta U_p^{\max}(k)$ for IC1–IC24 Shown in Fig. 4.7

IC No.	$\Delta U_p^{\min}(k)$	$\Delta U_p^{\max}(k)$
1	$\frac{-5E(k)-5R(k)+2}{10R(k)-22}$	$\frac{-5E(k)-5R(k)-2}{10R(k)-18}$
2	$\frac{-5E(k)-5R(k)+2}{10E(k)-22}$	$\frac{-5E(k)-5R(k)-2}{10E(k)-18}$
3	$\frac{5E(k)+5R(k)-2}{10R(k)+18}$	$\frac{5E(k)+5R(k)+2}{10R(k)+22}$
4	$\frac{5E(k)+5R(k)-2}{10E(k)+18}$	$\frac{5E(k)+5R(k)+2}{10E(k)+22}$
5	$\frac{-5E(k)-5R(k)+2}{10R(k)-22}$	1
6	$\frac{-5E(k)-5R(k)+2}{10E(k)-22}$	1
7	$\frac{-5E(k)+6}{5E(k)+5R(k)-10}$	$0.5R(k)+0.6$
8	$0.5E(k)-0.6$	$\frac{5R(k)+6}{5E(k)+5R(k)+10}$
9	-1	$\frac{5E(k)+5R(k)+2}{10R(k)+22}$
10	-1	$\frac{5E(k)+5R(k)+2}{10E(k)+22}$
11	$0.5R(k)-0.6$	$\frac{5E(k)+6}{5E(k)+5R(k)+10}$
12	$\frac{-5R(k)+6}{5E(k)+5R(k)-10}$	$0.5E(k)+0.6$
13	$\frac{-5E(k)-5R(k)+2}{10E(k)-22}$	$0.5R(k)+0.6$
14	$0.5E(k)-0.6$	$\frac{5E(k)+5R(k)+2}{10R(k)+22}$

TABLE 4.4 **(*Continued*)**

IC No.	$\Delta U_p^{\min}(k)$	$\Delta U_p^{\max}(k)$
15	$0.5R(k)-0.6$	$\dfrac{5E(k)+5R(k)+2}{10E(k)+22}$
16	$\dfrac{-5E(k)-5R(k)+2}{10R(k)-22}$	$0.5E(k)+0.6$
17	$0.5E(k)-0.6$	$\dfrac{5E(k)+5R(k)+2}{5R(k)+14}$
18	$0.5R(k)-0.6$	$\dfrac{5E(k)+5R(k)+2}{5E(k)+14}$
19	$\dfrac{-5E(k)-5R(k)+2}{5R(k)-14}$	$0.5E(k)+0.6$
20	$\dfrac{-5E(k)-5R(k)+2}{5E(k)-14}$	$0.5R(k)+0.6$
21	$\dfrac{5E(k)+5R(k)-2}{10R(k)+18}$	$\dfrac{5E(k)+5R(k)+2}{10R(k)+18}$
22	$\dfrac{5E(k)+5R(k)-2}{10E(k)+18}$	$\dfrac{5E(k)+5R(k)+2}{10E(k)+18}$
23	$\dfrac{-5E(k)-5R(k)+2}{10R(k)-18}$	$\dfrac{-5E(k)-5R(k)-2}{10R(k)-18}$
24	$\dfrac{-5E(k)-5R(k)+2}{10E(k)-18}$	$\dfrac{-5E(k)-5R(k)-2}{10E(k)-18}$

Let us discuss how to derive the analytical structure of a T2 fuzzy PI controller that is a little bit different from the one in the last section. The main difference is the type-reducer implementation—it uses the modified interval weighted-average algorithm instead of the interval weighted-average algorithm. The structure-deriving technique in the previous section can be fully utilized up to the step before type reduction. The center-of-sets type of reducer with the average defuzzifier produces

$$\Delta u(k) = k_{\Delta U}\,\Delta U(k) = \frac{1}{16}\sum_{p=1}^{16} k_{\Delta U}\,\Delta U_p(k) \tag{4.25}$$

Keeping Eq. (4.22) in mind, Eq. (4.25) can be written as

$$\Delta u(k) = \frac{1}{16}\sum_{p=1}^{16} k_{\Delta U} \frac{\varphi_1^p k_e e(k) + \varphi_2^p k_r r(k) + \varphi_3^p}{\psi_1^p k_e e(k) + \psi_2^p k_r r(k) + \psi_3^p} \tag{4.26}$$

Let

$$K_p(e(k),\ r(k)) = \frac{1}{16}\sum_{p=1}^{16} \frac{k_{\Delta U} k_r \varphi_2^p}{\psi_1^p k_e e(k) + \psi_2^p k_r r(k) + \psi_3^p}$$

$$K_i(e(k),\ r(k)) = \frac{1}{16}\sum_{p=1}^{16} \frac{k_{\Delta U} k_e \varphi_1^p}{\psi_1^p k_e e(k) + \psi_2^p k_r r(k) + \psi_3^p}$$

$$\delta(e(k),\ r(k)) = \frac{1}{16}\sum_{p=1}^{16} \frac{k_{\Delta U} \varphi_3^p}{\psi_1^p k_e e(k) + \psi_2^p k_r r(k) + \psi_3^p}$$

then

$$\Delta u(k) = K_p(e(k), r(k)) \cdot r(k) + K_i(e(k), r(k)) \cdot e(k) + \delta(e(k), r(k)) \tag{4.27}$$

which is a nonlinear PI controller in incremental form with variable proportional-gain $K_p(e(k),\ r(k))$, variable integral-gain $K_i(e(k),\ r(k))$, and variable control offset $\delta(e(k),\ r(k))$. The following conclusion is obvious.

THEOREM 4.2 The T2 fuzzy PI (or the corresponding PD) controller in this section is structurally equivalent to a nonlinear PI (or PD) controller with variable gains and a variable control offset.

When introducing the average center-of-sets type of reducer in Section 4.3, it was pointed out that it generated an output value that approximated the output value produced by the center-of-sets type of reducer using the interval weighted-average algorithm. As a demonstration, let us see the simple comparison example below.

Example 4.2 We construct two fuzzy controllers using the settings in Section 4.1. One fuzzy controller uses the center-of-sets type reducer implemented by the modified interval weighted-average method and the average defuzzifier, whereas the other employs the center-of-sets type of reducer involving the interval weighted-average algorithm plus the centroid defuzzifier. Their parameter settings are $\theta_1 = \theta_2 = 0.1,\ L_1 = L_2 = 1,\ b_1 = 1,\ b_2 = b_3 = 0,$ and $b_4 = -1$. Our goal is to compare output of these two fuzzy controllers.

The assumptions $b_1 = 1,\ b_2 = b_3 = 0,$ and $b_4 = -1$ are reasonable. Referring to Fig. 4.4, the controller output should be increased if the system output is in the region governed by rule R^1 and decreased in the area managed by rule R^4. It is sensible that the magnitudes of the increment and decrement are equal. In the regions related to the other two rules, no change to the controller output is necessary. One observes from Fig. 4.8 that only a marginal difference exists in controller output between these two fuzzy PI controllers whose configurations are exactly the same except the type of reducer implementation and defuzzifier. The maximum difference is 2.83%. The numbers and shapes of the ICs (not shown) are somewhat different, which are not important.

We wish to differentiate the role that the analytical structure plays from that of the 3D control surface does (e.g., Fig. 4.8) because some people might feel that viewing the 3D surface is a good way to analyze or design the fuzzy controller, T1 or T2, and that the analytical structure does not offer much more value. This is a misconception. While a 3D control surface can be conveniently generated by a computer program without knowing the analytical structure, the surface can play only a rather limited role. First of all, it works only for a fuzzy controller with one or two input variables. In contrast, analytical structure derivation methodology can, in principle, be extended to fuzzy controllers with three or more input variables. (The caveat for doing the three-variable cases is that they will be significantly more challenging than doing the two-variable cases because the ICs will be three-dimensional. Involving more than three variables will be even harder, although theoretically still possible.) Second, 3D surface's usefulness is mainly to permit gross visual inspection of the controller output, which might be helpful *after* the fuzzy controller is constructed. The 3D surface does not, and cannot, expose the underlying control algorithm (i.e., the mathematical input–output relationship) responsible for the surface. The only exception is that if the surface is a plane, the computer program (not human vision) will be able to determine this to be the case. Nevertheless, that the surface is a plane means a linear controller; but (almost) all the T2 fuzzy controllers are nonlinear controllers. A rigorous controller analysis or design cannot be meaningfully performed based on viewing the 3D surface. This is true even when the surface is a plane. A case in point is the linear PI or PD controller whose control surface is a plane. The slope of the plane is determined by the controller's two gain parameters. Designing the PI or PD controller means finding suitable gain values. Hundreds of research papers have been published, and are still being published, to develop new and better methods to compute the gain values, none of which utilizes 3D control plane viewing. The nonlinear control theory can be applied to analyzing or designing the fuzzy controllers using the analytical structures and the process is rigorous.

Another difference between these two fuzzy PI controllers is the way that the gains vary. Look at the central area formed by $[-L_1 + P_1, L_1 - P_1] \times [-L_2 + P_2, L_2 - P_2]$. According to Fig. 4.6, the variable gains of the fuzzy PI controller with the center-of-sets type of reducer implemented via the modified interval weighted-average method plus the average defuzzifier is symmetrical about the $E(k)$ axis, the $R(k)$ axis, and the origin of the input space, whereas the fuzzy PI controller with the center-of-sets type of reducer realized by the interval weighted-average method plus the centroid defuzzifier is symmetrical only about the origin (see Fig. 4.7). From a control standpoint, all these symmetries are sensible and useful for different control objectives.

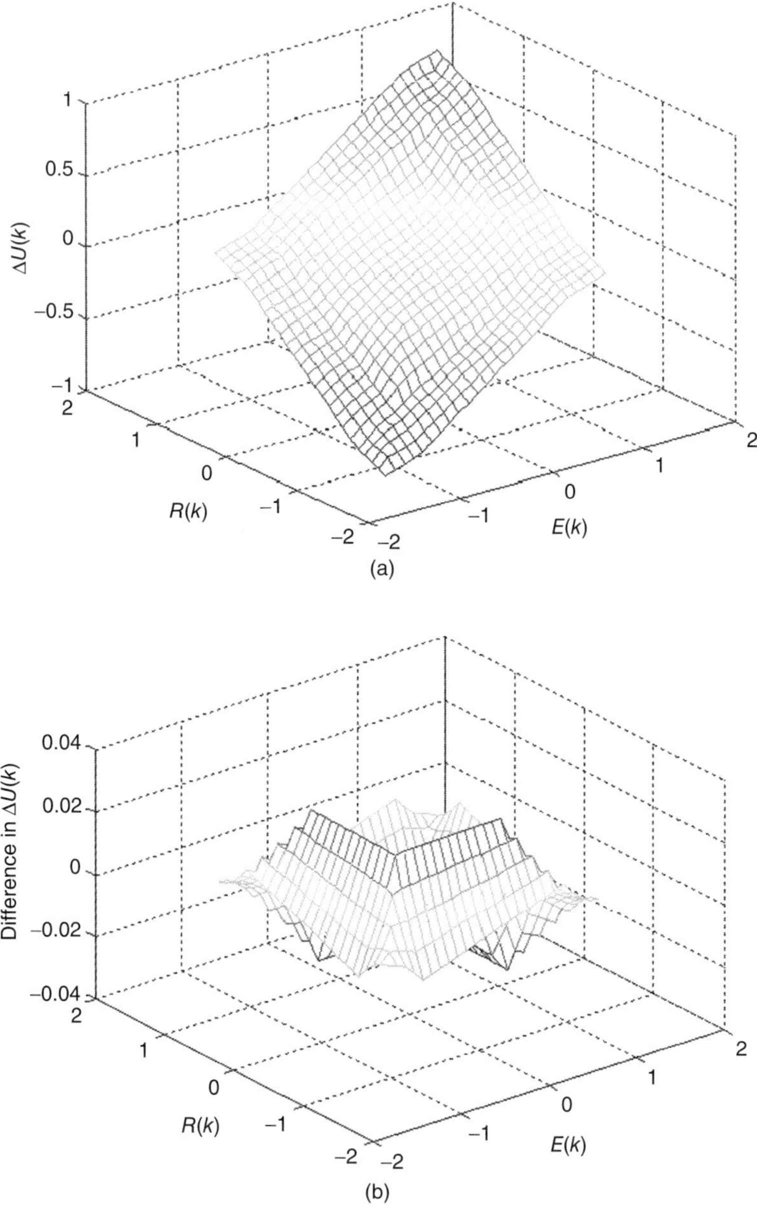

Figure 4.8 (a) Output of the T2 fuzzy PI controller using the center-of-sets type reducer implemented by the interval weighted-average method and the centroid defuzzifier. (b) Output difference between the controller uses the center-of-sets type reducer implemented by the modified interval weighted-average method and the average defuzzifier, and the controller uses the center-of-sets type reducer implemented by the interval weighted-average method and the centroid defuzzifier.

For a control developer wanting to explicitly know and analyze the analytical structure of his/her T2 fuzzy PI (or PD) control system, the center-of-sets type of reducer implemented by the modified interval weighted-average method plus the average defuzzifier may be a better choice because:

1. The analytical structure of the fuzzy PI and PD controllers can be derived through algebraic manipulations without requiring numerical calculation assistance from a computer. In contrast, a computer program must be made to aid the derivation process involving the interval weighted-average center-of-sets type of reducer.
2. The IC divisions are analytically derived and hence the resulting boundaries are precise. This, however, is not the case for the boundaries of the final ICs in the previous section because they are found through numerical calculations, leading to approximate boundaries whose errors become smaller as the number of points used in the calculations increases.
3. The IC divisions and analytical structure expressions can be determined in a general setting (the parameters do not have values), which is not true for the interval weighted-average center-of-sets type of reducer. For the latter controllers, the analytical structure expressions and the associated computed IC boundaries are valid only for the particular parameter values used in the calculations. For a different set of parameter values, the program has to be rerun to find ICs and the analytical structure.

Example 4.3 Find the variable proportional-gain $K_p(e(k), r(k))$of the fuzzy PI controller using the average center-of-sets type of reducer for IC1 in Fig. 4.6b.

Variable $K_p(e(k), r(k))$ is the coefficient of the $r(k)$. Therefore, one needs to first add the 16 coefficients of $r(k)$ in the numerators in Table 4.3. After algebraic simplifications, one obtains the following:

$$
\begin{aligned}
K_p(e(k),\ r(k)) = {} & \frac{k_{\Delta u} k_r (b_1 - b_2) L_1}{16 L_2} \\
& \times \left[\frac{2}{2L_1 - k_e e(k)} + \frac{1}{2L_1(1+\theta_2) - k_e e(k)} \right. \\
& + \frac{1}{2L_1(1-\theta_1) - k_e e(k)} + \frac{1}{2L_1(1-\theta_2) - k_e e(k)} \\
& + \frac{1}{2L_1(1+\theta_1) - k_e e(k)} + \frac{1}{4L_1(1+\theta_1+\theta_2) - 2k_e e(k)} \\
& + \frac{1}{4L_1(1-\theta_1+\theta_2) - 2k_e e(k)} + \frac{1}{4L_1(1+\theta_1-\theta_2) - 2k_e e(k)} \\
& \left. + \frac{1}{4L_1\left(1-\theta_1-\theta_2\right) - 2k_e e(k)} \right]
\end{aligned}
$$

which is the explicit variable proportional-gain expression sought. The result is consistent with Eq. (4.27) and Theorem 4.2.

Example 4.4 Find the variable proportional-gain $K_p(e(k), r(k))$ of the fuzzy PI controller using the average center-of-sets type of reducer for IC1 to IC16 in Fig. 4.6b when the parameter setting is $L_1 = L_2 = L$, $\theta_1 = \theta_2 = \theta$, $b_1 = b$, $b_4 = -b$, and $b_2 = b_3 = 0$.

Do again what we did in Example 4.3, but this time for all the 16 ICs instead of IC1 only. After simplifying the intermediate results, the final results are summarized in Table 4.5, which shows the variable proportional gain for the entire input space as opposed to just in one IC (e.g., Example 4.3). If you understand the method presented in this section and apply it correctly, you will have no problem arriving at the same expressions as those in Table 4.5 after some effort is made. The table provides a concrete case to illustrate Eq. (4.27) and Theorem 4.2.

4.6 MAMDANI FUZZY PI AND PD CONTROLLERS—CONFIGURATION 3[3]

4.6.1 Fuzzy PI Controller Configuration

We now explore the analytical structure of the fuzzy PI and PD controllers that have the same configurations as in Section 4.5 (i.e., T2 triangular input fuzzy sets, Mamdani fuzzy rules, the centroid type of reducer implemented via the modified interval weighted-average method, and the average defuzzifier) except they adopt T2 singleton fuzzy sets as output fuzzy sets as opposed to the T1 singleton fuzzy sets. The output fuzzy sets $\tilde{B}_m$ are shown in Fig. 4.9 where $0 \le \beta_m \le \alpha_m \le 1$ $(m = 1, 2, 3, 4)$.

The firing interval for each rule is

$$\mu_{\tilde{B}_1}(\Delta U) = [\underline{\mu}_{\tilde{B}_1}, \overline{\mu}_{\tilde{B}_1}] = [\min(\underline{\mu}_{\tilde{A}_{11}}, \underline{\mu}_{\tilde{A}_{21}}, \beta_1), \ \min(\overline{\mu}_{\tilde{A}_{11}}, \overline{\mu}_{\tilde{A}_{21}}, \alpha_1)] \tag{4.28}$$

$$\mu_{\tilde{B}_2}(\Delta U) = [\underline{\mu}_{\tilde{B}_2}, \overline{\mu}_{\tilde{B}_2}] = [\min(\underline{\mu}_{\tilde{A}_{11}}, \underline{\mu}_{\tilde{A}_{22}}, \beta_2), \ \min(\overline{\mu}_{\tilde{A}_{11}}, \overline{\mu}_{\tilde{A}_{22}}, \alpha_2)] \tag{4.29}$$

$$\mu_{\tilde{B}_3}(\Delta U) = [\underline{\mu}_{\tilde{B}_3}, \overline{\mu}_{\tilde{B}_3}] = [\min(\underline{\mu}_{\tilde{A}_{12}}, \underline{\mu}_{\tilde{A}_{21}}, \beta_3), \ \min(\overline{\mu}_{\tilde{A}_{12}}, \overline{\mu}_{\tilde{A}_{21}}, \alpha_3)] \tag{4.30}$$

$$\mu_{\tilde{B}_4}(\Delta U) = [\underline{\mu}_{\tilde{B}_4}, \overline{\mu}_{\tilde{B}_4}] = [\min(\underline{\mu}_{\tilde{A}_{12}}, \underline{\mu}_{\tilde{A}_{22}}, \beta_4), \ \min(\overline{\mu}_{\tilde{A}_{12}}, \overline{\mu}_{\tilde{A}_{22}}, \alpha_4)] \tag{4.31}$$

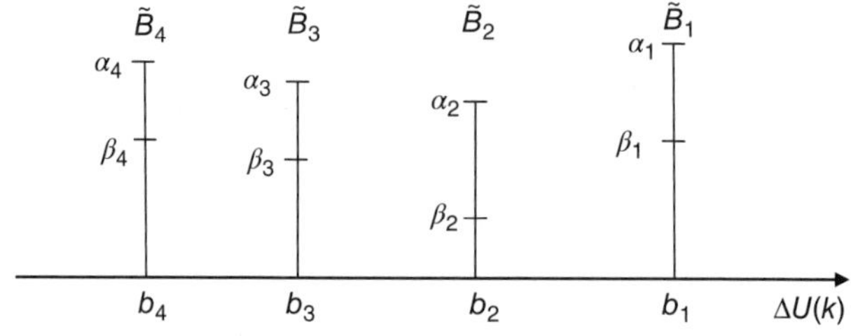

Figure 4.9 Primary membership functions of the T2 singleton fuzzy sets for $\Delta U(k)$.

[3]Part of the material in this section is adapted from Du and Ying (2008; © 2008, IEEE).

TABLE 4.5 Variable Proportional-gain $K_p(e(k), r(k))$ for Example 4.4

IC No.	$K_p(e(k), r(k))$
1 and 3	$\frac{k_{\Delta U}k_r b}{32}\left[\frac{4}{2L(1-\theta)-k_e\lvert e(k)\rvert}+\frac{4}{2L(1+\theta)-k_e\lvert e(k)\rvert}+\frac{1}{2L(1-2\theta)-k_e\lvert e(k)\rvert}+\frac{1}{2L(1+2\theta)-k_e\lvert e(k)\rvert}+\frac{6}{2L-k_e\lvert e(k)\rvert}\right]$
2 and 4	$\frac{k_{\Delta U}k_r b}{32}\left[\frac{4}{2L(1-\theta)-k_r\lvert r(k)\rvert}+\frac{4}{2L(1+\theta)-k_r\lvert r(k)\rvert}+\frac{1}{2L(1-2\theta)-k_r\lvert r(k)\rvert}+\frac{1}{2L(1+2\theta)-k_r\lvert r(k)\rvert}+\frac{6}{2L-k_r\lvert r(k)\rvert}\right]$
5 and 9	$\frac{k_{\Delta U}k_r b}{32}\left[\frac{2}{3L+2\theta L-2k_e\lvert e(k)\rvert+k_r\lvert r(k)\rvert}+\frac{4}{2L-k_e\lvert e(k)\rvert+k_r\lvert r(k)\rvert}+\frac{2}{L-2\theta L+k_r\lvert r(k)\rvert}+\frac{4}{3L+2\theta L-k_e\lvert e(k)\rvert}+\frac{1}{2L+2\theta L-k_e\lvert e(k)\rvert}+\frac{1}{L}\right]$
6, 7, 10 and 11	$\frac{k_{\Delta U}k_r b}{32}\left[\frac{2}{2L+k_e\lvert e(k)\rvert-k_r\lvert r(k)\rvert}+\frac{2}{4L+4\theta L-k_e\lvert e(k)\rvert-k_r\lvert r(k)\rvert}+\frac{2}{5L+6\theta L-k_e\lvert e(k)\rvert-2k_r\lvert r(k)\rvert}+\frac{4}{3L+2\theta L-k_r\lvert r(k)\rvert}+\frac{2}{2L+2\theta L-k_r\lvert r(k)\rvert}+\frac{2}{3L+2\theta L+k_e\lvert e(k)\rvert-2k_r\lvert r(k)\rvert}\right]$
8 and 12	$\frac{k_{\Delta U}k_r b}{32}\left[\frac{2}{5L+6\theta L-2k_e\lvert e(k)\rvert-k_r\lvert r(k)\rvert}+\frac{4}{4L+4\theta L-k_e\lvert e(k)\rvert-k_r\lvert r(k)\rvert}+\frac{4}{3L+2\theta L-k_e\lvert e(k)\rvert}+\frac{1}{2L+2\theta L-k_e\lvert e(k)\rvert}+\frac{2}{3L+2\theta L-k_r\lvert r(k)\rvert}+\frac{1}{L}\right]$
13 and 15	$\frac{k_{\Delta U}k_r b}{32}\left[\frac{4}{3L-2\theta L-k_e\lvert e(k)\rvert}+\frac{8}{3L+2\theta L-k_e\lvert e(k)\rvert}+\frac{2}{2L+2\theta L-k_e\lvert e(k)\rvert}+\frac{4}{3L+6\theta L-k_e\lvert e(k)\rvert}+\frac{1}{2L+4\theta L-k_e\lvert e(k)\rvert}+\frac{1}{2L-k_e\lvert e(k)\rvert}+\frac{1}{L(1+2\theta)}+\frac{1}{L(1-2\theta)}+\frac{2}{L}\right]$
14 and 16	$\frac{k_{\Delta U}k_r b}{32}\left[\frac{2}{3L-2\theta L-k_r\lvert r(k)\rvert}+\frac{4}{3L+2\theta L-k_r\lvert r(k)\rvert}+\frac{2}{2L+2\theta L-k_r\lvert r(k)\rvert}+\frac{2}{3L+6\theta L-k_r\lvert r(k)\rvert}+\frac{1}{2L+4\theta L-k_r\lvert r(k)\rvert}+\frac{1}{2L-k_r\lvert r(k)\rvert}\right]$

According to Eq. (4.16) and the centroid type of reducer, the result after type reduction and average defuzzification is

$$\Delta U_p(k) = \frac{\sum_{s=1}^{4}\mu_{\widetilde{B}_s} b_s}{\sum_{s=1}^{4}\mu_{\widetilde{B}_s}} \qquad p = 1, \cdots, 16 \tag{4.32}$$

where $\mu_{\widetilde{B}_s} = \underline{\mu}_{\widetilde{B}_s}$ or $\overline{\mu}_{\widetilde{B}_s}$ and b_s is the nonzero location of $\widetilde{B}_s$.

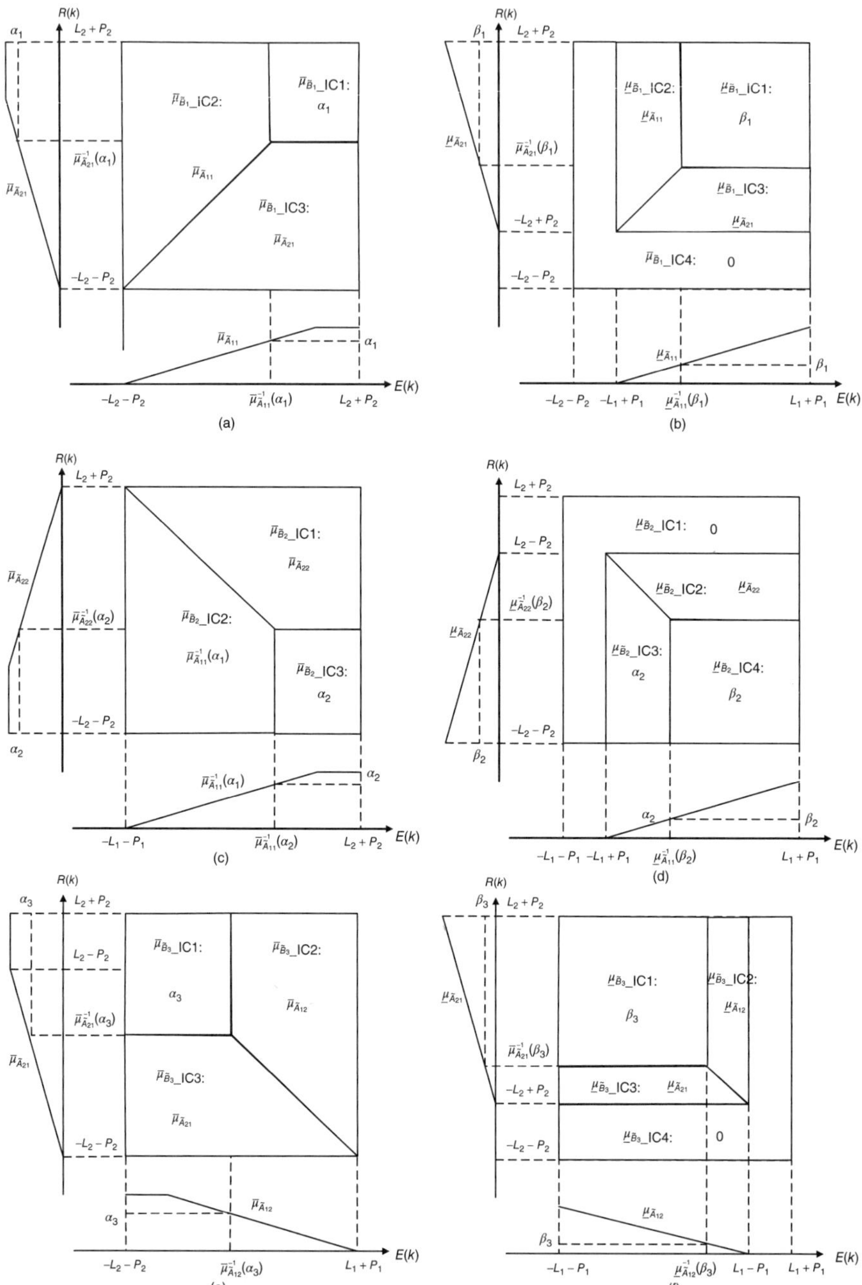

Figure 4.10 ICs for the upper and lower bounds of the four firing intervals. Figures (a)–(h) are corresponding to $\overline{\mu}_{\tilde{B}_1}$, $\underline{\mu}_{\tilde{B}_1}$, $\overline{\mu}_{\tilde{B}_2}$, $\underline{\mu}_{\tilde{B}_2}$, $\overline{\mu}_{\tilde{B}_3}$, $\underline{\mu}_{\tilde{B}_3}$, $\overline{\mu}_{\tilde{B}_4}$, and $\underline{\mu}_{\tilde{B}_4}$ in Eqs. (4.29)–(4.32), respectively.

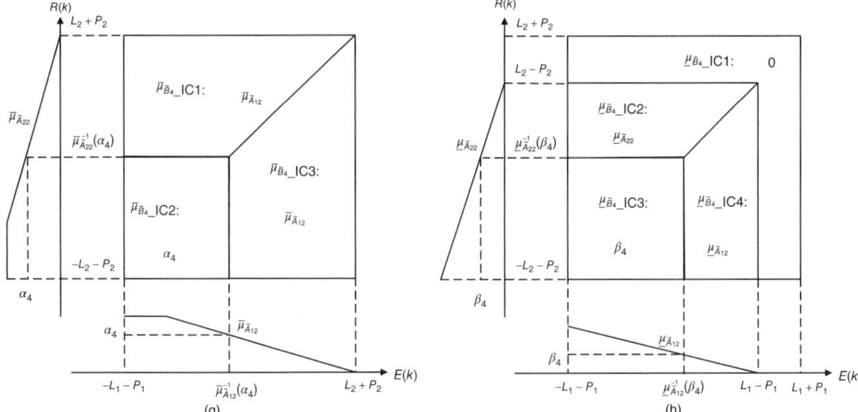

Figure 4.10 (*Continued*)

4.6.2 Method for Deriving the Analytical Structure

Similar to what was done with Eqs. (4.18)–(4.21) in Section 4.4 to divide the input space into ICs, we need to use Eqs. (4.29)–(4.32) to divide the input space into ICs. The results are shown in Fig. 4.10 where $\overline{\mu}_{\tilde{A}_{11}}^{-1}$, $\underline{\mu}_{\tilde{A}_{11}}^{-1}$, $\overline{\mu}_{\tilde{A}_{12}}^{-1}$, $\underline{\mu}_{\tilde{A}_{12}}^{-1}$, $\overline{\mu}_{\tilde{A}_{21}}^{-1}$, $\underline{\mu}_{\tilde{A}_{21}}^{-1}$, $\overline{\mu}_{\tilde{A}_{22}}^{-1}$, and $\underline{\mu}_{\tilde{A}_{22}}^{-1}$ are the inverse functions of the linear part of the corresponding membership functions. For instance, the result in Fig. 4.10a is a division for $\overline{\mu}_{\tilde{B}_1} = \min(\overline{\mu}_{\tilde{A}_{11}}, \overline{\mu}_{\tilde{A}_{21}}, \alpha_1)$. When $E(k)$ and $R(k)$ are in the region labeled as $\overline{\mu}_{\tilde{B}_1}$-IC1 in Fig. 4.10a, $\overline{\mu}_{\tilde{A}_{11}}$ and $\overline{\mu}_{\tilde{A}_{21}}$ are always larger than α_1. So $\overline{\mu}_{\tilde{B}_1} = \alpha_1$. Similarly, in the region labeled as $\overline{\mu}_{\tilde{B}_1}$-IC2 the value of the membership function $\overline{\mu}_{\tilde{A}_{11}}$ is always smaller than those of $\overline{\mu}_{\tilde{A}_{21}}$ and α_1. Thus, $\overline{\mu}_{\tilde{B}_1} = \overline{\mu}_{\tilde{A}_{11}}$. And in the region of $\overline{\mu}_{\tilde{B}_1}$-IC3, the value of the membership function $\overline{\mu}_{\tilde{A}_{21}}$ is always smaller than those of $\overline{\mu}_{\tilde{A}_{11}}$ and α_1, leading to $\overline{\mu}_{\tilde{B}_1} = \overline{\mu}_{\tilde{A}_{21}}$. In the common boundary of $\overline{\mu}_{\tilde{B}_1}$-IC2 and $\overline{\mu}_{\tilde{B}_1}$-IC3, the value of $\overline{\mu}_{\tilde{A}_{11}}$ is equal to that of $\overline{\mu}_{\tilde{A}_{21}}$, and both are smaller than α_1.

Like before, these eight figures must be superimposed so that the four rules are counted for at the same time. The number and shape of the final ICs after the superimposition depend on the shapes of the input fuzzy sets. Without loss of generality, we assume $L_1 > L_2$, $L_1 \times \theta_1 = L_2 \times \theta_2$, $\alpha_1 > \alpha_2$, $\alpha_3 > \alpha_4$, $\beta_1 > \beta_2$, and $\beta_3 > \beta_4$ and show the superimposition result—in total 83 ICs (Fig. 4.11). For brevity, the upper and lower limits of the four firing intervals for IC1–IC30 are listed in Table 4.6.

For each IC, putting the eight membership functions of the IC into Eq. (4.33) we obtain the analytical structure of the fuzzy PI controller for that IC. Like the other T2 fuzzy controllers in Sections 4.4 and 4.5, for every IC, $\Delta U_p(k)$ consists of a series of fractions, each of which shares structure similar to Eq. (4.22). As an example, Table 4.7 shows the analytical expressions of $\Delta U_p(k)$ for IC1.

TABLE 4.6 Upper and Lower Limits of Firing Intervals for Four Rules for IC1–IC30

	R^1		R^2		R^3		R^4	
IC No.	$\bar{\mu}_{\tilde{B}_1}$	$\underline{\mu}_{\tilde{B}_1}$	$\bar{\mu}_{\tilde{B}_3}$	$\underline{\mu}_{\tilde{B}_3}$	$\bar{\mu}_{\tilde{B}_4}$	$\underline{\mu}_{\tilde{B}_4}$	$\bar{\mu}_{\tilde{B}_2}$	$\underline{\mu}_{\tilde{B}_2}$
1	$\bar{\mu}_{\tilde{A}_{11}}$	0	α_3	β_3	$\underline{\mu}_{\tilde{A}_{21}}$	0	$\bar{\mu}_{\tilde{A}_{11}}$	0
2	$\bar{\mu}_{\tilde{A}_{11}}$	0	α_3	β_3	$\underline{\mu}_{\tilde{A}_{21}}$	0	$\underline{\mu}_{\tilde{A}_{21}}$	0
3	$\bar{\mu}_{\tilde{A}_{11}}$	$\underline{\mu}_{\tilde{A}_{11}}$	α_3	β_3	$\underline{\mu}_{\tilde{A}_{21}}$	0	$\underline{\mu}_{\tilde{A}_{21}}$	0
4	$\bar{\mu}_{\tilde{A}_{11}}$	$\underline{\mu}_{\tilde{A}_{11}}$	α_3	β_3	$\underline{\mu}_{\tilde{A}_{21}}$	0	$\bar{\mu}_{\tilde{A}_{11}}$	0
5	$\bar{\mu}_{\tilde{A}_{11}}$	$\underline{\mu}_{\tilde{A}_{11}}$	$\bar{\mu}_{\tilde{A}_{12}}$	β_3	$\underline{\mu}_{\tilde{A}_{21}}$	0	$\underline{\mu}_{\tilde{A}_{21}}$	0
6	$\bar{\mu}_{\tilde{A}_{11}}$	β_1	$\bar{\mu}_{\tilde{A}_{12}}$	β_3	$\underline{\mu}_{\tilde{A}_{21}}$	0	$\underline{\mu}_{\tilde{A}_{21}}$	0
7	$\bar{\mu}_{\tilde{A}_{11}}$	β_1	$\bar{\mu}_{\tilde{A}_{12}}$	$\underline{\mu}_{\tilde{A}_{12}}$	$\underline{\mu}_{\tilde{A}_{21}}$	0	$\underline{\mu}_{\tilde{A}_{21}}$	0
8	α_1	β_1	$\bar{\mu}_{\tilde{A}_{12}}$	$\underline{\mu}_{\tilde{A}_{12}}$	$\underline{\mu}_{\tilde{A}_{21}}$	0	$\underline{\mu}_{\tilde{A}_{21}}$	0
9	α_1	β_1	$\bar{\mu}_{\tilde{A}_{12}}$	$\underline{\mu}_{\tilde{A}_{12}}$	$\bar{\mu}_{\tilde{A}_{12}}$	0	$\underline{\mu}_{\tilde{A}_{21}}$	0
10	α_1	β_1	$\bar{\mu}_{\tilde{A}_{12}}$	0	$\underline{\mu}_{\tilde{A}_{21}}$	0	$\underline{\mu}_{\tilde{A}_{21}}$	0
11	α_1	β_1	$\bar{\mu}_{\tilde{A}_{12}}$	0	$\bar{\mu}_{\tilde{A}_{12}}$	0	$\underline{\mu}_{\tilde{A}_{21}}$	0
12	$\bar{\mu}_{\tilde{A}_{11}}$	0	α_3	β_3	$\underline{\mu}_{\tilde{A}_{21}}$	$\underline{\mu}_{\tilde{A}_{22}}$	$\bar{\mu}_{\tilde{A}_{11}}$	0
13	$\bar{\mu}_{\tilde{A}_{11}}$	0	$\bar{\mu}_{\tilde{A}_{21}}$	β_3	$\underline{\mu}_{\tilde{A}_{21}}$	$\underline{\mu}_{\tilde{A}_{22}}$	$\bar{\mu}_{\tilde{A}_{11}}$	0
14	$\bar{\mu}_{\tilde{A}_{11}}$	$\underline{\mu}_{\tilde{A}_{11}}$	α_3	β_3	$\underline{\mu}_{\tilde{A}_{21}}$	$\underline{\mu}_{\tilde{A}_{22}}$	$\bar{\mu}_{\tilde{A}_{11}}$	$\underline{\mu}_{\tilde{A}_{11}}$
15	$\bar{\mu}_{\tilde{A}_{11}}$	$\underline{\mu}_{\tilde{A}_{11}}$	$\bar{\mu}_{\tilde{A}_{21}}$	β_3	$\underline{\mu}_{\tilde{A}_{21}}$	$\underline{\mu}_{\tilde{A}_{22}}$	$\bar{\mu}_{\tilde{A}_{11}}$	$\underline{\mu}_{\tilde{A}_{11}}$
16	$\bar{\mu}_{\tilde{A}_{11}}$	$\underline{\mu}_{\tilde{A}_{11}}$	α_3	β_3	$\underline{\mu}_{\tilde{A}_{21}}$	$\underline{\mu}_{\tilde{A}_{22}}$	$\bar{\mu}_{\tilde{A}_{11}}$	$\underline{\mu}_{\tilde{A}_{22}}$
17	$\bar{\mu}_{\tilde{A}_{11}}$	$\underline{\mu}_{\tilde{A}_{11}}$	$\bar{\mu}_{\tilde{A}_{12}}$	β_3	$\underline{\mu}_{\tilde{A}_{21}}$	$\underline{\mu}_{\tilde{A}_{22}}$	$\bar{\mu}_{\tilde{A}_{11}}$	$\underline{\mu}_{\tilde{A}_{22}}$
18	$\bar{\mu}_{\tilde{A}_{11}}$	$\underline{\mu}_{\tilde{A}_{11}}$	$\bar{\mu}_{\tilde{A}_{12}}$	β_3	$\underline{\mu}_{\tilde{A}_{21}}$	$\underline{\mu}_{\tilde{A}_{22}}$	$\underline{\mu}_{\tilde{A}_{21}}$	$\underline{\mu}_{\tilde{A}_{22}}$
19	$\bar{\mu}_{\tilde{A}_{11}}$	β_1	$\bar{\mu}_{\tilde{A}_{12}}$	β_3	$\underline{\mu}_{\tilde{A}_{21}}$	$\underline{\mu}_{\tilde{A}_{22}}$	$\underline{\mu}_{\tilde{A}_{21}}$	$\underline{\mu}_{\tilde{A}_{22}}$
20	$\bar{\mu}_{\tilde{A}_{11}}$	β_1	$\bar{\mu}_{\tilde{A}_{12}}$	$\underline{\mu}_{\tilde{A}_{12}}$	$\underline{\mu}_{\tilde{A}_{21}}$	$\underline{\mu}_{\tilde{A}_{22}}$	$\underline{\mu}_{\tilde{A}_{21}}$	$\underline{\mu}_{\tilde{A}_{22}}$
21	$\bar{\mu}_{\tilde{A}_{11}}$	β_1	$\bar{\mu}_{\tilde{A}_{12}}$	$\underline{\mu}_{\tilde{A}_{12}}$	$\bar{\mu}_{\tilde{A}_{12}}$	$\underline{\mu}_{\tilde{A}_{22}}$	$\underline{\mu}_{\tilde{A}_{21}}$	$\underline{\mu}_{\tilde{A}_{22}}$
22	$\bar{\mu}_{\tilde{A}_{21}}$	β_1	$\bar{\mu}_{\tilde{A}_{12}}$	$\underline{\mu}_{\tilde{A}_{12}}$	$\bar{\mu}_{\tilde{A}_{12}}$	$\underline{\mu}_{\tilde{A}_{12}}$	$\underline{\mu}_{\tilde{A}_{21}}$	$\underline{\mu}_{\tilde{A}_{22}}$
23	α_1	β_1	$\bar{\mu}_{\tilde{A}_{12}}$	$\underline{\mu}_{\tilde{A}_{12}}$	$\bar{\mu}_{\tilde{A}_{12}}$	$\underline{\mu}_{\tilde{A}_{22}}$	$\underline{\mu}_{\tilde{A}_{21}}$	$\underline{\mu}_{\tilde{A}_{22}}$
24	α_1	β_1	$\bar{\mu}_{\tilde{A}_{12}}$	$\underline{\mu}_{\tilde{A}_{12}}$	$\bar{\mu}_{\tilde{A}_{12}}$	$\underline{\mu}_{\tilde{A}_{12}}$	$\underline{\mu}_{\tilde{A}_{21}}$	$\underline{\mu}_{\tilde{A}_{22}}$
25	α_1	β_1	$\bar{\mu}_{\tilde{A}_{12}}$	0	$\bar{\mu}_{\tilde{A}_{12}}$	0	$\underline{\mu}_{\tilde{A}_{21}}$	$\underline{\mu}_{\tilde{A}_{22}}$
26	$\bar{\mu}_{\tilde{A}_{21}}$	β_1	$\bar{\mu}_{\tilde{A}_{12}}$	0	$\bar{\mu}_{\tilde{A}_{12}}$	0	$\underline{\mu}_{\tilde{A}_{21}}$	$\underline{\mu}_{\tilde{A}_{22}}$
27	$\bar{\mu}_{\tilde{A}_{11}}$	0	$\bar{\mu}_{\tilde{A}_{21}}$	β_3	$\underline{\mu}_{\tilde{A}_{21}}$	β_4	$\bar{\mu}_{\tilde{A}_{11}}$	0
28	$\bar{\mu}_{\tilde{A}_{11}}$	$\underline{\mu}_{\tilde{A}_{11}}$	$\bar{\mu}_{\tilde{A}_{21}}$	β_3	$\underline{\mu}_{\tilde{A}_{21}}$	β_4	$\bar{\mu}_{\tilde{A}_{11}}$	$\underline{\mu}_{\tilde{A}_{11}}$
29	$\bar{\mu}_{\tilde{A}_{11}}$	$\underline{\mu}_{\tilde{A}_{11}}$	$\bar{\mu}_{\tilde{A}_{12}}$	β_3	$\underline{\mu}_{\tilde{A}_{21}}$	β_4	$\bar{\mu}_{\tilde{A}_{11}}$	$\underline{\mu}_{\tilde{A}_{22}}$
30	$\bar{\mu}_{\tilde{A}_{11}}$	$\underline{\mu}_{\tilde{A}_{11}}$	$\bar{\mu}_{\tilde{A}_{12}}$	β_3	$\underline{\mu}_{\tilde{A}_{21}}$	β_4	$\bar{\mu}_{\tilde{A}_{11}}$	β_2

TABLE 4.7 $\Delta U_p(k)$ for IC1

p	$\Delta U_p(k)$
1	$\dfrac{L_2\left(b_1+b_2\right)E(k)-L_1b_4R(k)+2L_1L_2\left[\left(\theta_1+0.5\right)\left(b_1+b_2\right)+\alpha_3b_3+\left(\theta_2+0.5\right)b_4\right]}{2L_2E(k)-L_1R(k)+2L_1L_2\left(2\theta_1+\theta_2+\alpha_3+1.5\right)}$
2	$\dfrac{\left(b_1+b_2\right)E(k)+2L_1\left[\left(\theta_1+0.5\right)\left(b_1+b_2\right)+\alpha_3b_3\right]}{2E(k)+2L_1\left(2\theta_1+\alpha_3+1\right)}$
3	$\dfrac{L_2\left(b_1+b_2\right)E(k)-L_1b_4R(k)+2L_1L_2\left[\left(\theta_1+0.5\right)\left(b_1+b_2\right)+\beta_3b_3+\left(\theta_2+0.5\right)b_4\right]}{2L_2E(k)-L_1R(k)+2L_1L_2\left(2\theta_1+\theta_2+\beta_3+1.5\right)}$
4	$\dfrac{\left(b_1+b_2\right)E(k)+2L_1\left[\left(\theta_1+0.5\right)\left(b_1+b_2\right)+\beta_3b_3\right]}{2E(k)+2L_1\left(2\theta_1+\beta_3+1\right)}$
5	$\dfrac{L_2b_1E(k)-L_1b_4R(k)+2L_1L_2\left[\left(\theta_1+0.5\right)b_1+\alpha_3b_3+\left(\theta_2+0.5\right)b_4\right]}{L_2E(k)-L_1R(k)+2L_1L_2\left(\theta_1+\theta_2+\alpha_3+1\right)}$
6	$\dfrac{b_1E(k)+2L_1\left[\left(\theta_1+0.5\right)b_1+\alpha_3b_3\right]}{E(k)+2L_1\left(\theta_1+\alpha_3+0.5\right)}$
7	$\dfrac{L_2b_1E(k)-L_1b_4R(k)+2L_1L_2\left[\left(\theta_1+0.5\right)b_1+\beta_3b_3+\left(\theta_2+0.5\right)b_4\right]}{L_2E(k)-L_1R(k)+2L_1L_2\left(\theta_1+\theta_2+\beta_3+1\right)}$
8	$\dfrac{b_1E(k)+2L_1\left[\left(\theta_1+0.5\right)b_1+\beta_3b_3\right]}{E(k)+2L_1\left(\theta_1+\beta_3+0.5\right)}$
9	$\dfrac{L_2b_2E(k)-L_1b_4R(k)+2L_1L_2\left[\left(\theta_1+0.5\right)b_2+\alpha_3b_3+\left(\theta_2+0.5\right)b_4\right]}{L_2E(k)-L_1R(k)+2L_1L_2\left(\theta_1+\theta_2+\alpha_3+1\right)}$
10	$\dfrac{b_2E(k)+2L_1\left[\left(\theta_1+0.5\right)b_2+\alpha_3b_3\right]}{E(k)+2L_1\left(\theta_1+\alpha_3+0.5\right)}$
11	$\dfrac{L_2b_2E(k)-L_1b_4R(k)+2L_1L_2\left[\left(\theta_1+0.5\right)b_2+\beta_3b_3+\left(\theta_2+0.5\right)b_4\right]}{L_2E(k)-L_1R(k)+2L_1L_2\left(\theta_1+\theta_2+\beta_3+1\right)}$

(*continued*)

TABLE 4.7 ***(Continued)***

p	$\Delta U_p(k)$
12	$\dfrac{b_2E(k) + 2L_1\left[\left(\theta_1 + 0.5\right)b_2 + \beta_3 b_3\right]}{E(k) + 2L_1\left(\theta_1 + \beta_3 + 0.5\right)}$
13	$\dfrac{b_4R(k) - 2L_2\left[\alpha_3 b_3 + \left(\theta_2 + 0.5\right)b_4\right]}{R(k) - 2L_2\left(\alpha_3 + \theta_2 + 0.5\right)}$
14	b_3
15	$\dfrac{b_4R(k) - 2L_2\left[\beta_3 b_3 + \left(\theta_2 + 0.5\right)b_4\right]}{R(k) - 2L_2\left(\beta_3 + \theta_2 + 0.5\right)}$
16	b_3

Without proof, the formal result similar to Theorems 4.1 and 4.2 is given below:

THEOREM 4.3 The T2 fuzzy PI (or the corresponding PD) controller in this section is structurally equivalent to a nonlinear PI (or PD) controller with variable gains and a variable control offset.

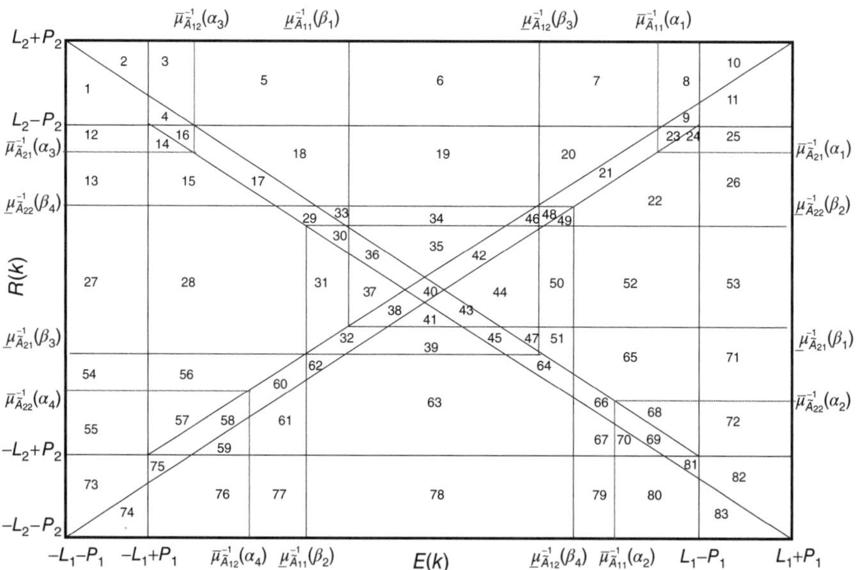

Figure 4.11 Overall input space division. Each region is labeled by an IC number.

4.7 MAMDANI FUZZY PI AND PD CONTROLLERS—CONFIGURATION 4[4]

4.7.1 Fuzzy PI Controller Configuration

The fuzzy controllers in this section are quite different from, and more complex than, those in Sections 4.4–4.6 as far as their configuration is concerned. More specifically, the configuration is extended to the following: Any kind and number of T2 fuzzy sets for the inputs, any type of T2 or T1 fuzzy sets for the output, Mamdani fuzzy rules, the center-of-sets type of reducer involving the interval weighted-average method and iterative KM algorithm, and the centroid defuzzifier. The complexity is mainly due to (1) nonlinear input fuzzy sets and (2) the iterative nature of the KM algorithm.

Let $E(k)$ be defined on $[L_1, R_1]$ that is divided into $N_1 - 1$ subintervals: $[S_1, S_2], \ldots, [S_i, S_{i+1}], \ldots, [S_{N_1-1}, S_{N_1}]$, $1 \leq i \leq N_1$. There are N_1 interval T2 fuzzy sets, $\widetilde{A}_{11}, \cdots, \widetilde{A}_{1i}, \cdots, \widetilde{A}_{1N_1}$. And $\widetilde{A}_{1i}$ is defined over $[S_{i-1}, S_{i+1}]$ and its membership value is zero everywhere else. Assume that (1) $\overline{\mu}_{\widetilde{A}_{1i}} \geq \underline{\mu}_{\widetilde{A}_{1i}}$ and (2) $\overline{\mu}_{\widetilde{A}_{1i}}$ (and $\underline{\mu}_{\widetilde{A}_{1i}}$) increases from 0, reaches its maximum, which can be a range of the same maximum, and then decreases to 0. Many of the widely used T2 fuzzy sets in the literature (e.g., the trapezoidal type) meet the assumptions. To illustrate the assumptions, Fig. 4.12 provides some example T2 fuzzy sets.

Likewise, let $R(k)$ be defined on $[L_2, R_2]$ that is divided into $N_2 - 1$ subintervals, namely $[M_1, M_2], \ldots, [M_j, M_{j+1}], \ldots, [M_{N_2-1}, M_{N_2}]$, over which N_2 interval T2 fuzzy sets meeting the same assumptions above are defined. Each of them is denoted $\widetilde{A}_{2j}$ $(1 \leq j \leq N_2)$.

A total of $N_1 \times N_2$ Mamdani fuzzy rules are used to cover all the possible combinations of the input fuzzy sets. Because of the way that the input fuzzy sets are defined, at any sampling instance, only two adjacent $\widetilde{A}_{1i}$ and two adjacent $\widetilde{A}_{2j}$ can

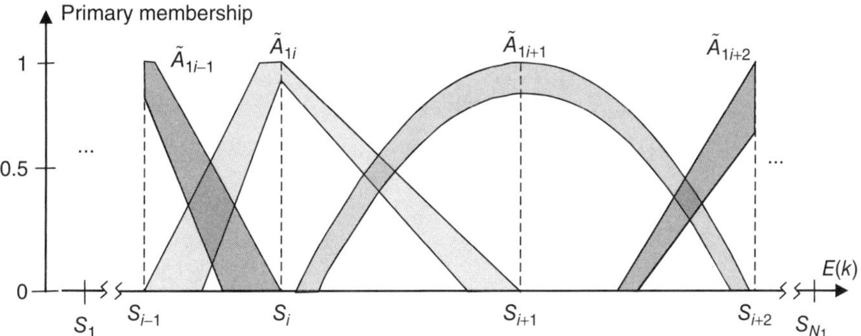

Figure 4.12 Example interval T2 fuzzy sets for $E(k)$.

[4]Part of the material in this section is adapted from Zhou and Ying (2011; © 2011, IEEE) and Zhou and Ying (2013; © 2013, IEEE).

be involved in fuzzifying $E(k)$ and $R(k)$, respectively. Without losing generality, we assume that $\widetilde{A}_{1i}$, $\widetilde{A}_{1i+1}$, $\widetilde{A}_{2j}$, and $\widetilde{A}_{2j+1}$ are the ones that are involved. Accordingly, the following four fuzzy rules are executed:

$$R^1: \text{If } E(k) \text{ is } \widetilde{A}_{1i} \text{ and } R(k) \text{ is } \widetilde{A}_{2j}, \text{ then } \Delta U(k) \text{ is } \widetilde{B}_{h(i,j)}.$$

$$R^2: \text{If } E(k) \text{ is } \widetilde{A}_{1i} \text{ and } R(k) \text{ is } \widetilde{A}_{2j+1}, \text{ then } \Delta U(k) \text{ is } \widetilde{B}_{h(i,j+1)}.$$

$$R^3: \text{If } E(k) \text{ is } \widetilde{A}_{1i+1} \text{ and } R(k) \text{ is } \widetilde{A}_{2j}, \text{ then } \Delta U(k) \text{ is } \widetilde{B}_{h(i+1,j)}.$$

$$R^4: \text{If } E(k) \text{ is } \widetilde{A}_{1i+1} \text{ and } R(k) \text{ is } \widetilde{A}_{2j+1}, \text{ then } \Delta U(k) \text{ is } \widetilde{B}_{h(i+1,j+1)}$$

where the Zadeh fuzzy AND operator is used, and $\widetilde{B}_{h(i,j)}$, $\widetilde{B}_{h(i,j+1)}$, $\widetilde{B}_{h(i+1,j)}$, and $\widetilde{B}_{h(i+1,j+1)}$ can be any continuous interval T2 fuzzy sets. These output fuzzy sets are indexed by integer subscripts whose values are computed using function $h(\cdot)$ so that the input and output fuzzy sets are mathematically linked. For example, one may simply use

$$h(i,j) = i + j$$

which, in a sense, links the input fuzzy sets with the output fuzzy set in a linear fashion, leading to the so-called linear fuzzy rules (Ying, 1993b). In this case, suppose that $\widetilde{A}_{12}$(i.e., $i = 2$) means $E(k)$ is "negative medium" and that $\widetilde{A}_{25}$ (i.e., $j = 5$) indicates $R(k)$ is "positive small." Then, $\widetilde{B}_{h(i,j)} = \widetilde{B}_7$ can be interpreted as the output being "positive large." Because $h(\cdot)$, chosen by the controller designer, can be any function, the fuzzy rules subsequently can be arbitrary in terms of input and output relation.

The firing intervals for the four rules are

$$\widetilde{F}^1 = [\underline{f}^1, \overline{f}^1] = [\min(\underline{\mu}_{\widetilde{A}_{1i}}, \underline{\mu}_{\widetilde{A}_{2j}}), \min(\overline{\mu}_{\widetilde{A}_{1i}}, \overline{\mu}_{\widetilde{A}_{2j}})] \tag{4.33}$$

$$\widetilde{F}^2 = [\underline{f}^2, \overline{f}^2] = [\min(\underline{\mu}_{\widetilde{A}_{1i}}, \underline{\mu}_{\widetilde{A}_{2j+1}}), \min(\overline{\mu}_{\widetilde{A}_{1i}}, \overline{\mu}_{\widetilde{A}_{2j+1}})] \tag{4.34}$$

$$\widetilde{F}^3 = [\underline{f}^3, \overline{f}^3] = [\min(\underline{\mu}_{\widetilde{A}_{1i+1}}, \underline{\mu}_{\widetilde{A}_{2j}}), \min(\overline{\mu}_{\widetilde{A}_{1i+1}}, \overline{\mu}_{\widetilde{A}_{2j}})] \tag{4.35}$$

$$\widetilde{F}^4 = [\underline{f}^4, \overline{f}^4] = [\min(\underline{\mu}_{\widetilde{A}_{1i+1}}, \underline{\mu}_{\widetilde{A}_{2j+1}}), \min(\overline{\mu}_{\widetilde{A}_{1i+1}}, \overline{\mu}_{\widetilde{A}_{2j+1}})] \tag{4.36}$$

The iterative KM algorithm links the firing intervals and the centroids for $\widetilde{B}_{h(i,j)}$, $\widetilde{B}_{h(i,j+1)}$, $\widetilde{B}_{h(i+1,j)}$, and $\widetilde{B}_{h(i+1,j+1)}$ to produce $\Delta U(k) = [\Delta U_l(k), \Delta U_r(k)]$, a T1 fuzzy set, which is then defuzzified by the centroid defuzzifier Eq. (3.68) to produce $\Delta U(k)$. The centroid for the output fuzzy set of R^s is denoted $[a^s, b^s]$, which can be computed either online or offline (note that $a^s = b^s$ when the output fuzzy set is T1). Then b^s and a^s are, respectively, arranged in the ascending orders and the results are represented by $b^{1*} \le b^{2*} \le b^{3*} \le b^{4*}$ and $a^{1*} \le a^{2*} \le a^{3*} \le a^{4*}$. Then relabel $\overline{f}^s$ and $\underline{f}^s$ so that they conform to $b^{1*} \le b^{2*} \le b^{3*} \le b^{4*}$ and $a^{1*} \le a^{2*} \le a^{3*} \le a^{4*}$, respectively, resulting in $\overline{f}^{1*}, \overline{f}^{2*}, \overline{f}^{3*}, \overline{f}^{4*}$ and $\underline{f}^{1*}, \underline{f}^{2*}, \underline{f}^{3*}, \underline{f}^{4*}$. According to Eq (3.63),

$$\Delta U_l(k) = \frac{\sum_{i=1}^{L} \overline{f}^{i*} a^i + \sum_{j=L+1}^{4} \underline{f}^{j*} a^j}{\sum_{i=1}^{L} \overline{f}^{i*} + \sum_{j=L+1}^{4} \underline{f}^{j*}} \tag{4.37}$$

$$\Delta U_r(k) = \frac{\sum_{i=1}^{R} \underline{f}^{i*} b^i + \sum_{j=R+1}^{4} \overline{f}^{j*} b^j}{\sum_{i=1}^{R} \underline{f}^{i*} + \sum_{j=R+1}^{4} \overline{f}^{j*}} \tag{4.38}$$

where integers L $(1 \leq L \leq 3)$ and R $(1 \leq R \leq 3)$ are switch points that depend on the input and output fuzzy sets and the values of $E(k)$ and $R(k)$, and hence vary with k. Finally

$$\Delta U(k) = 1/2(\Delta U_l(k) + \Delta U_r(k)) \tag{4.39}$$

4.7.2 Method for Deriving the Analytical Structure

Suppose arbitrarily that $\widetilde{A}_{1i}$, $\widetilde{A}_{1i+1}$, $\widetilde{A}_{2j}$, and $\widetilde{A}_{2j+1}$ are as shown in Fig. 4.13 with the mathematical definitions given in Table 4.8 (outside the intervals in the right column of Table 4.8, the membership values are 0).

The process of developing the ICs is the same as in the previous sections of this chapter (i.e., the process of generating Figs. 4.10 and 4.11). Without loss of generality, let $\eta_{j+1} \leq \lambda_{i+1} \leq \eta_j \leq \lambda_i$. Figure 4.14 shows the ICs where $\mu^{-1}_{\underline{\widetilde{A}}_{1i}}(\cdot)$, $\mu^{-1}_{\underline{\widetilde{A}}_{2j}}(\cdot)$, $\mu^{-1}_{\underline{\widetilde{A}}_{2j}}(\cdot)$, $\mu^{-1}_{\underline{\widetilde{A}}_{2j+1}}(\cdot)$, $\overline{\mu}^{-1}_{\widetilde{A}_{2j+1}}(\cdot)$, and $\mu^{-1}_{\underline{\widetilde{A}}_{1i+1}}(\cdot)$ are the inverse functions of the respective membership functions. It turns out that there are a total of 50 regions, labeled IC1–IC50. For each IC, $[\underline{f}^s, \overline{f}^s]$ can be determined. For example, for IC27, Eqs. (4.33)–(4.36) generate: $\underline{f}^1 = \min(\mu_{\underline{\widetilde{A}}_{1i}}, \mu_{\underline{\widetilde{A}}_{2j}}) = \mu_{\underline{\widetilde{A}}_{2j}}$, $\underline{f}^2 = \min(\mu_{\underline{\widetilde{A}}_{1i}}, \mu_{\underline{\widetilde{A}}_{2j+1}}) = \mu_{\underline{\widetilde{A}}_{1i}}$, $\underline{f}^3 = \min(\mu_{\underline{\widetilde{A}}_{1i+1}}, \mu_{\underline{\widetilde{A}}_{2j}}) = \mu_{\underline{\widetilde{A}}_{2j}}$, $\underline{f}^4 = \min(\mu_{\underline{\widetilde{A}}_{1i+1}}, \mu_{\underline{\widetilde{A}}_{2j+1}}) = \mu_{\underline{\widetilde{A}}_{2j+1}}$, $\overline{f}^1 = \min(\overline{\mu}_{\widetilde{A}_{1i}}, \overline{\mu}_{\widetilde{A}_{2j}}) = \overline{\mu}_{\widetilde{A}_{2j}}$, $\overline{f}^2 = \min(\overline{\mu}_{\widetilde{A}_{1i}}, \overline{\mu}_{\widetilde{A}_{2j+1}}) = \overline{\mu}_{\widetilde{A}_{1i}}$, $\overline{f}^3 = \min(\overline{\mu}_{\widetilde{A}_{1i+1}}, \overline{\mu}_{\widetilde{A}_{2j}}) = \overline{\mu}_{\widetilde{A}_{2j}}$, and $\overline{f}^4 = \min(\overline{\mu}_{\widetilde{A}_{1i+1}}, \overline{\mu}_{\widetilde{A}_{2j+1}}) = \overline{\mu}_{\widetilde{A}_{2j+1}}$. Note that the result can be different if the input fuzzy sets are different.

For the type reduction procedure, assume $b^{s*} = b^s$ and $a^{s*} = a^s$, and accordingly $\overline{f}^{s*} = \overline{f}^s$ and $\underline{f}^{s*} = \underline{f}^s$. For an IC put the eight membership functions resulting from the min( ) operations in the four rules and the centroids of the output fuzzy sets into the type reducer represented by Eqs. (4.37) and (4.38). One will obtain $\Delta U_R(k)$ and $\Delta U_L(k)$, which will lead to the analytical structure of the fuzzy controller via Eq. (4.39) for that particular IC if L in Eq. (4.37) and R in Eq. (4.38) are known. The exact values of L and R can always be computed as long as all the parameters of the input fuzzy sets, controller inputs, and the centroids of all the output fuzzy sets are numerically available. Hence, in principle for any specific fuzzy controller,

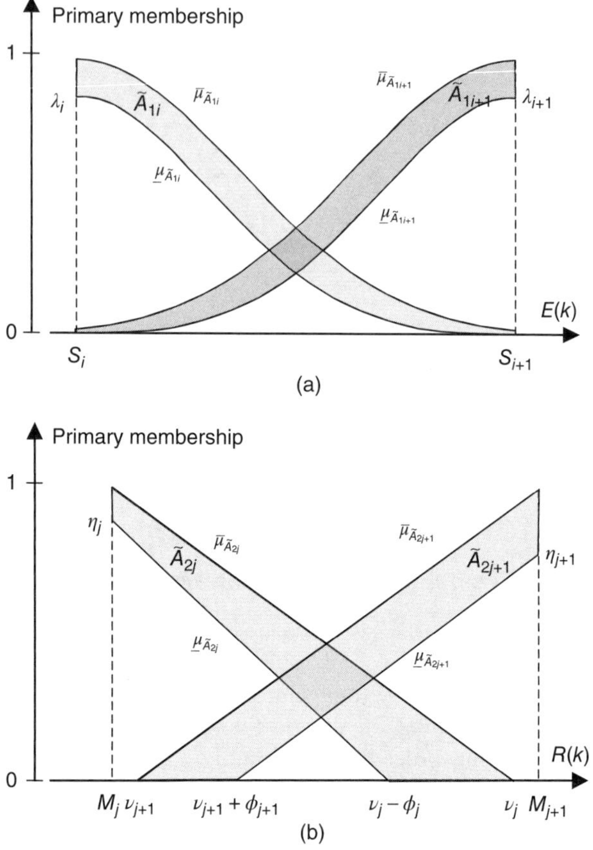

Figure 4.13 Example interval T2 fuzzy sets used for (a) $E(k)$ and (b) $R(k)$.

the structure can always be derived. Nevertheless, our interest is on a general controller configuration where these parameters are not assigned values. As a result, L and R do not have specific values, making the analytical structure derivation significantly more complicated. Note that $L = 1 \sim 3$ and $R = 1 \sim 3$ for the controller in this section. They lead to a total of $3 \times 3 = 9$ possible scenarios, each of which is called a case. Nine cases are the maximum number of cases that Eqs. (4.37) and (4.38) can produce for an IC. When $L = 1$ and $R = 1$, it is case 1, for which the corresponding structure of the T2 fuzzy controller is

$$\Delta U(k) = \frac{1}{2}\left(\frac{\bar{f}^1 a^1 + \underline{f}^2 a^2 + \underline{f}^3 a^3 + \underline{f}^4 a^4}{\bar{f}^1 + \underline{f}^2 + \underline{f}^3 + \underline{f}^4} + \frac{\underline{f}^1 b^1 + \bar{f}^2 b^2 + \bar{f}^3 b^3 + \bar{f}^4 b^4}{\underline{f}^1 + \bar{f}^2 + \bar{f}^3 + \bar{f}^4} \right) \tag{4.40}$$

TABLE 4.8 Mathematical Definitions of $\widetilde{A}_{1i}$, $\widetilde{A}_{1i+1}$, $\widetilde{A}_{2j}$, and $\widetilde{A}_{2j+1}$

Mathematical Definition	Interval
$\overline{\mu}_{\widetilde{A}_{1i}} = \exp\left\{-\dfrac{\left[E(k)-S_i\right]^2}{a_i^2}\right\}$	$[S_i, S_{i+1}]$
$\underline{\mu}_{\widetilde{A}_{1i}} = \lambda_i \exp\left\{-\dfrac{\left[E(k)-S_i\right]^2}{\sigma_i^2}\right\}$	$[S_i, S_{i+1}]$
$\overline{\mu}_{\widetilde{A}_{1i+1}} = \exp\left\{-\dfrac{\left[E(k)-S_{i+1}\right]^2}{a_{i+1}^2}\right\}$	$[S_i, S_{i+1}]$
$\underline{\mu}_{\widetilde{A}_{1i+1}} = \lambda_{i+1} \exp\left\{-\dfrac{\left[E(k)-S_{i+1}\right]^2}{\sigma_{i+1}^2}\right\}$	$[S_i, S_{i+1}]$
$\overline{\mu}_{\widetilde{A}_{2j}} = -\dfrac{R(k)}{\nu_j - M_j} + \dfrac{M_j}{\nu_j - M_j} + 1$	$[M_j, \nu_j]$
$\underline{\mu}_{\widetilde{A}_{2j}} = -\dfrac{\eta_j}{\nu_j - \phi_j - M_j} R(k) + \dfrac{M_j \eta_j}{\nu_j - \phi_j - M_j} + \eta_j$	$[M_j, \nu_j - \phi_j]$
$\overline{\mu}_{\widetilde{A}_{2j+1}} = \dfrac{1}{M_{j+1} - \nu_{j+1}} R(k) - \dfrac{M_{j+1}}{M_{j+1} - \nu_{j+1}} + 1$	$[\nu_{j+1}, M_{j+1}]$
$\underline{\mu}_{\widetilde{A}_{2j+1}} = \dfrac{\eta_{j+1}}{M_{j+1} - \nu_{j+1} - \phi_{j+1}} R(k) - \dfrac{M_{j+1}\eta_{j+1}}{M_{j+1} - \nu_{j+1} - \phi_{j+1}} + \eta_{j+1}$	$[\nu_{j+1} + \phi_{j+1}, M_{j+1}]$

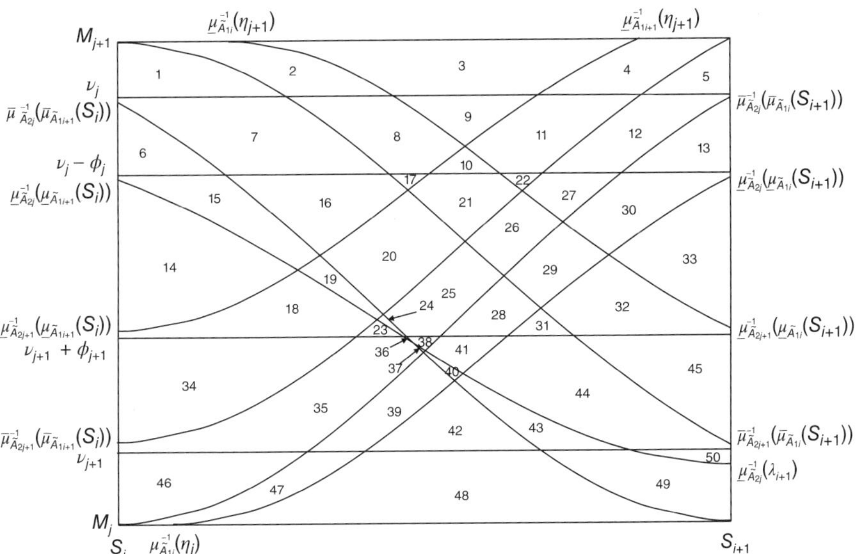

Figure 4.14 Input space is divided into 50 ICs.

The other eight cases can also be established by the type reducer (Table 4.9).

Because of the lack of the specific parameter values, the analytical structure of the fuzzy controller cannot be uniquely determined partially due to the multiple (i.e., nine) cases stemming from the unknown values of L and R. We now concentrate on a particular fuzzy controller with all the parameter values specified so that its analytical structure can be explicitly derived.

Example 4.5 Let us use IC27 just mentioned above and IC20 as examples.

For any point in this IC, one can derive nine different analytical structures for the nine cases. For instance, the analytical structure for case 1 is

$$\Delta U(k) = \frac{b^2}{\Omega_2} \exp\left\{-\left[\frac{E(k) - S_i}{a_i}\right]^2\right\} + \left(\frac{\Gamma_1}{\Omega_1} + \frac{\Gamma_2}{\Omega_2}\right) R(k) + \left(\frac{\Gamma_3}{\Omega_1} + \frac{\Gamma_4}{\Omega_2}\right) \tag{4.41}$$

where

$$\Omega_1 = 2\left(1 + \eta_j + 2\eta_{j+1} - \frac{M_j}{M_j - v_j} - \frac{\eta_j M_j}{M_j + \phi_j - v_j} - \frac{2\eta_{j+1} M_{j+1}}{M_{j+1} - v_{j+1} - \phi_{j+1}}\right)$$

$$+ 2\left(\frac{1}{M_j - v_j} + \frac{\eta_j}{M_j + \phi_j - v_j} + \frac{2\eta_{j+1}}{M_{j+1} - v_{j+1} - \phi_{j+1}}\right) R(k)$$

$$\Omega_2 = 2\left(2 + \eta_j - \frac{M_j}{M_j - v_j} - \frac{\eta_j M_j}{M_j + \phi_j - v_j} - \frac{M_{j+1}}{M_{j+1} - v_{j+1}}\right)$$

$$+ 2\left(\frac{1}{M_{j+1} - v_{j+1}} + \frac{1}{M_j - v_j} + \frac{\eta_j}{M_j + \phi_j - v_j}\right) R(k)$$

$$+ 2\exp\left\{-\left[\frac{E(k) - S_i}{a_i}\right]^2\right\}$$

$$\Gamma_1 = \frac{a^1}{M_j - v_j} + \frac{a^3 \eta_j}{M_j + \phi_j - v_j} + \frac{(a^2 + a^4)\eta_{j+1}}{M_{j+1} - v_{j+1} - \phi_{j+1}}$$

$$\Gamma_2 = \frac{b^1 \eta_j}{M_j + \phi_j - v_j} + \frac{b^3}{M_j - v_j} + \frac{b^4}{M_{j+1} - v_{j+1}}$$

TABLE 4.9 Structure of T2 Fuzzy Controller for Each of Nine Cases

Case No.	Switching Points	$\Delta U(k)$
1	$L=1, R=1$	$\frac{1}{2}\left(\frac{\bar{f}^1 a^1+\underline{f}^2 a^2+\underline{f}^3 a^3+\underline{f}^4 a^4}{\bar{f}^1+\underline{f}^2+\underline{f}^3+\underline{f}^4}+\frac{\underline{f}^1 b^1+\bar{f}^2 b^2+\bar{f}^3 b^3+\bar{f}^4 b^4}{\underline{f}^1+\bar{f}^2+\bar{f}^3+\bar{f}^4}\right)$
2	$L=1, R=2$	$\frac{1}{2}\left(\frac{\bar{f}^1 a^1+\underline{f}^2 a^2+\underline{f}^3 a^3+\underline{f}^4 a^4}{\bar{f}^1+\underline{f}^2+\underline{f}^3+\underline{f}^4}+\frac{\underline{f}^1 b^1+\underline{f}^2 b^2+\bar{f}^3 b^3+\bar{f}^4 b^4}{\underline{f}^1+\underline{f}^2+\bar{f}^3+\bar{f}^4}\right)$
3	$L=1, R=3$	$\frac{1}{2}\left(\frac{\bar{f}^1 a^1+\underline{f}^2 a^2+\underline{f}^3 a^3+\underline{f}^4 a^4}{\bar{f}^1+\underline{f}^2+\underline{f}^3+\underline{f}^4}+\frac{\underline{f}^1 b^1+\underline{f}^2 b^2+\underline{f}^3 b^3+\bar{f}^4 b^4}{\underline{f}^1+\underline{f}^2+\underline{f}^3+\bar{f}^4}\right)$
4	$L=2, R=1$	$\frac{1}{2}\left(\frac{\bar{f}^1 a^1+\bar{f}^2 a^2+\underline{f}^3 a^3+\underline{f}^4 a^4}{\bar{f}^1+\bar{f}^2+\underline{f}^3+\underline{f}^4}+\frac{\underline{f}^1 b^1+\bar{f}^2 b^2+\bar{f}^3 b^3+\bar{f}^4 b^4}{\underline{f}^1+\bar{f}^2+\bar{f}^3+\bar{f}^4}\right)$
5	$L=2, R=2$	$\frac{1}{2}\left(\frac{\bar{f}^1 a^1+\bar{f}^2 a^2+\underline{f}^3 a^3+\underline{f}^4 a^4}{\bar{f}^1+\bar{f}^2+\underline{f}^3+\underline{f}^4}+\frac{\underline{f}^1 b^1+\underline{f}^2 b^2+\bar{f}^3 b^3+\bar{f}^4 b^4}{\underline{f}^1+\underline{f}^2+\bar{f}^3+\bar{f}^4}\right)$
6	$L=2, R=3$	$\frac{1}{2}\left(\frac{\bar{f}^1 a^1+\bar{f}^2 a^2+\underline{f}^3 a^3+\underline{f}^4 a^4}{\bar{f}^1+\bar{f}^2+\underline{f}^3+\underline{f}^4}+\frac{\underline{f}^1 b^1+\underline{f}^2 b^2+\underline{f}^3 b^3+\bar{f}^4 b^4}{\underline{f}^1+\underline{f}^2+\underline{f}^3+\bar{f}^4}\right)$
7	$L=3, R=1$	$\frac{1}{2}\left(\frac{\bar{f}^1 a^1+\bar{f}^2 a^2+\bar{f}^3 a^3+\underline{f}^4 a^4}{\bar{f}^1+\bar{f}^2+\bar{f}^3+\underline{f}^4}+\frac{\underline{f}^1 b^1+\bar{f}^2 b^2+\bar{f}^3 b^3+\bar{f}^4 b^4}{\underline{f}^1+\bar{f}^2+\bar{f}^3+\bar{f}^4}\right)$
8	$L=3, R=2$	$\frac{1}{2}\left(\frac{\bar{f}^1 a^1+\bar{f}^2 a^2+\bar{f}^3 a^3+\underline{f}^4 a^4}{\bar{f}^1+\bar{f}^2+\bar{f}^3+\underline{f}^4}+\frac{\underline{f}^1 b^1+\underline{f}^2 b^2+\bar{f}^3 b^3+\bar{f}^4 b^4}{\underline{f}^1+\underline{f}^2+\bar{f}^3+\bar{f}^4}\right)$
9	$L=3, R=3$	$\frac{1}{2}\left(\frac{\bar{f}^1 a^1+\bar{f}^2 a^2+\bar{f}^3 a^3+\underline{f}^4 a^4}{\bar{f}^1+\bar{f}^2+\bar{f}^3+\underline{f}^4}+\frac{\underline{f}^1 b^1+\underline{f}^2 b^2+\underline{f}^3 b^3+\bar{f}^4 b^4}{\underline{f}^1+\underline{f}^2+\underline{f}^3+\bar{f}^4}\right)$

$$\Gamma_3 = a^1 + a^3\eta_j + (a^2 + a^4)\eta_{j+1} - \frac{a^1 M_j}{M_j - v_j} - \frac{a^3 \eta_j M_j}{M_j + \phi_j - v_j} - \frac{(a^2 + a^4)\eta_{j+1} M_{j+1}}{M_{j+1} - v_{j+1} - \phi_{j+1}}$$

$$\Gamma_4 = b^1\eta_j + b^3 + b^4 - \frac{b^1 \eta_j M_j}{M_j + \phi_j - v_j} + \frac{b^3 M_j}{v_j - M_j} + \frac{b^4 M_{j+1}}{v_{j+1} - M_{j+1}}$$

As another instance, the analytical structure for case 2 for IC20 is

$$\Delta U(k) = \frac{1}{\Omega_2}\left(b^2 \lambda_i \exp\left\{ -\left[\frac{E(k) - S_i}{\sigma_i}\right]^2 \right\} + b^4 \exp\left\{ -\left[\frac{E(k) - S_{i+1}}{a_{i+1}}\right]^2 \right\} \right) + \left(\frac{\Gamma_1}{\Omega_1} + \frac{\Gamma_2}{\Omega_2}\right) R(k) + \left(\frac{\Gamma_3}{\Omega_1} + \frac{\Gamma_4}{\Omega_2}\right) \quad (4.42)$$

where

$$\Omega_1 = 2\left(1 + \eta_j + 2\eta_{j+1} - \frac{M_j}{M_j - v_j} - \frac{\eta_j M_j}{M_j + \phi_j - v_j} - \frac{2\eta_{j+1} M_{j+1}}{M_{j+1} - v_{j+1} - \phi_{j+1}}\right) + 2\left(\frac{1}{M_j - v_j} + \frac{\eta_j}{M_j + \phi_j - v_j} + \frac{2\eta_{j+1}}{M_{j+1} - v_{j+1} - \phi_{j+1}}\right) R(k)$$

$$\Omega_2 = 2\left(1 + \eta_j - \frac{M_j}{M_j - v_j} - \frac{\eta_j M_j}{M_j + \phi_j - v_j}\right) + 2\left(\frac{1}{M_j - v_j} + \frac{\eta_j}{M_j + \phi_j - v_j}\right) R(k) + 2\lambda_i \exp\left\{ -\left[\frac{E(k) - S_i}{\theta_i}\right]^2 \right\} + \exp\left\{ -\left[\frac{E(k) - S_{i+1}}{a_{i+1}}\right]^2 \right\}$$

$$\Gamma_1 = \frac{a^1}{M_j - v_j} + \frac{a^3 \eta_j}{M_j + \phi_j - v_j} + \frac{(a^2 + a^4)\eta_{j+1}}{M_{j+1} - v_{j+1} - \phi_{j+1}}$$

$$\Gamma_2 = \frac{b^1 \eta_j}{M_j + \phi_j - v_j} + \frac{b^3}{M_j - v_j}$$

$$\Gamma_3 = a^1 + a^3\eta_j + (a^2 + a^4)\eta_{j+1} - \frac{a^1 M_j}{M_j - v_j} - \frac{a^3 \eta_j M_j}{M_j + \phi_j - v_j} - \frac{(a^2 + a^4)\eta_{j+1} M_{j+1}}{M_{j+1} - v_{j+1} - \phi_{j+1}}$$

$$\Gamma_4 = b^1\eta_j + b^3 + b^4 - \frac{b^1 \eta_j M_j}{M_j + \phi_j - v_j} - \frac{b^3 M_j}{M_j - v_j}$$

The analytical structures in Eqs. (4.41) and (4.42) are fundamentally different from those in Sections 4.4–4.6 because of $\exp(-\{[E(k) - S_i]/a_i\}^2)$ in Eq. (4.41) and $\exp(-\{[E(k) - S_i]/\sigma_i\}^2)$ and $\exp(-\{[E(k) - S_{i+1}]/a_{i+1}\}^2)$ in Eq. (4.42), which make the structures non-PI control type. The reasons behind the appearance of these two terms are the use of the nonlinear input fuzzy sets $\widetilde{A}_{1i}$ and $\widetilde{A}_{1i+1}$ (see Fig. 4.13a).

Clearly, the type of input fuzzy set can dictate the form of the analytical structure. As a result, a theorem similar to Theorems 4.1–4.3 cannot be established for the controllers in the present section. Nevertheless, if all the input fuzzy sets are restricted to piecewise linear fuzzy sets (e.g., triangular and/or trapezoidal), the following can be proved to be true.

THEOREM 4.4 The T2 fuzzy PI (or the corresponding PD) controller in this section is structurally equivalent to a nonlinear PI (or PD) controller with variable gains and a variable control offset if and only if all the input fuzzy sets are piecewise linear.

Example 4.6 below is a concrete (albeit indirect) illustration of the validity of this theorem. We wish to point out first that the IC distribution in the input space is determined by the controller parameters only, whereas the case distribution depends on both the controller parameters and the input values. Hence, the two distributions are not necessarily related (we have yet to find their possible connection—a future research topic). When the controller parameters are unknown (e.g., for a general configuration), it may be impossible to determine the boundaries between adjacent cases. If, on the other hand, the controller parameters are given, the boundaries can be easily determined quantitatively by a computer program written for this purpose.

A MATLAB program was developed to find: (1) an IC such that every point in it belonged to the same case, (2) an IC such that points in it belonged to more than one case, and (3) multiple ICs such that points in them belonged to the same case (the program is available for the reader to download). Any point is assigned to one and only one case. Which case is assigned to a point will be known if the controller parameters and the values of the input variables are specified. But when they are not specified, that is, when the controller configuration is general, the multiple possibilities arise. The reader is also reminded that two ICs having the same case may not have the same analytical structure. For the 50 ICs shown in Fig. 4.14, in theory there exist a total of maximum $9 \times 50 = 450$ (i.e., each IC is supposed to have all 9 cases) and minimum $1 \times 50 = 50$ (when each IC has only 1 case) possible analytical structures. The exact number of different analytical structures cannot be known because of the general configuration. A Mathematica program was written to derive all of the 450 analytical structures (not shown here) and found them all to be similar to either Eq. (4.41) or (4.42).

In summary, the total number of analytical structures is unknown unless all the controller parameters are given. Even when all the parameter values are spelled out, a computer program will most likely need to be written by the reader to determine which case is assigned to which point in the input space. Once this is carried out, the total number of cases will readily be known.

Example 4.6 Suppose that the universe of $E(k)$ of a controller meeting the configuration of this section is $[-9, 9]$ and is divided into three subintervals, $[-9, -3]$, $[-3, 3]$ and $[3, 9]$, over which four interval T2 fuzzy sets, $\widetilde{A}_{11}, \widetilde{A}_{12}, \widetilde{A}_{13}$, and $\widetilde{A}_{14}$, exist (Zhou and Ying, 2013). Likewise, $R(k)$ is in $[-6, 6]$ and is divided into three subintervals, $[-6, -2]$, $[-2, 2]$, and $[2, 6]$. Four interval T2 fuzzy sets, $\widetilde{A}_{21}, \widetilde{A}_{22}, \widetilde{A}_{23}$, and

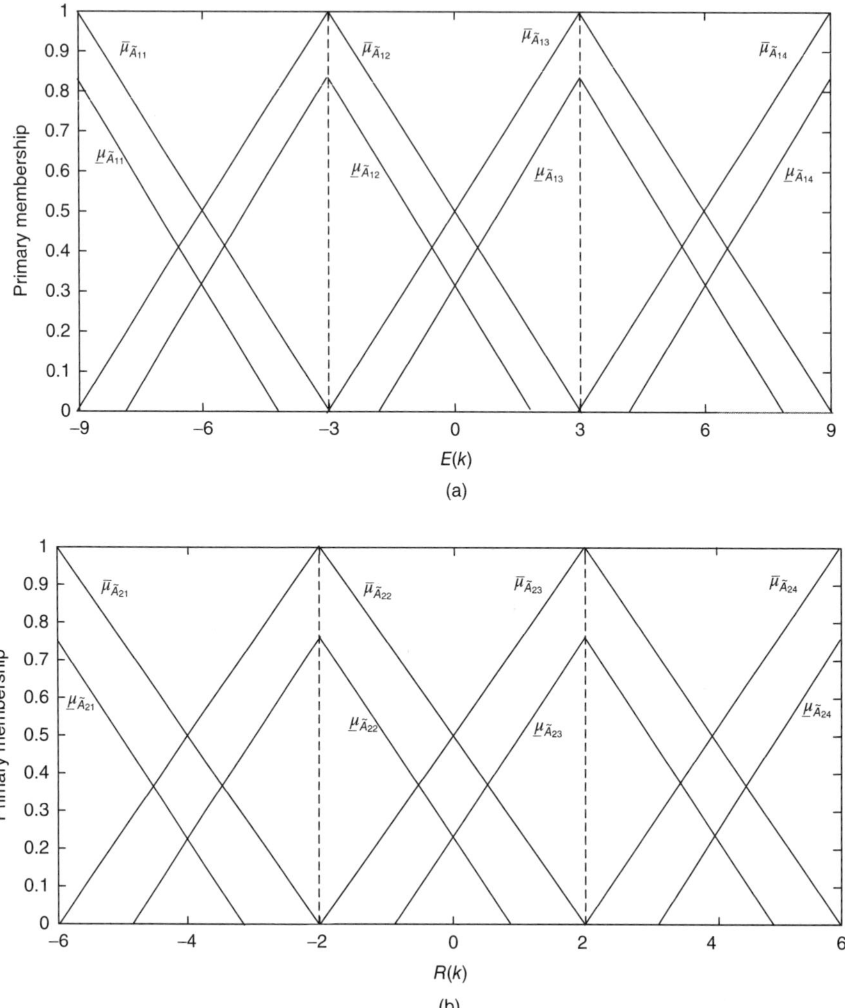

Figure 4.15 Example triangular interval T2 fuzzy sets for (a) $E(k)$ and (b) $R(k)$.

$\widetilde{A}_{24}$, are employed. These fuzzy sets are symmetrically triangular (Fig. 4.15). In the fuzzy rules, the input fuzzy sets and output fuzzy sets are linked by $h(i,j)=2i+j$. Derive the analytical structure of this fuzzy controller.

For brevity, we only derive the analytical structure for $-3 \le E(k) \le 3$ and $-2 \le R(k) \le 2$, and the structure for other subintervals can be obtained in a similar fashion. Because $-3 \le E(k) \le 3$, $\widetilde{A}_{12}$ and $\widetilde{A}_{13}$ are used, and owing to $-2 \le R(k) \le 2$, $\widetilde{A}_{22}$ and $\widetilde{A}_{23}$ are involved (accordingly $i=2,3$ and $j=2,3$). Four fuzzy rules involving these four fuzzy sets are executed. The mathematical definitions for nonzero values of $\widetilde{A}_{12}$, $\widetilde{A}_{13}$, $\widetilde{A}_{22}$, and $\widetilde{A}_{23}$ needed in the derivation are

TABLE 4.10 Mathematical Definitions for Parts of $\widetilde{A}_{12}, \widetilde{A}_{13}, \widetilde{A}_{22}$, and $\widetilde{A}_{23}$

Definition	Interval
$\overline{\mu}_{\widetilde{A}_{12}} = (-E(k)+3)/6$	$E(k) \in [-3, 3]$
$\underline{\mu}_{\widetilde{A}_{12}} = (-E(k)+2)/6$	$E(k) \in [-3, 2]$
$\overline{\mu}_{\widetilde{A}_{13}} = (E(k)+3)/6$	$E(k) \in [-3, 3]$
$\underline{\mu}_{\widetilde{A}_{13}} = (E(k)+2)/6$	$E(k) \in [-2, 3]$
$\overline{\mu}_{\widetilde{A}_{22}} = (-R(k)+2)/4$	$R(k) \in [-2, 2]$
$\underline{\mu}_{\widetilde{A}_{22}} = (-R(k)+1)/4$	$R(k) \in [-2, 1]$
$\overline{\mu}_{\widetilde{A}_{23}} = (R(k)+2)/4$	$R(k) \in [-2, 2]$
$\underline{\mu}_{\widetilde{A}_{23}} = (R(k)+1)/4$	$R(k) \in [-1, 2]$

listed in Table 4.10. The resulting output fuzzy sets are $\widetilde{B}_{h(2,2)} = \widetilde{B}_6$, $\widetilde{B}_{h(2,3)} = \widetilde{B}_7$, $\widetilde{B}_{h(3,2)} = \widetilde{B}_8$, and $\widetilde{B}_{h(3,3)} = \widetilde{B}_9$ [the subscripts are calculated by $h(i,j) = 2i + j$]. They can be of any shape. Whatever the shapes, their centroids are computable before the structure derivation. Suppose the centroids for the four output fuzzy sets are [4.8, 5.2], [5.3, 5.7], [5.8, 6.1], and [6.4, 6.6], respectively.

The input space covered by $[-3, 3] \times [-2, 2]$ must be divided into IC1–IC40 shown in Fig. 4.16. Because the primary membership functions of all the input fuzzy sets are symmetrical and piecewise linear (i.e., triangular), the 40 ICs subsequently show some symmetricities. Now that the controller configuration is specified, values of L and R can be determined. The procedure is as follows. Choose 301 points for $E(k)$: $-3, -2.98, \ldots, 2.98, 3$ and 201 points for $R(k)$: $-2, -1.98, \ldots, 1.98, 2$. Each combination of an $E(k)$ value and a $R(k)$ value has a case number. Figure 4.17 shows how the computer-calculated cases distribute over the 60,501 (301×201) points. It happens in this example that only cases 1, 2, 5, 6, and 9 show up in Fig. 4.17, and the other four of the nine cases never appear. Note that more points used leads to smoother case boundaries.

Superimposing Fig. 4.16 onto Fig. 4.17 produces Fig. 4.18, showing ICs and cases at the same time. As already pointed out above, one IC can have several different cases. For example, in IC35 cases 1, 2, 5, and 6 exist. Also, the locations of the ICs and the case distributions have no known relation. The location of an IC depends on the eight individual input space divisions owing to the min( ) operation, which, in turn, depend on the input fuzzy sets. A case (i.e., the values of L and R) is decided not only by all the input fuzzy sets but also by all the output fuzzy sets

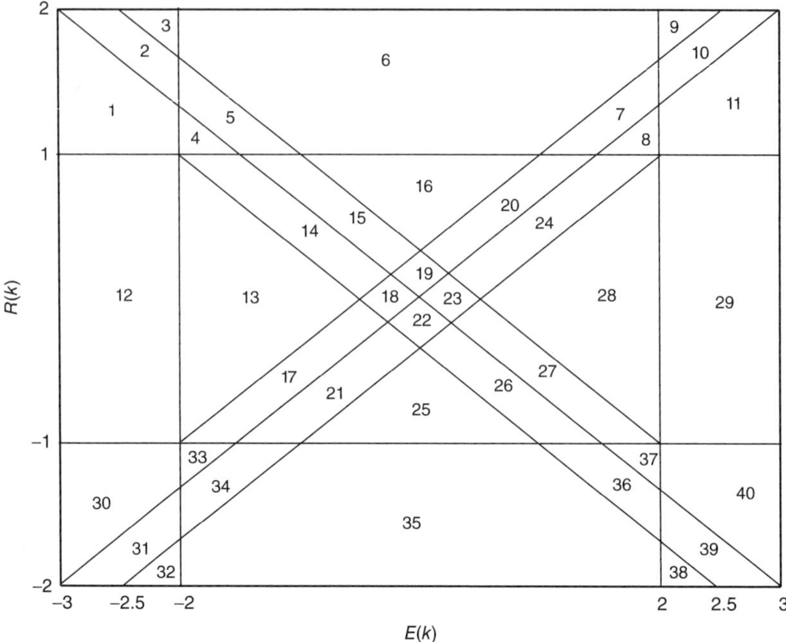

Figure 4.16 Dividing the input space covered by $[-3, 3] \times [-2, 2]$ in to ICs.

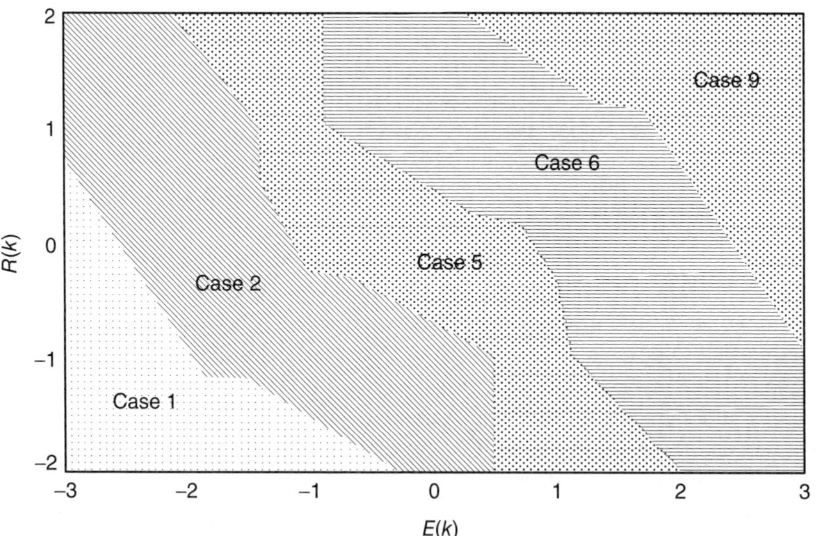

Figure 4.17 Case distribution in the region shown in Fig. 4.16.

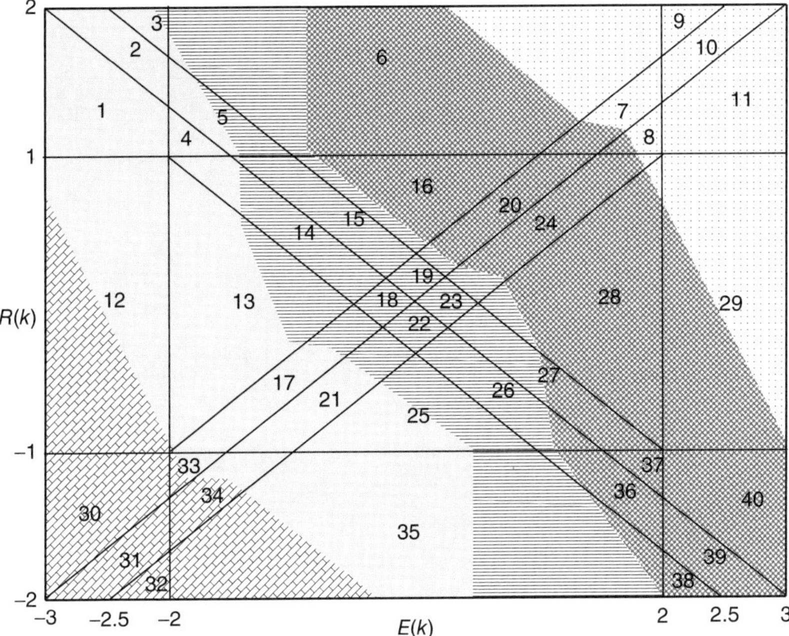

Figure 4.18 Result of superimposing Fig. 4.16 onto Fig. 4.17.

as well as the values of $E(k)$ and $R(k)$. Based on Fig. 4.18, one can use the iterative KM algorithm for the center-of-sets type of reducer and the centroid defuzzifier to derive the analytical structure for each IC that has one case or more than one case. Four ICs are selected, each of which has one case only, and their analytical structures are listed in Table 4.11.

The MATLAB program written for this example is available for the reader to download at this book's publisher's website.

4.8 TSK FUZZY PI AND PD CONTROLLERS—CONFIGURATION 5[5]

4.8.1 Fuzzy PI Controller Configuration

The fuzzy controllers in this section differ from those in the last section because they use (1) TSK fuzzy rules, not Mamdani rules, and (2) T2 singleton fuzzy sets for the output. All the other components are exactly the same (i.e., any type and number of T2 fuzzy sets for the inputs, the interval weighted-average method and the iterative KM algorithm for the center-of-sets type of reducer, and the centroid defuzzifier). Like before, suppose only the following four TSK fuzzy rules (out of

[5]Part of the material in this section is adapted from Zhou and Ying (2012; © 2012, IEEE).

TABLE 4.11 Analytical Structures of T2 Fuzzy Controller in Four Selected ICs

IC No.	$\Delta U(k)$
1	$\frac{127}{150+40E(k)+30R(k)}E(k)+\left[\frac{171}{300+80E(k)+60R(k)}+\frac{1}{12}\right]R(k)$ $+\left[\frac{933}{300+80E(k)+60R(k)}+\frac{149}{60}\right]$
11	$-\frac{101}{210-40E(k)}E(k)+\frac{9}{210-40E(k)}R(k)+\left[\frac{1073}{210-40E(k)}+\frac{33}{10}\right]$
18	$\frac{127}{180+40E(k)}E(k)+\left[\frac{3}{52+16E(k)}-\frac{11}{120}\right]R(k)+\left[\frac{1089}{360+80E(k)}+\frac{27}{10}\right]$
30	$\frac{127}{210-40E(k)}E(k)+\frac{3}{84-16E(k)}R(k)+\left[\frac{63}{21-4E(k)}+\frac{12}{5}\right]$

a total of $N_1 \times N_2$ fuzzy rules) are executed at time k:

R^1: If $E(k)$ is $\widetilde{A}_{1i}$ and $R(k)$ is $\widetilde{A}_{2j}$, then $\Delta U(k) = C_i^j + D_i^j E(k) + E_i^j R(k)$.

R^2: If $E(k)$ is $\widetilde{A}_{1i}$ and $R(k)$ is $\widetilde{A}_{2j+1}$, then $\Delta U(k) = C_i^{j+1} + D_i^{j+1} E(k) + E_i^{j+1} R(k)$.

R^3: If $E(k)$ is $\widetilde{A}_{1i+1}$ and $R(k)$ is $\widetilde{A}_{2j}$, then $\Delta U(k) = C_{i+1}^j + D_{i+1}^j E(k) + E_{i+1}^j R(k)$.

R^4: If $E(k)$ is $\widetilde{A}_{1i+1}$ and $R(k)$ is $\widetilde{A}_{2j+1}$, then $\Delta U(k) = C_{i+1}^{j+1} + D_{i+1}^{j+1} E(k) + E_{i+1}^{j+1} R(k)$.

where the Zadeh fuzzy AND operator is used. Intervals C_i^j, C_i^{j+1}, C_{i+1}^j, C_{i+1}^{j+1}, D_i^j, D_i^{j+1}, D_{i+1}^j, D_{i+1}^{j+1}, E_i^j, E_i^{j+1}, E_{i+1}^j, and E_{i+1}^{j+1} are interval T1 fuzzy sets:

$$C_i^j = [c_i^j - \varpi_i^j, c_i^j + \varpi_i^j] \tag{4.43}$$

$$D_i^j = [d_i^j - \gamma_i^j, d_i^j + \gamma_i^j] \tag{4.44}$$

$$E_i^j = [e_i^j - \varepsilon_i^j, e_i^j + \varepsilon_i^j] \tag{4.45}$$

where c_i^j, d_i^j, and e_i^j denote the centers (i.e., means) of C_i^j, D_i^j, and E_i^j, respectively, whereas ϖ_i^j, γ_i^j, and ε_i^j denote, respectively, the spreads of C_i^j, D_i^j, and E_i^j [E_i^j is an interval that is unrelated to the input variable $E(k)$]. These fuzzy sets are indexed by integer superscripts and subscripts so that the input fuzzy sets and rule consequents are mathematically linked. We use the interval coefficients because (1) there may be situations in which one only knows the ranges for the coefficients in the rule consequents, and (2) a single-valued coefficient is a special case of the interval T1 fuzzy set.

The firing intervals are computed in the same way as in Eqs. (4.33)–(4.36). Because C_i^j, D_i^j, and E_i^j are intervals and $E(k)$ and $R(k)$ are numbers, the computing result for the consequent of rule R^s is an interval:

$$\Delta U^s(k) = [\Delta U_l^s(k), \Delta U_r^s(k)] \tag{4.46}$$

where

$$\Delta U_l^1(k) = c_i^j + d_i^j E(k) + e_i^j R(k) - \varpi_i^j - \gamma_i^j |E(k)| - \varepsilon_i^j |R(k)| \tag{4.47}$$

$$\Delta U_l^2(k) = c_i^{j+1} + d_i^{j+1} E(k) + e_i^{j+1} R(k) - \varpi_i^{j+1} - \gamma_i^{j+1} |E(k)| - \varepsilon_i^{j+1} |R(k)| \tag{4.48}$$

$$\Delta U_l^3(k) = c_{i+1}^j + d_{i+1}^j E(k) + e_{i+1}^j R(k) - \varpi_{i+1}^j - \gamma_{i+1}^j |E(k)| - \varepsilon_{i+1}^j |R(k)| \tag{4.49}$$

$$\Delta U_l^4(k) = c_{i+1}^{j+1} + d_{i+1}^{j+1} E(k) + e_{i+1}^{j+1} R(k) - \varpi_{i+1}^{j+1} - \gamma_{i+1}^{j+1} |E(k)| - \varepsilon_{i+1}^{j+1} |R(k)| \tag{4.50}$$

and

$$\Delta U_r^1(k) = c_i^j + d_i^j E(k) + e_i^j R(k) + \varpi_i^j + \gamma_i^j |E(k)| + \varepsilon_i^j |R(k)| \tag{4.51}$$

$$\Delta U_r^2(k) = c_i^{j+1} + d_i^{j+1} E(k) + e_i^{j+1} R(k) + \varpi_i^{j+1} + \gamma_i^{j+1} |E(k)| + \varepsilon_i^{j+1} |R(k)| \tag{4.52}$$

$$\Delta U_r^3(k) = c_{i+1}^j + d_{i+1}^j E(k) + e_{i+1}^j R(k) + \varpi_{i+1}^j + \gamma_{i+1}^j |E(k)| + \varepsilon_{i+1}^j |R(k)| \tag{4.53}$$

$$\Delta U_r^4(k) = c_{i+1}^{j+1} + d_{i+1}^{j+1} E(k) + e_{i+1}^{j+1} R(k) + \varpi_{i+1}^{j+1} + \gamma_{i+1}^{j+1} |E(k)| + \varepsilon_{i+1}^{j+1} |R(k)| \tag{4.54}$$

The interval arithmetic used here is described in Mendel (2001, Theorem 7–4, p. 228). Similar to the process in the last section, $\Delta U_l^s(k)$ and $\Delta U_r^s(k)$ need to be arranged in their respective ascending orders: $\Delta U_l^{1*}(k) \le \Delta U_l^{2*}(k) \le \Delta U_l^{3*}(k) \le \Delta U_l^{4*}(k)$ and $\Delta U_r^{1*}(k) \le \Delta U_r^{2*}(k) \le \Delta U_r^{3*}(k) \le \Delta U_r^{4*}(k)$ [note that $\Delta U_l^s(k)$ does not necessarily correspond to $\Delta U_l^{s*}(k)$; this also holds true for the relationship between $\Delta U_r^{s*}(k)$ and $\Delta U_r^s(k)$]. Then arrange $\bar{f}^s$ and $\underline{f}^s$ to correspond to $\Delta U_l^{1*}(k) \le \Delta U_l^{2*}(k) \le \Delta U_l^{3*}(k) \le \Delta U_l^{4*}(k)$ and $\Delta U_r^{\bar{1}*}(k) \le \Delta U_r^{2*}(k) \le \Delta U_r^{3*}(k) \le \Delta U_r^{4*}(k)$, respectively, leading to $\bar{f}^{1*}, \bar{f}^{2*}, \bar{f}^{3*}, \bar{f}^{4*}$ and $\underline{f}^{1*}, \underline{f}^{2*}, \underline{f}^{3*}, \underline{f}^{4*}$. Finally

$$\Delta U_l(k) = \frac{\sum_{i=1}^{L} \bar{f}^{i*} \Delta U_l^{i*}(k) + \sum_{j=L+1}^{4} \underline{f}^{j*} \Delta U_l^{j*}(k)}{\sum_{i=1}^{L} \bar{f}^{i*} + \sum_{j=L+1}^{4} \underline{f}^{j*}} \tag{4.55}$$

$$\Delta U_r(k) = \frac{\sum_{i=1}^{R} \underline{f}^{i*} \Delta U_r^{i*}(k) + \sum_{j=R+1}^{4} \bar{f}^{j*} \Delta U_r^{j*}(k)}{\sum_{i=1}^{R} \underline{f}^{i*} + \sum_{j=R+1}^{4} \bar{f}^{j*}} \tag{4.56}$$

$$\Delta U(k) = \tfrac{1}{2}[\Delta U_l(k) + \Delta U_r(k)] \tag{4.57}$$

4.8.2 Deriving the Analytical Structure

Suppose that $\widetilde{A}_{1i}$, $\widetilde{A}_{1i+1}$, $\widetilde{A}_{2j}$, and $\widetilde{A}_{2j+1}$ are as shown in Fig. 4.13. Due to the fact that the T2 TSK controllers differ from the T2 Mamdani controllers only in the rule consequent, the technique for determining the min( ) operation outcome for the TSK controllers is identical to that for the Mamdani controllers. This is because Eqs. (4.33)–(4.36) only involve the input fuzzy sets. Assume $\Delta U_l^{s*}(k) = \Delta U_l^s(k)$ and $\Delta U_r^{s*}(k) = \Delta U_r^s(k)$ and, accordingly, $\overline{f}^{s*} = \overline{f}^s$ and $\underline{f}^{s*} = \underline{f}^s$. The rest of the derivation steps are identical to those presented in the last section since the same type of reducer and defuzzifier are employed. Due to the highly similar derivation processes between the Mamdani and TSK controllers, they share the same two issues mentioned in the last section: (1) multiple cases can exist in one IC, and (2) the analytical structure of the TSK fuzzy controller for an IC can be attained only if L and R are known. For any specific TSK fuzzy controller the exact values of L and R can always be computed as long as all the parameters of the input fuzzy sets, controller inputs, and the coefficients in the rule consequents are numerically available. For a general controller configuration, however, these parameter values are not specified. We know that only one of the cases can be for any point in the input space, but do not know what it is until all the controller parameter values are available.

In the previous section, the input space region of $[S_i, S_{i+1}] \times [M_j, M_{j+1}]$ is divided into 50 ICs (Fig. 4.14). The ICs are valid for the TSK controllers, and we will use them to illustrate the analytical structure derivation process for the TSK controllers. Taking IC27, which is also used as an example in the last section, one can derive 9 different analytical structures for the 9 cases. For case 1, the analytical structure of the TSK controller found by a Mathematica program is

$$
\begin{aligned}
\Delta U(k) = {} & \frac{\Omega_1}{P}E(k) + \frac{\Omega_2}{P}R(k) + \frac{\Omega_3}{P}|E(k)| + \frac{\Omega_4}{P}|R(k)| + \frac{\Omega_5 + \Omega_7}{P}E^2(k) \\
& + \frac{\Omega_6}{P}E(k)R(k) + \frac{\Omega_8}{P}E(k)|R(k)| + \frac{\Omega_{10}}{P}|E(k)|R(k) + \frac{\Omega_{12}}{P}|E(k)||R(k)| \\
& + \frac{\Omega_9 + \Omega_{11}}{P}R^2(k) + \frac{\Omega_{13} + \Omega_{15}}{P}E^2(k)R(k) + \frac{\Omega_{14} + \Omega_{16}}{P}E(k)R^2(k) \\
& + \frac{\Omega_{18} + \Omega_{20}}{P}|E(k)|R^2(k) + \frac{\Omega_{17}}{P}R^3(k) + \frac{\Omega_{19}}{P}|R^3(k)| \\
& + \frac{\Omega}{P}\exp\left\{-\left[\frac{E(k) - S_i}{a_i}\right]^2\right\} + \frac{\Omega_{21}}{P}
\end{aligned}
\tag{4.58}
$$

All the coefficients are provided in Appendix 4A because they are quite messy. Clearly, this controller is not of the PI type because of the $E(k)R(k)$, $|E(k)|R(k)$, $R^2(k)$, $\exp(-\{[E(k) - S_i]/a_i\}^2)$, and the like terms in the numerators of Eq. (4.58). Also note that all the numerators in Eq. (4.58) are constants for the Ω_j.

The program also produces the analytical structures corresponding to the other 8 cases as well as the analytical structure of every combination of an IC and a case (there are a total of $9 \times 50 = 450$ of such combinations). An inspection of these 450 structures reveals that they are all in the form of Eq. (4.58). Only a subset of these 450 expressions correspond to the underlying analytical structures being sought (every point in the input space can have only 1 case, not all the 9 cases). Because the controller is in a general setting without specific values for its parameters, one is unable to know which structures belong to that subset.

From Section 4.4 through this section, we have discussed how to derive the analytical structure of various T2 fuzzy controllers. All of them have two input variables. Note that the structure-deriving methods are not restricted to two input variables. Their overreaching principle is applicable to the T2 fuzzy controllers involving more than two input variables. Nevertheless, for such controllers the derivation task will be (much) more challenging. The biggest difficulty is to divide the n-dimension input space ($n \geq 3$) into a number of n-dimensional ICs. Take $n = 3$ as an example. One must properly divide the 3D input space into many 3D ICs so that Zadeh AND operation can be carried out for each of the fuzzy rules (at least $2^3 = 8$ rules). Also, for all the 8 rules, there will be $2 \times 2^3 = 16$ individual divisions that will be superimposed to generate overall 3D ICs, as opposed to $2 \times 2^2 = 8$ individual divisions (e.g., Fig. 4.5). The number of switching points will also be greater, that is, $L = 1 \sim 7$ and $R = 1 \sim 7$, resulting in a total of $7 \times 7 = 49$ different cases. Obviously, if the dimension is higher than 3, obtaining the ICs will be even more challenging. Fortunately, 2 or 3 input variables are usually sufficient for many important control applications.

4.9 ANALYZING THE DERIVED ANALYTICAL STRUCTURES[6]

The analytical structure of the various T2 fuzzy PI and PD controllers derived in Sections 4.4–4.8 reveal these fuzzy controllers to be equivalent to nonlinear PI or PD controllers with variable gains and a variable control offset. Since the turn of the 1990s the analytical structure of a variety of T1 fuzzy PI and PD controllers has been studied, and their structures are widely known to be nonlinear PI and PD controllers with variable gains, usually without the variable offset term (Ying, 2000). The results in the preceding sections offer an unprecedented opportunity to more insightfully and rigorously examine the differences between the T2 and T1 fuzzy controllers from a control theory standpoint.

We now focus on the T2 fuzzy PI controller in Section 4.5 and analyze characteristics of its analytical structure with respect to that of the corresponding T1 fuzzy PI controller. By corresponding T1 controller, we mean the T1 controller that a T2 controller degenerates to when its T2 components degenerate to T1 types. Obviously, every T2 controller has its corresponding T1 controller. Consequently, the principle of the comparison work applies to the rest of the T2 controllers in this chapter and beyond.

[6]Part of the material in this section is adapted from Du and Ying (2010; © 2010, IEEE).

The derivation in Section 4.5.2 is general in that the controller parameters can be any values. For an easier and more insightful analysis, let us suppose, relative to the parameter values set in Fig. 4.6b, that

$$\begin{aligned} L_1 &= L_2 = L \qquad \theta_1 = \theta_2 = \theta \qquad P_1 = P_2 = P \\ b_1 &= b \qquad b_4 = -b \qquad b_2 = b_3 = 0 \end{aligned} \tag{4.59}$$

where $b > 0$. These assumptions are not restrictive as the analysis is applicable to any other parameter value settings. Note that when $L_1 = L_2$, the rectangular overall region in Fig. 4.6b becomes a square. Due to Eq. (4.59), the expressions for the variable gains and the variable control offset are substantially simplified. The variable proportional gain, $K_p(e(k), r(k))$, of the T2 fuzzy PI controller is already listed IC by IC in Table 4.5. The expressions for the integral gain, $K_i(e(k), r(k))$, and the offset are provided in Tables 4.12 and 4.13, respectively. The mathematical structures of the variable integral gain are similar to those of the variable proportional gain shown in Table 4.5. These parameter settings will be used to conduct the following analysis. For brevity, only the fuzzy PI controllers will be used without involvement of the fuzzy PD controllers and only $K_p(e(k), r(k))$ will be discussed [the result is applicable to $K_i(e(k), r(k)]$ due to the mathematical similarity between $K_p(e(k), r(k))$ and $K_i(e(k), r(k))$.

4.9.1 Structural Connection with the Corresponding T1 Fuzzy PI Controller

The T2 fuzzy PI controller contains its corresponding T1 fuzzy PI controller as a special case when $\theta_1 = \theta_2 = 0$. Therefore, $K_p(e(k), r(k))$ of the corresponding T1 fuzzy PI controller is obtained by letting $\theta = 0$ in Table 4.5:

$$K_p(e(k), r(k)) = \begin{cases} \dfrac{k_{\Delta U} k_r b}{2\left(2L - k_e\left|e(k)\right|\right)} & \text{IC1 and IC3} \\[2ex] \dfrac{k_{\Delta U} k_r b}{2(2L - k_r|r(k)|)} & \text{IC2 and IC4} \end{cases}$$

Note that when $\theta = 0$, IC5–IC16 no longer exists (Fig. 4.19). Incidentally, the proportional-gain expression for the T1 fuzzy controller is exactly the same as that derived for the T1 fuzzy controller with the same configuration, which was studied before the era of T2 fuzzy control (e.g., Ying et al., 1990). This fact also indirectly validates the correctness of the T2 structure derivation results. Obviously, the proportional gains of the T2 and T1 PI controllers share similar mathematical structures.

To better present the similarities, how the proportional gains vary with $e(n)$ and $r(n)$ in IC1–IC4 are plotted in Fig. 4.20 where without loss of generality, $\theta = 0.3$, $L = b = 1$, and $k_e = k_r = k_{\Delta U} = 1$. The characteristics of the variable gains indeed look quite similar. The control surfaces of these T2 and T1 controllers under this

TABLE 4.12 Variable Integral-gain $K_i(e(k), r(k))$ when T2 Controller Uses Parameter Values Given in Eq. (4.59)

IC No.	$K_i(e(k), r(k))$
1 and 3	$\frac{k_{\Delta U}k_e b}{32}\left[\frac{4}{2L(1-\theta)-k_e\vert e(k)\vert}+\frac{4}{2L(1+\theta)-k_e\vert e(k)\vert}+\frac{1}{2L(1-2\theta)-k_e\vert e(k)\vert}+\frac{1}{2L(1+2\theta)-k_e\vert e(k)\vert}+\frac{6}{2L-k_e\vert e(k)\vert}\right]$
2 and 4	$\frac{k_{\Delta U}k_e b}{32}\left[\frac{4}{2L(1-\theta)-k_r\vert r(k)\vert}+\frac{4}{2L(1+\theta)-k_r\vert r(k)\vert}+\frac{1}{2L(1-2\theta)-k_r\vert r(k)\vert}+\frac{1}{2L(1+2\theta)-k_r\vert r(k)\vert}+\frac{6}{2L-k_r\vert r(k)\vert}\right]$
5, 8, 9 and 12	$\frac{k_{\Delta U}k_e b}{32}\left[\frac{2}{2L-k_e\vert e(k)\vert+k_r\vert r(k)\vert}+\frac{2}{5L+6\theta L-2k_e\vert e(k)\vert-k_r\vert r(k)\vert}+\frac{2}{3L+2\theta L-2k_e\vert e(k)\vert+k_r\vert r(k)\vert}+\frac{2}{4L+4\theta L-k_e\vert e(k)\vert-k_r\vert r(k)\vert}+\frac{2}{2L+2\theta L-k_e\vert e(k)\vert}+\frac{4}{3L+2\theta L-k_e\vert e(k)\vert}\right]$
6 and 10	$\frac{k_{\Delta U}k_e b}{32}\left[\frac{2}{3L+2\theta L+k_e\vert e(k)\vert-2k_r\vert r(k)\vert}+\frac{4}{2L+k_e\vert e(k)\vert-k_r\vert r(k)\vert}+\frac{2}{L+2\theta L-k_e\vert e(k)\vert}+\frac{4}{3L+2\theta L-k_r\vert r(k)\vert}+\frac{1}{2L+2\theta L-k_r\vert r(k)\vert}+\frac{1}{L}\right]$
7 and 11	$\frac{k_{\Delta U}k_e b}{32}\left[\frac{2}{5L+6\theta L-k_e\vert e(k)\vert-2k_r\vert r(k)\vert}+\frac{4}{4L+4\theta L-k_e\vert e(k)\vert-k_r\vert r(k)\vert}+\frac{4}{3L+2\theta L-k_e\vert e(k)\vert}+\frac{1}{2L+2\theta L-k_r\vert r(k)\vert}+\frac{2}{3L+2\theta L-k_r\vert r(k)\vert}+\frac{1}{L}\right]$
13 and 15	$\frac{k_{\Delta U}k_e b}{32}\left[\frac{2}{3L-2\theta L-k_e\vert e(k)\vert}+\frac{4}{3L+2\theta L-k_e\vert e(k)\vert}+\frac{2}{2L+2\theta L-k_e\vert e(k)\vert}+\frac{2}{3L+6\theta L-k_e\vert e(k)\vert}+\frac{1}{2L+4\theta L-k_e\vert e(k)\vert}+\frac{1}{2L-k_e\vert e(k)\vert}\right]$
14 and 16	$\frac{k_{\Delta U}k_e b}{32}\left[\frac{4}{3L-2\theta L-k_r\vert r(k)\vert}+\frac{8}{3L+2\theta L-k_r\vert r(k)\vert}+\frac{2}{2L+2\theta L-k_r\vert r(k)\vert}+\frac{4}{3L+6\theta L-k_r\vert r(k)\vert}+\frac{1}{2L+4\theta L-k_r\vert r(k)\vert}+\frac{1}{2L-k_r\vert r(k)\vert}+\frac{1}{L(1+2\theta)}+\frac{1}{L(1-2\theta)}+\frac{2}{L}\right]$

TABLE 4.13 Variable Offset $\delta(e(k), r(k))$ When T2 Controller Uses Parameter Values Given in Eq. (4.59)

IC No.	$\delta(e(k), r(k))$
1, 2, 3, and 4	0
5	$\frac{k_{\Delta U}b}{32}\left\{3-2\theta+\frac{4L}{3L+2L\theta-k_r[r(k)]}+\frac{8L(1-2\theta)}{3L+2L\theta-k_e[e(k)]}+\frac{L(1-6\theta)}{2L+2L\theta-k_e[e(k)]}+\frac{2L(1-6\theta)}{2L+k_r[r(k)]-k_e[e(k)]}+\frac{2L(3-2\theta)}{4L+4L\theta-k_e[e(k)]-k_r[r(k)]}+\frac{2L(1-2\theta)}{5L+6L\theta-2k_e[e(k)]-k_r[r(k)]}+\frac{2L(1-2\theta)}{L-2L\theta+k_r[r(k)]}-\frac{8L\theta}{3L+2L\theta-2k_e[e(k)]+k_r[r(k)]}\right\}$
6	$\frac{k_{\Delta U}b}{32}\left\{3-2\theta+\frac{4L}{3L+2L\theta-k_e[e(k)]}+\frac{8L(1-2\theta)}{3L+2L\theta-k_r[r(k)]}+\frac{L(1-6\theta)}{2L+2L\theta-k_r[r(k)]}+\frac{2L(1-6\theta)}{2L+k_e[e(k)]-k_r[r(k)]}+\frac{2L(3-2\theta)}{4L+4L\theta-k_e[e(k)]-k_r[r(k)]}+\frac{2L(1-2\theta)}{5L+6L\theta-k_e[e(k)]-2k_r[r(k)]}+\frac{2L(1-2\theta)}{L-2L\theta+k_e[e(k)]}-\frac{8L\theta}{3L+2L\theta+k_e[e(k)]-2k_r[r(k)]}\right\}$
7	$\frac{k_{\Delta U}b}{32}\left\{1+2\theta-\frac{2L(1+2\theta)}{2L-k_e[e(k)]-k_r[r(k)]}-\frac{L(1+2\theta)}{2L+2L\theta-k_r[r(k)]}-\frac{2L(1+2\theta)}{3L+2L\theta-k_e[e(k)]-2k_r[r(k)]}+\frac{2L(1+2\theta)}{4L+4L\theta+k_e[e(k)]-k_r[r(k)]}+\frac{2L(1+2\theta)}{3L+2L\theta-k_e[e(k)]}\right\}$
8	$\frac{k_{\Delta U}b}{32}\left\{\frac{2L(1+2\theta)}{2L+k_e[e(k)]+k_r[r(k)]}+\frac{L(1+2\theta)}{2L+2L\theta+k_e[e(k)]}-\frac{2L(1+2\theta)}{3L+2L\theta-k_r[r(k)]}+\frac{2L(1+2\theta)}{3L+2L\theta+2k_e[e(k)]+k_r[r(k)]}-\frac{2L(1+2\theta)}{4L+4L\theta+k_e[e(k)]-k_r[r(k)]}-1-2\theta\right\}$
9	$\frac{k_{\Delta U}b}{32}\left\{2\theta-3-\frac{8L(1-2\theta)}{3L+2L\theta+k_e[e(k)]}-\frac{2L(1-6\theta)}{4L+4L\theta+2k_e[e(k)]}-\frac{4L}{3L+2L\theta+k_r[r(k)]}-\frac{2L(1-6\theta)}{2L+k_e[e(k)]-k_r[r(k)]}-\frac{2L(3-2\theta)}{4L+4L\theta+k_e[e(k)]+k_r[r(k)]}-\frac{2L(1-2\theta)}{5L+6L\theta+2k_e[e(k)]+k_r[r(k)]}+\frac{8L\theta}{3L+2L\theta+2k_e[e(k)]-k_r[r(k)]}+\frac{2L(1-2\theta)}{2L\theta-L+k_r[r(k)]}\right\}$

TABLE 4.13 (*Continued*)

IC No.	$\delta(e(k),\ r(k))$
10	$\frac{k_{\Delta U}b}{32}\left\{2\theta-3-\frac{8L(1-2\theta)}{3L+2L\theta+k_r[r(k)]}-\frac{2L(1-6\theta)}{4L+4L\theta+2k_r[r(k)]}-\frac{4L}{3L+2L\theta+k_e[e(k)]}-\frac{2L(1-6\theta)}{2L-k_e[e(k)]+k_r[r(k)]}-\frac{2L(3-2\theta)}{4L+4L\theta+k_e[e(k)]+k_r[r(k)]}-\frac{2L(1-2\theta)}{5L+6L\theta+k_e[e(k)]+2k_r[r(k)]}+\frac{8L\theta}{3L+2L\theta+2k_r[r(k)]-k_e[e(k)]}+\frac{2L(1-2\theta)}{2L\theta-L+k_e[e(k)]}\right\}$
11	$\frac{k_{\Delta U}b}{32}\left\{\frac{2L(1+2\theta)}{2L+k_e[e(k)]+k_r[r(k)]}-\frac{2L(1+2\theta)}{3L+2L\theta-k_e[e(k)]}+\frac{2L(1+2\theta)}{4L+4L\theta+2k_r[r(k)]}+\frac{2L(1+2\theta)}{3L+2L\theta+k_e[e(k)]+2k_r[r(k)]}-\frac{2L(1+2\theta)}{4L+4L\theta-k_e[e(k)]+k_r[r(k)]}-1-2\theta\right\}$
12	$\frac{k_{\Delta U}b}{32}\left\{1+2\theta-\frac{2L(1+2\theta)}{2L-k_e[e(k)]-k_r[r(k)]}-\frac{2L(1+2\theta)}{4L+4L\theta-2k_e[e(k)]}-\frac{2L(1+2\theta)}{3L+2L\theta-2k_e[e(k)]-k_r[r(k)]}+\frac{2L(1+2\theta)}{4L+4L\theta-k_e[e(k)]+k_r[r(k)]}+\frac{2L(1+2\theta)}{3L+2L\theta+k_r[r(k)]}\right\}$
13	$\frac{k_{\Delta U}b}{32}\left\{4+\frac{2L(1-6\theta)}{3L-2L\theta-k_e[e(k)]}+\frac{4L(1-2\theta)}{3L+2L\theta-k_e[e(k)]}-\frac{4L\theta}{2L-k_e[e(k)]}+\frac{2L(1+2\theta)}{3L+6L\theta-k_e[e(k)]}-\frac{4L\theta}{2L+2L\theta-k_e[e(k)]}\right\}$
14	$\frac{k_{\Delta U}b}{32}\left\{4+\frac{2L(1-6\theta)}{3L-2L\theta-k_r[r(k)]}+\frac{4L(1-2\theta)}{3L+2L\theta-k_r[r(k)]}+\frac{2L(1+2\theta)}{3L+6L\theta-k_r[r(k)]}-\frac{4L\theta}{2L-k_r[r(k)]}-\frac{4L\theta}{2L+2L\theta-k_r[r(k)]}\right\}$
15	$\frac{k_{\Delta U}b}{32}\left\{\frac{4L\theta}{2L+k_e[e(k)]}-\frac{2L(1-6\theta)}{3L-2L\theta+k_e[e(k)]}-\frac{4L(1+2\theta)}{3L+2L\theta+k_e[e(k)]}-4+\frac{4L\theta}{2L+2L\theta+k_e[e(k)]}-\frac{2L(1+2\theta)}{3L+6L\theta+k_e[e(k)]}\right\}$
16	$\frac{k_{\Delta U}b}{32}\left\{\frac{4L\theta}{2L+k_r[r(k)]}-\frac{2L(1-6\theta)}{3L-2L\theta+k_r[r(k)]}-4+\frac{4L\theta}{2L+2L\theta+k_r[r(k)]}-\frac{4L(1-2\theta)}{3L+2L\theta+k_r[r(k)]}-\frac{2L(1+2\theta)}{3L+6L\theta+k_r[r(k)]}\right\}$

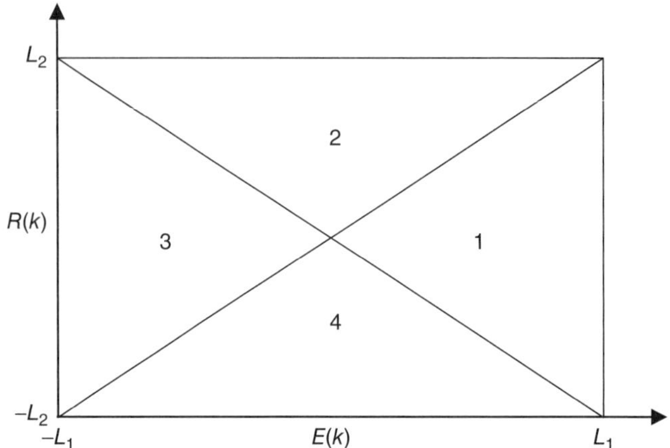

Figure 4.19 ICs for the T1 fuzzy PI controller comparable to the T2 fuzzy PI controller when $\theta_1 = \theta_2 = 0$.

parameter setting are also provided (Fig. 4.21), which are quite similar. It should be understood that the similarities in the variable gains and control surface are independent of the values of the parameters involved. These similarities should not come as a surprise. After all, the T2 fuzzy PI controller is equal to the average of two T1 fuzzy PI controllers [i.e., $\Delta U_j^{\min}(n)$ represents one controller and $\Delta U_j^{\max}(n)$ represents another and their average is due to the centroid defuzzifier].

Analytically speaking, the use of the T2 input fuzzy sets makes infinitely many different (but somewhat similar) nonlinear PI controllers as opposed to only one controller of the same kind in the case of the T1 fuzzy controllers. The characteristics of the gains are parameterized by θ and are adjustable by it (this is the reason why there exist infinite versions of similar nonlinear PI controllers). We now analyze how θ influences the characteristics of the variable gains.

4.9.2 Characteristics of the Variable Gains of the T2 Fuzzy PI Controller

Our analysis will focus on the T2 fuzzy PI controller in IC1–IC4 only. This is the most important region for the controller operation because (1) it contains the system's equilibrium point $(k_e e(k), k_r r(k)) = (0, 0)$, and (2) when the controller operates outside of these ICs, at least one of the two input variables is fuzzified by the flat portions of the T2 fuzzy sets (i.e., when their membership functions are either 0 or 1; see Fig. 4.2). Furthermore, one only needs to analyze the proportional gain because the integral gain is proportional to it due to

$$\frac{K_p(e(k), r(k))}{K_i(e(k), r(k))} = \frac{k_r}{k_e} \tag{4.60}$$

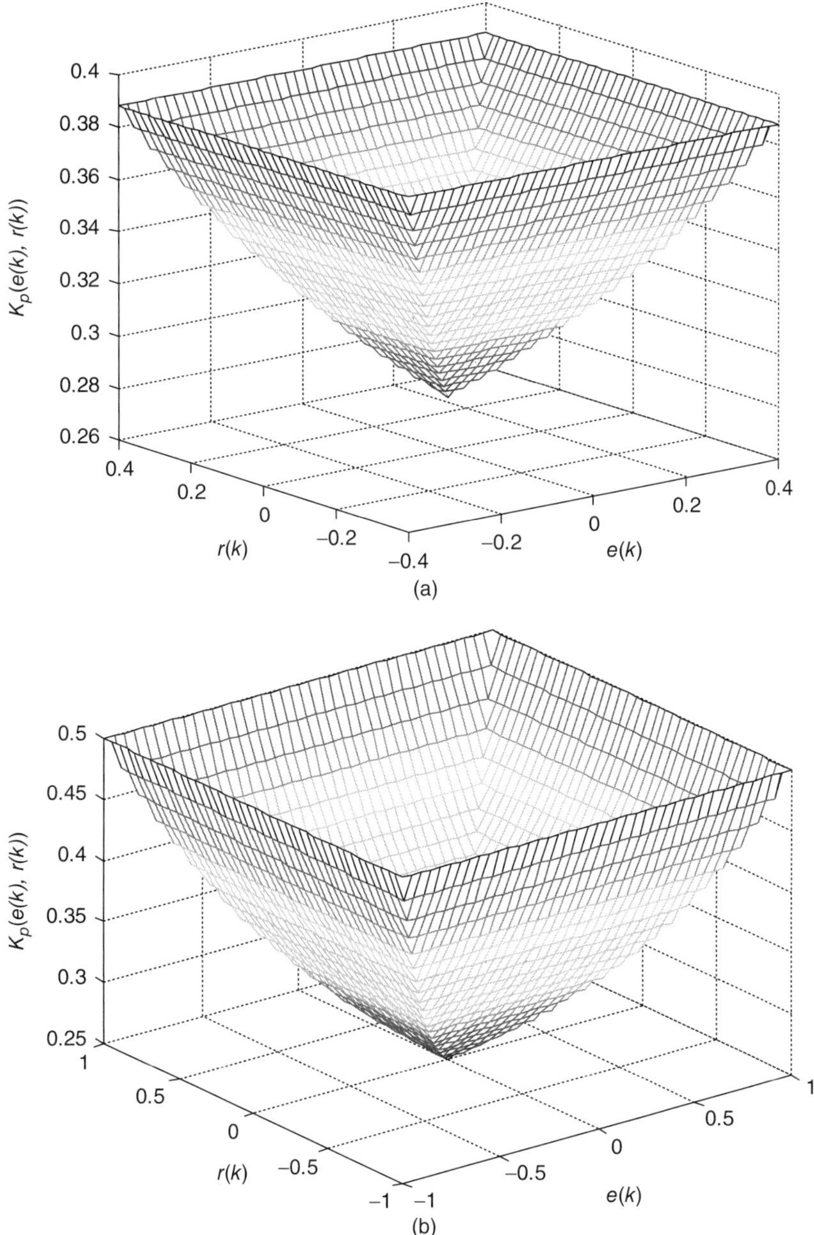

Figure 4.20 Three-dimensional plots of the variable proportional gain of the (a) T2 fuzzy PI controller, and that of its corresponding (b) T1 fuzzy PI controller when $\theta = 0.3$, $L = b = 1$, and $k_e = k_r = k_{\Delta U} = 1$.

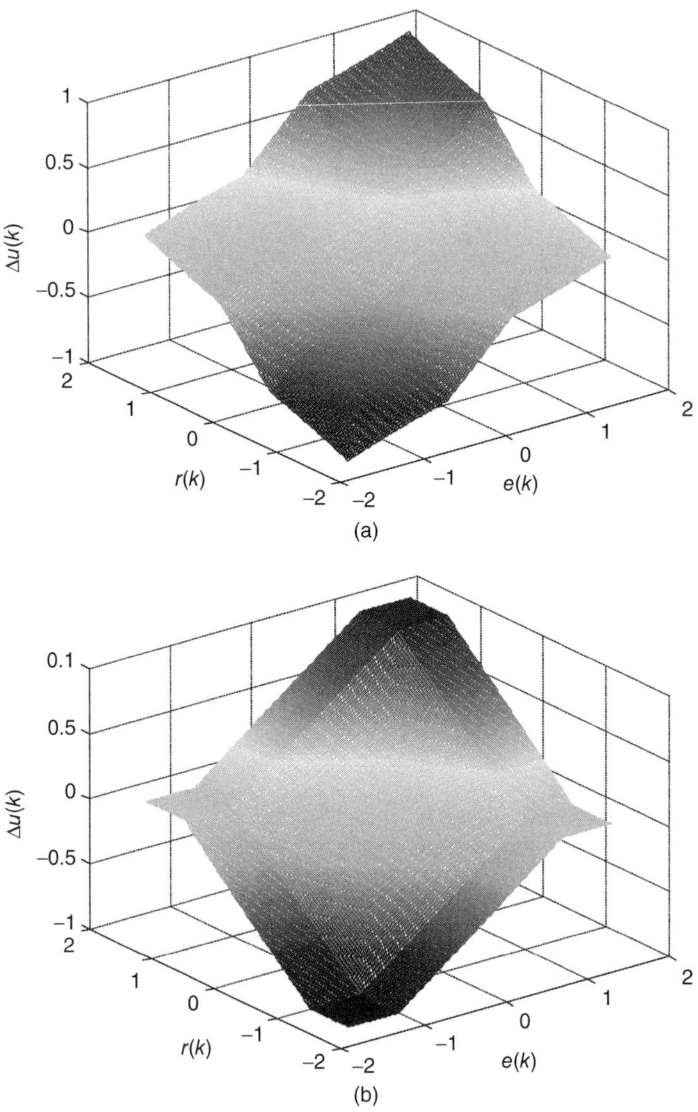

Figure 4.21 Control surface of the (a) T2 fuzzy PI controller and that of its corresponding (b) T1 fuzzy PI controller when $\theta = 0.3$, $L = b = 1$, and $k_e = k_r = k_{\Delta U} = 1$.

Moreover, note that (1) the gain expression for IC1 and IC3 is the same as that for IC2 and IC4 if $k_e|e(k)|$ and $k_r|r(k)|$ exchange their positions; and (2) the gain expressions for IC1 and IC3 are symmetrical with respect to the $r(k)$ axis, and those for IC2 and IC4 are symmetrical with respect to the $e(k)$ axis. Hence, for the T2 controller, it suffices to study the proportional-gain variation characteristics for IC1, which is

$$K_p(e(k),\ r(k)) = \frac{k_{\Delta U} k_r b}{32} \left[\frac{4}{2L(1-\theta) - k_e e(k)} + \frac{4}{2L(1+\theta) - k_e e(k)} + \frac{1}{2L(1-2\theta) - k_e e(k)} + \frac{1}{2L(1+2\theta) - k_e e(k)} + \frac{6}{2L - k_e e(k)} \right] \tag{4.61}$$

The valid range for θ is $[0, 0.5)$ (Fig. 4.2).

Property 4.1 The size of the area occupied by IC1–IC4 decreases with the increase of θ.

Furthermore, θ determines the size of each IC, including IC1–IC4. The larger the value of θ, the smaller the area of IC1–IC4. More specifically, each side of the square formed by IC1–IC4 is $2L - 4L\theta$.

Property 4.2 Keeping θ constant, the proportional gain monotonically increases with the increase of $e(k)$. The maximal proportional-gain $K_{max}(\theta)$ and the minimal proportional-gain $K_{min}(\theta)$ are reached when $k_e e(k) = L - 2\theta L$ and $k_e e(k) = 0$, respectively:

$$K_{max}(\theta) = \frac{k_{\Delta U} k_r b}{32L} \left(\frac{4}{1+4\theta} + \frac{1}{1-2\theta} + \frac{1}{1+6\theta} + \frac{6}{1+2\theta} + 4 \right) \tag{4.62}$$

$$K_{min}(\theta) = \frac{k_{\Delta U} k_r b}{32L} \left(\frac{4}{1-\theta^2} + \frac{1}{1-4\theta^2} + 3 \right) \tag{4.63}$$

This property can be understood from Eq. (4.61). When $e(k)$ increases, the values of all the five fraction expressions in Eq. (4.61) will monotonically increase, leading to a higher proportional gain. This can also be seen from the 3D plot in Fig. 4.20a. It is trivial to determine the maximal gain and the minimal gain as stated in Eqs. (4.62) and (4.63). Note that $k_e e(k) = L - 2\theta L$ indicates that $k_e e(k)$ reaches one of the boundaries of IC1 (Fig. 4.6b).

Property 4.3 Minimal proportional-gain $K_{min}(\theta)$ increases with θ monotonically, whereas $K_{max}(\theta)$ does not, leading to a nonmonotonic relationship for the gain ratio $K_{max}(\theta)/K_{min}(\theta)$ with respect to θ.

It is obvious that $K_{min}(\theta)$ increases as the denominators of the two fraction expressions in Eq. (4.63) increase with θ. In order to prove that $K_{max}(\theta)$ is nonmonotonic, solve $dK_{max}{}^{(\theta)}/_{\theta} = 0$ and find that $K_{max}(\theta)$ achieves its minimum 0.387 when $\theta = 0.272$. That means $K_{max}(\theta)$ decreases when θ is in [0, 0.272] and increases when θ belongs to [0.272, 0.5]. Figure 4.22a shows how $K_{max}(\theta)$ and $K_{min}(\theta)$ vary with θ. Subsequently, $K_{max}(\theta)/K_{min}(\theta)$ shows a nonmonotonic relationship with respect to θ, which is plotted in Fig. 4.22b. It can be calculated that the minimum ratio 134.1% takes place when $\theta = 0.383$ and the ratio becomes 200% either when $\theta = 0$ or when θ approaches to 0.5. Importantly though, the area of IC1–IC4 becomes smaller and smaller as θ goes closer and closer to 0.5. Thus, too large a θ may not necessarily be desirable.

As a reference, it can be computed on the basis of the above analytical structure for the T1 fuzzy PI controller; the maximal and minimal gains are $K_{\max}(0) = K_{\Delta U}K_r b/(2L)$ and $K_{\min}(0) = K_{\Delta U}K_r b/(4L)$, respectively. Consequently, the gain ratio range is [100%, 200%], meaning the variable gain can be anywhere between the minimal gain and up to two times of it.

These three properties are investigated for the variable proportional gain in IC1. They hold true for IC3 by letting $k_e e(k)$ be $k_e|e(k)|$ where $k_e e(k) \in [-L+2\theta L, 0]$. They are also valid for IC2 and IC4—just replace $K_e e(k)$ by $k_r r(k)$ for IC2 or by $k_r|r(k)|$ for IC4.

The variable integral gain shares the same three properties due to Eq. (4.60).

With these three properties, the T2 and T1 fuzzy PI controllers can be compared in terms of gain characteristics. Property 1 says that the size of the area occupied by IC1–IC4 decreases with the increase of θ. Since the T1 fuzzy PI controller has the minimum value of θ, that is, $\theta = 0$, the area size of the T1 fuzzy PI (or PD) controller is always larger than that of the T2 fuzzy PI controller when $\theta \neq 0$. Property 2 indicates the changing trend of the variable gains for the T2 fuzzy PI controller with respect to the input variables. Because the T1 controller is the special case of the T2 controller, the variable gains of the T1 controller have the same tendency. For the T1 fuzzy PI controller, the minimal proportional gain is

$$K_{\min}(0) = \frac{K_{\Delta U}K_r b}{4L}$$

when $k_e e(k) = 0$ and the maximal proportional gain is

$$K_{\max}(0) = \frac{k_{\Delta U}k_r b}{2L}$$

Property 3 indicates that the minimal proportional-gain gain of the T1 fuzzy controller is always smaller than that of the T2 fuzzy controller. The gain ratio of the T1 fuzzy controller is always greater than or equal to that of the T2 fuzzy controller.

This line of analyses is directly applicable to the related T2 and T1 fuzzy PD controllers and can be extended to the other ICs of the T2 fuzzy controllers.

4.10 DESIGN GUIDELINES FOR THE T2 FUZZY PI AND PD CONTROLLERS[7]

The derived analytical structures in Sections 4.4–4.8 not only enable the insightful, precise analyses conducted in the previous section but also make it possible to empower design of the controllers. We now discuss design of the T2 fuzzy PI

[7]Part of the material in this section is adapted from Du and Ying (2010; © 2010, IEEE).

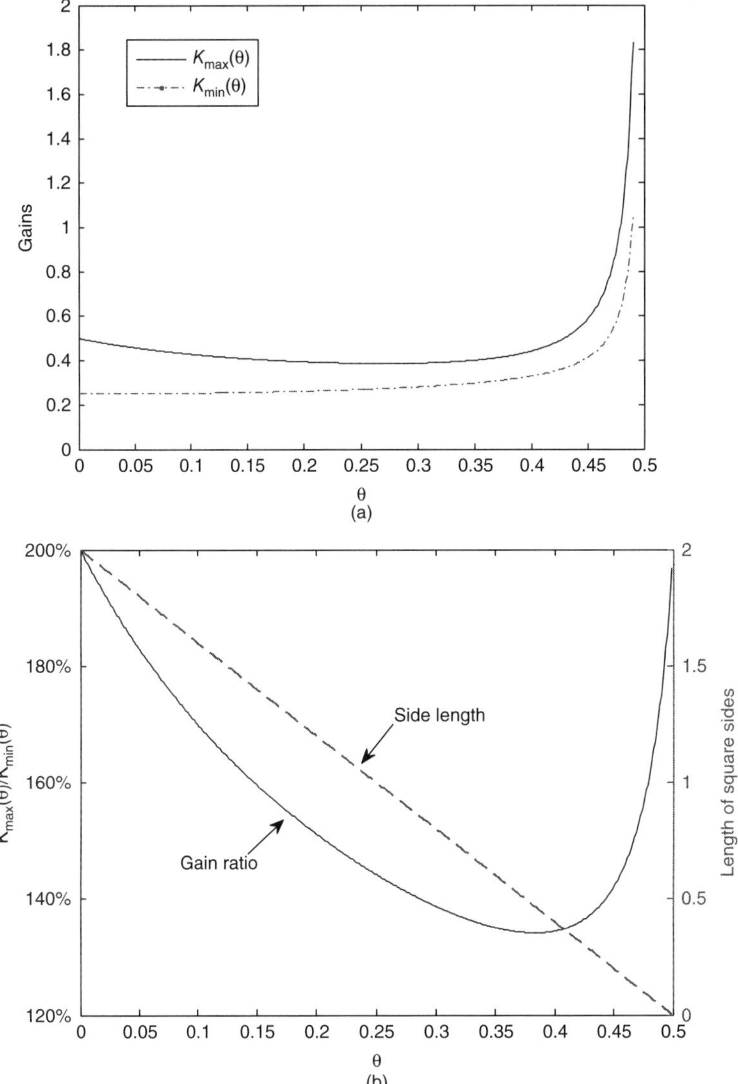

Figure 4.22 (a) Variations of $K_{max}(\theta)$ and $K_{min}(\theta)$ with θ (up to 0.49) for the T2 fuzzy PI controller, and (b) how the gain ratio (solid line) and the side length of the square area formed by IC1 to IC4 (dotted line) vary with θ.

(or PD) control systems, with a focus on the controller configuration in Sections 4.5. This discussion has relevance to the other T2 fuzzy PI and PD controllers in this book and beyond.

Even though the T2 fuzzy PI and PD controllers in Section 4.5 are relatively simple in terms of the numbers of input fuzzy sets (only 2 fuzzy sets per input variable and they are all linear fuzzy sets), output fuzzy sets (4 singleton fuzzy sets), and fuzzy rules (only 4 of them), the number of design parameters is still as many as 11, very high if compared to the most popular controller in the world—the PID controller that has only 3. The parameters are θ_1, θ_2, L_1, L_2, k_e, k_r, $k_{\Delta U}$, b_1, b_2, b_3, and b_4. Manually tuning all of them in a trial-and-error fashion is impractical. Design guidelines below may reduce the burden.

4.10.1 Determination of θ_1 and θ_2 Values

As a benefit of the analytical structure derivation and gain characteristics analysis, one is now able to treat the T2 fuzzy PI and PD controllers as nonlinear controllers with variable gains, rather than blackbox controllers. Because of the availability of the analytical structures, the roles that θ_1 and θ_2 play can now be clearly understood from a control theory standpoint as opposed to vague and subjective measures of uncertainties of the T2 input fuzzy sets treated from the angle of linguistic knowledge representation (e.g., the larger the θ_1 or θ_2, the more uncertainty in the input fuzzy sets are captured and represented). Furthermore, the system analysis or design techniques in the well-established nonlinear control theory are in principle applicable now. These include local, global, or bounded-input bounded-output stability, Lyapunov stability analysis, phase plane analysis, and describing function analysis. To reduce the number of design parameters, we suggest to choosing $\theta_1 = \theta_2 = \theta$ for most applications. Compared to the T1 fuzzy controllers, θ represents an extra degree of freedom for the system developer. Depending on control requirements, one may choose different gain ratios by using different values of θ (Fig. 4.22b). If a higher ratio is desired, use of a smaller θ may be appropriate. For certain nonlinear systems, a higher gain ratio may be beneficial to control performance. For instance, the analytical structure and gain variation characteristics of the T1 Mamdani fuzzy PI controller whose gain ratio was inherently fixed at 200% was studied and the controller was successfully applied to the real-time control of mean arterial pressure in postoperative open-heart patients during their recovery in the Cardiac Surgical Intensive Care Unit (Ying et al., 1992). Whether a higher gain ratio is always desirable for T2 fuzzy control remains an open question and is worth further investigation.

As pointed out earlier, the larger the θ, the smaller the square area formed by IC1–IC4 (Fig. 4.22b). Recall that each side of the square is $2L - 4L\theta$ and when θ approaches to 0.5, the square will be close to disappearing. Too small a square may not be desirable. Thus, the value of θ should be appropriately selected for the system to be controlled. Because this issue is being discussed in the context of nonlinear control, a formula is not expected to be derived that is capable of calculating the proper value of θ for any given system. A certain amount of experimentation (e.g., computer simulation) seems to be inevitable in order to find the optimal θ for a specific system. It is possible that a number of different combinations of θ values and square area sizes will produce the same or similar performances that are all satisfactory to the developer. This scenario is universally encountered throughout designs of conventional control systems, linear or nonlinear.

4.10.2 Determination of the Remaining Nine Parameter Values

The remaining parameters are $L_1, L_2, k_e, k_r, k_{\Delta U}, b_1, b_2, b_3$, and b_4; and L_1 and k_e are related, as are L_2 and k_r. In a practical application, L_1 and L_2 represent the ranges of the input physical variables and hence are specified. The ranges should be used as L_1 and L_2. If the actual ranges are unknown (e.g., in a study with a hypothetical setting), it is appropriate to set L_1 and L_2 at arbitrary levels, say 1. This is because k_e and k_r are allowed to take any values, which enables one to always find suitable values for these two parameters regardless of the values of L_1 and L_2. For many control needs, letting $b_2 = b_3 = 0$ and $b_1 = -b_4$ is proper (Fig. 4.4). Since $k_{\Delta U}$, b_1, and b_4 are related (they always appear as a product), one should fix b_1 and b_4 first (e.g., let $b_1 = -b_4 = 1$). What remains to be determined are k_e, k_r, and $k_{\Delta U}$.

Based on the analytical structures derived above, they can be determined using the technique developed in Ying (1994b) for the related T1 fuzzy PI and PD controllers. The basic idea is this: At the equilibrium point [i.e., when $e(k) = r(k) = 0$], both the proportional gain and integral gain of the T2 fuzzy PI controller become fixed. As an example, from the above assumptions about the parameters (e.g., $L_1 = L_2 = 1$), the proportional gain for IC1–IC4 given in Table 4.5 becomes

$$K_p(0,0) = \frac{k_{\Delta U} k_r}{64}\left(\frac{4}{1-\theta} + \frac{4}{1+\theta} + \frac{1}{1-2\theta} + \frac{1}{1+2\theta} + 6\right)$$

Owing to Eq. (4.60),

$$K_i(0,0) = \frac{k_{\Delta U} k_e}{64}\left(\frac{4}{1-\theta} + \frac{4}{1+\theta} + \frac{1}{1-2\theta} + \frac{1}{1+2\theta} + 6\right)$$

Now, apply the linear PI controller to the system to be controlled. The system can be as complex as being nonlinear, time varying, and with time delay. Worse (but more realistically), its mathematical model is unknown. Tune the proportional gain and integral gain of the linear PI controller to achieve a reasonable system output performance (e.g., the output is merely stable) such that the proportional gain and integral gain at that time are K_p^* and K_i^*, respectively. This can usually be achieved rather easily and quickly because it is well known that this can be done for a linear PI controller. The initial values of k_e, k_r, and $k_{\Delta U}$ (and θ, if desired) can be determined by using

$$\frac{k_{\Delta U} k_r}{64}\left(\frac{4}{1-\theta} + \frac{4}{1+\theta} + \frac{1}{1-2\theta} + \frac{1}{1+2\theta} + 6\right) = K_p^*$$

$$\frac{k_{\Delta U} k_r}{64}\left(\frac{4}{1-\theta} + \frac{4}{1+\theta} + \frac{1}{1-2\theta} + \frac{1}{1+2\theta} + 6\right) = K_i^*$$

There are three (or four, if θ is included) unknowns in two equations. Hence the solution is not unique, and indeed there are infinite sets of solutions. One solution can be selected that uses the initial values of the parameters to control the system, after which they can be fine tuned to achieve the desired system performance.

This technique is applicable to all the T2 fuzzy PI (and PD) controllers in this chapter because their analytical structures are known—nonlinear PI (and PD) controllers with variable gains plus a variable control offset, which are similar to the fuzzy PI controller in the present Section.

4.11 SUMMARY

Analytical structure-deriving techniques for four different classes of interval T2 Mamdani fuzzy controllers and one type of interval T2 TSK fuzzy controllers, all of which employ Zadeh AND operator (see Table 4.1), have been presented. Regardless of the controller configurations, one common step in the structure derivation is the necessity of dividing the input space into a number of ICs thanks to the min( ) operation. The number of ICs is much more for the T2 controllers than for the comparable T1 fuzzy controllers because every input fuzzy set for the T2 controller has two membership functions (i.e., the upper and lower membership functions) instead of just one as in the case of the T1 fuzzy sets used by the T1 controllers (Section 4.9.1 shows an example). This difference alone makes the derivation and subsequent analysis of the analytical structure of the T2 controllers more complicated.

Table 4.14 summarizes the analytical structure results obtained by the five structure-deriving techniques for the five classes of the T2 controllers covered in this chapter. The results for the controller configurations 1–3 are all general—the T2 fuzzy controllers are equivalent to the nonlinear PI (or PD) controllers with variable gains plus a variable control offset. Only part of the result for the controller configuration 4 is general—the T2 fuzzy controllers are equivalent to nonlinear PI (or PD) controllers with variable gains plus a variable control offset if and only if all the input fuzzy sets are piecewise linear. The results for the remaining controllers in configuration 4 and all the controllers in configuration 5 are not general. Whether or not a general conclusion can be attained for a class of controllers depends on the nature of controllers' components. The input fuzzy sets play an important role—when they are linear or piecewise linear (e.g., the triangular type and the trapezoidal type), it is very likely that a general conclusion can be made.

For controller configurations 4 and 5 where the iterative KM algorithm is used in the center-of-sets type of reducer, an additional derivation step is necessary, namely to determine the distribution and boundaries of the case numbers in the input space.

A T2 fuzzy controller uses only one of the two fuzzy AND operators, Zadeh or product. Deriving the analytical structure of a fuzzy controller, T1 or T2, with the product AND operator is, relatively speaking, much easier regardless of shape of input fuzzy sets. This is because the multiplication operation is carried out without needing to first divide the input space into ICs, which is required by the min( ) operation. The two operators lead to substantially different analytical structures.

TABLE 4.14 Summary of Derived Analytical Structures for Five T2 Fuzzy Controller Classes Given in Table 4.4

T2 Controller Configuration	Section	Analytical Structure
1	4.4	General conclusion: nonlinear PI (or PD) controllers with variable gains and a variable control offset
2	4.5	General conclusion: nonlinear PI (or PD) controllers with variable gains and a variable control offset
3	4.6	General conclusion: nonlinear PI (or PD) controllers with variable gains and a variable control offset
4	4.7	General conclusion: nonlinear PI (or PD) controllers with variable gains and a variable control offset if and only if all the input fuzzy sets are piecewise linear. The analytical structure of the remaining T2 controllers depends on the nature of the controller components.
5	4.8	The analytical structure depends on the nature of the controller components. No general conclusion can be drawn.

Roughly speaking, using the product AND operator for a T2 controller will make its analytical structure contain terms such as $E^2(k)$, $R^2(k)$, and $E(k)R(k)$ in its numerator even when piecewise linear input fuzzy sets are employed. Consequently, for this situation, the analytical structure usually cannot be presented as a nonlinear PI or PD controller. A logical conclusion is that one operator cannot be said to be better or worse than the other. They offer different benefits for different applications. It is up to the controller developer to decide which AND operator to use.

Revealing the analytical structure of a T2 fuzzy controller with the min() operator is a relatively new research direction. There are not many additional results in literature besides those for the five controller configurations in this chapter and some other configurations dealt with in Chapter 5. One of the newest controller configurations and its associated analytical structure results are given in Zhou and Ying (2013), in which a technique is developed for deriving the analytical structure of a broad class of typical two-input interval T2 Mamdani fuzzy controllers. A theorem is proved that the T2 fuzzy controllers with the piecewise linear input fuzzy sets are the sum of two nonlinear PI (or PD) controllers plus a variable control offset, each of which is with variable gains. The theorem represents a necessary and sufficient condition. This result differs substantially from all the existing analytical structure results about T2 and T1 fuzzy controllers in that it is the only known structure at this point in time that consists of the sum of two PI (or PD) controllers.

The principle underlying the structure-deriving techniques in this book is applicable to most, if not all, two-input interval T2 fuzzy controllers with Zadeh AND operator that have not been investigated in the literature. It may also be applicable to the two-input fuzzy controllers with general T2 fuzzy sets. No report exists that covers a three-input T2 fuzzy controller. Effort in that direction can be rewarding; but the exploration can be substantially more challenging than dealing with the two-input controllers. It is our view that developing new analytical structure derivation techniques can be technically fruitful.

Based on the analytical structure results in this book as well as those in the literature, it is observed that the use of the piecewise linear T2 input fuzzy sets is essential if one wants the nonlinear PI or PD type of control. When other types of input fuzzy sets are employed, the analytical structure will most likely be just nonlinear controllers with seemingly complex mathematical formulations that have no apparent structural similarity to any conventional controller. Hence, it is advantageous to employ the piecewise linear input fuzzy sets (mainly trapezoidal or triangular fuzzy sets), if one wants a structurally understandable T2 controller.

As an important benefit of revealing the analytical structure of a T2 fuzzy controller, the fuzzy controller can now be treated rigorously as a nonlinear controller, rather than as a blackbox controller. The roles that various parameters play, such as the footprints of uncertainty of the input fuzzy sets, can be clearly understood from a control theory standpoint as opposed to the vague and subjective angle of linguistic knowledge representation. Furthermore, the structure information can be used to facilitate control system design, an almost vacuous field so far. Realizing this potential is an interesting and rewarding research topic.

In summary, the new analytical structure results in this book together with the results in the literature represent a stepping stone for research that explores the structure of other T2 fuzzy controllers, paving the way for eventual realization of all the significant benefits pointed out in Section 4.1.

APPENDIX 4A

Equation (4.58) is

$$\begin{aligned}\Delta U(k) = {} & \frac{\Omega_1}{P}E(k) + \frac{\Omega_2}{P}R(k) + \frac{\Omega_3}{P}|E(k)| + \frac{\Omega_4}{P}|R(k)| + \frac{\Omega_5+\Omega_7}{P}E^2(k) \\ & + \frac{\Omega_6}{P}E(k)R(k) + \frac{\Omega_8}{P}E(k)|R(k)| + \frac{\Omega_{10}}{P}|E(k)|R(k) + \frac{\Omega_{12}}{P}|E(k)||R(k)| \\ & + \frac{\Omega_9+\Omega_{11}}{P}R^2(k) + \frac{\Omega_{13}+\Omega_{15}}{P}E^2(k)R(k) + \frac{\Omega_{14}+\Omega_{16}}{P}E(k)R^2(k) \\ & + \frac{\Omega_{18}+\Omega_{20}}{P}|E(k)|R^2(k) + \frac{\Omega_{17}}{P}R^3(k) + \frac{\Omega_{19}}{P}|R^3(k)| \\ & + \frac{\Omega}{P}\exp\left\{-\left[\frac{E(k)-S_j}{a_i}\right]^2\right\} + \frac{\Omega_{21}}{P}\end{aligned}$$

where the coefficients are

$$P = \Lambda_1 + \Lambda_2 R(k) + \Lambda_3 R^2(k) + \Lambda_4 \exp\left\{ -\left[\frac{E(k) - S_j}{a_i}\right]^2 \right\} + \Lambda_5 R(k) \exp\left\{ -\left[\frac{E(k) - S_j}{a_i}\right]^2 \right\}$$

$$\Omega = \Omega_{22} E(k) + \Omega_{23} R(k) + \Omega_{24} |E(k)| + \Omega_{25} |R(k)| + \Omega_{26} E^2(k) + \Omega_{27} E(k) R(k) + \Omega_{28} E^2(k) + \Omega_{29} E(k) |R(k)| + \Omega_{31} |E(k)| R(k) + \Omega_{33} |E(k)| |R(k)| + \left(\Omega_{30} + \Omega_{32}\right) R^2(k) + \Omega_{34}$$

$$\Omega_1 = \Upsilon_3 \Gamma_1 + \Theta_1 \Upsilon_1$$

$$\Omega_2 = \Upsilon_3 \Gamma_2 + \Theta_2 \Upsilon_1 + \Gamma_{13} \Upsilon_4 + \Theta_{13} \Upsilon_2$$

$$\Omega_3 = \Upsilon_3 \Gamma_3 + \Theta_3 \Upsilon_1$$

$$\Omega_4 = \Upsilon_3 \Gamma_4 + \Theta_4 \Upsilon_1$$

$$\Omega_5 = \Upsilon_3 \Gamma_5 + \Theta_5 \Upsilon_1$$

$$\Omega_6 = \Upsilon_3 \Gamma_6 + \Theta_6 \Upsilon_1 + \Gamma_1 \Upsilon_4 + \Theta_1 \Upsilon_2$$

$$\Omega_7 = \Upsilon_3 \Gamma_7 + \Theta_7 \Upsilon_1$$

$$\Omega_8 = \Upsilon_3 \Gamma_8 + \Theta_8 \Upsilon_1$$

$$\Omega_9 = \Upsilon_3 \Gamma_9 + \Theta_9 \Upsilon_1 + \Gamma_2 \Upsilon_4 + \Theta_2 \Upsilon_2$$

$$\Omega_{10} = \Upsilon_3 \Gamma_{10} + \Theta_{10} \Upsilon_1 + \Gamma_3 \Upsilon_4 + \Theta_3 \Upsilon_2$$

$$\Omega_{11} = \Upsilon_3 \Gamma_{11} + \Theta_{11} \Upsilon_1 + \Gamma_4 \Upsilon_4 + \Theta_4 \Upsilon_2$$

$$\Omega_{12} = \Upsilon_3 \Gamma_{12} + \Theta_{12} \Upsilon_1$$

$$\Omega_{13} = \Gamma_5 \Upsilon_4 + \Theta_5 \Upsilon_2$$

$$\Omega_{14} = \Gamma_6 \Upsilon_4 + \Theta_6 \Upsilon_2$$

$$\Omega_{15} = \Gamma_7 \Upsilon_4 + \Theta_7 \Upsilon_2$$

$$\Omega_{16} = \Gamma_8 \Upsilon_4 + \Theta_8 \Upsilon_2$$

$$\Omega_{17} = \Gamma_9 \Upsilon_4 + \Theta_9 \Upsilon_2$$

$$\Omega_{18} = \Gamma_{10} \Upsilon_4 + \Theta_{10} \Upsilon_2$$

$$\Omega_{19} = \Gamma_{11} \Upsilon_4 + \Theta_{11} \Upsilon_2$$

$$\Omega_{20} = \Gamma_{12} \Upsilon_4 + \Theta_{12} \Upsilon_2$$

$$\Omega_{21} = \Upsilon_3 \Gamma_{13} + \Theta_{13} \Upsilon_1$$

$$\Omega_{22} = 2\Gamma_1 + \Upsilon_1 d_i^{j+1}$$

$$\Omega_{23} = 2\Gamma_2 + \Upsilon_1 e_i^{j+1}$$

$$\Omega_{24} = 2\Gamma_3 + \Upsilon_1 \gamma_i^{j+1}$$

$$\Omega_{25} = 2\Gamma_4 + \Upsilon_1 \varepsilon_i^{j+1}$$

$$\Omega_{26} = 2\Gamma_5$$

$$\Omega_{27} = 2\Gamma_6 + \Upsilon_2 d_i^{j+1}$$

$$\Omega_{28} = 2\Gamma_7$$

$$\Omega_{29} = 2\Gamma_8$$

$$\Omega_{30} = 2\Gamma_9 + \Upsilon_2 e_i^{j+1}$$

$$\Omega_{31} = 2\Gamma_{10} + \Upsilon_2 \gamma_i^{j+1}$$

$$\Omega_{32} = 2\Gamma_{11} + \Upsilon_2 \varepsilon_i^{j+1}$$

$$\Omega_{33} = 2\Gamma_{12}$$

$$\Omega_{34} = 2\Gamma_{13}$$

$$\Lambda_1 = \Upsilon_1 \Upsilon_3$$

$$\Lambda_2 = \Upsilon_1 \Upsilon_4 + \Upsilon_2 \Upsilon_3$$

$$\Lambda_3 = \Upsilon_2 \Upsilon_4$$

$$\Lambda_4 = 2\Upsilon_1$$

$$\Lambda_5 = 2\Upsilon_2$$

$$\Upsilon_1 = 2 + \frac{2M_j}{v_j - M_i} + 2\eta_j - \frac{2\eta_j M_j}{M_j + \phi_j - v_j} + 4\eta_{j+1} - \frac{4\eta_{j+1} M_{j+1}}{M_{j+1} - v_{j+1} - \phi_{j+1}}$$

$$\Upsilon_2 = \frac{2}{M_j - v_j} + \frac{2\eta_j}{M_j - v_j - \phi_j} + \frac{4\eta_{j+1}}{M_{j+1} - v_{j+1} - \phi_{j+1}}$$

$$\Upsilon_3 = 4 + \frac{2M_j}{v_j - M_j} + \frac{2M_{j+1}}{M_{j+1} - v_{j+1}} + 2\eta_j \frac{2\eta_j M_j}{M_j + \phi_j - v_j}$$

$$\Upsilon_4 = \frac{2M_j}{M_j - v_j} + \frac{2}{M_{j+1} - v_{j+1}} + \frac{2\eta_j}{M_j + v_j + \phi_j}$$

$$\Gamma_1 = \left(d_{i+1}^{j+1} d_i^{j+1}\right)\left(\eta_{j+1} - \frac{\eta_{j+1} M_{j+1}}{M_{j+1} - \phi_{j+1} - v_{j+1}}\right) + \frac{d_i^j v_j}{v_j - M_j} + \frac{d_{i+1}^j \eta_j \eta_j \left(\phi_j - v_j\right)}{M_j + \phi_j - v_j}$$

$$\Gamma_2 = \frac{\left[c_{i+1}^{j+1} - \varpi_{i+1}^{j+1} - \varepsilon_{i+1}^{j+1} + c_i^{j+1} - \varpi_i^{j+1} + \left(e_{i+1}^{j+1} + e_i^{j+1}\right)\left(-\phi_{j+1} - v_{j+1}\right)\eta_{j+1}\right]}{M_{j+1} - v_{j+1} - \phi_{j+1}}$$

$$+\frac{c_i^j - \varpi_i^j - e_i^j \nu_j}{M_j - \nu_j} + \frac{\left(c_{i+1}^j - \varpi_{i+1}^j\right)\eta_j}{M_j - \nu_j - \phi_j} + \frac{e_{i+1}^j \eta_j \left(\phi_j - \nu_j\right)}{M_j + \phi_j - \nu_j}$$

$$\Gamma_3 = \gamma_{i+1}^{j+1} - \gamma_i^{j+1}\left(\eta_{j+1} - \frac{\eta_{j+1}M_{j+1}}{M_{j+1} - \phi_{j+1} - \nu_{j+1}}\right) - \gamma_i^j\left(1 + \frac{M_j}{\nu_j - M_j}\right)$$

$$-\gamma_{i+1}^j\left(\eta_j - \frac{\eta_j M_j}{M_j + \phi_j - \nu_j}\right)$$

$$\Gamma_4 = -\varepsilon_i^j\left(1 + \frac{M_j}{\nu_j - M_j}\right) - \varepsilon_{i+1}^j\left(\eta_j - \frac{n_j M_j}{M_j + \phi_j - \nu_j}\right)$$

$$-\varepsilon_i^{j+1}\left(\eta_{j+1} - \frac{\eta_{j+1}M_{j+1}}{M_{j+1} - \nu_{j+1} - \phi_{j+1}}\right)$$

$$\Gamma_5 = 0$$

$$\Gamma_6 = \frac{\eta_{j+1}\left(d_{i+1}^{j+1} + d_i^{j+1}\right)}{M_{j+1} - \nu_{j+1} - \phi_{j+1}} - \frac{d_i^j}{\nu_j - M_j} + \frac{d_{i+1}^j \eta_j}{M_j - \nu_j - \phi_j}$$

$$\Gamma_7 = 0$$

$$\Gamma_8 = 0$$

$$\Gamma_9 = \frac{\left(e_{i+1}^{j+1} + e_i^{j+1}\right)\eta_{j+1}}{M_{j+1} - \nu_{j+1} - \phi_{j+1}} - \frac{e_i^j}{\nu_j - M_j} + \frac{e_{i+1}^j \eta_j}{M_j - \nu_j - \phi_j}$$

$$\Gamma_{10} = \frac{\left(-\gamma_{i+1}^{j+1} + \gamma_i^{j+1}\right)\eta_{j+1}}{M_{j+1} - \nu_{j+1} - \phi_{j+1}} + \frac{\gamma_i^j}{\nu_j - M_j} + \frac{\gamma_{i+1}^j \eta_j}{M_j - \nu_j - \phi_j}$$

$$\Gamma_{11} = \frac{\varepsilon_i^j}{\nu_j - M_j} + \frac{\varepsilon_{i+1}^j \eta_j}{M_j - \nu_j - \phi_j} + \frac{\varepsilon_{i+1}^j \eta_{j+1}}{M_{j+1} - \nu_{j+1} - \phi_{j+1}}$$

$$\Gamma_{12} = 0$$

$$\Gamma_{13} = \left(c_{i+1}^{j+1} - \varpi_{i+1}^{j+1} - \varepsilon_{i+1}^{j+1} + c_i^{j+1} - \varpi_i^{j+1}\right)\left(\eta_{j+1} - \frac{\eta_{j+1}M_{j+1}}{M_{j+1} - \phi_{j+1} - \nu_{j+1}}\right)$$

$$+\left(c_i^j - \varpi_i^j\right)\left(1 + \frac{M_j}{\nu_j - M_j}\right) + \left(c_{i+1}^j - \varpi_{i+1}^j\right)\left(\eta_j - \frac{\eta_j M_j}{M_j - \phi_j - \nu_j}\right)$$

$$\Theta_1 = \frac{-\nu_{j+1}d_{i+1}^{j+1}}{M_{j+1} - \nu_{j+1}} + \frac{d_i^j \eta_j \left(\phi_j - \nu_j\right)}{M_j + \phi_j - \nu_j} + \frac{d_{i+1}^j \nu_j}{\nu_j - M_j}$$

$$\Theta_2 = \frac{\varpi_{i+1}^{j+1} + c_{i+1}^{j+1} + \varepsilon_{i+1}^{j+1} - \nu_{j+1}e_{i+1}^{j+1}}{M_{j+1} - \nu_{j+1}} + \frac{\left(\varpi_i^j + c_i^j\right)\eta_j + e_i^j \eta_j \left(\phi_j - \nu_j\right)}{M_j + \phi_j - \nu_j}$$

$$+\frac{\varpi_{i+1}^{j}+c_{i+1}^{j}-e_{i+1}^{j}v_j}{v_j-M_j}$$

$$\Theta_3=\frac{-v_{j+1}\gamma_{i+1}^{j+1}}{M_{j+1}-v_{j+1}}+\frac{(\phi_j-v_j)\,\gamma_i^j\eta_j}{M_j+\phi_j-v_j}+\frac{v_j\gamma_{i+1}^{j}}{v_j-M_j}$$

$$\Theta_4=\frac{\varepsilon_i^j\eta_j\,(\phi_j-v_j)}{M_j+\phi_j-v_j}+\frac{\varepsilon_{i+1}^{j}v_j}{v_j-M_j}$$

$$\Theta_5=0$$

$$\Theta_6=\frac{d_{i+1}^{j+1}}{M_{j+1}-v_{j+1}}+\frac{d_i^j\eta_j}{M_j-v_j+\phi_j}+\frac{d_{i+1}^{j}}{v_j-M_j}$$

$$\Theta_7=0$$

$$\Theta_8=0$$

$$\Theta_9=\frac{e_{i+1}^{j+1}}{M_{j+1}-v_{j+1}}+\frac{e_i^j\eta_j}{M_j-v_j+\phi_j}+\frac{e_{i+1}^{j}}{v_j-M_j}$$

$$\Theta_{10}=\frac{\gamma_{i+1}^{j+1}}{M_{j+1}-v_{j+1}}+\frac{\gamma_i^j\eta_j}{M_j-v_j+\phi_j}+\frac{\gamma_{i+1}^{j}}{v_j-M_j}$$

$$\Theta_{11}=\frac{\varepsilon_i^j\eta_j}{M_j-v_j+\phi_j}+\frac{\varepsilon_{i+1}^{j}}{v_j-M_j}$$

$$\Theta_{12}=0$$

$$\Theta_{13}=\frac{-v_{j+1}\left(\varpi_{i+1}^{j+1}+c_{i+1}^{j+1}+\varepsilon_{i+1}^{j+1}\right)}{M_{j+1}-v_{j+1}}+\frac{(\phi_j-v_j)\left(\varpi_i^j+c_i^j\right)\eta_j}{M_j+\phi_j-v_j}$$

$$+\frac{v_j\left(\varpi_{i+1}^{j}+c_{i+1}^{j}\right)}{v_j-M_j}$$

CHAPTER 5

Analysis of Simplified Interval Type-2 Fuzzy PI and PD Controllers

5.1 INTRODUCTION

An interval type-2 (IT2) fuzzy set (FS) can be thought of as a collection of many embedded T1 FSs, and a T2 fuzzy logic controller (FLC) may, therefore, be conceptually treated as a collection of many (embedded) T1 FLCs whose crisp output is obtained by aggregating the outputs of all the embedded T1 FLCs (see Section 1.5). Due to the extra degree of freedoms, a T2 FLC has the *potential* to outperform a T1 FLC under certain conditions. This chapter focuses on the design, properties demonstration via simulations/experiments, and theoretical analysis of interval IT2 (IT2) fuzzy proportional plus integral (PI) or proportional plus derivative (PD) controllers.

The structure of a typical IT2 fuzzy PI and PD controller is shown in Fig. 3.4, where the input signals are the feedback error $E(k)$ and the change of error $R(k)$ and the output signal is the change in control signal $\Delta u(k)$ (PI) or the control signal $u(k)$ (PD). As mentioned in the preview of properties of interval IT2 (IT2) FLCs in Section 1.7, one feature is a smoother control surface than that of a T1 FLC, especially around the origin. Consequently, experimental results show that small disturbances around the steady-state value will not result in significant control signal changes so there are less oscillations (Wu and Tan, 2006). However, the potential performance improvement provided by IT2 FLCs comes with the cost of higher computational power needed to implement the Karnik–Mendel (KM) type reducer. To reduce the computational burden while preserving the advantages of IT2 FLCs, two approaches may be considered: (1) faster type reduction methods, such as the algorithms presented in Section 2.3.6, and (2) a simpler architecture that reduces the computing requirement. The feasibility of the second approach is demonstrated in this chapter. The key idea is to only replace some critical T1 FSs by IT2 sets. A procedure that uses genetic algorithm to evolve an IT2 FLC that is robust enough to cope well with the uncertainties while having minimum

Introduction to Type-2 Fuzzy Logic Control: Theory and Applications, First Edition.
Jerry M. Mendel, Hani Hagras, Woei-Wan Tan, William W. Melek, and Hao Ying.

computational cost is proposed. Experimental results that establish the feasibility of the simplified IT2 FLC structure are also presented.

Having demonstrated the properties of IT2 FLC via simulations and experimental results, a theoretical study of IT2 FLC is presented. The motivation is to establish more general results as the simulation and experimental analysis cannot be generalized and applied into other systems due to the limitation of case studies. First, the KM type reducer is analyzed in order to devise a methodology for partitioning the input space such that the output for an IT2 fuzzy logic system may be explicitly related to the firing strength of all rules. With this breakthrough, the analytical structure for T1 FLCs can be extended to analyze IT2 FLCs.

The chapter is organized as follows: The simplified IT2 FLC is presented in Section 5.2, while a computational cost comparison and the potential reduction in computing requirements provided by the simplified IT2 FLC is illustrated in Section 5.2.3. Two type-1 FLCs and two type-2 FLCs with different degrees of freedom are designed in Section 5.2.5, and their abilities to handle modeling uncertainties are compared using a coupled-tank liquid-level control system. Section 5.2.6 discusses the performances of the proposed architecture. Next, a methodology for deriving the analytical structure for a case of IT2 PI/PD FLCs, with the configuration defined in Section 5.3.1, is presented. The algorithm delineated in Section 5.3.2 leverages on a property of the KM type reducer that constrains the switch points to the location of the consequent sets to extend the analytical structure technique for T1 FLCs. Section 5.3.3 then presents five examples that illustrate how the analytical structure framework may be applied to explain why an IT2 fuzzy controller has the potential to better achieve the conflicting aims of fast rise time and small overshoot. Finally, conclusions are drawn in Section 5.4.

5.2 SIMPLIFIED TYPE-2 FLCs: DESIGN, COMPUTATION, AND PERFORMANCE[1]

Extensive experiments (Wu and Tan, 2006; Hagras, 2007; Hagras and Wagner, 2012) have demonstrated that IT2 FLCs have the potential to provide better control performances as they are better able to handle disturbances, noises, and unmodeled dynamics. The reason is that they usually have smoother control surfaces around the $(E(k), R(k)) = (0, 0)$ point (Wu and Tan, 2006; Wu, 2012a). However, the robust performance of IT2 FLCs may be achieved at the cost of higher computational requirements (Wu, 2013). To balance the control performance and computational cost, we may employ a simplified IT2 FLC, in which IT2 FSs are only used near the steady state and T1 FSs are used elsewhere. An example of such an IT2 FLC is shown in Fig. 5.1. This section presents the design procedure and compares the computational cost and control performance of the simplified IT2 FLC with T1 FLCs and traditional IT2 FLCs.

[1]Much of the material is taken directly from Wu and Tan (2006; © 2006, Elsevier).

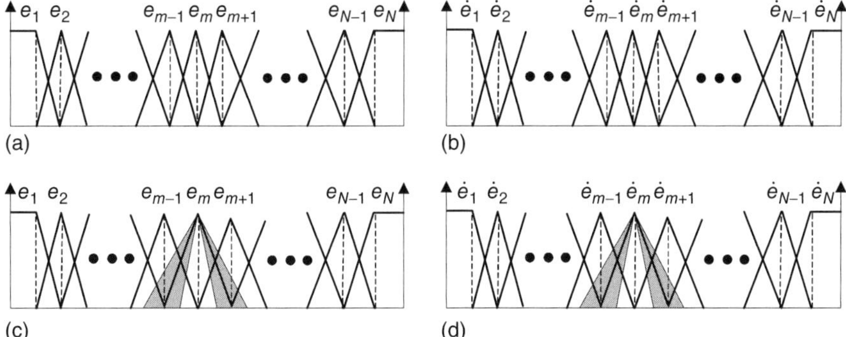

Figure 5.1 Example FSs of a baseline T1 FLC and a simplified IT2 FLC. (a) The N FSs for input, $E(k)$, of the T1 FLC, where e_i is the center of the ith fuzzy set; (b) The N FSs for input, $R(k)$, of the T1 FLC, where $\dot{e}_i$ is the center of the ith fuzzy set; (c) FSs of $E(k)$ of the simplified IT2 FLC, where the middle one is an IT2 FS and all the others are T1 FSs; and (d) FSs of $R(k)$ of the simplified IT2 FLC, where the middle one is an IT2 FS and all the others are T1 FSs (Wu and Tan, 2006; © 2006, Elsevier).

5.2.1 Structure of a Simplified IT2 FLC

A simplified IT2 FLC may be designed by gradually replacing T1 FSs by their IT2 counterparts until the resulting IT2 FLC meets the robustness requirements, starting with the FSs that characterize the region around steady state. Since the computational cost will increase significantly when the number of IT2 FSs increases, as few IT2 FSs as possible should be introduced. For a PI-like FLC, the response near steady state is determined mainly by the control surface around the $(E(k), R(k)) = (0, 0)$ origin, which is governed by the middle FSs of the error $E(k)$ and change of error $R(k)$ input domains. Hence, the procedure for designing a simplified IT2 FLC is:

1. Design a baseline T1 FLC through simulation on a nominal model.
2. Change the most important input FS to IT2 FS. For the two inputs of a PI-like FLC, the change of error, $R(k)$, is more susceptible to noises so the FS corresponding to zero $R(k)$ is changed to IT2 first, as illustrated in Fig. 5.1d.
3. If the IT2 FLC designed above cannot cope well with the actual plant, the FS associated with zero, $E(k)$, is also changed to IT2 FS, as illustrated in Fig. 5.1c.
4. If the resulting IT2 FLC is still not robust enough, more IT2 FSs may be introduced starting from the middle of each input domain and gradually moving toward the limits of the domain. Another criterion is to use IT2 FSs to characterize the operating region that needs a smoother control surface.

An FLC designed by the proposed procedure has two parts—a T1 part and an IT2 part. Different fuzzy partitions will be activated when the state of the plant is in different operating regions. During the transient stage, the FLC behaves like a T1 FLC since no IT2 FSs are fired. When the output approaches the set point, IT2 FSs will be fired and the plant is controlled by an IT2 FLC. Smoother control signals will be generated, which help to eliminate oscillations.

5.2.2 Output Computation

Consider a simplified IT2 FLC that has N^2 rules with crisp consequents y_i, $i = 1, 2, \ldots, N^2$, where N is the number of fuzzy sets for each of the two input domains. Suppose the first M $(0 < M < N^2)$ rules contain only T1 FSs in the antecedent, and the remaining $N^2 - M$ rules have at least one IT2 FS in the antecedent. There will be M firing levels (f_i, $i = 1, 2, \ldots, M$) and $N^2 - M$ firing intervals ($\widetilde{f}_i$, $i = M + 1, M + 2, \ldots, N^2$). Assuming product sum fuzzy inference and the result of the fuzzy inference engine is processed using center-of-sets type reduction (see Section 3.3.2.4), the output of the simplified IT2 FLC may be expressed as

$$\begin{aligned}
\dot{u} &= \frac{\sum_{i=1}^{M} u_i f_i + \sum_{i=M+1}^{N^2} u_i \widetilde{f}_i}{\sum_{i=1}^{M} f_i + \sum_{i=M+1}^{N^2} \widetilde{f}_i} \\
&= \frac{\beta + \sum_{i=M+1}^{N^2} u_i \widetilde{f}_i}{\alpha + \sum_{i=M+1}^{N^2} \widetilde{f}_i} \\
&= \frac{\beta}{\alpha} + \frac{\sum_{i=M+1}^{N^2} u_i \widetilde{f}_i - \frac{\beta}{\alpha} \sum_{i=M+1}^{N^2} \widetilde{f}_i}{\alpha + \sum_{i=M+1}^{N^2} \widetilde{f}_i} \\
&= \frac{\beta}{\alpha} + \frac{\sum_{i=M+1}^{N^2} (u_i - \frac{\beta}{\alpha}) \widetilde{f}_i}{\alpha + \sum_{i=M+1}^{N^2} \widetilde{f}_i}
\end{aligned} \tag{5.1}$$

where $\alpha = \sum_{i=1}^{M} f_i$ and $\beta = \sum_{i=1}^{M} u_i f_i$. Define u_i' and $\widetilde{f}_{N^2+1}$ as

$$u_i' = \begin{cases} u_i - \frac{\beta}{\alpha} & i = M + 1, M + 2, \ldots, N^2 \\ 0 & i = N^2 + 1 \end{cases}$$

$$\widetilde{f}_{N^2+1} = \alpha$$

then Eq. (5.1) can be further simplified to

$$\dot{u} = \frac{\beta}{\alpha} + \frac{\sum_{i=M+1}^{N^2+1} u_i' \widetilde{f}_i}{\sum_{i=M+1}^{N^2+1} \widetilde{f}_i} \tag{5.2}$$

The second term on the right-hand side of Eq. (5.2) can be calculated by the Karnik–Mendel algorithms or their enhancements described in Section 2.3.6.

5.2.3 Computational Cost

From the discussion in the previous section, the simplified IT2 FLC has reduced computational cost when compared with traditional IT2 FLCs. The lower computing load is due to a reduction in the number of membership grades that need to be computed and the reduction in the size of the type reduction problem.

Example 5.1 Consider T1 and IT2 fuzzy sets with triangular membership functions. Let a, b, c be the apexes of the triangular membership for the T1 fuzzy set and the upper/lower membership function of the IT2 fuzzy set. The membership grade of an input, x, in the T1 fuzzy set or the upper/lower membership of the IT2 fuzzy set may similarly be expressed as

$$f_i(x) = \begin{cases} \dfrac{b-x}{b-a} & a < x < b \\ \dfrac{c-x}{c-b} & b < x < c \\ 0 & \text{otherwise} \end{cases} \tag{5.3}$$

This equation involves two subtraction and one division operations. If the simplified IT2 FLC replaces M IT2 fuzzy sets by T1 fuzzy sets, there will be a reduction of $2M$ subtraction and M division operations because Eq. (5.3) needs to be executed once for a T1 set and twice for an IT2 set.

As a further study, the computational cost of simplified IT2 FLCs is quantitatively compared with a T1 FLC and a traditional IT2 FLC. All FLCs have two inputs, $E(k)$ and $R(k)$. Each input has N equally spaced FSs, and the output domain has N^2 numbers. The T1 FLC is denoted as FLC_1, and its FSs are shown in Figs. 5.1a and 5.1b. The first simplified IT2 FLC is FLC_{2s}, where only the middle FS of $R(k)$ is an IT2 FS, as shown in Figs. 5.1a and 5.1d. The second simplified IT2 FLC is FLC_{2m}, where the middle FSs of both $E(k)$ and $R(k)$ are IT2 FSs, as shown in Figs. 5.1c and 5.1d. The traditional IT2 FLC is denoted as FLC_{2f}, whose all FSs are IT2 FSs.

We partitioned the domain of $E(k)$, $[-1,\ 1]$, into 101 equally distributed points e_i; and 101 $\dot{e}_i$'s were generated in the same way. Thus, all possible combinations of e_i and $\dot{e}_i$ yielded 10,201 input pairs. Computational cost was evaluated by comparing the time needed to find outputs for these 10,201 inputs. All the experiments were done by Matlab on a 996-MHz computer with 256 MB of RAM and Windows XP. The standard KM algorithm described in Section 2.3.6 was used to implement the type reduction step. Table 5.1 shows the results for different values of N. Observe that the computational cost of the simplified structure is less than half of a traditional IT2 FLC, and the saving is much more significant when N is small.

TABLE 5.1 Computational Cost of the Four FLCs

N\FLC	FLC_1 (s)	FLC_{2s}(s)	FLC_{2m} (s)	FLC_{2f} (s)
3	1.2	2.0	6.7	10.4
5	1.6	2.5	5.1	10.3
7	2.3	3.7	5.0	12.0
9	3.3	5.6	6.6	15.0
11	4.6	8.6	9.5	19.6

5.2.4 Genetic Tuning of FLC

The task of tuning the parameters of an FLC is vital for ensuring its performance. Genetic algorithm (GA), a general-purpose search algorithm that uses principles inspired by natural population genetics to evolve solutions to problems, is a common strategy used for tuning an FLC. The tuning method fits the membership functions of the fuzzy sets by minimizing an error function defined using a set of evaluation input–output data.

A GA is a general-purpose search algorithm based on the mechanics of natural selection and genetics (Holland, 1975; Goldberg, 1989). Figure 5.2 shows the flowchart of a basic GA. The key idea behind the search process is inspired by the natural evolution of biological creatures, where the fittest among a group of artificial entities survive to form a new generation together with those that are produced through gene exchange. Given an initial population of candidate solutions that are represented as strings called chromosome, GAs operate in cycles known as

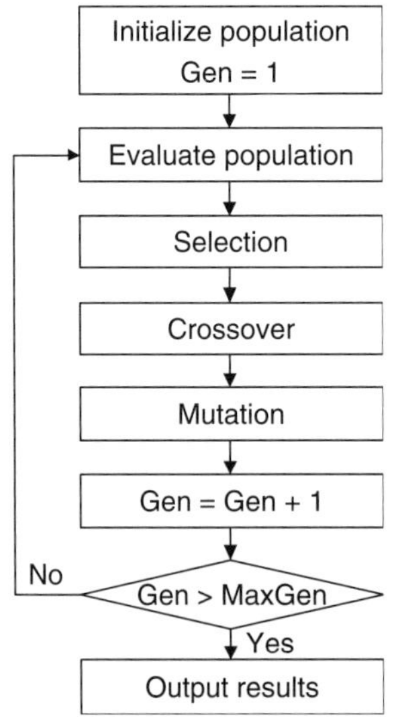

Figure 5.2 Flowchart of a basic GA (Wu and Tan, 2006; © 2006, Elsevier).

generations. For each generation, the fitness of all individuals with respect to the optimization task is evaluated via a scalar objective function (fitness function), and the result of this evaluation is used to drive the Darwinian selection process. A certain percentage of every generation is also produced randomly via genetic operators such as crossover, and mutation to provide genetic diversity, and thus increasing the probability of finding the global minimal. As a result of this evolutionary cycle of selection, crossover, and mutation, more and more suitable solutions to the optimization problem emerge within the population. In the next subsection, the performance of the simplified IT2 FLCs and T1 FLCs tuned via a GA-based optimization approach is studied and compared.

5.2.5 Performance

5.2.5.1 Coupled-Tank System The control performance of the simplified IT2 FLC is evaluated in this subsection via simulation and experimental results obtained from a coupled-tank system shown in Fig. 3.5a and repeated as Fig. 5.3. It consists of two small tower-type tanks mounted above a reservoir that stores the water. Water is pumped into the top of each tank by two independent pumps, and the levels of water are measured by two capacitive-type probe sensors. Each tank is fitted with an outlet, at the side near the base. Raising the baffle between the two tanks allows water to flow between them. The amount of water that returns to the reservoir is approximately proportional to the square root of the height of the water in the tank, which is the main source of nonlinearity in the system.

The dynamics of the coupled-tank apparatus can be described by the following nonlinear differential equations:

$$A_1 \frac{dH_1}{dt} = Q_1 - \alpha_1 \sqrt{H_1} - \alpha_3 \sqrt{H_1 - H_2} \tag{5.4a}$$

$$A_2 \frac{dH_2}{dt} = Q_2 - \alpha_2 \sqrt{H_2} + \alpha_3 \sqrt{H_1 - H_2} \tag{5.4b}$$

where A_1, A_2 are the cross-sectional areas of tank 1, 2; H_1, H_2 are the liquid levels in tank 1, 2; Q_1, Q_2 are the volumetric flow rate (cm^3/sec) of pump 1, 2; $\alpha_1, \alpha_2, \alpha_3$ are the proportionality constants corresponding to the $\sqrt{H_1}$, $\sqrt{H_2}$ and $\sqrt{H_1 - H_2}$ terms. We assume $H_1 \geq H_2$, which is always satisfied in the experiments.

The coupled-tank apparatus can be configured as a second-order single-input single-output system by turning off pump 2 and using pump 1 to control the water level in tank 2. Since pump 2 is turned off, Q_2 equals zero and Eq. (5.4b) reduces to

$$A_2 \frac{dH_2}{dt} = -\alpha_2 \sqrt{H_2} + \alpha_3 \sqrt{H_1 - H_2} \tag{5.5}$$

Equations (5.4a) and (5.5) are used to construct a simulation model of the coupled tank for the GA to evaluate the fitness of the candidate solutions. The parameters

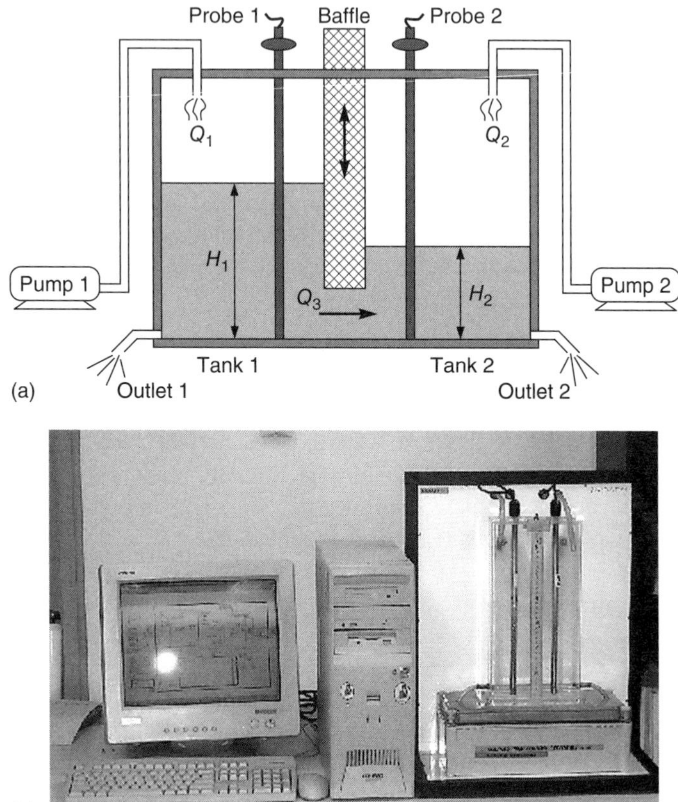

Figure 5.3 Coupled-tank liquid-level control system: (a) schematic diagram and (b) experimental setup (Wu and Tan, 2006; © 2006, Elsevier).

used in the simulation model are

$$A_1 = A_2 = 36.52 \text{ cm}^2$$
$$\alpha_1 = \alpha_2 = 5.6186$$
$$\alpha_3 = 10$$

The area of the tank was measured manually while the discharge coefficients (α_1, α_2, and α_3) were found by measuring the time taken for a predetermined change in the water levels to occur. Although the DC power source can supply between 0 and 5 V to the pumps, the maximum control signal was capped at 4.906 V, which corresponds to an input flow rate of about 75cm^3/sec. To compensate for the pump dead zone, the minimum control signal was chosen to be 1.646 V. A sampling period of 1 sec was employed.

5.2.5.2 *Structure of the FLCs* Four FLCs (FLC_{13}, FLC_{15}, FLC_{2s}, and FLC_{2f}) are tuned by GA. Among them, FLC_{13} and FLC_{15} are T1 FLCs, FLC_{2s} is a simplified IT2 FLC, and FLC_{2f} is a traditional IT2 FLC. FLC_{15} has five FSs in each input domain and the output domain is described by nine singleton fuzzy sets as shown in Table 5.2. FLC_{13}, FLC_{2s}, and FLC_{2f} have three FSs in each input domain, which are labeled as *N*, *Z*, and *P*. The output space ($\dot{u}$) is characterized by five singleton fuzzy sets: *NB*, *NS*, *Z*, *PS*, and *PB*. The rule base is shown in Table 5.3. The various FS operations adopted in the experiments are the sum-product inference engine, center-of-sets type reducer, and height defuzzifier.

5.2.5.3 *GA Optimization* All 4 FLCs are tuned by GA. First, the chromosome coding scheme is described. Figure 5.4 shows the general structure of the chromosome used by the GA to optimize the parameters of the FLCs, and the actual length of the chromosome varies according to the number of FLC parameters that needs to be optimized. The domain for the feedback error input, $E(k)$, of FLC_{13} is partitioned by the 3 FSs illustrated in Fig. 5.5. The T1 fuzzy sets are defined by the 3 points N_e, Z_e, and P_e. Similarly, the 3 points that define the 3 sets for the $R(k)$ domain are $N_{\dot{e}}$, $Z_{\dot{e}}$, and $P_{\dot{e}}$. Another 5 numbers are needed to define the fuzzy singletons used to characterized the output domain. The chromosome for FLC_{13}, therefore, has a total of 11 parameters. For FLC_{15}, each chromosome comprises 19 genes because 5 parameters are needed to determine the T1 FSs for each input and 9 parameters for the fuzzy singleton consequents. The first 11 genes for the simplified IT2 FLC, FLC_{2s}, are common with the parameters of FLC_{13}. The next two genes in the chromosome determine the FOU of the only IT2 FS used to partition the $R(k)$ domain of FLC_{2s}. They define the amount by which the T1 FS is shifted ($d_{\dot{e}_{2l}}$ and $d_{\dot{e}_{2r}}$ in

TABLE 5.2 Rule Base of FLC_{15}

$E(k)\backslash R(k)$	$\dot{e}_1$	$\dot{e}_2$	$\dot{e}_3$	$\dot{e}_4$	$\dot{e}_5$
e_1	$\dot{u}_1$	$\dot{u}_2$	$\dot{u}_3$	$\dot{u}_4$	$\dot{u}_5$
e_2	$\dot{u}_2$	$\dot{u}_3$	$\dot{u}_4$	$\dot{u}_5$	$\dot{u}_6$
e_3	$\dot{u}_3$	$\dot{u}_4$	$\dot{u}_5$	$\dot{u}_6$	$\dot{u}_7$
e_4	$\dot{u}_4$	$\dot{u}_5$	$\dot{u}_6$	$\dot{u}_7$	$\dot{u}_8$
e_5	$\dot{u}_5$	$\dot{u}_6$	$\dot{u}_7$	$\dot{u}_8$	$\dot{u}_9$

TABLE 5.3 Rule Base of FLC_{13}, FLC_{2s}, and FLC_{2f}

$E(k)\backslash R(k)$	$N_{\dot{e}}$	$Z_{\dot{e}}$	$P_{\dot{e}}$
N_e	NB	NS	Z
Z_e	NS	Z	PS
P_e	Z	PS	PB

1	2	3	4	5	6	7	8	9	10	11	12	13	14	15	16	17	18	19
Ne	Ze	Pe	$N\dot{e}$	$Z\dot{e}$	$P\dot{e}$	NB	NS	Z	PS	PB	$d_{\dot{e}2l}$	$d_{\dot{e}2r}$	d_{e2l}	d_{e2r}	$d_{\dot{e}1r}$	$d_{\dot{e}3l}$	d_{e1r}	d_{e3l}
Sub-chromosome of e			Sub-chromosome of $\dot{e}$			Sub-chromosome of $\dot{u}$												

Figure 5.4 GA coding scheme of the FLCs (Wu and Tan, 2006; © 2006, Elsevier).

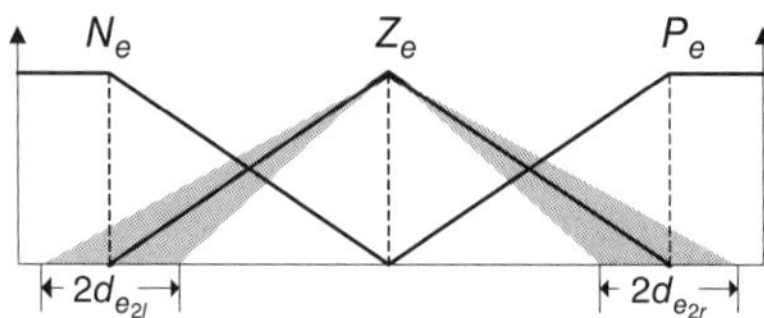

Figure 5.5 Example FSs of $E(k)$ (Wu and Tan 2006; © 2006, Elsevier).

TABLE 5.4 Plants Used to Assess Fitness of Candidate Solutions

	Plant I	Plant II	Plant III	Plant IV
$A_1 = A_2$ (cm^2)		36.52		
$\alpha_1 = \alpha_2$		5.6186		
α_3	10	10	10	8
Set point (cm)	$0 \rightarrow 15$	$0 \rightarrow 22.5 \rightarrow 7.5$	$0 \rightarrow 15$	$0 \rightarrow 15$
Transport delay (s)	0	0	2	0

Fig. 5.5) to generate the FOU of the IT2 FS. In the case of FLC$_{2f}$, the input domains are partitioned by 6 IT2 FSs so the chromosome is exactly the 19 genes shown in Fig. 5.4.

The fitness of each chromosome in the GA population is assessed by subjecting the simulation model of the liquid-level process to step inputs. As the goal is to explore the four FLCs' ability to handle modeling uncertainties, each candidate solution is used to control plants with the four sets of parameters shown in Table 5.4. The integral of time absolute error (ITAE) obtained for each of the four plants are added together and used to evaluate the fitness of the FLCs:

$$F = \sum_{i=1}^{4} \alpha_i \left[\sum_{j=1}^{N_i} j * e_i(j) \right] \tag{5.6}$$

where α_i is the weight corresponding to the ITAE of the ith plant, and $N_i = 200$ is the number of sampling instants. There is a need to introduce α_i because the ITAE of the second plant is usually several times bigger than that of other plants. To ensure that the ITAE of the four plants can be reduced with equal emphasis, α_2 is defined as $\frac{1}{3}$ and the other weights are unity.

The GA consisted of 200 chromosomes in each generation and optimization is terminated after 100 generations. The crossover rate was 0.8 and the mutation rate

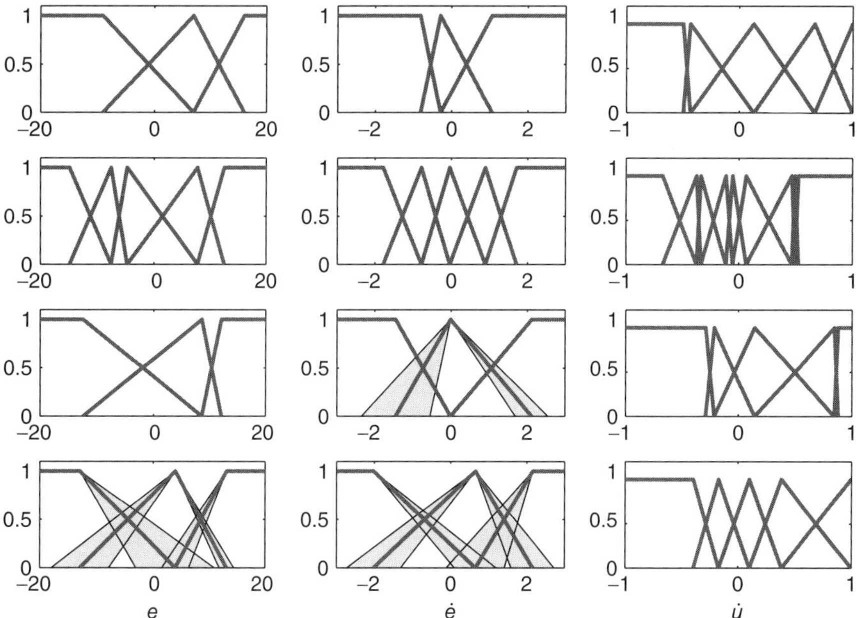

Figure 5.6 FSs of the four FLCs (FLC_{13}, FLC_{15}, FLC_{2s}, and FLC_{2f}, from the top to the bottom (Wu and Tan, 2006; © 2006, Elsevier).

was 0.1. After the crossover and mutation process, the genes in each subchromosome may not remain in the proper order after crossover and mutation in the sense that the center of the type-2 set corresponding to N_e may be larger than that of Z_e. Every subchromosome is, therefore, sorted before fitness evaluation is performed. The FSs of FLC_{13}, FLC_{15}, FLC_{2s}, and FLC_{2f} evolved by GA are shown in Fig. 5.6. The parameters are listed in Tables 5.5 and 5.6.

5.2.5.4 Experimental Results The control performances of the four FLCs are compared through simulation and experiments. As pointed out in Teo et al. (1998), the volumetric flow rate of the pumps in the coupled-tank apparatus is nonlinear and the system has nonzero transport delay. Besides, the data read by the sensor is noisy. These characteristics are not accurately captured by the model used by the GA to optimize the FLC parameters. Hence, the ability of the four FLCs to handle modeling uncertainties can be ascertained by examining their control performances on the actual plant. Figures 5.7 and 5.8 show the step responses for different set points and the corresponding control signals. All four FLCs can handle the uncertainties introduced by the pump nonlinearity and the unmodeled transport delay.

To further test the FLCs, the flow rate between the two tanks was reduced by lowering the baffle separating the two tanks. This change gave rise to a system with slower dynamics. In addition, the difference in liquid level between the two tanks was larger at steady state. The corresponding step responses and the control signals are shown in Fig. 5.9. Although the simulated step responses of the four

TABLE 5.5 FSs of FLC_{13}, FLC_{2s}, and FLC_{2f}

	$N_{\dot{e}}$	$y_{\dot{e}}$	$P_{\dot{e}}$
		(a) FSs of $E(k)$	
FLC_1	−9.0611	6.9846	16.0539
FLC_{2s}	−12.4578	8.6232	12.1405
	−12.9137	3.9722	13.1283
FLC_{2f}	$d_{e_{1r}} = 7.0388$	$d_{e_{2l}} = 5.0656, d_{e_{2r}} = 1.2868$	$d_{e_{3l}} = 2.4127$

	$N_{\dot{e}}$	$y_{\dot{e}}$	$P_{\dot{e}}$
		(b) FSs of $R(k)$	
FLC_1	−0.8093	−0.2884	1.0538
		−0.0119	
FLC_{2s}	−1.4505	$d_{\dot{e}_{2l}} = 0.9002, d_{\dot{e}_{2r}} = 0.4327$	2.1192
	−2.0186	0.6459	2.1534
FLC_{2f}	$d_{\dot{e}_{1r}} = 0.5479$	$d_{\dot{e}_{2l}} = 0.7091, d_{\dot{e}_{2r}} = 0.5697$	$d_{\dot{e}_{3l}} = 0.7644$

	NB	NS	Z	PS	PB
			(c) FSs of $\dot{u}$		
FLC_1	−0.4985	−0.4362	0.1282	0.6613	0.9998
FLC_{2s}	−0.2906	−0.2130	0.1422	0.8490	0.8817
FLC_{2f}	−0.3967	−0.1702	0.1002	0.3802	0.9978

FLCs are satisfactory, the experimental results corresponding to the two T1 FLCs (FLC_{13} and FLC_{15}) exhibit large oscillations. The two IT2 FLCs have the ability to eliminate these oscillations quickly, and the liquid level reaches its desired height at steady state. Notice that the performances of the two IT2 FLCs are similar, though they have a different number of IT2 FSs.

Lastly, the ability of the four FLCs to deal with transport delay was studied, and the results are shown in Fig. 5.10. Once again, both IT2 FLCs outperform their T1 counterparts. The performances of the two IT2 FLCs are similar, though they have a different number of IT2 FSs. This suggests that some IT2 FSs may not be necessary, and the computational cost may be reduced without sacrificing robustness by using T1 FSs in place of some IT2 FSs.

5.2.6 Discussions

Observe from Figs. 5.7–5.10 that all four FLCs gave similar simulation results. However, the simulation and experimental results obtained using the IT2 FLCs

TABLE 5.6 FSs of FLC_{15}

e_1	e_2	e_3	e_4	e_5
		(a) FSs of $E(k)$		
−14.8778	−7.5460	−4.7217	7.7783	12.5710

$\dot{e}_1$	$\dot{e}_2$	$\dot{e}_3$	$\dot{e}_4$	$\dot{e}_5$
		(b) FSs of $R(k)$		
−1.7824	−0.7799	−0.0387	0.8896	1.7115

$\dot{u}_1$	$\dot{u}_2$	$\dot{u}_3$	$\dot{u}_4$	$\dot{u}_5$	$\dot{u}_6$	$\dot{u}_7$	$\dot{u}_8$	$\dot{u}_9$
				(c) FSs of $\dot{u}$				
−0.6755	−0.3771	−0.3381	−0.1142	−0.0543	0.0645	0.4632	0.4921	0.5194

generally concur more with each other than the simulation and experimental results generated using the T1 FLCs. A quantitative measure of the performances of the four FLCs was generated by calculating the ITAE for all cases that were studied. The result is presented in Fig. 5.11. The plot is scaled such that the ITAEs for the step responses obtained using FLC_{13} is 100%. Consequently, a smaller number in Fig. 5.11 translates to a lower ITAE, and therefore better performance. Observe that the experimental performances achieved by the two IT2 FLCs are better than that of the T1 FLCs. Most notably, FLC_{2s} outperforms FLC_{15} even though FLC_{15} has six more design parameters.

The control surfaces of the four FLCs are shown in Fig. 5.12. The control surfaces of the two IT2 FLCs are smoother than that of FLC_{13} around the origin $[(E(k), R(k)) = (0, 0)]$, which is why the IT2 FLCs are more robust. Note that the control surface of FLC_{2s} is similar to that of FLC_{2f}, even though FLC_{2f} has more IT2 FSs. The control surfaces provide further evidence that there will not be significant performance deterioration when the proposed simplified IT2 FLC is used in place of a traditional IT2 FLC where all the input sets are IT2 FSs.

In common with the standard practice in GA-based tuning of FLC parameters, a statistical evaluation was conducted by repeatedly performing the several optimization exercises. Five versions of each controller (FLC_{13}, FLC_{15}, FLC_{2s}, and FLC_{2f}) were evolved and tested on the practical plant. Most of the T1 FLCs performed poorly. The step responses either had long settling times or exhibited persistent oscillations. FLC_{13} and FLC_{15} presented in the previous section actually have the best performance from among the various T1 FLCs. Several IT2 FLCs from different runs were also tested on the actual plant. The experimental results did not differ from those presented here. This is another indication of the superior ability of IT2

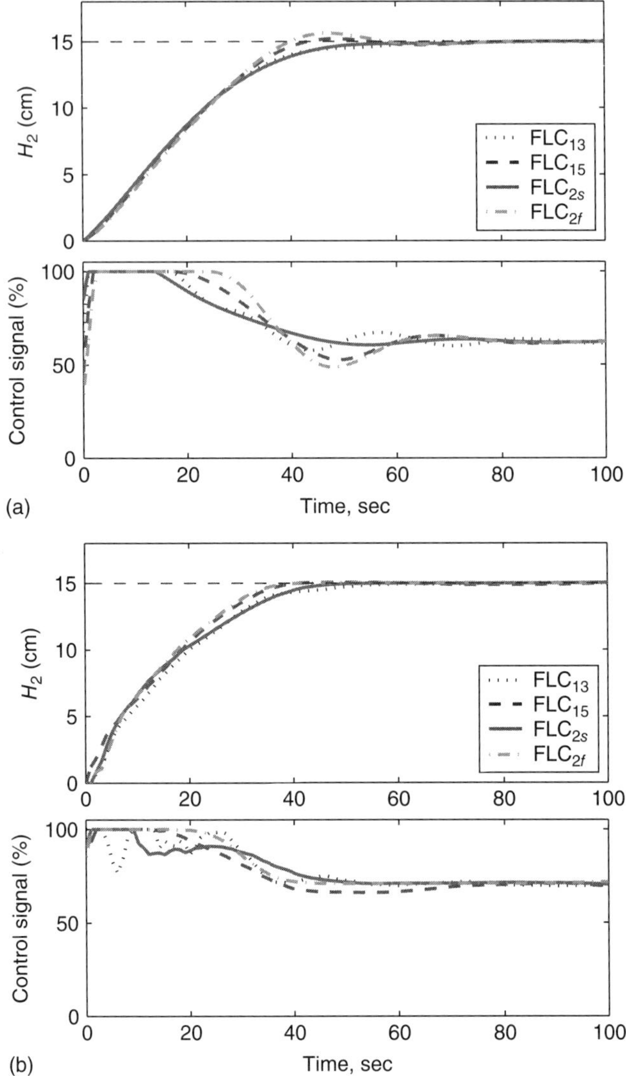

Figure 5.7 Step responses when the set point was 15 cm: (a) simulation results and (b) experiment results (Wu and Tan, 2006; © 2006, Elsevier).

FLCs to tolerate more modeling uncertainties. When a simulation model is used to evaluate the GA candidate solutions, the IT2 FLCs will have a higher probability of performing well on the actual plant.

With the simplified architecture, the computational cost of resulting simplified IT2 FLCs is much lower than that of a traditional IT2 FLC. The time taken by the GA to evolve the four FLCs is shown in Table 5.7. The data was obtained using a 996-MHz computer with 256 MB of RAM. A 10,000-step simulation [the set

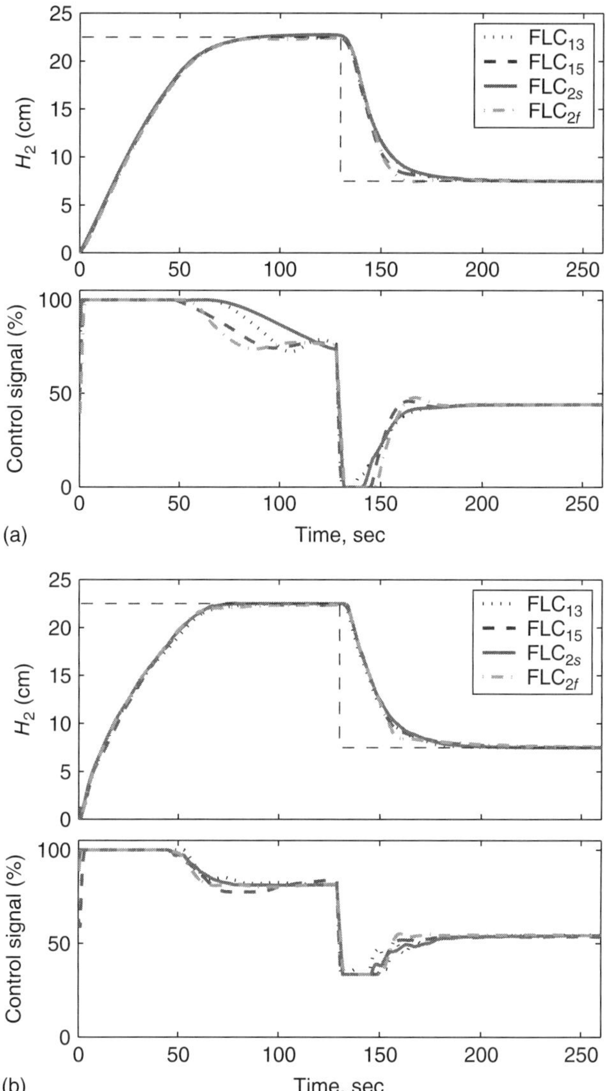

Figure 5.8 Step responses when the set point was changed. (a) simulation results and (b) experiment results (Wu and Tan, 2006; © 2006, Elsevier).

point is $15 + 10\sin(i/50)$, where $i = 1, 2, \ldots, 10,000$ is the time instant], using the evolved FLCs was also run on the same computer and the computation time is shown in Table 5.7. The results indicate that the computational cost of FLC_{2s} is much lower when compared with that of FLC_{2f}. These results suggest that the simplified IT2 FLC is more suitable for real-time implementation. It enables computational cost to be reduced without a degradation in the control performance and the ability to handle modeling uncertainties.

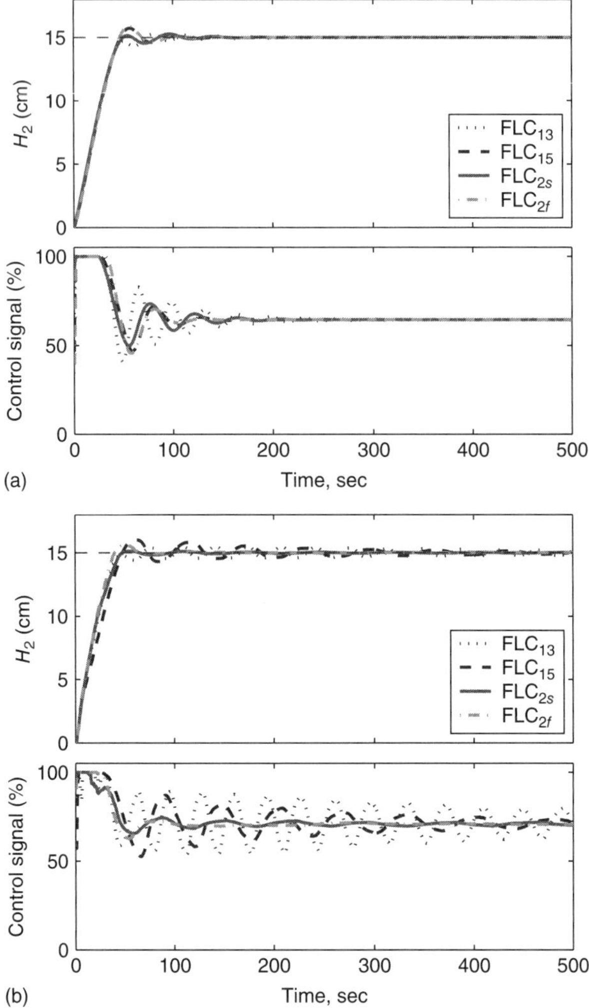

Figure 5.9 Step responses when the baffle was lowered: (a) simulation results and (b) experiment results (Wu and Tan, 2006; © 2006, Elsevier).

In summary, the simulation and experimental results indicate the smoother control surface around the $(E(k), R(k)) = (0, 0)$ region provided by an IT2 FLC is a potential source for performance improvement.

The remaining parts of the chapter focus on a theoretical study of a class of IT2 FLCs in order to systematically establish the unique features of IT2 FLCs that provide the potential of better performance. Using the technique for analysis introduced in Chapter 4, the focus is to derive the equivalent nonlinear proportional, derivative, and integral gains and study how their variations with $E(k)$ and $R(k)$ equip the IT2 FLC with the potential to yield better performance.

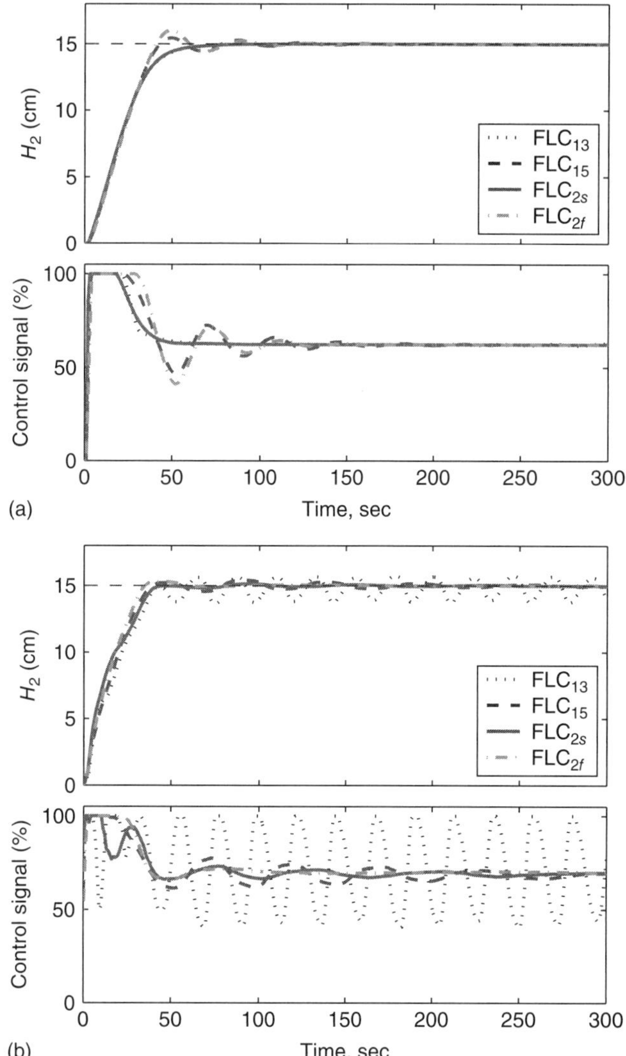

Figure 5.10 Step responses when there was a 2-sec transport delay: (a) simulation results and (b) experiment results (Wu and Tan, 2006; © 2006, Elsevier).

5.3 ANALYTICAL STRUCTURE OF INTERVAL T2 FUZZY PD AND PI CONTROLLER

5.3.1 Configuration of Interval T2 Fuzzy PD and PI Controller[2]

IT2 fuzzy PI and PD controllers are introduced in Section 4.3, and the general structure is shown in Fig. 4.1. The two inputs are $E(n) = K_e e(n) = K_e(\mathrm{SP}(n) - y(n))$

[2]Much of the material is taken directly from Nie and Tan, 2012; © 2012, IEEE).

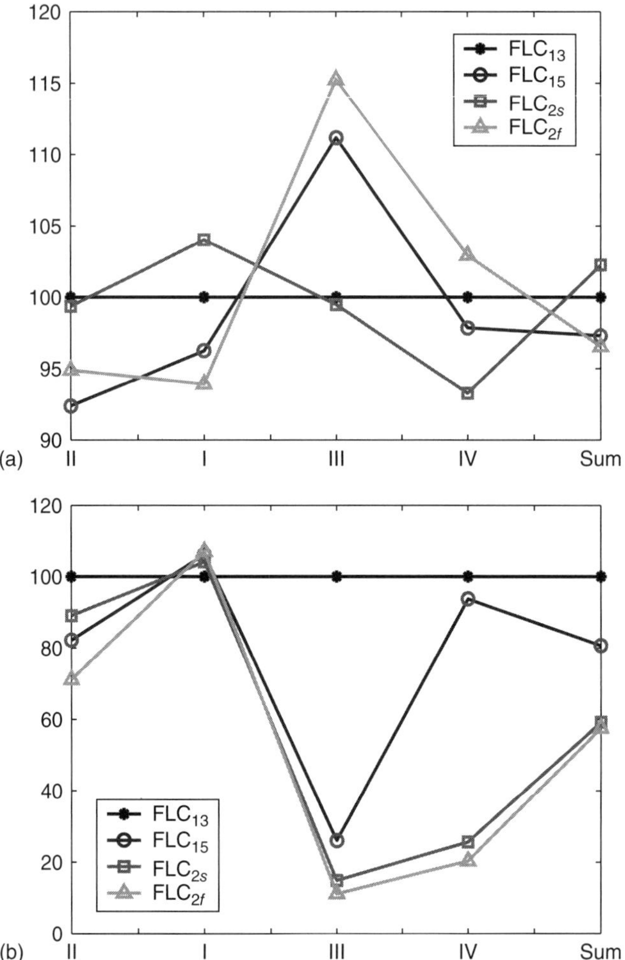

Figure 5.11 Comparison of four FLCs on four plants: (a) simulation results and (b) experiment results. I, II, III, and IV in horizontal axis stands for Plant I, Plant II, Plant III, and Plant IV in Table 5.4, respectively. Sum means the sum of the ITAEs on the four plants (Wu and Tan, 2006; © 2006, Elsevier).

[Eq. (4.7)] and $R(n) = K_r r(n) = K_r(e(n) - e(n-1))$ (Eq. 4.8), where SP(n) is the reference signal, $y(n)$ is the output of the closed-loop system, K_e and K_r are the scaling constants. Then, the output of the controller may be defined as

$$\Delta u(n) = f(E(n), R(n)) \tag{5.7}$$

The output $\Delta u(n)$ may be interpreted directly as the control signal or as the rate of change in the actuation signal. Depending on how the output of the fuzzy controller is defined, it may be interpreted as a fuzzy PD or a fuzzy PI controller.

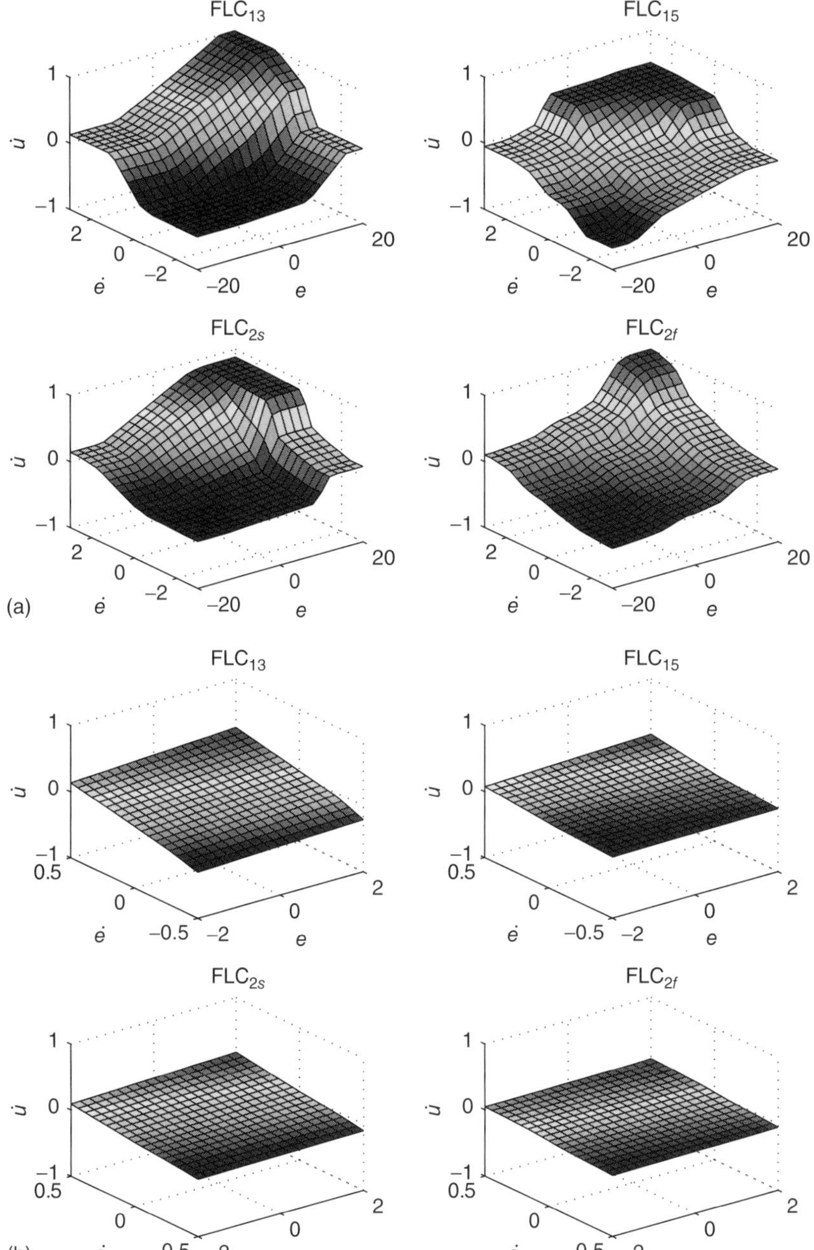

Figure 5.12 Control surfaces of the four FLCs: (a) complete control surfaces and (b) control surfaces near the origin (Wu and Tan, 2006; © 2006, Elsevier).

TABLE 5.7 Comparison of Computational Cost

Item\FLC	FLC_{13}	FLC_{15}	FLC_{2s}	FLC_{2f}
GA tuning (sec)	500	550	1200	4500
Simulation (sec)	1.28	1.40	3.19	11.70

For this study, two IT2 FSs are used to partition the space of each input: *EN* and *EP* for $E(n)$, *RN* and *RP* for $R(n)$. By shifting the membership of two symmetrical T1 FSs horizontally by the amount of θ_1, the upper and lower membership of *EN* and *EP*, $\overline{\mathbf{EN}}$, $\underline{\mathbf{EN}}$, $\overline{\mathbf{EP}}$, and $\underline{\mathbf{EP}}$ in Fig. 5.13a, can be obtained (refer to Chapter 4). In the same way, θ_2 represents the amount by which the upper and lower membership of *RN* and *RP* are shifted to obtain $\overline{\mathbf{RN}}$, $\underline{\mathbf{RN}}$, $\overline{\mathbf{RP}}$, and $\underline{\mathbf{RP}}$ in Fig. 5.13b. Based on whether its value depends on the inputs, the lower bound or upper bound of any antecedent set can be decomposed into dependent part or independent segment (0 or 1). For example,

$$\overline{\mathbf{EN}} = \begin{cases} -\dfrac{1}{2L_1}E(n) + 0.5 + \theta_1 & \text{for} \quad -L_1 + P_1 \leq E(n) \leq L_1 + P_1 \\ 0 & \text{for} \quad E(n) \geq L_1 + P_1 \\ 1 & \text{for} \quad E(n) \leq -L_1 + P_1 \end{cases} \tag{5.8}$$

where $\overline{\mathbf{EN}}$ is the segment of the membership function of $\overline{\mathbf{EN}}$ that is linearly related with the input, while the membership grade of the other two segments of $\overline{\mathbf{EN}}$ is fixed at 0 or 1. Eq. (5.8) reveals the relationship between the membership grade and the input.

For fuzzy systems that partition the input space using IT2 FSs shown in Figs. 5.13a and 5.13b, a commonly used rule base comprised of the following IF–THEN statements:

- Rule 1: IF $E(n)$ is positive AND $R(n)$ is positive THEN $\Delta u(n)$ is H_1.
- Rule 2: IF $E(n)$ is positive AND $R(n)$ is negative THEN $\Delta u(n)$ is H_2.
- Rule 3: IF $E(n)$ is negative AND $R(n)$ is positive THEN $\Delta u(n)$ is H_3.
- Rule 4: IF $E(n)$ is negative AND $R(n)$ is negative THEN $\Delta u(n)$ is H_4.

where H_1, H_2, H_3, and H_4 are the four singleton consequent FSs as illustrated in Fig. 5.14.

The result of matching the input signals to the fuzzy sets in the antecedent part of the fuzzy rules is an IT1 set called the firing set. Using the minimum t-norm operator, the firing sets for each rule are as follows:

$$R_1 = [\underline{R}_1, \overline{R}_1] = [\min(\underline{\mathbf{EP}}, \underline{\mathbf{RP}}), \min(\overline{\mathbf{EP}}, \overline{\mathbf{RP}})] \quad \text{for } \Delta u(n) = H_1 \tag{5.9}$$

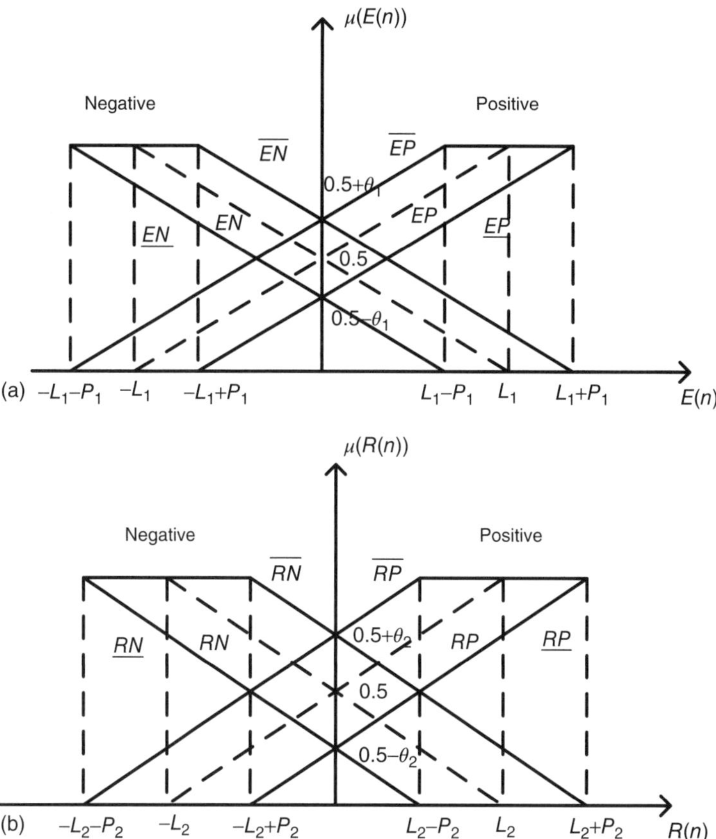

Figure 5.13 IT2 antecedent FSs: (a) IT2 FSs *EN* and *EP* for the input $E(n)$ ($P_1 = 2L_1\theta_1$). (b) IT2 FSs *RN* and *RP* for the input $R(n)$ ($P_2 = 2L_2\theta_2$) (Nie and Tan, 2012; © 2012, IEEE).

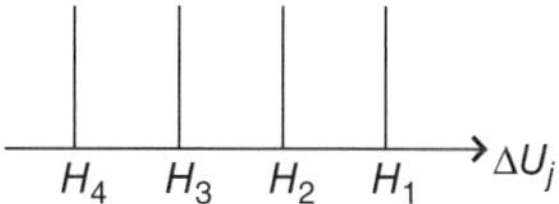

Figure 5.14 Illustration of the singleton consequent FSs of the IT2 fuzzy PD controller (Nie and Tan, 2012; © 2012, IEEE).

$$R_2 = [\underline{R}_2, \overline{R}_2] = [\min(\underline{\mathbf{EP}}, \underline{\mathbf{RN}}), \min(\overline{\mathbf{EP}}, \overline{\mathbf{RN}})] \quad \text{for } \Delta u(n) = H_2 \tag{5.10}$$

$$R_3 = [\underline{R}_3, \overline{R}_3] = [\min(\underline{\mathbf{EN}}, \underline{\mathbf{RP}}), \min(\overline{\mathbf{EN}}, \overline{\mathbf{RP}})] \quad \text{for } \Delta u(n) = H_3 \tag{5.11}$$

$$R_4 = [\underline{R}_4, \overline{R}_4] = [\min(\underline{\mathbf{EN}}, \underline{\mathbf{RN}}), \min(\overline{\mathbf{EN}}, \overline{\mathbf{RN}})] \quad \text{for } \Delta u(n) = H_4 \tag{5.12}$$

Using the wavy-slice representation theorem (Chapter 2, Theorem 2.2), the IT2 FS formed by the fuzzy inference engine may be viewed as the collection of all of its embedded IT1 FSs. Hence, the output of the inference engine may be type reduced into an IT1 set comprising of the centroids of all embedded T1 FSs:

$$\Delta u_j = \frac{R_1^* * H_1 + R_2^* * H_2 + R_3^* * H_3 + R_4^* * H_4}{R_1^* + R_2^* + R_3^* + R_4^*} \tag{5.13}$$

where R_i^* is a value within the lower and upper bound of the firing set for the ith rule, R_i. In summary, the type-reduced set $\Delta u_{\mathrm{TR}}(n)$ may be expressed mathematically as $[\Delta u_j^{\min}, \Delta u_j^{\max}]$, where $\Delta u_j^{\min}$ and $\Delta u_j^{\max}$ are, respectively, the smallest and largest centroid of all the possible embedded T1 FS. Lastly, using height defuzzification, the crisp output of the IT2 FLC is

$$\Delta u(n) = \frac{\Delta u_j^{\min} + \Delta u_j^{\max}}{2} \tag{5.14}$$

As described in Chapter 2, Section 2.3.4, the upper and lower bound of the type-reduced set, $\Delta u_j^{\min}$ and $\Delta u_j^{\max}$, may be expressed as the centroids of two unique embedded type-1 sets each of which involves only one switch between the lower and upper MF of the IT2 fuzzy set produced by the fuzzy inference engine (Mendel, 2001). The position of each switch point depends on the values of the singleton consequent sets. Hence, unlike a T1 fuzzy controller where the partitions of the input space are independent of the consequent sets, there is a need for the following assumptions in order to simplify the derivation of the analytical structure of IT2 fuzzy PD controller: (1) The rule base is symmetrical. In other words, $H_2 = H_3$. (2) $H_1, H_2 = H_3$, and H_4 are equally spaced as in Fig. 5.15. The assumption $H_4 < H_3 = H_2 < H_1$ is made based on the observation that the output of a linear PD controller increases when either input signal increases.

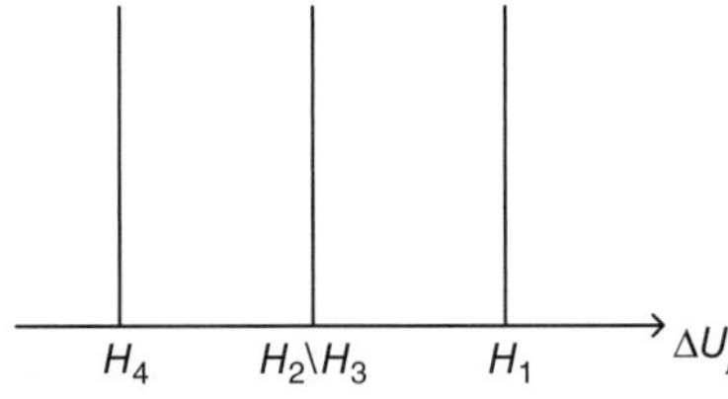

Figure 5.15 Singleton consequent FSs of IT2 fuzzy PD controller (Nie and Tan, 2012; © 2012, IEEE).

5.3.2 Analysis of the Karnik–Mendel Type-Reduced IT2 Fuzzy PD Controller[3]

The analytical structure of an FLC may be established by deriving the mathematical relationship between the inputs and the output. Similar to the case of a T1 FLC, the main concept used to determine the input–output relationship in an IT2 fuzzy PD controller is to specify the firing strength by dividing the input space into regions and to replace each firing strength with its corresponding mathematical expression. This subsection discusses the strategy that will enable the derivation of analytical expressions for the lower and upper bounds of the type reduced set, $\Delta u_j^{\min}$ and $\Delta u_j^{\max}$.

Recalling (2.66) and (2.67) in Chapter 2, the expressions for the type-reduced sets are

$$c_l(L) = \frac{\sum_{i=1}^{L} \overline{\mu}_i x_i + \sum_{i=L+1}^{m} \underline{\mu}_i x_i}{\sum_{i=1}^{L} \overline{\mu}_i + \sum_{i=L+1}^{m} \underline{\mu}_i} \tag{5.15}$$

$$c_r(R) = \frac{\sum_{i=1}^{R} \underline{\mu}_i x_i + \sum_{i=R+1}^{m} \overline{\mu}_i x_i}{\sum_{i=1}^{R} \underline{\mu}_i + \sum_{i=R+1}^{m} \overline{\mu}_i} \tag{5.16}$$

For the IT2 FLC described in Section 5.3.1, the type-reduced set may be constructed from Eqs. (5.15) and (5.16) by making the following substitutions:

$$\overline{\mu}_i = \overline{R}_j, \qquad \underline{\mu}_i = \underline{R}_j, \qquad x_i = H_j \tag{5.17}$$

where $\overline{R}_j$ and $\underline{R}_j$ are the upper and lower bound of the firing set associated with the jth rule in Eqs. (5.9)–(5.12). Furthermore, since it is assumed that $H_4 < H_3 = H_2 < H_1$, Eqs. (5.15) and (5.16) may be re-expressed as

$$\Delta u_j^{\min} = \frac{\sum_{i=1}^{L-1} \underline{R}_i H_i + \sum_{i=L}^{4} \overline{R}_i H_i}{\sum_{i=1}^{L-1} \underline{R}_i + \sum_{i=L}^{4} \overline{R}_i} \tag{5.18}$$

$$\Delta u_j^{\max} = \frac{\sum_{i=1}^{R-1} \overline{R}_i H_i + \sum_{i=R}^{4} \underline{R}_i H_i}{\sum_{i=1}^{R-1} \overline{R}_i + \sum_{i=R}^{4} \underline{R}_i} \tag{5.19}$$

These expressions indicate that $\Delta u_j^{\min}$ and $\Delta u_j^{\max}$ may be expressed as the average of all singleton consequents weighted by the lower or upper bound of the firing strength and an embedded T1 FS. Once the switch points L in Eq. (5.15) and R in Eq. (5.16) are known, then the problem of analyzing the structure of an IT2

[3]Much of the material is taken directly from (Nie and Tan 2012; © 2012, IEEE).

fuzzy PD controller reduces to a T1 system that can be studied using the following well-established techniques (Ying, 2000):

1. Partition the input space into regions by applying the minimum t-norm to the antecedent membership functions.
2. Determine the specific expression for each firing strength, respectively, in the corresponding subregion of each embedded T1 FLC.

The above discussion establishes that the key enabling theory for extending analytical structure analysis from T1 to IT2 FLC is a method for identifying the partitioning lines that correspond to a change in the switch point. In the case of the IT2 fuzzy PD controller, the location property given in Table 2.2 shows that the conditions for finding the switch points L and R are as follows:

$$H_L \leq \Delta u_j^{\min} < H_{L-1} \tag{5.20}$$

$$H_R \leq \Delta u_j^{\max} < H_{R-1} \tag{5.21}$$

Example 5.2 Consider the IT2 FLC described in Section 5.3.1. The positions of the switch points L and R may be obtained by analyzing the inequalities in Eqs. (5.20) and (5.21), which state the constraint that L and R must be positioned at one of the three singleton consequent sets.

In this case, there are three unique consequent sets ($H_1, H_2 = H_3$, and H_4). Combined with the conditions from Eqs. (5.20) and (5.21) that $H_L \leq \Delta u_j^{\min}$ and $H_R \leq \Delta u_j^{\max}$, L and R can assume one of only two values, that is, $H_L = \{H_4, H_2 = H_3\}$ and $H_R = \{H_4, H_2 = H_3\}$. Consequently, the lower bound of the type-reduced set may be computed using one of the following two equations:

$$\Delta u_{j1}^{\min} = \frac{\overline{R}_4 * H_4 + \underline{R}_3 * H_3 + \underline{R}_2 * H_2 + \underline{R}_1 * H_1}{\overline{R}_4 + \underline{R}_3 + \underline{R}_2 + \underline{R}_1} \tag{5.22}$$

$$\Delta u_{j2}^{\min} = \frac{\overline{R}_4 * H_4 + \overline{R}_3 * H_3 + \overline{R}_2 * H_2 + \underline{R}_1 * H_1}{\overline{R}_4 + \overline{R}_3 + \overline{R}_2 + \underline{R}_1} \tag{5.23}$$

Similarly, the upper bound of the type-reduced set may be derived using one of the following two equations:

$$\Delta u_{j1}^{\max} = \frac{\underline{R}_4 * H_4 + \underline{R}_3 * H_3 + \underline{R}_2 * H_2 + \overline{R}_1 * H_1}{\underline{R}_4 + \underline{R}_3 + \underline{R}_2 + \overline{R}_1} \tag{5.24}$$

$$\Delta u_{j2}^{\max} = \frac{\underline{R}_4 * H_4 + \overline{R}_3 * H_3 + \overline{R}_2 * H_2 + \overline{R}_1 * H_1}{\underline{R}_4 + \overline{R}_3 + \overline{R}_2 + \overline{R}_1} \tag{5.25}$$

The example illustrates that the only two possible expressions for the lower bound of the type-reduced set, $\Delta u_j^{\min}$, for the IT2 FLC defined in Section 5.3.1 may be written as:

1. Mode 1: When $H_4 \leq \Delta u_j^{\min} \leq H_2 = H_3 \Longleftrightarrow$ The left switch point L coincides with H_4:

$$\Delta u_j^{\min} = \Delta u_{j1}^{\min} = \frac{\overline{R}_4 * H_4 + \underline{R}_3 * H_3 + \underline{R}_2 * H_2 + \underline{R}_1 * H_1}{\overline{R}_4 + \underline{R}_3 + \underline{R}_2 + \underline{R}_1} \tag{5.26}$$

2. Mode 2: When $H_2 = H_3 \leq \Delta u_j^{\min} \leq H_1 \Longleftrightarrow$ The left switch point L coincides with $H_2 = H_3$:

$$\Delta u_j^{\min} = \Delta u_{j2}^{\min} = \frac{\overline{R}_4 * H_4 + \overline{R}_3 * H_3 + \overline{R}_2 * H_2 + \underline{R}_1 * H_1}{\overline{R}_4 + \overline{R}_3 + \overline{R}_2 + \underline{R}_1} \tag{5.27}$$

By comparing the expressions of modes 1 and 2 in Eqs. (5.26) and (5.27), their properties can be generalized as:

1. The weight associated with H_1 is always the lower bound of the firing set for rule 1. Similarly, H_4 is weighted by the upper bound of the firing set for rule 4. Consequently, the weight on H_1 and H_4 is independent of the switch points L and R.
2. In mode 1, $H_2 = H_3$ are weighted by the lower bounds $\underline{R}_2$ and $\underline{R}_3$, while the corresponding upper bounds $\overline{R}_2$ and $\overline{R}_3$ are used to weight $H_2 = H_3$ in mode 2. The condition when the switch point changes from the position of $H_2 = H_3$ to H_1 and vice versa can be established as:

$$\Delta u_j^{\min} = \Delta u_{j1}^{\min} = \Delta u_{j2}^{\min} = H_2 = H_3 \tag{5.28}$$

By replacing $\Delta u_{j1}^{\min}$ and $\Delta u_{j2}^{\min}$ with their corresponding expressions in Eqs. (5.26) and (5.27), the above equation may be written as

$$\frac{\overline{R}_4 * H_4 + \underline{R}_3 * H_3 + \underline{R}_2 * H_2 + \underline{R}_1 * H_1}{\overline{R}_4 + \underline{R}_3 + \underline{R}_2 + \underline{R}_1} = \frac{\overline{R}_4 * H_4 + \overline{R}_3 * H_3 + \overline{R}_2 * H_2 + \underline{R}_1 * H_1}{\overline{R}_4 + \overline{R}_3 + \overline{R}_2 + \underline{R}_1} = H_2 = H_3$$

$$\Longleftrightarrow \overline{R}_4(H_4 - H_2) = \underline{R}_1(H_2 - H_1)$$
$$\text{or} \quad \overline{R}_4(H_4 - H_3) = \underline{R}_1(H_3 - H_1) \tag{5.29}$$

Due to the assumption that the three consequent sets are equally spaced and $H_4 < H_3 = H_2 < H_1$, the condition derived above may be simplified to

$$\overline{R}_4 = \underline{R}_1 \tag{5.30}$$

Further, the subregions where Eq. (5.20) is used to calculate the left end point of the type-reduced set (mode 1) should satisfy the condition

$$\Delta u_{j1}^{\min} \leq H_2 = H_3 \Longleftrightarrow \overline{R}_4 \geq \underline{R}_1 \tag{5.31}$$

and the areas where the IT2 fuzzy PD controller operate in mode 2, that is, the output is defined by Eq. (5.21), can be found using the following equality:

$$\Delta u_{j2}^{\min} \geq H_2 = H_3 \Longleftrightarrow \overline{R}_4 \leq \underline{R}_1 \tag{5.32}$$

The first property shows that the firing strength of rules 1 and 4 used to calculate $\Delta u_j^{\min}$ is independent of the switch point L, and the firing strength of rules 1 and 4 is always governed by the lower and upper bound of the firing set, that is, $\underline{R}_1$ and $\overline{R}_4$. Furthermore, the boundary defined by the conditions in Eqs. (5.31) and (5.32) also depends on $\underline{R}_1$ and $\overline{R}_4$. Based on this observation, the first step in partitioning the input space in order to derive a closed-form firing level is performed by considering the outcomes of the minimum t-norm operations for rules 1 and 4. Next, the relative firing strength of rules 1 and 4 in each subspace is compared in order to determine whether the IT2 fuzzy PD controller will operate in modes 1 and 2. In the subregions under modes 1 and 2, the embedded type 1 set is completely defined by Eqs. (5.26) and (5.27) so the partitions can be found using existing technique, that is, ascertaining the minimum of the lower or upper bound of the firing sets. In summary, an algorithm to derive the partitions of the input space by $\Delta u_j^{\min}$ can be generalized as:

Step 1: The firing strength of rule 1 ($\underline{R}_1$) and rule 4 ($\overline{R}_4$) can be specified by dividing the input space via the outcomes of the minimum t-norm operations for rules 1 and 4, that is,

$$\underline{R}_1 = \min\{\underline{\mathbf{EP}}, \underline{\mathbf{RP}}\} \tag{5.33}$$

$$\overline{R}_4 = \min\{\overline{\mathbf{EN}}, \overline{\mathbf{RN}}\} \tag{5.34}$$

Step 2: The partitions obtained using $\underline{R}_1$ and $\overline{R}_4$ in step 1 is further subdivided into the following two groups, which correspond to one of the two possible operating modes:

$$\text{Mode 1:} \quad \overline{R}_4 > \underline{R}_1 \tag{5.35}$$

$$\text{Mode 2:} \quad \overline{R}_4 < \underline{R}_1 \tag{5.36}$$

Step 3: To specify the firing strength of rules 2 and 3 by dividing the corresponding regions for modes 1 and 2 using Eqs. (5.26) and (5.27), respectively. Under mode 1, the firing strength of rules 2 and 3 is given by the lower MF. Hence, partitioning can be achieved from the outcomes of the following minimum t-norm operations:

$$\underline{R}_2 = \min\{\underline{\mathbf{EP}}, \underline{\mathbf{RN}}\} \tag{5.37}$$

$$\underline{R}_3 = \min\{\underline{\mathbf{EN}}, \underline{\mathbf{RP}}\} \tag{5.38}$$

For the regions where the IT2 fuzzy PD controller is operating in mode 2, the following minimum t-norm operations should be used to divide the input space:

$$\overline{R}_2 = \min\{\overline{\mathbf{EP}}, \overline{\mathbf{RN}}\} \tag{5.39}$$

$$\overline{R}_3 = \min\{\overline{\mathbf{EN}}, \overline{\mathbf{RP}}\} \tag{5.40}$$

Step 4: Superimpose all the partitions obtained by considering the mode switch and the minimum t-norm operations.

Similarly, the firing strength in the equation for deriving the right end point $\Delta u_j^{\max}$ can be specified. The procedures to derive the analytical structure of the IT2 fuzzy PD controller can be generalized in Fig. 5.16. Applying the algorithm in Fig. 5.16, the partitions of the input space for $\Delta u_j^{\min}$ and $\Delta u_j^{\max}$ were derived (Nie and Tan, 2012). Since the output of an IT2 FLC is the average of the two endpoints, superimposing the partitions by $\Delta u_j^{\min}$ and $\Delta u_j^{\max}$ yields the 35 partitions shown in Fig. 5.17, which may be used to analyze an IT2 fuzzy PD controller. Each of the 35 partitions correspond to unique groups of input conditions (ICs).

5.3.3 Analysis of the IT2 Fuzzy PD Controller

This section presents examples that illustrate how the ICs or partitions may be useful for analyzing the characteristics of IT2 FLCs. The central idea is to obtain a closed-form equation for the output of the IT2 FLC in a particular region by replacing each firing strength in Eqs. (5.18) and (5.19) with the relevant expressions. Properties of the IT2 fuzzy PD controller can then be analyzed using the equations.

Example 5.3 ***Interpreting IT2 PD FLC as Nonlinear PD Controllers***
Consider the IT2 FLC defined in Section 5.3.1. For any input pairs that satisfy IC1 and lie in partition 1 of Fig. 5.17, the switch points, L and R, correspond to

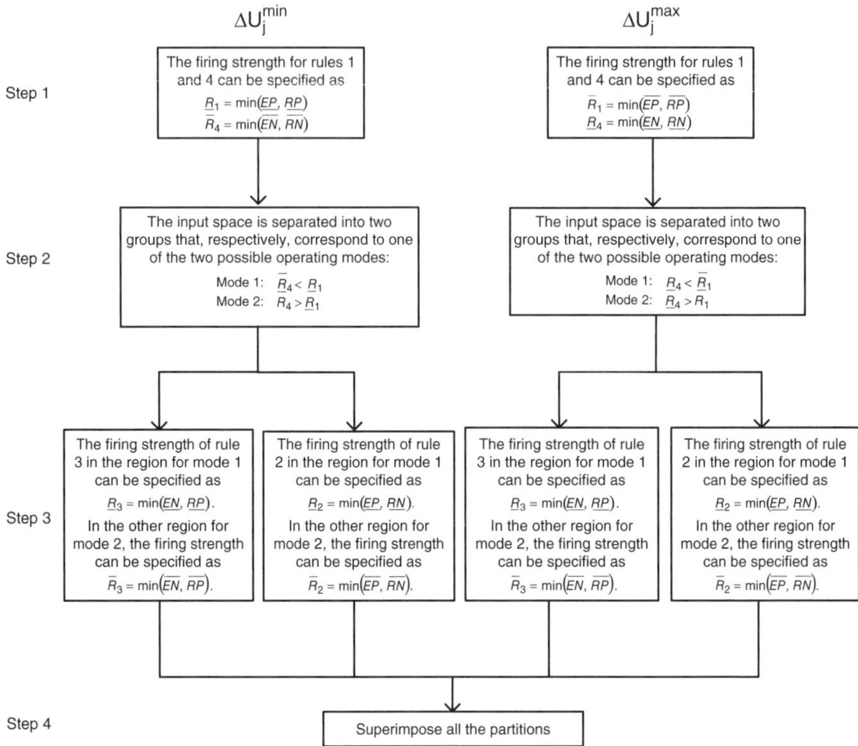

Figure 5.16 Flowchart of the algorithm to derive the analytical structure of IT2 fuzzy PD controller (Nie and Tan 2012; © 2012, IEEE).

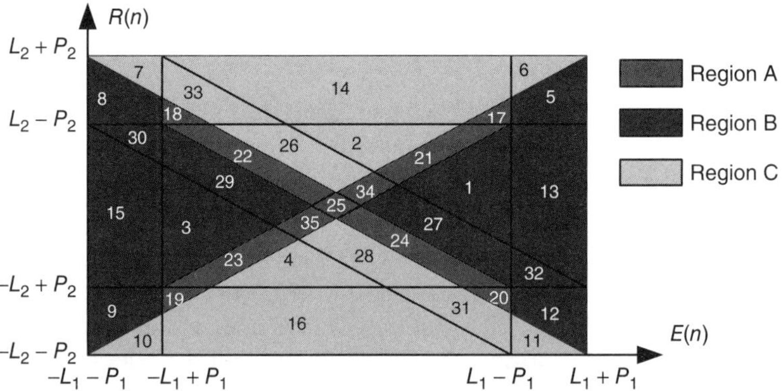

Figure 5.17 Partition of input space by IT2 FLC when $\theta_1 < \theta_2$ (Nie and Tan, 2012; © 2012, IEEE).

the locations of the consequent sets $H_3 = H_2$ and H_1, respectively. Consequently, the firing strengths in Eq. (5.18) for $\Delta u_j^{\min}$ should be

$$R_4 = R_3 = \overline{EN} = -\frac{1}{2L_1}E(n) + 0.5 + \theta_1 \tag{5.41}$$

$$R_2 = \overline{RN} = -\frac{1}{2L_2}R(n) + 0.5 + \theta_2 \tag{5.42}$$

$$R_1 = \underline{RP} = \frac{1}{2L_2}R(n) + 0.5 - \theta_2 \tag{5.43}$$

Substituting the above equations into Eqs. (5.23) and (5.25), the closed-form expression for $\Delta u_j^{\min}$ and $\Delta u_j^{\max}$ are as follows:

$$\begin{aligned}
\Delta u_{j\text{IC}1}^{\min} &= \frac{R_4^* * H_4 + R_3^* * H_3 + R_2^* * H_2 + R_1^* * H_1}{R_4^* + R_3^* + R_2^* + R_1^*} \\
&= \frac{\overline{EN} * H_4 + \overline{EN} * H_3 + \overline{RN} * H_2 + \underline{RP} * H_1}{\overline{EN} + \overline{EN} + \overline{RN} + \underline{RP}} \\
&= \frac{\begin{array}{c}[-(1/2L_1)E(n) + 0.5 + \theta_1](H_3 + H_4) + [-(1/2L_2)R(n) + 0.5 + \theta_2]H_2 \\ +[(1/2L_2)R(n) + 0.5 - \theta_2]H_1\end{array}}{\begin{array}{c}2[-(1/2L_1)E(n) + 0.5 + \theta_1] + [-(1/2L_2)R(n) + 0.5 + \theta_2] \\ +[(1/2L_2)R(n) + 0.5 - \theta_2]\end{array}} \\
&= \frac{-L_2(H_3 + H_4)E(n) + L_1(H_1 - H_2)R(n)}{4L_1L_2(1 + \theta_1) - 2L_2E(n)} \\
&\quad + \frac{L_1L_2[(0.5 + \theta_1)(H_3 + H_4) + (0.5 - \theta_2)H_1 + (0.5 + \theta_2)H_2]}{2L_1L_2(1 + \theta_1) - L_2E(n)} \\
&= K_p^1E(n) + K_d^1R(n) + \delta^1
\end{aligned} \tag{5.44}$$

$$\begin{aligned}
\Delta u_{j\text{IC}1}^{\max} &= \frac{R_4^* * H_4 + R_3^* * H_3 + R_2^* * H_2 + R_1^* * H_1}{R_4^* + R_3^* + R_2^* + R_1^*} \\
&= \frac{\underline{EN} * H_4 + \underline{EN} * H_3 + \underline{RN} * H_2 + \overline{RP} * H_1}{\underline{EN} + \underline{EN} + \underline{RN} + \overline{RP}} \\
&= \frac{\begin{array}{c}[-(1/2L_1)E(n) + 0.5 - \theta_1](H_3 + H_4) + [-(1/2L_2)R(n) + 0.5 - \theta_2]H_2 \\ +[(1/2L_2)R(n) + 0.5 + \theta_2]H_1\end{array}}{\begin{array}{c}2[-(1/2L_1)E(n) + 0.5 - \theta_1] + [-(1/2L_2)R(n) + 0.5 - \theta_2] \\ +[(1/2L_2)R(n) + 0.5 + \theta_2]\end{array}} \\
&= \frac{-L_2(H_3 + H_4)E(n) + L_1(H_1 - H_2)R(n)}{4L_1L_2(1 - \theta_1) - 2L_2E(n)}
\end{aligned}$$

$$+\frac{L_1L_2[(0.5-\theta_1)(H_3+H_4)+(0.5+\theta_2)H_1+(0.5-\theta_2)H_2]}{2L_1L_2(1-\theta_1)-L_2E(n)} \tag{5.45}$$

Example 5.3 demonstrates that the closed-form equations for the bounds of the type-reduced set may be expressed in terms of $E(n)$ and $R(n)$. Consequently, each subregion may be interpreted as a nonlinear PI/PD controller of the following form:

$$\Delta u_{j\mathrm{IC}q} = K_p^q E(n) + K_d^q R(n) + \delta^q \tag{5.46}$$

where $\Delta u_{j\mathrm{IC}q}$ is the output of IC q, K_p^q is the corresponding proportional gain, K_d^q is the derivative gain, and δ^q is the offset. The equivalent proportional-gains K_p^q and derivative-gains K_d^q for the 35 partitions are shown in Nie and Tan (2012) and reproduced as Tables 5.8 and 5.9. In the next example, the analytical structure and equivalent proportional and derivative gains will be used to investigate the potential advantages introduced by nonequal FOU.

Example 5.4 ***Potential Benefits of Using Unequal FOU*** Assume an IT2 FLC that has the structure described in Section 5.3.1, but with the additional constraints that $H_2 = H_3 = 0$ and the FOU size for both the $E(k)$ and $R(k)$ are equal, that is, $\theta_1 = \theta_2$. By following the procedure described in Fig. 5.16, the partitions in Fig. 5.18 can be obtained.

A comparison of Figs. 5.17 and 5.18 reveals that nonequal FOU, that is, $\theta_1 \neq \theta_2$, introduces additional partitions that are collectively grouped as region A in Fig. 5.17. An examination of Tables 5.8 and 5.9 reveals that the equivalent proportional and integral gains for input pairs in region A, except for IC 17–20, are functions of both $E(n)$ and $R(n)$, for example, $K_p = f\{E(n), R(n)\}$. In contrast, the equivalent proportional and derivative gains of the IT2 FLC for the remaining

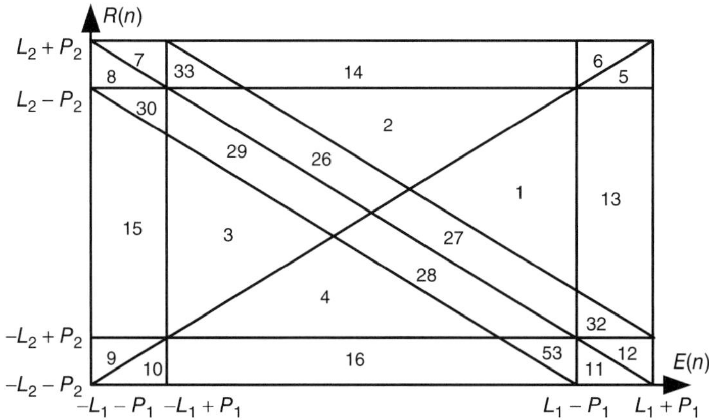

Figure 5.18 Partition of the input space by the IT2 FLC when $\theta_1 = \theta_2$ (Nie and Tan, 2012; © 2012, IEEE).

TABLE 5.8 Gains for External Subregions

IC No.	K_p	K_d	δ
5	$\frac{H_1}{8L_1(1+\theta_1)-4E(n)}$	$\frac{L_1H_1}{8L_1L_2(1+\theta_1)-4L_2E(n)}$	$-\frac{L_1(\theta_1+\theta_2)H_1}{4L_1(1+\theta_1)-2E(n)}+0.5H_1$
6	$\frac{L_2H_1}{8L_1L_2(1+\theta_2)-4L_1R(n)}$	$\frac{H_1}{8L_2(1+\theta_2)-4R(n)}$	$-\frac{L_2(\theta_1+\theta_2)H_1}{4L_2(1+\theta_2)-2R(n)}+0.5H_1$
7	$\frac{H_1}{4L_1}$	$\frac{L_1H_1}{4L_1L_2(1-\theta_1+\theta_2)-2L_1R(n)-2L_2E(n)}$	$\frac{L_1L_2(0.5+\theta_2)H_4}{2L_1L_2(1-\theta_1+\theta_2)-L_1R(n)-L_2E(n)}+0.5(0.5+\theta_1)H_1$
8	$\frac{L_2H_1}{4L_1L_2(1-\theta_2+\theta_1)+2L_1R(n)+2L_2E(n)}$	$\frac{H_1}{4L_2}$	$\frac{L_1L_2(0.5+\theta_1)H_1}{2L_1L_2(1-\theta_2+\theta_1)+L_1R(n)+L_2E(n)}+0.5(0.5+\theta_2)H_4$
9	$\frac{H_1}{8L_1(1+\theta_1)+4E(n)}$	$\frac{L_1H_1}{8L_1L_2(1+\theta_1)+4L_2E(n)}$	$\frac{L_1H_1(\theta_1+\theta_2)}{4L_1(1+\theta_1)+2E(n)}+0.5H_4$
10	$\frac{H_1L_2}{8L_1L_2(1+\theta_2)+4L_1R(n)}$	$\frac{H_1}{8L_2(1+\theta_2)+4R(n)}$	$\frac{L_2(\theta_1+\theta_2)H_1}{4L_2(1+\theta_2)+2R(n)}+0.5H_4$
11	$\frac{H_1}{4L_1}$	$\frac{L_1H_1}{4L_1L_2(1-\theta_1+\theta_2)+2L_2E(n)+2L_1R(n)}$	$\frac{L_1L_2(0.5+\theta_2)H_1}{2L_1L_2(1-\theta_1+\theta_2)+L_2E(n)+L_1R(n)}+0.5(0.5+\theta_1)H_4$
12	$\frac{L_2H_1}{4L_1L_2(1+\theta_1-\theta_2)-2L_2E(n)-2L_1R(n)}$	$\frac{H_1}{4L_2}$	$\frac{L_1L_2(0.5+\theta_1)H_4}{2L_1L_2(1+\theta_1-\theta_2)-L_2E(n)-L_1R(n)}+0.5(0.5+\theta_2)H_1$
13	$\frac{H_1}{8L_1(1+\theta_1)-4E(n)}$	$\frac{L_1H_1}{8L_1L_2(1+\theta_1)-4L_2E(n)}+\frac{H_1}{4L_2}$	$-\frac{L_1(\theta_1+\theta_2)H_1}{4L_1(1+\theta_1)-2E(n)}+0.5(0.5+\theta_2)H_1$
14	$\frac{L_2H_1}{8L_1L_2(1+\theta_2)-4L_1R(n)}+\frac{H_1}{4L_1}$	$\frac{H_1}{8L_2(1+\theta_2)-4R(n)}$	$-\frac{L_2(\theta_1+\theta_2)H_1}{4L_2(1+\theta_2)-2R(n)}+0.5(0.5+\theta_1)H_1$
15	$\frac{H_1}{8L_1(1+\theta_1)+4E(n)}$	$\frac{L_1H_1}{8L_1L_2(1+\theta_1)+4L_2E(n)}+\frac{H_1}{4L_2}$	$\frac{L_1(\theta_1+\theta_2)H_1}{4L_1(1+\theta_1)+2E(n)}+0.5(0.5+\theta_2)H_4$

(continued)

TABLE 5.8 (*Continued*)

IC No.	K_p	K_d	δ
16	$\frac{L_2H_1}{8L_1L_2(1+\theta_2)+4L_1R(n)}+\frac{H_1}{4L_1}$	$\frac{H_1}{8L_2(1+\theta_2)+4R(n)}$	$\frac{L_2\theta_1+\theta_2)H_1}{4L_2(1+\theta_2)+2R(n)}+0.5(0.5+\theta_1)H_4$
17	$\frac{H_1}{4L_1}+\frac{H_1}{8L_1(1+\theta_1)-4E(n)}$	$\frac{L_1H_1}{8L_1L_2(1+\theta_1)-4L_2E(n)}$	$-\frac{L_1(\theta_1+\theta_2)H_1}{4L_1(1+\theta_1)-2E(n)}+0.5(0.5+\theta_1)H_1$
18	$\frac{H_1}{4L_1(1.5-\theta_1)+2E(n)}+\frac{L_2H_1}{4L_1L_2(1-\theta_2+\theta_1)+2L_1R(n)+2L_2E(n)}$	$\frac{L_1H_1}{4L_1L_2(1.5-\theta_1)+2L_2E(n)}$	$\frac{L_1L_2(0.5+\theta_1)H_1}{2L_1L_2(1-\theta_2+\theta_1)+L_1R(n)+L_2E(n)}+\frac{L_1H_4(\theta_1+\theta_2)}{2L_1(1.5-\theta_1)+E(n)}$
19	$\frac{H_1}{8L_1(1+\theta_1)+4E(n)}+\frac{H_1}{4L_1}$	$\frac{L_1H_1}{8L_1L_2(1+\theta_1)+4L_2E(n)}$	$\frac{L_1(\theta_1+\theta_2)H_1}{4L_1(1+\theta_1)+2E(n)}+0.5(0.5+\theta_1)H_4$
20	$\frac{H_1}{4L_1(1.5-\theta_1)-2E(n)}+\frac{L_2H_1}{4L_1L_2(1+\theta_1-\theta_2)-2L_2E(n)-2L_1R(n)}$	$\frac{L_1H_1}{4L_1L_2(1.5-\theta_1)-2L_2E(n)}$	$\frac{L_1L_2(0.5+\theta_1)H_4}{2L_1L_2(1+\theta_1-\theta_2)-L_2E(n)-L_1R(n)}+\frac{L_1(\theta_1+\theta_2)H_1}{2L_1(1.5-\theta_1)-E(n)}$
30	$\frac{H_1}{4L_1(1.5+\theta_1-2\theta_2)+2E(n)}$	$\frac{L_1H_1}{4L_1L_2(1.5+\theta_1-2\theta_2)+2L_2E(n)}+\frac{H_1}{4L_2}$	$\frac{L_1(\theta_1+\theta_2)H_1}{2L_1(1.5+\theta_1-2\theta_2)+E(n)}+0.5(0.5+\theta_2)H_4$
31	$\frac{H_1}{4L_1}+\frac{L_2H_1}{4L_1L_2(1.5-2\theta_1+\theta_2)+2L_1R(n)}$	$\frac{H_1}{4L_2(1.5-2\theta_1+\theta_2)+R(n)}$	$\frac{L_2(\theta_1+\theta_2)H_1}{2L_2(1.5-2\theta_1+\theta_2)+R(n)}+0.5(0.5+\theta_1)H_4$
32	$\frac{H_1}{4L_1(1.5+\theta_1-2\theta_2)-2E(n)}$	$\frac{L_1H_1}{4L_1L_2(1.5+\theta_1-2\theta_2)-2L_2E(n)}+\frac{H_1}{4L_2}$	$\frac{L_1H_4(\theta_1+\theta_2)}{2L_1(1.5+\theta_1-2\theta_2)-E(n)}+0.5(0.5+\theta_2)H_1$
33	$\frac{L_2H_1}{4L_1L_2(1.5-2\theta_1+\theta_2)-2L_1R(n)}+\frac{H_1}{4L_1}$	$\frac{L_2H_1}{4L_1L_2(1.5-2\theta_1+\theta_2)-2L_1R(n)}$	$\frac{L_2H_4(\theta_1+\theta_2)}{2L_2(1.5-2\theta_1+\theta_2)-R(n)}+0.5(0.5+\theta_1)H_1$

TABLE 5.9 Gains for Internal Subregions

IC No.	K_p	K_d	δ
1	$\frac{H_1}{8L_1(1+\theta_1)-4E(n)}+\frac{H_1}{8L_1(1-\theta_1)-4E(n)}$	$\frac{L_1H_1}{8L_1L_2(1+\theta_1)-4L_2E(n)}+\frac{L_1H_1}{8L_1L_2(1-\theta_1)-4L_2E(n)}$	$\frac{L_1(\theta_1+\theta_2)H_4}{4L_1(1+\theta_1)-2E(n)}+\frac{L_1(\theta_1+\theta_2)H_1}{4L_1(1-\theta_1)-2E(n)}$
2	$\frac{L_2H_1}{8L_1L_2(1+\theta_2)-4L_1R(n)}+\frac{L_2H_1}{8L_1L_2(1-\theta_2)-4L_1R(n)}$	$\frac{H_1}{8L_2(1+\theta_2)-4R(n)}+\frac{H_1}{8L_1L_2(1-\theta_2)-4R(n)}$	$\frac{L_2(\theta_1+\theta_2)H_4}{4L_2(1+\theta_2)-2R(n)}+\frac{L_2(\theta_1+\theta_2)H_1}{4L_2(1-\theta_2)-2R(n)}$
3	$\frac{H_1}{8L_1(1-\theta_1)+4E(n)}+\frac{H_1}{8L_1(1+\theta_1)+4E(n)}$	$\frac{L_1H_1}{8L_1L_2(1-\theta_1)+2L_2E(n)}+\frac{L_1H_1}{8L_1L_2(1+\theta_1)+4L_2E(n)}$	$\frac{L_1(\theta_1+\theta_2)H_4}{4L_1(1-\theta_1)+2E(n)}+\frac{L_1(\theta_1+\theta_2)H_1}{4L_1(1+\theta_1)+2E(n)}$
4	$\frac{L_2H_1}{8L_1L_2(1-\theta_2)+4L_1R(n)}+\frac{L_2H_1}{8L_1L_2(1+\theta_2)+4L_1R(n)}$	$\frac{H_1}{8L_2(1-\theta_2)+4R(n)}+\frac{H_1}{8L_2(1+\theta_2)+4R(n)}$	$\frac{L_2(\theta_1+\theta_2)H_4}{4L_2(1-\theta_2)+2R(n)}+\frac{L_2(\theta_1+\theta_2)H_1}{4L_2(1+\theta_2)+2R(n)}$
21[a]	$\frac{H_1}{8L_1(1+\theta_1)-4E(n)}+\frac{L_2H_1}{8L_1L_2(1-\theta_2)-4L_1R(n)}$	$\frac{L_1H_1}{8L_1L_2(1+\theta_1)-4L_2E(n)}+\frac{H_1}{8L_2(1-\theta_2)-4R(n)}$	$\frac{L_1(\theta_1+\theta_2)H_4}{4L_1(1+\theta_1)-2E(n)}+\frac{L_2(\theta_1+\theta_2)H_1}{4L_2(1-\theta_2)-2R(n)}$

(continued)

TABLE 5.9 (Continued)

IC No.	K_p	K_d	δ
22[a]	$\frac{L_2H_1}{4L_1L_2(2-\theta_1-\theta_2)+2L_2E(n)-2L_1R(n)} + + \frac{L_2H_1}{4L_1L_2(2-3\theta_2+\theta_1)+2L_2E(n)-2L_1R(n)}$	$\frac{L_1H_1}{4L_1L_2(2-\theta_1-\theta_2)+2L_2E(n)-2L_1R(n)} + \frac{L_1H_1}{4L_1L_2(2-3\theta_2+\theta_1)+2L_2E(n)-2L_1R(n)}$	$\frac{L_1L_2(\theta_1+\theta_2)H_4}{2L_1L_2(2-\theta_1-\theta_2)+L_2E(n)-L_1R(n)} + \frac{L_1L_2(\theta_1+\theta_2)H_1}{2L_1L_2(2-3\theta_2+\theta_1)+L_2E(n)-L_1R(n)}$
23[a]	$\frac{L_2H_1}{8L_1L_2(1-\theta_2)+4L_1R(n)} + \frac{H_1}{8L_1(1+\theta_1)+4E(n)}$	$\frac{H_1}{8L_2(1-\theta_2)+4R(n)} + \frac{L_1H_1}{8L_1L_2(1+\theta_1)+4l_2E(n)}$	$\frac{L_2(\theta_1+\theta_2)H_4}{4L_2(1-\theta_2)+2R(n)} + \frac{L_1(\theta_1+\theta_2)H_1}{4L_1(1+\theta_1)+2E(n)}$
24[a]	$\frac{L_2H_1}{4L_1L_2(2+\theta_1-3\theta_2)+2L_1R(n)-2L_2E(n)} + \frac{L_2H_1}{4L_1L_2(2-\theta_1-\theta_2)+2L_1R(n)-2L_2E(n)}$	$\frac{L_1H_1}{4L_1L_2(2+\theta_1-3\theta_2)+2L_1R(n)-2L_2E(n)} + \frac{L_1H_1}{4L_1L_2(2-\theta_1-\theta_2)+2L_1R(n)-2L_2E(n)}$	$\frac{L_1L_2(\theta_1+\theta_2)H_4}{2L_1L_2(2+\theta_1-3\theta_2)+L_1R(n)-L_2E(n)} + \frac{L_1L_2(\theta_1+\theta_2)H_1}{2L_1L_2(2-\theta_1-\theta_2)+L_1R(n)-L_2E(n)}$
25[a]	$\frac{L_2H_1}{4L_1L_2(2+\theta_1-3\theta_2)+2L_1R(n)-2L_2E(n)} + \frac{L_2H_1}{4L_1L_2(2-3\theta_2+\theta_1)+2L_2E(n)-2L_1R(n)}$	$\frac{L_1H_1}{4L_1L_2(2+\theta_1-3\theta_2)+2L_1R(n)-2L_2E(n)} + \frac{L_1H_1}{4L_1L_2(2-3\theta_2+\theta_1)+2L_2E(n)-2L_1R(n)}$	$\frac{L_1L_2(\theta_1+\theta_2)H_4}{2L_1L_2(2+\theta_1-3\theta_2)+L_1R(n)-L_2E(n)} + \frac{L_1L_2(\theta_1+\theta_2)H_1}{2L_1L_2(2-3\theta_2+\theta_1)+L_2E(n)-L_1R(n)}$
26	$\frac{L_2H_1}{8L_1L_2(1-\theta_1)-4L_1R(n)} + \frac{L_2H_1}{8L_1L_2(1-\theta_2)-4L_1R(n)}$	$\frac{H_1}{8L_2(1-\theta_1)-4R(n)} + \frac{H_1}{8L_2(1-\theta_2)-4R(n)}$	$\frac{L_2(\theta_1+\theta_2)H_4}{4L_2(1-\theta_1)-2R(n)} + \frac{L_2(\theta_1+\theta_2)H_1}{4L_2(1-\theta_2)-2R(n)}$

27	$\frac{H_1}{8L_1(1-\theta_2)-4E(n)}+\frac{H_1}{8L_1(1-\theta_1)-4E(n)}$	$\frac{L_1H_1}{8L_1L_2(1-\theta_2)-4L_2E(n)}+\frac{L_1H_1}{8L_1L_2(1-\theta_1)-4L_2E(n)}$	$\frac{L_1(\theta_1+\theta_2)H_4}{4L_1(1-\theta_2)-2E(n)}+\frac{L_1(\theta_1+\theta_2)H_1}{4L_1(1-\theta_1)-2E(n)}$
28	$\frac{L_2H_1}{8L_1L_2(1-\theta_2)+4L_1R(n)}+\frac{L_2H_1}{8L_1L_2(1-\theta_1)+4L_1R(n)}$	$\frac{H_1}{8L_2(1-\theta_2)+4R(n)}+\frac{H_1}{8L_2(1-\theta_1)+4R(n)}$	$\frac{L_2(\theta_1+\theta_2)H_4}{4L_2(1-\theta_2)+2R(n)}+\frac{L_2(\theta_1+\theta_2)H_1}{4L_2(1-\theta_1)+2R(n)}$
29	$\frac{H_1}{8L_1(1-\theta_1)+4E(n)}+\frac{H_1}{8L_1L_2(1-\theta_2)+4E(n)}$	$\frac{L_1H_1}{8L_1L_2(1-\theta_1)+4L_2E(n)}+\frac{L_1H_1}{8L_1L_2(1-\theta_2)+4L_2E(n)}$	$\frac{L_1(\theta_1+\theta_2)H_4}{4L_1(1-\theta_1)+2E(n)}+\frac{L_1(\theta_1+\theta_2)H_1}{4L_1(1-\theta_2)+2E(n)}$
34[a]	$\frac{H_1}{8L_1(1-\theta_2)-4E(n)}+\frac{L_2H_1}{8L_1L_2(1-\theta_2)-4L_1R(n)}$	$\frac{L_1H_1}{8L_1L_2(1-\theta_2)-4L_2E(n)}+\frac{H_1}{8L_2(1-\theta_2)-4R(n)}$	$\frac{L_1(\theta_1+\theta_2)H_4}{4L_1(1-\theta_2)-2E(n)}+\frac{L_2(\theta_1+\theta_2)H_1}{4L_2(1-\theta_2)-2R(n)}$
35[a]	$\frac{L_2H_1}{8L_1L_2(1-\theta_2)+4L_1R(n)}+\frac{L_1H_1}{8L_1L_2(1-\theta_2)+4L_2E(n)}$	$\frac{H_1}{8L_2(1-\theta_2)+4R(n)}+\frac{H_1}{8L_1(1-\theta_2)+4E(n)}$	$\frac{L_2(\theta_1+\theta_2)H_4}{4L_2(1-\theta_2)+2R(n)}+\frac{L_1(\theta_1+\theta_2)H_1}{4L_1(1-\theta_2)+2E(n)}$

[a] Region A comprises IC conditions 21, 22, 23, 24, 25, 34, and 35.

partitions are functions of either $E(n)$ or $R(n)$ only, that is, $K_p = f\{E(n)\}$ or $K_p = f\{R(n)\}$.

Figure 5.17 shows that partition 25 in region A is around the $(E(n), R(n)) = (0, 0)$ point. Since the equivalent proportional and derivative gains in region A are functions of both the inputs $E(n)$ and $R(n)$, the use of nonequal FOU ($\theta_1 \neq \theta_2$) may provide the potential for a smoother control surface around the zero feedback error point. Consequently, these regions may help to eliminate the overshoot despite greater control efforts provided by the IT2 fuzzy PD controller.

Example 5.5 ***Comparative Study of IT2 FLC with T1 FLC*** We consider the IT2 FLC with the structure described in Section 5.3.1, but with the additional conditions that the FOU sizes for both inputs are equal ($\theta_1 = \theta_2$, $H_2 = H_3 = 0$ and $H_4 = -H_1$). Due to the symmetry in the resulting IT2 fuzzy PD controller, it is sufficient to compare the characteristics of the output signals of the IT2 FLC and T1 FLC in IC1 and IC27. Again using the equivalent proportional and derivative gains tabulated in Tables 5.8 and 5.9 as well as the assumption that $\theta_1 = \theta_2$, the output signals of the IT2 fuzzy PD controllers for input pairs in IC1 and IC27 are

$$\begin{aligned}
\Delta u_{j\mathrm{IC1}} &= \left[\frac{H_1}{8L_1(1+\theta_1) - 4E(n)} + \frac{H_1}{8L_1(1-\theta_1) - 4E(n)}\right] E(n) \\
&\quad + \left[\frac{L_1H_1}{8L_1L_2(1+\theta_1) - 4L_2E(n)} + \frac{L_1H_1}{8L_1L_2(1-\theta_1) - 4L_2E(n)}\right] R(n) + \delta_1 \\
&= \left[\frac{1}{8L_1L_2(1+\theta_1) - 4L_2E(n)}\right. \\
&\quad \left. + \frac{1}{8L_1L_2(1-\theta_1) - 4L_2E(n)}\right] (L_1R(n) + L_2E(n))H_1 + \delta_1
\end{aligned} \tag{5.47}$$

$$\begin{aligned}
\Delta u_{j\mathrm{IC27}} &= \left[\frac{H_1}{8L_1(1-\theta_1) - 4E(n)} + \frac{H_1}{8L_1(1-\theta_1) - 4E(n)}\right] E(n) \\
&\quad + \left[\frac{L_1H_1}{8L_1L_2(1-\theta_1) - 4L_2E(n)} + \frac{L_1H_1}{8L_1L_2(1-\theta_1) - 4L_2E(n)}\right] R(n) \\
&= \left[\frac{1}{8L_1L_2(1-\theta_1) - 4L_2E(n)}\right. \\
&\quad \left. + + \frac{1}{8L_1L_2(1-\theta_1) - 4L_2E(n)}\right] (L_1R(n) + L_2E(n))H_1
\end{aligned} \tag{5.48}$$

The behavior of this IT2 FLC is compared with a T1 FLC with the antecedent sets shown in Figs. 5.19a and 5.19b. The antecedent sets for the T1 FLC are constructed by replacing every IT2 FS with a T1 FS such that both the IT2 and T1 FLC have the same input space. By partitioning the input space into the four regions shown

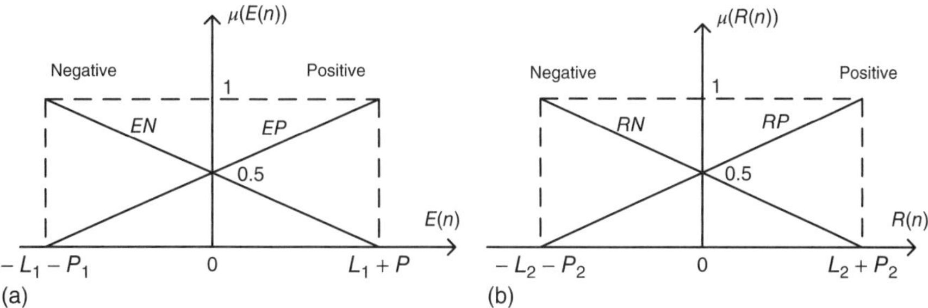

Figure 5.19 (a) T1 FSs *EN* and *EP* (solid lines) as antecedent sets for the input $E(n)$. (b) T1 FSs *RN* and *RP* (solid lines) as antecedent sets for the input $R(n)$ (Nie and Tan, 2012; © 2012, IEEE).

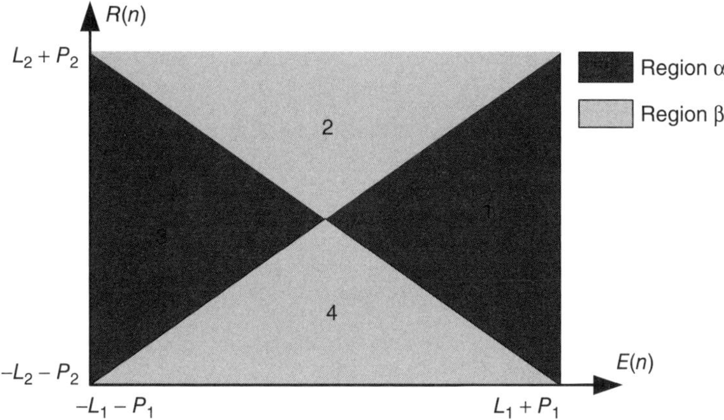

Figure 5.20 Partitions of the input space by T1 FLC (Nie and Tan, 2012; © 2012, IEEE).

in Fig. 5.20 (Ying, 2000), the type 1 FLC may essentially be interpreted as four nonlinear PD controllers with the following structural equation:

$$\Delta u_{T1,j\mathrm{IC}h} = k_p^h E(n) + k_d^h R(n) \tag{5.49}$$

It may be observed by comparing Fig. 5.17 with Fig. 5.20 that the subregions IC1 and IC27 in the input space of IT2 fuzzy PD controller are a subset of the subregion IC1 for its T1 counterpart. Hence, the comparative study may be performed by using the equivalent proportional- and derivative-gains expressions (k_p^q, k_d^q) for the T1 fuzzy PD controller to construct the output signal. For input pairs that satisfy IC1, the output signal of the T1 FLC may be expressed as

$$\begin{aligned}\Delta u_{T1,j\mathrm{IC}1} &= \frac{L_1(1+2\theta_1)R(n) + L_2(1+2\theta_2)E(n)}{4L_1L_2(1+2\theta_1)(1+2\theta_2) - 2L_2(1+2\theta_2)E(n)}H_1 \\ &= \frac{1}{4L_1L_2(1+2\theta_1) - 2L_2E(n)}(L_1R(n) + L_2E(n))H_1\end{aligned}$$

$$= \left[\frac{1}{8L_1L_2(1+2\theta_1) - 4L_2E(n)} + \frac{1}{8L_1L_2(1+2\theta_1) - 4L_2E(n)} \right] (L_1R(n) + L_2E(n))H_1 \tag{5.50}$$

Examining the structure of Eqs. (5.47), (5.48), and (5.50) reveals that conclusions about the relative size of the output signals for the IT2 FLC and T1 FLC depend the following inequalities:

$$\frac{1}{8L_1L_2(1-\theta_1) - 4L_2E(n)} > \frac{1}{8L_1L_2(1+2\theta_1) - 4L_2E(n)}$$

$$\frac{1}{8L_1L_2(1+\theta_1) - 4L_2E(n)} > \frac{1}{8L_1L_2(1+2\theta_1) - 4L_2E(n)}$$

For inputs that satisfy IC1, the $L_1R(n) + L_2E(n)$ term in Eqs. (5.47) and (5.50) is positive because $E(n)$ and $R(n)$ are positive in IC1 while L_1 and L_2 are parameters of the FOU so they must be positive. Furthermore, from Table 5.9,

$$\delta_1 = \frac{L_1(\theta_1+\theta_2)H_4}{4L_1(1+\theta_1) - 2E(n)} + \frac{L_1(\theta_1+\theta_2)H_1}{4L_1(1-\theta_1) - 2E(n)} \tag{5.51}$$

As $\theta_1 = \theta_2$ and $H_1 > 0$ (H_1 is larger than $H_2 = H_3$, which is assumed to be 0), $H_1L_1(\theta_1+\theta_2) = 2H_1L_1\theta_1 > 0$. Finally, invoking the assumption that $H_4 = -H_1$,

$$\delta_1 = 2H_1L_1\theta_1 \left[\frac{1}{4L_1(1-\theta_1) - 2E(n)} - \frac{1}{4L_1(1+\theta_1) - 2E(n)} \right] > 0$$

Consequently, $\Delta u_{jIC1} > \Delta u_{jIC1} > 0$ in IC1 and their difference increases as θ_1 increases because θ_1 appears in the denominator. Likewise, $L_1R(n) + L_2E(n) > 0$ in IC27 so the magnitude of the control effort provided by IT2 fuzzy PD controller for the same input pair is bigger. The difference in the size of the control signal also increases as $\theta_1 = \theta_2$ increases.

The discussions in Examples 5.4 and 5.5 provide an interesting theoretical foundation for explaining the characteristics of an IT2 FLC observed from the experimental study in Section 5.2.5.4. The potential of an IT2 FLC to outperform a T1 FLC by providing fast rise time and small overshoot may be summarized as:

1. As demonstrated by Example 5.4, the equivalent proportional and derivative gains in region A are functions of both the inputs $E(n)$ and $R(n)$. As region A comprises the $(E(n), R(n)) = (0, 0)$ point, this property provides the potential for the IT2 FLC to achieve smoother surface thereby reducing the oscillation amplitude. More importantly, the area of region A depends on the value of $|\theta_1 - \theta_2|$.

2. Example 5.5 indicates that the IT2 FLC may produce control efforts that are larger in magnitude, compared to its T1 counterpart, for the same input pairs. Greater control effort offers the potential to decrease the rise time. The amount by which the control effort can be enlarged depends on the value of the FOU.

While the above results indicate that the IT2 FLC has the potential to mitigate the amount of compromise between fast rise time and small overshoot, the trade-off between these two important control performance indicators still exits. This is because θ_1 and θ_2 need to be large for the IT2 FLC to produce bigger control efforts that result in fast rise time. However, a small overshoot requires the size of region A to be enlarged by setting a large absolute value $|\theta_1 - \theta_2|$. The two conditions cannot be achieved simultaneously. In the last example of the chapter, a numerical study is presented here to further illustrate the above properties of an IT2 fuzzy PD controller gained by the analysis in the previous sections.

Example 5.6 Consider the coupled tank described in Section 5.2.5.1. The configurations of IT2 fuzzy PI controller and T1 fuzzy PI controller are identical to those in Section 5.3.1 where $K_r = K_e = 1, K_{\Delta U} = 75$. The predefined antecedent sets of T1 fuzzy PI controller for Error ($E(n)$) and Rate ($R(n)$) are shown in Figs. 5.21a and 5.21b. The singleton consequent sets are predefined as $H_2 = H_3 = 0, H_4 = -H_1$. The two parameters θ_{Error} and θ_{Rate} are defined as the distance between the upper bound and the lower bound for every antecedent sets when α-cut is 1.

Analysis in the last subsection shows that IT2 fuzzy PD controller can outperform its T1 counterpart in rise time, overshoot, and disturbance rejection. To substantiate the theoretical study, the following three cases are simulated:

1. *Case 1* The parameters of IT2 fuzzy PD controller are optimized as $\theta_{\text{Error}} = 5$, $\theta_{\text{Rate}} = 2$, and $H_1 = 8$ using a genetic algorithm with ITAE as the fitness function. As shown in Fig. 5.22a, the response obtained using the IT2 FLC has comparative rise time with the T1 case but exhibits smaller overshoot and is less oscillatory. Figure 5.22b shows the error versus rate trajectory and the trajectory for the IT2 FLC is much smoother when it is near Error = 0.
2. *Case 2* In terms of rise time, these two FLCs are compared by choosing $H_1 = 0.08$. With a small H_1, their difference is more obvious in rise time. Figure 5.23a shows the step responses for case 2, while the error versus rate trajectory is shown in Fig. 5.23b. The IT2 FLC achieves larger convergence rate and less rise time as it can provide larger control effort than the T1 case.
3. *Case 3* By keeping $H_1 = 8$, the results for the plant with random disturbances are compared with those in case 1 to show their disturbance rejection ability. As shown in Figs. 5.24a and 5.24b, more oscillation is caused by disturbance in the T1 FLC while IT2 FLC exhibits similar performance with the one in case 1.

In terms of overshoot, rise time, and disturbance rejection, the IT2 FLC can outperform its T1 counterpart, which is identical to the theoretical analysis in the last

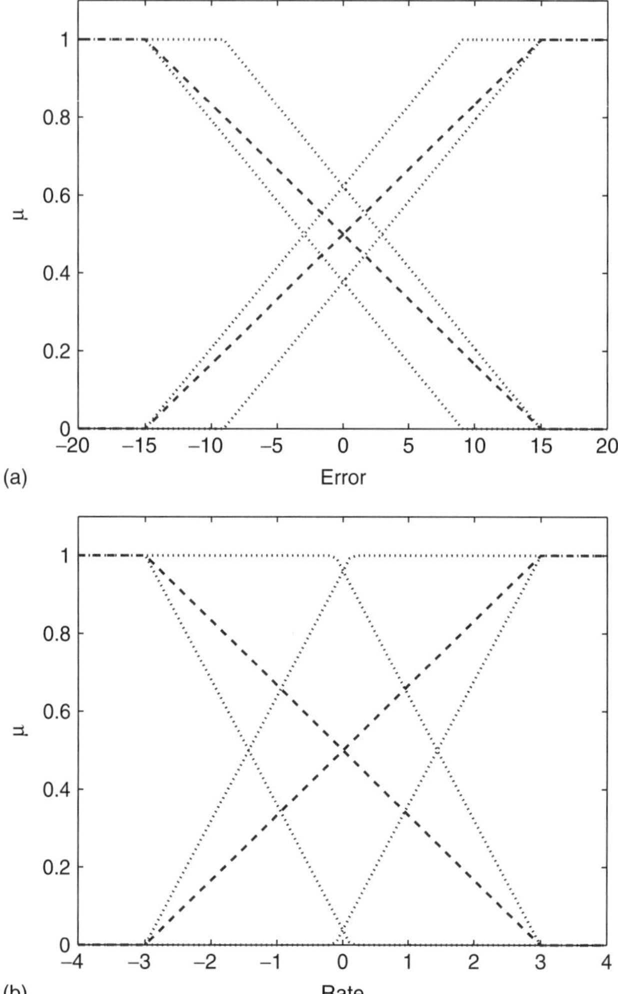

Figure 5.21 IT2 antecedent FSs: (a) antecedent sets of error and (b) antecedent sets of rate (the dashed line for T1 FLC, the dotted line for IT2 FLC) (Nie and Tan, 2012; © 2012, IEEE).

subsection. To find out how much the IT2 FLC can outperform its T1 case, further study is done by gradually increasing the value of H_1 and comparing the value of their ITAE. Figure 5.25 shows that the improvement in control performance (as measured by the ITAE) provided by the IT2 FLC over the T1 case increases as H_1 increases. The increase of this rate becomes slower when H_1 is increased beyond some value. Figures 5.26 and 5.27 show the control surface of the T1 FLC and the IT2 FLC, respectively. From the differences in the control surfaces of the T1 and IT2 FLCs shown in Fig. 5.28, it can be observed that for most of the input pairs the

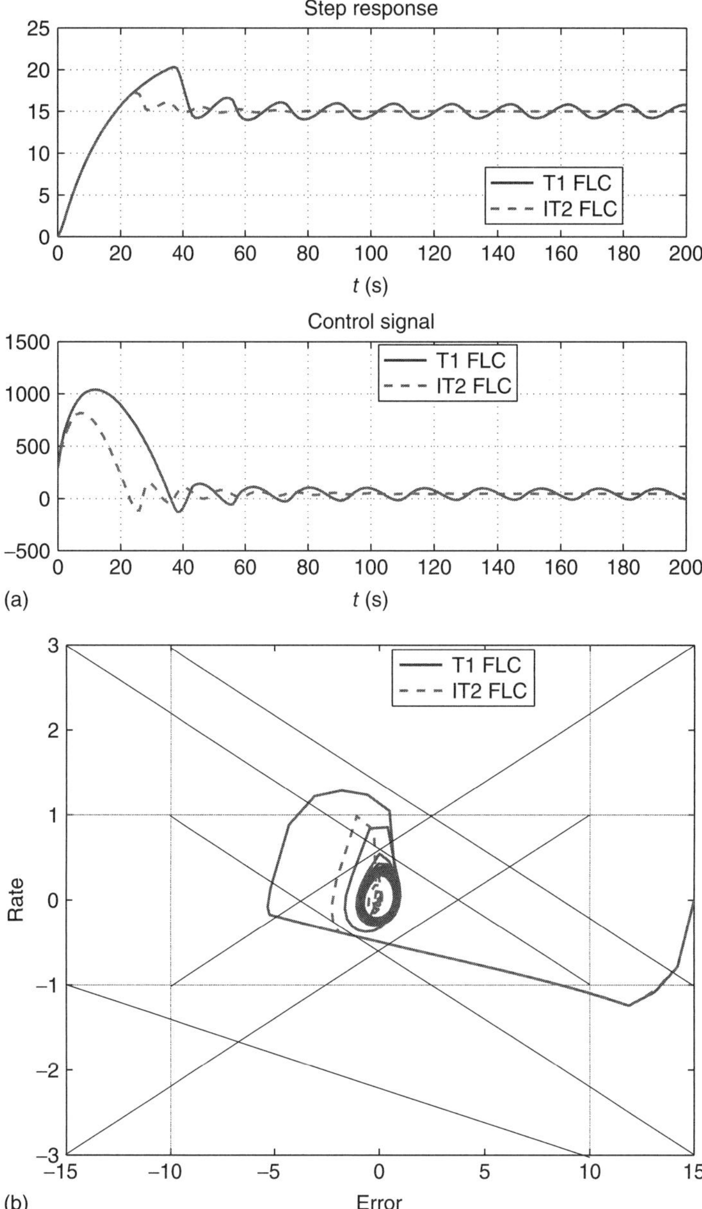

Figure 5.22 Case 1: (a) The output of the system using T1 FLC and IT2 FLC (the solid line for T1 FLC, the dashed line for IT2 FLC). (b) The trajectory of error and rate (the solid line for T1 FLC, the dashed line for IT2 FLC) (Nie and Tan, 2012; © 2012, IEEE).

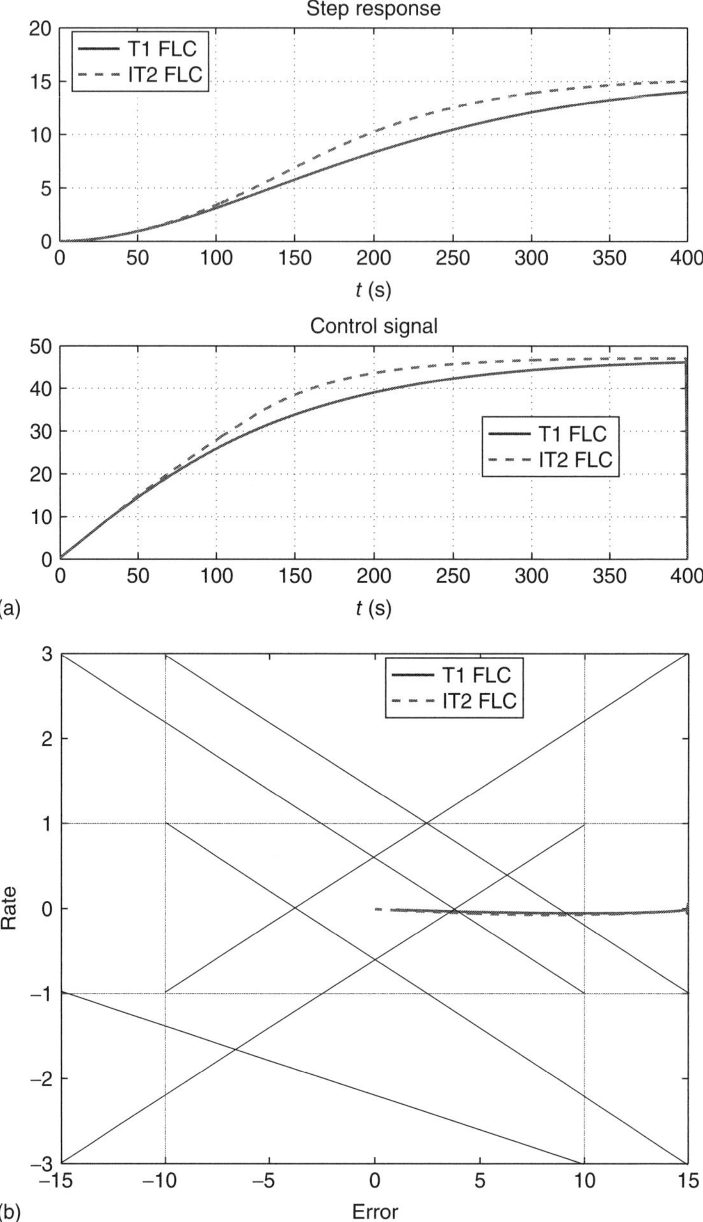

Figure 5.23 Case 2: (a) The output of the system using T1 FLC and IT2 FLC (the solid line for T1 FLC, the dashed line for IT2 FLC). (b) The trajectory of error and rate (the solid line for T1 FLC, the dashed line for IT2 FLC) (Nie and Tan, 2012; © 2012, IEEE).

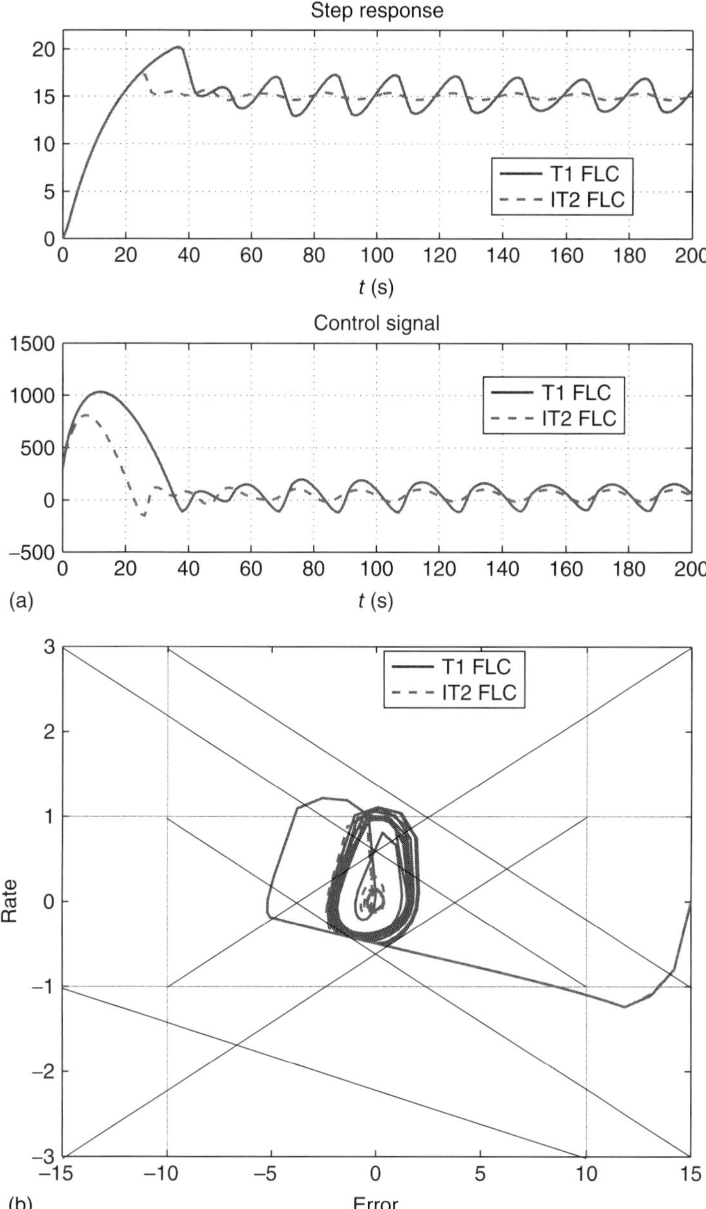

Figure 5.24 Case 3: (a) The output of the system using T1 FLC and IT2 FLC (the solid line for T1 FLC, the dashed line for IT2 FLC). (b) The trajectory of error and rate (the solid line for T1 FLC, the dashed line for IT2 FLC) (Nie and Tan, 2012; © 2012, IEEE).

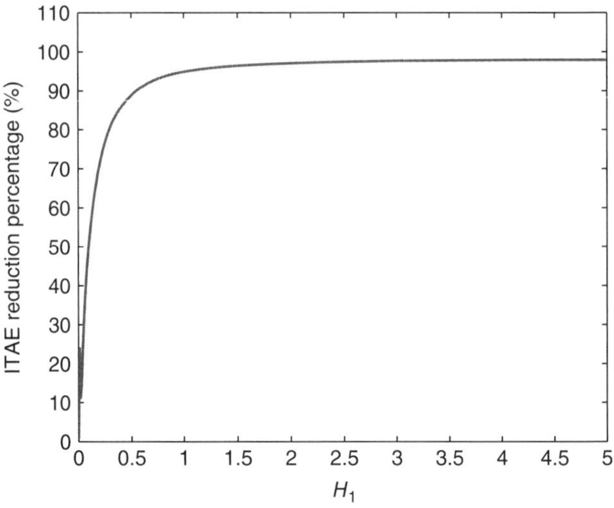

Figure 5.25 ITAE difference percentage: $\frac{\text{ITAE for T1 FLC} - \text{ITAE for IT2 FLC}}{\text{ITAE for T1 FLC}} \times 100\%$ (Nie and Tan, 2012; © 2012, IEEE).

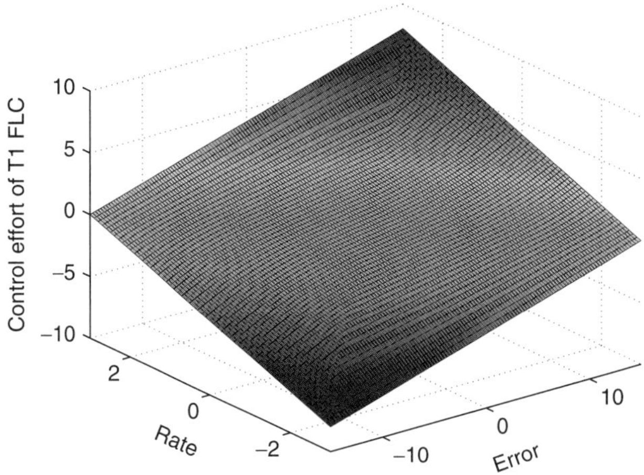

Figure 5.26 Control surface produced by the T1 FLC ($H_1 = 8$) (Nie and Tan, 2012; © 2012, IEEE).

IT2 FLC generates control efforts that are larger in magnitude, which is consistent with property 2.

5.4 CONCLUSIONS

In this chapter, a simplified type-2 FLC, which is more suitable for real-time control, is introduced, and a type-2 FLC with simplified structure is designed for a

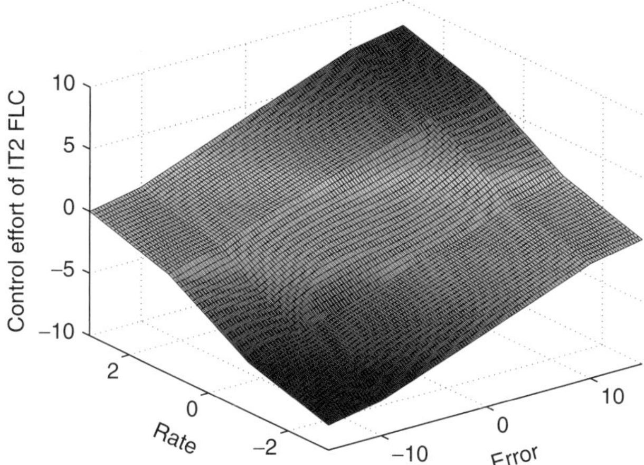

Figure 5.27 Control surface produced by the IT2 FLC ($H_1 = 8$) (Nie and Tan, 2012; © 2012, IEEE).

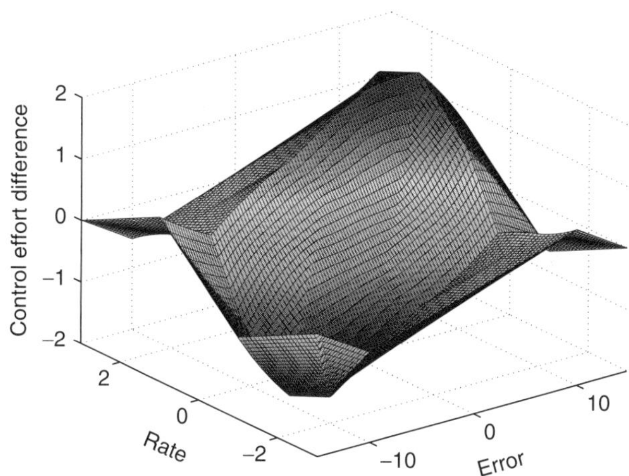

Figure 5.28 Surface difference between the IT2 FLC and the T1 FLC ($H_1 = 8$) (Nie and Tan 2012; © 2012, IEEE).

coupled-tank liquid-level control process. Experimental results show that the simplified type-2 FLC outperforms a T1 FLC. Analysis also indicates there will be at least 50% reduction in computational cost if the simplified type-2 FLC is used in place of a traditional type-2 FLC. Crucially, the experimental results indicate that the simplified structure retains the ability to alleviate the trade-off between fast rise time and small overshoot, which is a main characteristic of a full-fledged IT2 FLC. To establish more general results, the analytical structure of a special class of IT2 fuzzy PD and PI controllers that uses the KM iterative algorithm for type reduction is presented. The theoretical structure leads to closed-form equations for the output signal. Finally, examples are provided to illustrate how the derived analytical structure may be used for theoretical quantification of the control surface in order to establish the potential advantages of the IT2 FLC over the T1 FLC.

CHAPTER 6

On the Design of IT2 TSK FLCs

6.1 INTRODUCTION

In this chapter, we use the TSK model structure for the design of IT2 FLCs. Our approach is based on rigorous mathematical analyses for the design of IT2 FLCs, a development that plays a key role in this systematic design and analyses. Some sample Matlab codes for the examples are available online on the Wiley website. The reader is expected to have some basic knowledge about Matlab LMI or CVX toolboxes.[1]

Here, we use TSK, a well-known mathematical framework for the analysis and design of FLCs that was proposed in Japan in the 1980's. It enables a mathematical formulation of FLCs that is very suitable for analytical design (TSK FLCs are described in Section 3.2.3 for T1 and Section 3.3.3 for IT2). Throughout this chapter, the TSK model structure is used for analysis and design of IT2 FLCs.

The organization of this chapter is as follows: Section 6.2 provides preliminaries for IT2 TSK FLCs. Section 6.3 presents a new inference engine for control design, which is used in the entire chapter. Section 6.4 presents stability analyses for IT2 TSK FLCs. Section 6.5 presents a practical approach for the design of adaptive IT2 TSK FLCs for robot manipulators. Section 6.6 presents the design of adaptive control with applications to robot manipulators. Section 6.7 presents robust control design. Section 6.8 presents a summary of this chapter, and the appendix includes proofs.

6.2 PRELIMINARIES[2]

We first present the rule structures for discrete and continuous T1 TSK FLCs to familiarize the reader with these two control systems. Most of the control

[1]CVX is a modeling system based on Matlab for solving convex optimization problems (Grant and Boyd, 2008, 2011).

[2]Much of the material in this section is taken directly from Biglarbegian et al. (2010; © 2010, IEEE).

Introduction to Type-2 Fuzzy Logic Control: Theory and Applications, First Edition.
Jerry M. Mendel, Hani Hagras, Woei-Wan Tan, William W. Melek, and Hao Ying.

development is given for discrete systems. However, since comparable results for continuous systems can be presented, we avoid repetition and present the results for a continuous system when applicable.

6.2.1 Discrete T1 TSK FLC: Rules and Firing Level

In this subsection, we first introduce a rule structure of a discrete T1 TSK FLC. The sth rule, shown as R^s, is expressed as follows:

$$R^s\text{: If } x(k) \text{ is } F_1^s \text{ and } x(k-1) \text{ is } F_2^s \text{ and } \cdots \text{ and } x(k-p+1) \text{ is } F_p^s\text{, then}$$
$$u_s = c_1^s x(k) + \cdots + c_p^s x(k-p+1) \tag{6.1}$$

where $s = 1, \ldots, M$, F_i^s represents the T1 FS of input state i in rule s, namely, $x(k-i)$, $c_1^s, \ldots, c_p^s$ are the coefficients of the output function, u_s is the output of the sth rule, and M is the number of rules. Additionally, the state vector, $\boldsymbol{x}$, is defined as

$$\boldsymbol{x} = [x(k), x(k-1), \ldots, x(k-p+1)]^T \tag{6.2}$$

The firing strength of the sth rule, $f^s(\boldsymbol{x})$, is given by

$$f^s(\boldsymbol{x}) = \mu_{F_1^s}(x(k)) * \cdots * \mu_{F_p^s}(x(k-p+1)) \tag{6.3}$$

where $*$ is a t-norm operator.

6.2.2 Continuous T1 TSK FLC: Rules and Firing Level

Similar to the discrete case, the sth rule for a continuous system can be written as follows:

$$R^s\text{: If } x_1(t) \text{ is } F_1^s \text{ and } x_2(t) \text{ is } F_2^s \text{ and } \cdots x_p(t) \text{ is } F_p^s\text{, then}$$
$$u_s = c_1^s x_1(t) + \cdots + c_p^s x_p(t) \tag{6.4}$$

where $x_i(t)$ is the ith input to the controller; similar to the discrete case, $s = 1, \ldots, M$, F_i^s represents the T1 FS of input state i in rule s, $c_1^s, \ldots, c_p^s$ are the coefficients of the output function, u_s is the output of the sth rule, and M is the number of rules. Additionally, the state vector, $\boldsymbol{x}$, is defined as

$$\boldsymbol{x} = [x_1(t), x_2(t), \ldots, x_p(t)]^T \tag{6.5}$$

The firing strength of the sth rule, $f^s(\boldsymbol{x})$, is given as

$$f^s(\boldsymbol{x}) = \mu_{F_1^s}(x_1(t)) * \cdots * \mu_{F_p^s}(x_p(t))) \tag{6.6}$$

6.2.3 T1 TSK FLC Output

The output of a T1 TSK FLC for both discrete and continuous models is given by

$$U_{\text{T1 TSK}}(\boldsymbol{x}) = \frac{\sum_{s=1}^{M} f^s(\boldsymbol{x})u_s}{\sum_{s=1}^{M} f^s(\boldsymbol{x})} \tag{6.7}$$

Note that to use (6.7) for a discrete system, $f^s(\boldsymbol{x})$ and u_s in Eqs. (6.3) and (6.1), respectively, are required. For a continuous system, $f^s(\boldsymbol{x})$, in Eq. (6.6) and u_s, expressed by Eq. (6.4), are used.

6.2.4 Discrete IT2 TSK FLC: Rules and Firing Interval

As in the T1 TSK FLC, we first introduce a rule structure of a discrete IT2 TSK FLC. The sth rule structure of a IT2 TSK, where antecedents are IT2 FS and consequents are crisp numbers, is expressed as (Mendel, 2001)

$$R^s\text{: If } x(k) \text{ is } \tilde{F}_1^s \text{ and } x(k-1) \text{ is } \tilde{F}_2^s \text{ and } \cdots \text{ and } x(k-p+1) \text{ is } \tilde{F}_p^s, \text{ then}$$
$$u_s = c_1^s x(k) + \cdots + c_p^s x(p-k+1) \tag{6.8}$$

where $s = 1, \ldots, M$, $\tilde{F}_i^s$ represents the IT2 FS of input state i in rule s, $c_1^s, \ldots, c_p^s$ are the coefficients of the output function for rule s (and hence are crisp numbers, i.e., type-0 FSs), u_s is the output of the sth rule, and M is the number of rules. The above rules allow us to model the uncertainties encountered in the antecedents. In an IT2 TSK model, lower and upper firing strengths of the sth rule, $\underline{f}^s(\boldsymbol{x})$ and $\overline{f}^s(\boldsymbol{x})$, are given by

$$\underline{f}^s(\boldsymbol{x}) = \underline{\mu}_{\tilde{F}_1^s}(x(k)) * \cdots * \underline{\mu}_{\tilde{F}_p^s}(x(k-p+1)) \tag{6.9}$$

$$\overline{f}^s(\boldsymbol{x}) = \overline{\mu}_{\tilde{F}_1^s}(x(k)) * \cdots * \overline{\mu}_{\tilde{F}_p^s}(x(k-p+1)) \tag{6.10}$$

where $\underline{\mu}_{\tilde{F}_i^s}$ and $\overline{\mu}_{\tilde{F}_i^s}$ represent the ith ($i = 1, \ldots, p$) lower and upper membership functions of rule s, respectively.

6.2.5 Continuous IT2 TSK FLC: Rules and Firing Interval

Similar to the discrete case, the sth rule for a continuous system is written as

$$R^s\text{: If } x_1(t) \text{ is } \tilde{F}_1^s \text{ and } x_2(t) \text{ is } \tilde{F}_2^s \text{ and } \cdots \text{ and } x_p(t) \text{ is } \tilde{F}_p^s, \text{ then}$$
$$u_s = c_1^s x_1(t) + \cdots + c_p^s x_p(t) \tag{6.11}$$

The lower and upper firing strengths of the sth rule are given by

$$\underline{f}^s(\boldsymbol{x}) = \underline{\mu}_{\tilde{F}_1^s}(x_1(t)) * \cdots * \underline{\mu}_{\tilde{F}_p^s}(x_p(t)) \tag{6.12}$$

$$\overline{f}^s(\boldsymbol{x}) = \overline{\mu}_{\tilde{F}_1^s}(x_1(t)) * \cdots * \overline{\mu}_{\tilde{F}_p^s}(x_p(t)) \tag{6.13}$$

6.2.6 IT2 TSK FLC Output

Using the KM algorithms introduced in Section 2.3.6 produces the final output of the IT2 TSK model for both discrete and continuous models as given by (Mendel, 2001)

$$U_{\text{TSK/A2-C0}}(\boldsymbol{x}) = [u_l(\boldsymbol{x}), u_r(\boldsymbol{x})] = \int_{f^1 \in [\underline{f}^1, \overline{f}^1]} \cdots \int_{f^M \in [\underline{f}^M, \overline{f}^M]} 1 \Bigg/ \frac{\sum_{i=1}^{M} f^s(\boldsymbol{x}) u_s}{\sum_{i=1}^{M} f^s(\boldsymbol{x})} \tag{6.14}$$

where u_s for the discrete model is given by the consequent portion of Eq. (6.8) and for the continuous model is given by the consequent portion of Eq. (6.11). The firing strengths, $\underline{f}^s(\boldsymbol{x})$ and $\overline{f}^s(\boldsymbol{x})$, for the discrete model are given by Eqs. (6.9) and (6.10), respectively, and for the continuous model are given by Eqs. (6.12) and (6.13).

$U_{\text{TSK/A2-C0}}$ is an interval type-1 set and depends only on its left and right end points u_l,u_r, which can be computed using the iterative KM algorithms, similar to the type reduction method explained in Section 3.3.3.2. Its final output is given as

$$U_{\text{output}}(\boldsymbol{x}) = \frac{u_l(\boldsymbol{x}) + u_r(\boldsymbol{x})}{2} \tag{6.15}$$

The final output given by Eq. (6.15) does not have a closed-form expression, as can be seen from Eq. (6.14).

Another method to compute U_{output} is to use WM UBs, which are fully explained in Section 3.4. For the purpose of rigorous analysis and design of control systems, it is required to have a closed-form expression for the controller. Doing so allows for mathematical analyses and investigation of system properties. In addition, for real time applications and specifically in fast dynamics, having a closed-form expression is desired. The most adopted IT2 FLCs use KM algorithms as well as WM UBs. As was shown in Section 2.3.6, the KM algorithms do not provide a closed form. WM UBs provide a closed form, yet they may not be suitable for control design due to their complex structure. Hence, an alternative approach (Biglarbegian et al., 2008, 2010) that parallels the WM UBs but has a simpler structure is proposed and used for the purpose of implementing the control methodologies developed here. In the next section, we introduce this simpler and novel inference engine for control design.

6.3 NOVEL INFERENCE ENGINE FOR CONTROL DESIGN[3]

Given the need to have a closed-form solution for control design and analysis, we introduce a new inference engine that enables us to simply express the output of any IT2 TSK FLC. This inference engine was introduced in Biglarbegian et al. (2008,

[3]Much of the material in this section is taken directly from Biglarbegian et al. (2010; © 2010, IEEE).

2010) and has been used successfully in several control applications and rigorous analyses of IT2 TSK FLCs. This inference engine, called m–n IT2 FLC, has a simple closed-form structure, shown to be effective in analysis, design, and, real-time implementation of IT2 FLCs. In this chapter, the m–n formula is adopted for the the design and analyses of IT2 FLCs. The m–n inference engine is (Biglarbegian et al., 2010)

$$U_{m-n}(\boldsymbol{x}) = m\frac{\sum_{s=1}^{M} \underline{f}^s(\boldsymbol{x})u_s}{\sum_{s=1}^{M} \underline{f}^s(\boldsymbol{x})} + n\frac{\sum_{s=1}^{M} \overline{f}^s(\boldsymbol{x})u_s}{\sum_{s=1}^{M} \overline{f}^s(\boldsymbol{x})} \tag{6.16}$$

where u_s is the output of each rule; for the discrete model Eq. (6.8) is used and for the continuous model Eq. (6.11) is used; $\underline{f}^s(\boldsymbol{x})$ and $\overline{f}^s(\boldsymbol{x})$ for discrete models are given by Eqs. (6.9) and (6.10), and for continuous models Eqs. (6.12) and (6.13), respectively; and m and n are two free parameters that will be chosen by the designer to satisfy the design requirements.

$U_{m-n}(\boldsymbol{x})$ is a simplified form of the WM UBs; as is proven in the Appendix.

6.4 STABILITY OF IT2 TSK FLCs

This section presents rigorous stability analyses of IT2 TSK FLCs. T1 TSK fuzzy logic systems have been shown to be universal approximators (Ying, 2000) and can model nonlinear plants (Tanaka and Sano, 1995; Wang et al., 1996). As a result, we investigate the stability of IT2 TSK FLCs that use T1 TSK for modeling plants to reduce the complexity of the final stability conditions of IT2 TSK FLCs. We develop the stability analysis for a discrete model, using a similar approach and then present the stability analysis for continuous models.

6.4.1 Stability of Discrete IT2 TSK FLC

First, the structure of a discrete T1 TSK FLC (where T1 TSK is used in the structures of both plant and controller) is reviewed. Next, the T1 TSK FLC is replaced with a discrete IT2 TSK FLC.

6.4.1.1 T1 TSK FLC The general sth rule for the plant, R_Q^s, can be expressed as (Tanaka and Sano, 1994)

$$R_Q^s\text{: If } x(k) \text{ is } Q_1^s \text{ and } \cdots \text{ and } x(k-p+1) \text{ is } Q_p^s, \text{ then}$$
$$\boldsymbol{x}_s(k+1) = \boldsymbol{A}_s\boldsymbol{x} + \boldsymbol{b}_s\boldsymbol{u}(k), s = 1, 2, \ldots, r \tag{6.17}$$

where $\boldsymbol{A}_s \in \mathbb{R}^{n\times n}$, $\boldsymbol{b}_s \in \mathbb{R}^{n\times m}$, $\boldsymbol{u}(k) \in \mathbb{R}^m$ (controller output), and Q_i^s represents a T1 FS of the ith input state of rule s, $\boldsymbol{x}_s(k+1)$ is the output of each rule, $\boldsymbol{x}$ is the state vector given by Eq. (6.2), and r is the number of rules. The output of the system, $\boldsymbol{x}(k+1)$, which is the weighted average of each rule is given by

$$\boldsymbol{x}(k+1) = \frac{\sum_{s=1}^{r} f^s(\boldsymbol{x})\{\boldsymbol{A}_s\boldsymbol{x} + \boldsymbol{b}_s\boldsymbol{u}(k)\}}{\sum_{s=1}^{r} f^s(\boldsymbol{x})} \tag{6.18}$$

in which $f^s(\boldsymbol{x})$ is calculated using Eq. (6.3) where $Q_1^s, Q_2^s, \ldots, Q_p^s$ should be used as antecedents.

The sth control rule, R_c^s, is (Tanaka and Sano, 1995)

$$R_c^s\text{: If } x(k) \text{ is } C_1^s \text{ and } \cdots \text{ and } x(k-p+1) \text{ is } C_p^s\text{, then}$$
$$\boldsymbol{u}_s(k) = \boldsymbol{F}_s\boldsymbol{x},\ s = 1, 2, \ldots, r \tag{6.19}$$

and $\boldsymbol{F}_s$ is the sth feedback gain matrix, and C_i^s represents the T1 FS of input state i of rule s. Assuming the number of the rules for the controller is r, $\boldsymbol{u}(k)$, is thus given by (Tanaka and Sano, 1995)

$$\boldsymbol{u}(k) = \frac{\sum_{s=1}^{r} f^s(\boldsymbol{x})\boldsymbol{F}_s\boldsymbol{x}}{\sum_{s=1}^{r} f^s(\boldsymbol{x})} \tag{6.20}$$

To calculate $f^s(\boldsymbol{x})$ use Eq. (6.3) in which $C_1^s, C_2^s, \ldots, C_p^s$ should be used as antecedents. To obtain a closed loop of T1 TSK FLC, one needs to substitute Eq. (6.20) into (6.18). Now, we review the framework of IT2 TSK FLC.

6.4.1.2 IT2 TSK FLC To develop IT2 TSK FLC, we replace C_i^s with their IT2 FS counterparts, expressed as $\tilde{C}_i^s$. Thus, $\boldsymbol{u}(k)$, for the IT2 TSK FLC is now given by

$$\boldsymbol{u}(k) = m_c \frac{\sum_{s=1}^{r} \underline{f}^s(\boldsymbol{x})\boldsymbol{F}_s\boldsymbol{x}}{\sum_{s=1}^{r} \underline{f}^s(\boldsymbol{x})} + n_c \frac{\sum_{s=1}^{r} \overline{f}^s(\boldsymbol{x})\boldsymbol{F}_s\boldsymbol{x}}{\sum_{s=1}^{r} \overline{f}^s(\boldsymbol{x})} \tag{6.21}$$

where m_c and n_c are controller tuning parameters; and, $\underline{f}^s(\boldsymbol{x})$ and $\overline{f}^s(\boldsymbol{x})$ are given by Eqs. (6.9) and (6.10), respectively.

By substituting Eq. (6.21) into (6.18), $\boldsymbol{x}(k+1)$ can be expressed as

$$\boldsymbol{x}(k+1) = \frac{\sum_{i,j,l=1}^{r} g_{ijl}\boldsymbol{G}_{ijl}}{\sum_{i,j,l=1}^{r} g_{ijl}}\boldsymbol{x} \tag{6.22}$$

where

$$g_{ijl} = f^i(\boldsymbol{x})\underline{f}^j(\boldsymbol{x})\overline{f}^l(\boldsymbol{x}) \tag{6.23}$$

$$\boldsymbol{G}_{ijl} = \boldsymbol{A}_i + m_c\boldsymbol{b}_i\boldsymbol{F}_j + n_c\boldsymbol{b}_i\boldsymbol{F}_l \tag{6.24}$$

In the following we show the properties of $\sum_{i,j,l=1}^{r} \boldsymbol{G}_{ijl}$ that help us obtain easier stability conditions. $\sum_{i,j,l=1}^{r} \boldsymbol{G}_{ijl}$ can be most clearly expressed as

$$\sum_{i,j,l=1}^{r} \boldsymbol{G}_{ijl} = \sum_{i=1}^{r} \boldsymbol{G}_{iii} + \sum_{i\neq j}^{r} \sum_{j=1}^{r} \boldsymbol{G}_{ijj} + \sum_{i=1}^{r} \sum_{j\neq l}^{r} \sum_{l=1}^{r} \boldsymbol{G}_{ijl} \tag{6.25}$$

as well as

$$\begin{aligned}\sum_{i\neq j}^{r} \sum_{j=1}^{r} \boldsymbol{G}_{ijj} &= \sum_{i<j}^{r} \sum_{j=1}^{r} \boldsymbol{G}_{ijj} + \sum_{i>j}^{r} \sum_{j=1}^{r} \boldsymbol{G}_{ijj} \\ &= \sum_{i<j}^{r} \sum_{j=1}^{r} \boldsymbol{G}_{ijj} + \sum_{t<p}^{r} \sum_{t=1}^{r} \boldsymbol{G}_{ptt} = 2\sum_{i<j}^{r} \sum_{j=1}^{r} \left[\frac{\boldsymbol{G}_{ijj} + \boldsymbol{G}_{jii}}{2}\right]\end{aligned} \tag{6.26}$$

and

$$\begin{aligned}\sum_{i=1}^{r} \sum_{j\neq l}^{r} \sum_{l=1}^{r} \boldsymbol{G}_{ijl} &= \sum_{i=1}^{r} \sum_{j<l}^{r} \sum_{l=1}^{r} \boldsymbol{G}_{ijl} + \sum_{i=1}^{r} \sum_{j>l}^{r} \sum_{l=1}^{r} \boldsymbol{G}_{ijl} \\ &= \sum_{i=1}^{r} \sum_{j<l}^{r} \sum_{l=1}^{r} \boldsymbol{G}_{ijl} + \sum_{i=1}^{r} \sum_{p>t}^{r} \sum_{l=1}^{r} \boldsymbol{G}_{ipt} \\ &= \sum_{i=1}^{r} \sum_{j<l}^{r} \sum_{l=1}^{r} \boldsymbol{G}_{ijl} + \sum_{i=1}^{r} \sum_{l>j}^{r} \sum_{l=1}^{r} \boldsymbol{G}_{ilj} \\ &= 2\sum_{i=1}^{r} \sum_{j<l}^{r} \sum_{l=1}^{r} \left[\frac{\boldsymbol{G}_{ijl} + \boldsymbol{G}_{ilj}}{2}\right]\end{aligned} \tag{6.27}$$

Next, we define $\boldsymbol{H}_t$ and $\boldsymbol{v}_t$, respectively, as

$$\boldsymbol{H}_t \equiv \begin{cases} \dfrac{\boldsymbol{G}_{ijl} + \boldsymbol{G}_{ilj}}{2} & t = i + r\left(j - 1 + \dfrac{(l-1)(l-2)}{2}\right) \quad \text{and} \quad j < l \\ \dfrac{\boldsymbol{G}_{ijj} + \boldsymbol{G}_{jii}}{2} & t = i + \dfrac{j(j-1)}{2} + \dfrac{r^2(r-1)}{2} \quad \text{and} \quad j = l, i < j \\ \boldsymbol{G}_{iii} & t = i + \dfrac{i(i-1)}{2} + \dfrac{r^2(r-1)}{2} \quad \text{and} \quad i = j = l \end{cases} \tag{6.28}$$

$$\boldsymbol{v}_t \equiv \begin{cases} 2g_{ijl} & t = i + r(j - 1 + \dfrac{(l-1)(l-2)}{2}) \quad \text{and} \quad j < l \\ 2g_{ijj} & t = i + \dfrac{j(j-1)}{2} + \dfrac{r^2(r-1)}{2} \quad \text{and} \quad j = l, i < j \\ g_{iii} & t = i + \dfrac{i(i-1)}{2} + \dfrac{r^2(r-1)}{2} \quad \text{and} \quad i = j = l \end{cases} \tag{6.29}$$

The number of $\boldsymbol{H}_t$ matrices given in Eq. (6.28) is $\frac{r(r^2+1)}{2}$. Thus, Eq. (6.22) can be expressed in a compact form as

$$\boldsymbol{x}(k+1) = \frac{\sum_{t=1}^{[r(r^2+1)]/2} v_t \boldsymbol{H}_t}{\sum_{t=1}^{[r(r^2+1)]/2} v_t} \boldsymbol{x} \tag{6.30}$$

The sufficient condition for system (6.30) to be globally asymptotically stable is that a common positive-definite matrix $\boldsymbol{P}$ exists such that (Wang et al., 1996)

$$\boldsymbol{H}_t^T \boldsymbol{P} \boldsymbol{H}_t - \boldsymbol{P} < \boldsymbol{0} \tag{6.31}$$

where $t = 1, 2, \ldots, [r(r^2+1)]/2$ linear matrix inequalities (LMIs) in Eq. (6.31) are the final stability conditions of the discrete IT2 TSK FLC. To ensure closed-loop system stability, one can verify the feasibility of those LMIs using the Matlab LMI toolbox or the CVX. Therefore, if a positive-definite $\boldsymbol{P}$ exists satisfying Eq. (6.31) for $t = 1, 2, \ldots, [r(r^2+1)]/2$, then the closed-loop system will be asymptotically stable.

Note that without using the properties of $\sum_{i,j,l=1}^{r} \boldsymbol{G}_{ijl}$ (shown earlier), the number of LMIs that need to be checked is r^3, but by using the properties of $\sum_{i,j,l=1}^{r} \boldsymbol{G}_{ijl}$, the number of LMIs to be satisfied becomes $[r(r^2+1)]/2$, which is fewer than r^3, that is, $[r(r^2+1)]/2 < r^3$.

6.4.2 Stability of Continuous IT2 TSK FLC

Similar to the discrete case, we use T1 TSK and IT2 FLC to model continuous plant and controller, respectively. The rule structure for the continuous model is expressed as follows:

The sth rule for the plant, R_Q^s, can be expressed as (Wang et al., 1996)

$$\begin{aligned} &R_Q^s\text{: If } x_1(t) \text{ is } Q_1^s \text{ and } \cdots \text{ and } x_p(t) \text{ is } Q_p^s\text{, then} \\ &\dot{\boldsymbol{x}}_s(t) = \boldsymbol{A}_s \boldsymbol{x} + \boldsymbol{b}_s \boldsymbol{u}(t), s = 1, 2, \cdots, r \end{aligned} \tag{6.32}$$

where $\boldsymbol{A}_s \in \mathbb{R}^{n\times n}$, $\boldsymbol{b}_s \in \mathbb{R}^{n\times m}$, $\boldsymbol{u} \in \mathbb{R}^m$ (controller output), and Q_i^s represents a T1 FS of the ith input state of rule s, $\dot{\boldsymbol{x}}_s(t)$ is the output of each rule, $\boldsymbol{x}$ is the state vector given by Eq. (6.5), and r is the number of rules. We use Eq. (6.21) as the controller output. Substituting Eq. (6.21) into (6.32) and using the weighted average of all the rules to find the total output, $\dot{\boldsymbol{x}}$, yields

$$\dot{\boldsymbol{x}} = \frac{\sum_{i,j,l=1}^{r} g_{ijl} \boldsymbol{G}_{ijl}}{\sum_{i,j,l=1}^{r} g_{ijl}} \boldsymbol{x} \tag{6.33}$$

For stability of continuous models, very similar methodology can be used. As was shown earlier for the discrete case, we can show Eq. (6.33) to be expressed as

$$\dot{\boldsymbol{x}} = \frac{\sum_{t=1}^{[r(r^2+1)]/2} \upsilon_t \boldsymbol{H}_t}{\sum_{t=1}^{[r(r^2+1)]/2} \upsilon_t} \boldsymbol{x} \tag{6.34}$$

Note that $\dot{\boldsymbol{x}}$ is the derivative of the state vector given by Eq. (6.5). Also observe the similarity of discrete Eq. (6.30) and continuous Eq. (6.34) dynamics. It was shown in Wang et al. (1996) that the sufficient conditions for the stability of (6.34) are

$$\boldsymbol{H}_t^T \boldsymbol{P} + \boldsymbol{P}\boldsymbol{H} < \boldsymbol{0} \tag{6.35}$$

where $t = 1, 2, \ldots, [r(r^2 + 1)]/2$. To ensure closed-loop system stability of the continuous IT2 TSK FLC, we have to solve the LMIs given by Eq. (6.35) for $t = 1, 2, \ldots, [r(r^2 + 1)]/2$. If a positive-definite $\boldsymbol{P}$ exists that satisfies all the LMIs in Eq. (6.35), then the closed-loop system will be asymptotically stable. The feasibility of those LMIs can be verified using the Matlab LMI toolbox or the CVX (similar to the discrete case).

6.4.3 Examples

Example 6.1 ***Car parking (the code for this example can be found online on the Wiley website of the textbook)*** This example presents the control design for the problem of parking a car and was adopted from Biglarbegian et al. (2010). The control problem is steering the car from a starting point to a final position (destination) without backward movement. The position and orientation of the car with respect to the coordinate frame is shown in Fig. 6.1, where x_0 is the yaw angle and x_1 is the vertical position of the car rear side. The control design problem is to steer the car from an initial position to the parking position at $x_0 = x_1 = 0$.

As discussed earlier, for the plant (car) a T1 TS model was used. We now redesign and replace the controller with a IT2 TSK FLC. The membership functions for modeling the plant and the controller are shown in Fig. 6.2.

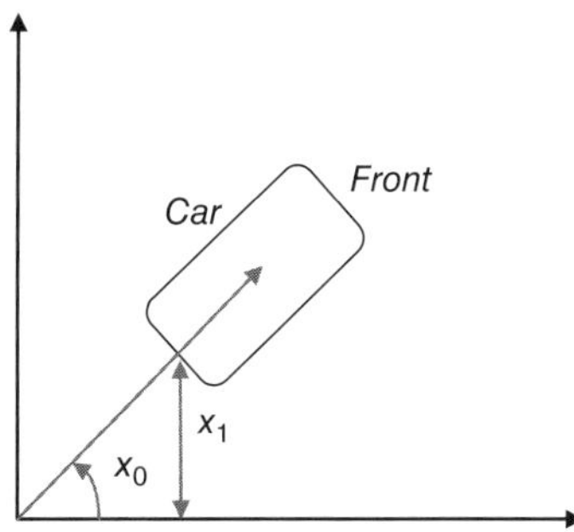

Figure 6.1 Robot position and orientation (Biglarbegian et al. 2010; © 2010, IEEE).

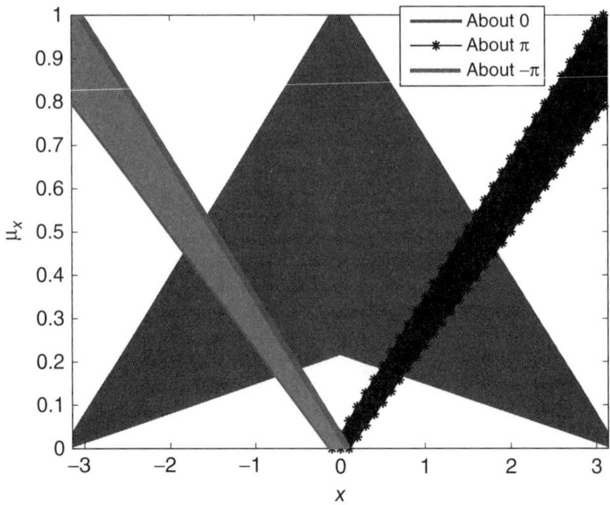

Figure 6.2 Membership functions (Biglarbegian et al., 2010; © 2010, IEEE).

The state vector, $\boldsymbol{x}(k)$, is defined as

$$\boldsymbol{x}(k) \equiv [x_0(k), x_1(k)]^T \tag{6.36}$$

As was shown in Tanaka and Sano (1995), to obtain an approximate T1 model of the plant (car), the dynamics of the car can be simplified around 0 and $\pm\pi$ angles. Thus, the expressions such as about "0" and "about $\pm\pi$" are used in the definition of the membership functions to model the plant and the controller for this example. These membership functions are given as follows (Biglarbegian et al., 2010; Tanaka and Sano, 1995):

Plant[4]

Rule 1: If $x_0(k)$ is "about 0," then $\boldsymbol{x}(k+1) = \boldsymbol{A}_1\boldsymbol{x}(k) + \boldsymbol{b}_1\boldsymbol{u}(k)$.
Rule 2: If $x_0(k)$ is "about π or $-\pi$," then $\boldsymbol{x}(k+1) = \boldsymbol{A}_2\boldsymbol{x}(k) + \boldsymbol{b}_2\boldsymbol{u}(k)$.

Control

Rule 1: If $x_0(k)$ is "about $\tilde{0}$," then $u(k) = \boldsymbol{f}_1\boldsymbol{x}(k)$.
Rule 2: If $x_0(k)$ is "about $\tilde{\pi}$ or $-\tilde{\pi}$," then $u(k) = \boldsymbol{f}_2\boldsymbol{x}(k)$.

[4] x_0 is given in radians.

The numerical values of the plant and controllers are as follows:

$$A_1 = \begin{bmatrix} 1 & 0 \\ 1 & 1 \end{bmatrix} \qquad A_2 = \begin{bmatrix} 1 & 0 \\ 0.003183 & 1 \end{bmatrix} \tag{6.37}$$

$$b_1 = \begin{bmatrix} 0.357143 \\ 1 \end{bmatrix} \qquad b_2 = \begin{bmatrix} 0.357143 \\ 1 \end{bmatrix} \tag{6.38}$$

$$f_1 = \begin{bmatrix} -0.4212 & -0.02933 \end{bmatrix} \qquad f_2 = \begin{bmatrix} -0.0991 & -0.00967 \end{bmatrix} \tag{6.39}$$

We choose $m = 3.2$ and $n = -2.2$. Since the number of the rules is 2, the number of required LMIs to be satisfied for stability is 5. We can now compute the $\boldsymbol{H}_i$ matrices according to Eq. (6.28).

$$\boldsymbol{H}_1 = \begin{bmatrix} 0.8495 & -0.0105 \\ 1 & 1 \end{bmatrix} \qquad \boldsymbol{H}_2 = \begin{bmatrix} 0.9071 & -0.0070 \\ 1 & 1 \end{bmatrix}$$

$$\boldsymbol{H}_3 = \begin{bmatrix} 0.9071 & -0.0070 \\ 0.5016 & 1 \end{bmatrix} \qquad \boldsymbol{H}_4 = \begin{bmatrix} 0.9071 & -0.0070 \\ 0.0032 & 1 \end{bmatrix}$$

$$\boldsymbol{H}_5 = \begin{bmatrix} 0.9646 & -0.0035 \\ 1 & 1 \end{bmatrix} \tag{6.40}$$

Using Matlab LMI toolbox, matrix $\boldsymbol{P}$ is thus given as

$$\boldsymbol{P} = \begin{bmatrix} 699.6386 & 57.3766 \\ 57.3766 & 11.7997 \end{bmatrix} \tag{6.41}$$

It is easy to verify that $\boldsymbol{P}$ satisfies the stability conditions. Details are as follows:

$$\boldsymbol{H}_1^T\boldsymbol{P}\boldsymbol{H}_1 - \boldsymbol{P} = \begin{bmatrix} -85.369 & -3.659 \\ -3.659 & -1.125 \end{bmatrix} < \mathbf{0},$$

$$\boldsymbol{H}_2^T\boldsymbol{P}\boldsymbol{H}_2 - \boldsymbol{P} = \begin{bmatrix} -8.077 & 1.650 \\ 1.650 & -0.765 \end{bmatrix} < \mathbf{0}$$

$$\boldsymbol{H}_3^T\boldsymbol{P}\boldsymbol{H}_3 - \boldsymbol{P} = \begin{bmatrix} -68.788 & -4.033 \\ -4.033 & -0.765 \end{bmatrix} < \mathbf{0},$$

$$\boldsymbol{H}_4^T\boldsymbol{P}\boldsymbol{H}_4 - \boldsymbol{P} = \begin{bmatrix} -123.637 & -9.714 \\ -9.714 & -0.765 \end{bmatrix} < \mathbf{0}$$

$$\boldsymbol{H}_5^T\boldsymbol{P}\boldsymbol{H}_5 - \boldsymbol{P} = \begin{bmatrix} -48.296 & -4.325 \\ -4.325 & -0.3880 \end{bmatrix} < \mathbf{0} \tag{6.42}$$

Therefore, the system is asymptotically stable.

Example 6.2 ***Control of a Chaotic System (the code for this example can be found online on the Wiley website of the textbook)*** In this example, adopted from Biglarbegian et al. (2010), we develop a IT2 TSK FLC and apply it to a nonlinear plant with chaotic behavior. The plant considered in this example is known as Chua's electric circuit (Chua et al., 1986). The Chua circuit consists of one inductor (L), two capacitors (C_1, C_2), one linear resistor (R), and one piecewise linear resistor [$g(v_{c1})$] and can be modeled as follows (Biglarbegian et al., 2010; Wang and Tanaka, 1996):

$$\dot{v}_{c1} = \frac{1}{C_1}\left[\frac{1}{R}(v_{c2} - v_{c1}) - g(v_{c1})\right] + u_1 \tag{6.43}$$

$$\dot{v}_{c2} = \frac{1}{C_2}\left[\frac{1}{R}(v_{c1} - v_{c2}) + i_L\right] + u_2 \tag{6.44}$$

$$\dot{i}_L = \frac{1}{L}(-v_{c2} - R_0 i_L) + u_3 \tag{6.45}$$

where $g(v_{c1})$ is given by

$$g(v_{c1}) = \begin{cases} G_b v_{c1} + (G_a - G_b)E & v_{c1} \geq E \\ G_a v_{c1} & -E < v_{c1} < E \\ G_b v_{c1} - (G_a - G_b)E & v_{c1} \leq -E \end{cases} \tag{6.46}$$

and v_{c1}, v_{c2}, i_L are state variables, G_a, G_b, E represent the characteristics of the resistor, and u_1, u_2, u_3 are the controls; see Chua et al. (1986, 1993) for more information.

First, we designate the state vector $\boldsymbol{x}(t) \equiv [x_1(t), x_2(t), x_3(t)]^T$ where $x_1 = v_{c1}$, $x_2 = v_{c2}$, $x_3 = i_L$. The values for the parameters used in this example are $R = 1.4286$, $R_0 = 0\Omega$, $C_1 = 0.1$, $C_2 = 0.2$, $L = 0.1429$, $G_a = -2$, $G_b = 0.1$, and $E = 1$. The next step is to identify the membership functions that are shown in Fig. 6.3.

The plant and control rules are as follows:

Plant Rules

Rule 1: If $x_1(t)$ is M_1, then $\dot{\boldsymbol{x}} = \boldsymbol{A}_1\boldsymbol{x}(t) + \boldsymbol{b}\boldsymbol{u}(t)$.

Rule 2: If $x_1(t)$ is M_2, then $\dot{\boldsymbol{x}} = \boldsymbol{A}_2\boldsymbol{x}(t) + \boldsymbol{b}\boldsymbol{u}(t)$.

Control Rules

Control Rule 1: If $x_1(t)$ is $\tilde{M}_1$, then $\boldsymbol{u}(t) = \boldsymbol{F}_1\boldsymbol{x}(t)$.

Control Rule 2: If $x_1(t)$ is $\tilde{M}_2$, then $\boldsymbol{u}(t) = \boldsymbol{F}_2\boldsymbol{x}(t)$.

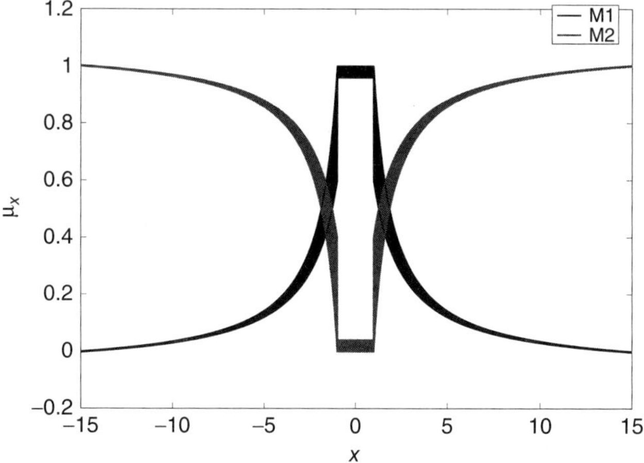

Figure 6.3 Membership functions (Biglarbegian et al., 2010; © 2009, IEEE).

where

$$\boldsymbol{A}_1 = \begin{bmatrix} 5.7143 & 14.2857 & 0 \\ 0.7143 & -0.7143 & 0.5 \\ 0 & -7 & 0 \end{bmatrix} \qquad \boldsymbol{A}_2 = \begin{bmatrix} -12.0190 & 14.2857 & 0 \\ 0.7143 & -0.7143 & 0.5 \\ 0 & -7 & 0 \end{bmatrix} \tag{6.47}$$

$$\boldsymbol{F}_1 = \begin{bmatrix} -33.3333 & -31.6202 & -1.7961 \\ 24.2702 & 0.0167 & -1.9808 \\ 1.7961 & 8.4808 & -0.3333 \end{bmatrix}$$

$$\boldsymbol{F}_2 = \begin{bmatrix} 3.0667 & 21.4379 & -3.7158 \\ -28.7879 & 0.0167 & -20.2722 \\ 3.7158 & 26.7722 & -0.3333 \end{bmatrix} \tag{6.48}$$

and $\boldsymbol{b}$ is a 3×3 identity matrix. Choose $m = n = 0.8$ and solve for the LMIs given in Eq. (6.35). The common $\boldsymbol{P}$ matrix that satisfies those LMIs is given as

$$\boldsymbol{P} = \begin{bmatrix} 2.2240 & 0.0112 & 0.1701 \\ 0.0112 & 2.5747 & -0.0198 \\ 0.1701 & -0.0198 & 2.0743 \end{bmatrix} \tag{6.49}$$

Since the LMIs are feasible, there is a positive-definite matrix that satisfies Eq. (6.35), the closed-loop system stability is guaranteed. It is also easy to verify that $\boldsymbol{P}$ satisfies the stability conditions. Details are as follows:

$$\boldsymbol{H}_1^T\boldsymbol{P} + \boldsymbol{P}\boldsymbol{H}_1 = \begin{bmatrix} -80.8608 & 9.3356 & -4.0332 \\ 9.3356 & -4.2432 & 0.4534 \\ -4.0332 & 0.4534 & -3.0280 \end{bmatrix} < \boldsymbol{0}$$

$$H_2^T P + PH_2 = \begin{bmatrix} -159.740 & 9.137 & -7.050 \\ 9.137 & -4.243 & 0.453 \\ -7.050 & 0.453 & -3.028 \end{bmatrix} < \mathbf{0}$$

$$H_3^T P + PH_3 = \begin{bmatrix} -209.974 & 21.278 & -9.433 \\ 21.278 & -4.612 & 0.570 \\ -9.433 & 0.570 & -3.085 \end{bmatrix} < \mathbf{0}$$

$$H_4^T P + PH_4 = \begin{bmatrix} -120.300 & 9.236 & -5.542 \\ 9.237 & -4.243 & 0.453 \\ -5.541 & 0.453 & -3.028 \end{bmatrix} < \mathbf{0}$$

$$H_5^T P + PH_5 = \begin{bmatrix} -30.626 & -2.805 & -1.650 \\ -2.805 & -3.874 & 0.337 \\ -1.650 & 0.337 & -2.971 \end{bmatrix} < \mathbf{0} \tag{6.50}$$

Therefore, the system is asymptotically stable.

6.5 DESIGN OF ADAPTIVE IT2 TSK FLC[5]

This section presents a methodology for the design of adaptive IT2 TSK FLC with application to robot manipulators. The structure of the proposed adaptive FLC depends on the PD FLC. First, we identify the structure of the PD FLC, presenting the membership functions, rules, and inference mechanism, in the subsections. Then, we use these for the design of the adaptive FLC.

6.5.1 Rule Bases

Rules play a key role in any FLC and obtaining effective rule bases is very important. Designers usually use their expertise or their intuition to develop rules, or they refer to some well-known rule bases, which is our choice in this section. We adopt the MacVicar–Whelan rule base for our design as it has been shown to be effective for tracking (Yager and Filev, 1994).

First, error, e, and the rate of change of the error, Δe, are defined as

$$e \equiv r - y \tag{6.51}$$

$$\Delta e \equiv e(k) - e((k-1)) \tag{6.52}$$

where r is the set point, y is the output, and k is an integer. We define the error vector, $\boldsymbol{e}$, as

$$\boldsymbol{e} = [e, \Delta e]^T \tag{6.53}$$

[5]Much of the material in this section is taken directly from Biglarbegian et al. (2009; © 2009, IEEE).

For the PD-type FLC, the rules are defined as follows:

$$R^s\text{: If } e \text{ is } \tilde{F}_1^s \text{ and } \Delta e \text{ is } \tilde{F}_2^s\text{, then } u_s = c_1^s e + c_2^s \Delta e \tag{6.54}$$

where $s = 1, \ldots, M$, $\tilde{F}_i^s$ represents the IT2 fuzzy set of input state i of rule s, c_1^s and c_2^s are the coefficients of the output function for rule s (and are crisp numbers), u_s is the output of the controller, with M the number of rules.

Lower firing level, $\underline{f}^s$, and upper firing level, $\overline{f}^s$, are given as follows:

$$\underline{f}^s(e) = \underline{\mu}_{\tilde{F}_1^s}(e_1) * \underline{\mu}_{\tilde{F}_2^s}(\Delta e) \tag{6.55}$$

$$\overline{f}^s(e) = \overline{\mu}_{\tilde{F}_1^s}(e_1) * \overline{\mu}_{\tilde{F}_2^s}(\Delta e) \tag{6.56}$$

where $\underline{\mu}_{\tilde{F}_i^s}$ and $\overline{\mu}_{\tilde{F}_i^s}$ represent the ith ($i = 1, 2$) lower and upper membership functions of rule s, respectively, and $*$ is a t-norm operator.

To determine the control action, u, a general MacVicar–Whelan rule base (Chopra et al., 2005) uses e and Δe. Although different rule bases can be defined, we present a system with only nine rules.

The nine-rule system is introduced in Table 6.1 where NB, ZE, and PB, represent negative big, zero, and positive big, respectively.

6.5.2 Membership Functions

Figures 6.4 shows some suggested membership functions for systems with nine rules.

The inputs to the membership functions shown above are within $[-1, 1]$. Hence, the values of e and Δe must be mapped onto $[-1, 1]$, which is done using scaling factors and introduced next.

6.5.3 Control Structure

The structure of the PD FLCs is shown in Fig. 6.5 where r is the input, y is the output to the closed loop, and e is the difference between the two. The inputs to the PD are e_n and Δe_n, and the corresponding output is u. Scaling factors are α_e and $\alpha_{\Delta e}$ map e and Δe onto $[-1, 1]$, use the following relationships:

$$e_n = \gamma_e . e \tag{6.57}$$

$$\Delta e_n = \gamma_{\Delta e} . \Delta e \tag{6.58}$$

TABLE 6.1 Rule Base for System with Nine Rules

$\Delta e/e$	**NB**	**ZE**	**PB**
NB	NB	NB	ZE
ZE	NB	ZE	PB
PB	ZE	PB	PB

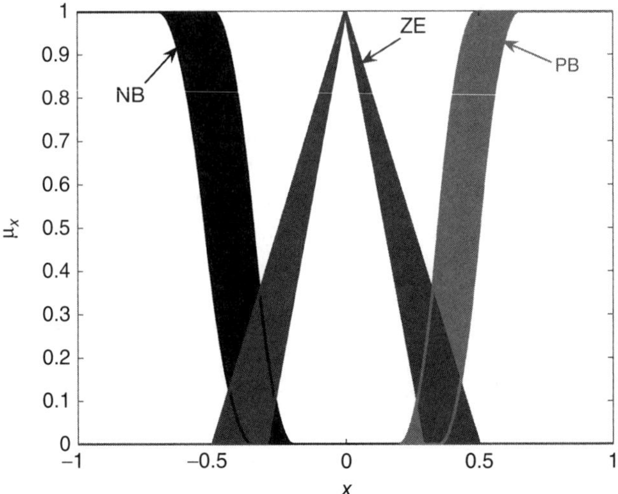

Figure 6.4 Membership functions for e and Δe of the proposed IT2 TSK FLCs for a system with nine rules (Biglarbegian et al., 2009; © 2009, IEEE).

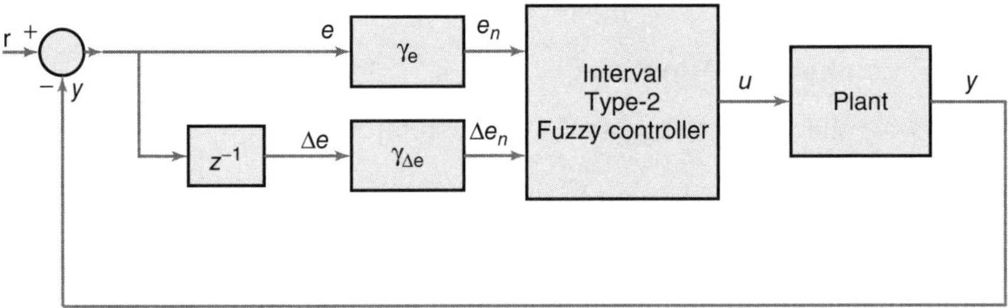

Figure 6.5 Structure of the IT2 PD FLCs (Bigalarbegian et al., 2009; © 2009, IEEE).

With the inference engine introduced earlier, the controller output, is given by the following expression:

$$u = m\frac{\sum_{s=1}^{M} \underline{f}^s(e)u_s}{\sum_{s=1}^{M} \underline{f}^s(e)} + n\frac{\sum_{s=1}^{M} \overline{f}^s(e)u_s}{\sum_{s=1}^{M} \overline{f}^s(e)} \tag{6.59}$$

6.5.4 Control Design

The next step is to determine the control parameters. There are a multitude of methods to find the consequent parameters of the IT2 TSK FLCs. To determine how to find the consequent parameters, the nine rules of Table 6.1 must first be considered:

If e is NB $(e = -1)$, then

Rule 1. If Δe is NB, $\Delta e = -1$: $u_1 = -c_1^1 - c_2^1 = -1$.
Rule 2. If Δe is ZE, $\Delta e = 0$: $u_2 = -c_1^2 = -1$.
Rule 3. If Δe is PB, $\Delta e = 1$: $u_3 = -c_1^3 + c_2^3 = 0$.

If e_n is ZE $(e_n = 0)$, then

Rule 4. If Δe_n is NB, $\Delta e_n = -1$: $u_4 = -c_2^4 = -1$.
Rule 5. If Δe_n is ZE, $\Delta e_n = 0$: $u_5 = 0$.
Rule 6. If Δe_n is PB, $\Delta e_n = 1$: $u_6 = c_2^6 = 1$.

If e is PB $(e = 1)$, then

Rule 7. If Δe is NB, $\Delta e = -1$: $u_7 = c_1^7 - c_2^7 = 0$.
Rule 8. If Δe is ZE, $\Delta e = 0$: $u_8 = c_1^8 = 1$.
Rule 9. If Δe is PB, $\Delta e = 1$: $u_9 = c_1^9 + c_2^9 = 1$.

From the above conditions we can solve for c_1^i and c_2^i, to find the consequent parameters. These constraints do not determine all the parameters, and we impose a condition on them so that they belong to [0, 1]. Depending on the design requirement, the designer can choose the free variables.

To tune the control parameters m, n, α_e, the following steps are recommended:

- To tune α_e, e_n is required to fall into $[-1, 1]$. Tuning $\alpha_{\Delta e}$ requires trial and error to achieve good transient response.
- To tune the parameters of the controller, m and n, it is best to start with small gains because high gains result in overshoot and even instability, that is, $0 < m, n <= 1$. If the design aspects are not met, start increasing m and/or n.

6.5.5 Control Performance

In this section, we use some standard metrics to compare the output of a PD-type IT2 TSK FLC on a nonlinear plant. These metrics are rise time, t_r, settling time, t_s, percent overshoot, OS, steady-state output, y_{ss}, and steady-state error, e_{ss}. The following plant is used for our investigation:

$$\dot{y}(t) = u(t) - y(t) - 0.8y^2(t) \tag{6.60}$$

We develop IT2 as well as T1 FLC and compare their outputs. Simulations were performed in Simulink.

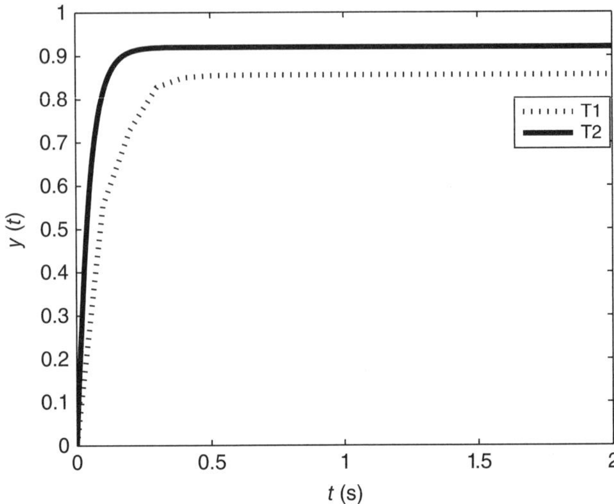

Figure 6.6 Step response of T1 and IT2 FLCs (Biglarbegian et al., 2009; © 2009, IEEE).

TABLE 6.2 T1 and IT2 FLCs Performance

	$t_r(s)$	$t_s(s)$	OS	y_{ss}	e_{ss}
T_1	0.227	0.249	0	0.856	0.144
IT_2	0.110	0.139	0	0.920	0.080

6.5.5.1 PD FLC

The results of the T1 and IT2 FLCs are shown in Fig. 6.6 where both controllers have output-tracking errors. However, the IT2 controller outperforms its T1 counterpart by providing a significantly faster response as well as reducing steady-state error. The transient response characteristics of both controllers are shown in Table 6.2, which verifies the enhanced performance of the IT2 controller. The values of m and n were chosen as 1.5 and 0.5, respectively; γ_e and $\gamma_{\Delta e}$ are selected to be 10 and 1.

6.6 ADAPTIVE CONTROL DESIGN WITH APPLICATION TO ROBOT MANIPULATORS[6]

In this section, we design adaptive IT2 TSK FLCs for modular and reconfigurable robots (MRR). These manipulators assume multiple configurations and thus can be used in several tasks. Because of the flexibility these manipulators offer, they are popular tools in automation industry for creating cost-effective solutions.

[6]Much of the material in this section is taken directly from Biglarbegian et al. (2011; © 2011, IEEE).

We first present the governing equations of motion for these manipulators. Then, we design controllers for a trajectory tracking problem, and finally present some experimental results. The dynamics of robot manipulators with p joints is given by Lewis et al. (1995) as

$$\boldsymbol{M}(\boldsymbol{q})\ddot{\boldsymbol{q}} + \boldsymbol{V}_m(\boldsymbol{q},\dot{\boldsymbol{q}})\dot{\boldsymbol{q}} + \boldsymbol{F}(\dot{\boldsymbol{q}}) + \boldsymbol{G}(\boldsymbol{q}) + \boldsymbol{\tau}_d = \boldsymbol{\tau} \tag{6.61}$$

where $\boldsymbol{M}(\boldsymbol{q})$ is the inertia matrix, $\boldsymbol{V}_m(\boldsymbol{q},\dot{\boldsymbol{q}})$ contains the Coriolis terms, and $\boldsymbol{F}(\dot{\boldsymbol{q}})$, $\boldsymbol{G}(\boldsymbol{q})$, $\boldsymbol{\tau}_d$ represent friction, gravity, disturbances, respectively, where $\boldsymbol{\tau}$ is the control variable, and, finally, $\boldsymbol{q}$ is the robot joint parameter.

The following properties that hold for manipulators are (Lewis et al., 1999):

- $\boldsymbol{M}(\boldsymbol{q})$ is symmetric, positive-definite matrix and bounded from above, that is, $||\boldsymbol{M}(\boldsymbol{q})|| \leq M_B$.
- $||\boldsymbol{V}_m(\boldsymbol{q},\dot{\boldsymbol{q}})|| < V_B||\dot{\boldsymbol{q}}||$.
- $\boldsymbol{M}(\boldsymbol{q}) - 2\boldsymbol{V}_m(\boldsymbol{q},\dot{\boldsymbol{q}})$, is skew-symmetric.
- $||\boldsymbol{F}(\dot{\boldsymbol{q}})|| \leq F_B||\dot{\boldsymbol{q}}|| + K_B$ and $||\boldsymbol{G}(\boldsymbol{q})|| \leq G_B$.
- The bound on disturbances are known, that is, $||\boldsymbol{\tau}_d|| \leq D_B$.

We use these properties for control design.

6.6.1 Tracking Control

In tracking, we require each joint of the robot to follow a desired trajectory. The control objective is to minimize the error as a result of varying dynamics and disturbances.

Define the tracking error, $\boldsymbol{e}$, and filtered tracking error, $\boldsymbol{r}$, as

$$\boldsymbol{e} \equiv \boldsymbol{q}_d - \boldsymbol{q} \tag{6.62}$$

$$\boldsymbol{r} \equiv \dot{\boldsymbol{e}} + \boldsymbol{\Lambda e} \tag{6.63}$$

where $\boldsymbol{\Lambda} \in \mathbb{R}^{p\times p}$ is a positive-definite design matrix.

We now solve for $\boldsymbol{q}$ from Eq. (6.62) and $\dot{\boldsymbol{e}}$ from Eq. (6.63):

$$\boldsymbol{q} = \boldsymbol{q}_d - \boldsymbol{e} \tag{6.64}$$

$$\dot{\boldsymbol{e}} = \boldsymbol{r} - \boldsymbol{\Lambda e} \tag{6.65}$$

First and second derivatives of Eq. (6.64) as well as the first derivative of Eq. (6.65) are given as follows, respectively:

$$\dot{\boldsymbol{q}} = \dot{\boldsymbol{q}}_d - \dot{\boldsymbol{e}} \tag{6.66}$$

$$\ddot{\boldsymbol{q}} = \ddot{\boldsymbol{q}}_d - \ddot{\boldsymbol{e}} \tag{6.67}$$

$$\ddot{\boldsymbol{e}} = \dot{\boldsymbol{r}} - \boldsymbol{\Lambda}\dot{\boldsymbol{e}} \tag{6.68}$$

Substituting $\dot{e}$ in Eq. (6.65) into Eq. (6.66) gives

$$\dot{q} = \dot{q}_d - r + \Lambda e \quad (6.69)$$

Similarly, substituting $\ddot{e}$ in Eq. (6.68) into Eq. (6.67) gives

$$\ddot{q} = \ddot{q}_d - \dot{r} + \Lambda \dot{e} \quad (6.70)$$

Using Eqs. (6.69) and (6.70), the dynamics of the robot [Eq. (6.61)] can be written as[7]

$$M\dot{r} = M[\ddot{q}_d + \Lambda \dot{e}] + V_m[\dot{q}_d + \Lambda e] - V_m r + F(\dot{q}) + G(q) + \tau_d - \tau \quad (6.71)$$

Absorbing the first, second, fourth, and fifth terms of the right hand of Eq. (6.71) into f, we will have

$$f = M[\ddot{q}_d + \Lambda \dot{e}] + V_m[\dot{q}_d + \Lambda e] + F(\dot{q}) + G(q) \quad (6.72)$$

Using the expression for f given in Eq. (6.72), the dynamics of the robot is expressed in a more compact form as follows:

$$M\dot{r} = -V_m r + f + \tau_d - \tau \quad (6.73)$$

where f has nonlinearity and unmodeled dynamics. For reconfigurable robots, the dynamic parameters will change when a robot arm is reconfigured; therefore, developing adaptive control techniques that are robust is necessary to handle these changing dynamics.

6.6.2 Control Structure

The structure of the IT2 TSK FLC is shown in Fig. 6.7, where q_d is the reference trajectory, $\dot{q}_d$ is the derivative of the reference trajectory, q is the MRR output, u_e is the auxiliary output (which will be explained later), u_{Fuzzy} is the IT2 FLC output, u_{PD} is the PD controller output, and e and $\dot{e}$ are the error and error rate, respectively.

The proposed controller has the following form:

$$\tau = u_{\text{PD}} + u_{\text{Fuzzy}} + u_e \quad (6.74)$$

where $u_{\text{PD}} = K_{\text{PD}} r$, and $u_{\text{Fuzzy}} = g(e, \dot{e})$, and u_e is the auxiliary control (which will be explained in detail later).

[7] M and V_m are short for $M(q)$ and $V_m(q, \dot{q})$, respectively.

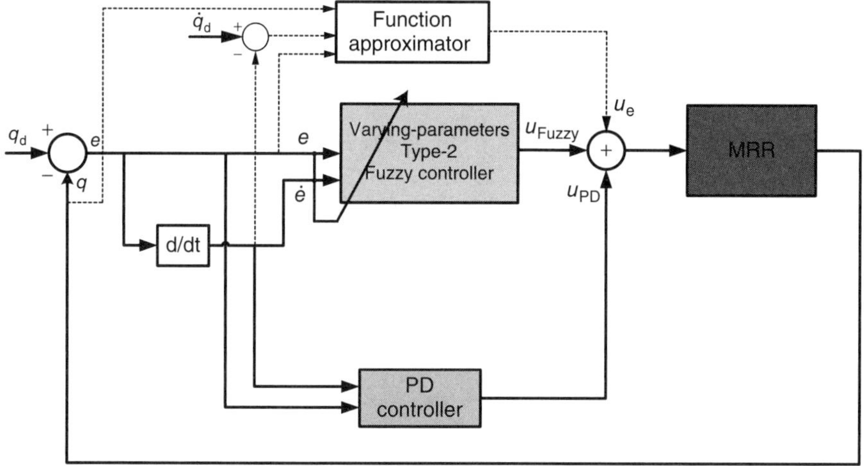

Figure 6.7 Control structure (Biglarbegian et al., 2011; © 2011, IEEE).

We now define the following terms that enable us to express the IT2 FLC in a matrix format for control design:

$$\underline{\boldsymbol{\phi}} \equiv \begin{bmatrix} m_1\left(\frac{\underline{f}_1^1}{\sum_{i=1}^M \underline{f}_1^i}\right) e_1 & \cdots & m_1\left(\frac{\underline{f}_1^M}{\sum_{i=1}^M \underline{f}_1^i}\right) e_1 & m_1\left(\frac{\underline{f}_1^1}{\sum_{i=1}^M \underline{f}_1^i}\right) \dot{e}_1 & \cdots & m_1\left(\frac{\underline{f}_1^M}{\sum_{i=1}^M \underline{f}_1^i}\right) \dot{e}_1 \\ m_2\left(\frac{\underline{f}_2^1}{\sum_{i=1}^M \underline{f}_2^i}\right) e_2 & \cdots & m_2\left(\frac{\underline{f}_2^M}{\sum_{i=1}^M \underline{f}_2^i}\right) e_2 & m_2\left(\frac{\underline{f}_2^1}{\sum_{i=1}^M \underline{f}_2^i}\right) \dot{e}_2 & \cdots & m_2\left(\frac{\underline{f}_2^M}{\sum_{i=1}^M \underline{f}_2^i}\right) \dot{e}_2 \\ \vdots & & \vdots & \vdots & & \vdots \\ m_p\left(\frac{\underline{f}_p^1}{\sum_{i=1}^M \underline{f}_p^i}\right) e_p & \cdots & m_p\left(\frac{\underline{f}_p^M}{\sum_{i=1}^M \underline{f}_p^i}\right) e_p & m_p\left(\frac{\underline{f}_p^1}{\sum_{i=1}^M \underline{f}_p^i}\right) \dot{e}_p & \cdots & m_p\left(\frac{\underline{f}_p^M}{\sum_{i=1}^M \underline{f}_p^i}\right) \dot{e}_p \end{bmatrix}$$

and

$$\overline{\boldsymbol{\phi}} \equiv \begin{bmatrix} n_1\left(\frac{\overline{f}_1^1}{\sum_{i=1}^M \overline{f}_1^i}\right) e_1 & \cdots & n_1\left(\frac{\overline{f}_1^M}{\sum_{i=1}^M \overline{f}_1^i}\right) e_1 & n_1\left(\frac{\overline{f}_1^1}{\sum_{i=1}^M \overline{f}_1^i}\right) \dot{e}_1 & \cdots & n_1\left(\frac{\overline{f}_1^M}{\sum_{i=1}^M \overline{f}_1^i}\right) \dot{e}_1 \\ n_2\left(\frac{\overline{f}_2^1}{\sum_{i=1}^M \overline{f}_2^i}\right) e_2 & \cdots & n_2\left(\frac{\overline{f}_2^M}{\sum_{i=1}^M \overline{f}_2^i}\right) e_2 & n_2\left(\frac{\overline{f}_2^1}{\sum_{i=1}^M \overline{f}_2^i}\right) \dot{e}_2 & \cdots & n_2\left(\frac{\overline{f}_2^M}{\sum_{i=1}^M \overline{f}_2^i}\right) \dot{e}_2 \\ \vdots & & \vdots & \vdots & & \vdots \\ n_p\left(\frac{\overline{f}_p^1}{\sum_{i=1}^M \overline{f}_p^i}\right) e_p & \cdots & n_p\left(\frac{\overline{f}_p^M}{\sum_{i=1}^M \overline{f}_p^i}\right) e_p & n_p\left(\frac{\overline{f}_p^1}{\sum_{i=1}^M \overline{f}_p^i}\right) \dot{e}_p & \cdots & n_p\left(\frac{\overline{f}_p^M}{\sum_{i=1}^M \overline{f}_p^i}\right) \dot{e}_p \end{bmatrix}$$

as well,

$$X \equiv (\underline{\boldsymbol{\phi}} + \overline{\boldsymbol{\phi}}) \tag{6.75}$$

and $\boldsymbol{\Theta}$ contains the consequent TSK parameters

$$\boldsymbol{\Theta} \equiv \left[c_1^1, \, \ldots \, , c_1^M \quad , c_2^1, \, \ldots \, , c_2^M\right]^T \tag{6.76}$$

Using Eqs. (6.75) and (6.76), the IT2 TSK FLC output is simply written as

$$\boldsymbol{u}_{\text{Fuzzy}} = \boldsymbol{g}(\boldsymbol{e}, \dot{\boldsymbol{e}}) = \boldsymbol{X}\Theta \tag{6.77}$$

As we will see later in the design, the compact form of the FLC output given by Eq. (6.77) helps to develop the adaptive controller.

Using Eqs. (6.62) and (6.77), the dynamics of the robot are given as

$$\boldsymbol{M}\dot{\boldsymbol{r}} = -\boldsymbol{V}_m\boldsymbol{r} - \boldsymbol{K}_{\text{PD}}\boldsymbol{r} - \boldsymbol{X}\Theta - \boldsymbol{u}_e + \boldsymbol{f} + \boldsymbol{\tau}_d \tag{6.78}$$

We assume the desired trajectory and its derivatives are bounded, that is, $\|\boldsymbol{q}_d\| \leq q_d$, $\|\dot{\boldsymbol{q}}_d\| \leq \dot{q}_d$, and $\|\ddot{\boldsymbol{q}}_d\| \leq \dot{q}_d$ and use this assumption in the next step for control design.

We now state an important theorem that is used to prove the stability of the controller, from which we will derive the adaptive law as well.

THEOREM 6.1 If for a nonlinear system $\dot{\boldsymbol{x}} = \boldsymbol{f}(\boldsymbol{x}) + \boldsymbol{d}(t)$, there exists a Lyapunov function $V(x, t)$ with continuous partial derivatives such that for x in a compact set $\boldsymbol{S} \subset \mathbb{R}^n$

$$V(x, t) > 0 \tag{6.79}$$

and

$$\dot{V}(x, t) < 0 \quad \text{for} \quad \|\boldsymbol{x}\| > R \tag{6.80}$$

for some $R > 0$ such that the ball of radius of R is contained in $\boldsymbol{S}$, then the system is uniformly ultimately bounded (UUB), and the norm of the state is bounded to within a neighborhood of R (Lewis et al., 1999).

Therefore, we need to find a Lyapunov function and show $\dot{V} < 0$ along the robot trajectories. Considering the following Lyapunov function

$$V = \tfrac{1}{2}\boldsymbol{r}^T\boldsymbol{M}\boldsymbol{r} + \int_0^t \boldsymbol{\Theta}^T\boldsymbol{F}\Theta \, dt \tag{6.81}$$

$\dot{V}$ is expressed as

$$\dot{V} = \boldsymbol{r}^T\boldsymbol{M}\dot{\boldsymbol{r}} + \tfrac{1}{2}\boldsymbol{r}^T\dot{\boldsymbol{M}}\boldsymbol{r} + \boldsymbol{\Theta}^T\boldsymbol{F}\Theta \tag{6.82}$$

Using (6.78), $\dot{V}$ can be written simply as

$$\dot{V} = \tfrac{1}{2}\boldsymbol{r}^T(\dot{\boldsymbol{M}} - 2\boldsymbol{V}_m)\boldsymbol{r} - \boldsymbol{\Theta}^T(\boldsymbol{X}^T\boldsymbol{r} - \boldsymbol{F}\Theta) - \boldsymbol{r}^T\boldsymbol{K}_{PD}\boldsymbol{r} + \boldsymbol{r}^T\boldsymbol{\phi} \tag{6.83}$$

where $\boldsymbol{\phi}$ is given by

$$\boldsymbol{\phi} \equiv \boldsymbol{M}[\ddot{\boldsymbol{q}}_d + \boldsymbol{\Lambda}\dot{\boldsymbol{e}}] + \boldsymbol{F}(\dot{\boldsymbol{q}}) + \boldsymbol{G}(\boldsymbol{q}) + \boldsymbol{V}_m[\dot{\boldsymbol{q}}_d + \boldsymbol{\Lambda}\boldsymbol{e}] - \boldsymbol{u}_e + \boldsymbol{\tau}_d \tag{6.84}$$

Assume a function approximator, termed $\boldsymbol{u}_e$, exists that can approximate $\boldsymbol{V}_m[\dot{\boldsymbol{q}}_d + \boldsymbol{\Lambda}\boldsymbol{e}]$. This approximator is the auxiliary control that was defined in Eq. (6.74).

As well, define

$$\boldsymbol{\epsilon} \equiv \boldsymbol{V}_m[\dot{\boldsymbol{q}}_d + \boldsymbol{\Lambda}\boldsymbol{e}] - \boldsymbol{u}_e \tag{6.85}$$

where $\|\boldsymbol{\epsilon}\| \leq \epsilon_n$. We make use of this definition in the following analysis.

Since $\boldsymbol{r}^T\boldsymbol{\phi} \leq \|\phi\|.\|\boldsymbol{r}\|$, we will have

$$\boldsymbol{r}^T\boldsymbol{\phi} \leq \|\boldsymbol{M}[\ddot{\boldsymbol{q}}_d + \boldsymbol{\Lambda}\dot{\boldsymbol{e}}] + \boldsymbol{F}(\dot{\boldsymbol{q}}) + \boldsymbol{G}(\boldsymbol{q}) + \boldsymbol{\epsilon} + \boldsymbol{\tau}_d\|.\|\boldsymbol{r}\| \tag{6.86}$$

The upper bound on $\|\boldsymbol{\phi}\|$ is calculated as

$$\begin{aligned} \|\boldsymbol{\phi}\| &\leq \|\boldsymbol{M}[\ddot{\boldsymbol{q}}_d + \boldsymbol{\Lambda}\dot{\boldsymbol{e}}]\| + \|\boldsymbol{F}(\dot{\boldsymbol{q}})\| + \|\boldsymbol{G}(\boldsymbol{q})\| + \|\boldsymbol{\tau}_d\| \\ &\leq \|\boldsymbol{M}\ddot{\boldsymbol{q}}_d\| + \|\boldsymbol{M}\Lambda\dot{\boldsymbol{e}}\| + \|\boldsymbol{F}(\dot{\boldsymbol{q}})\| + \|\boldsymbol{G}(\boldsymbol{q})\| + \|\boldsymbol{\epsilon}\| + \|\boldsymbol{\tau}_d\| \end{aligned} \tag{6.87}$$

Using the upper bounds of the robot parameters, as well as boundedness of the desired trajectory and its derivatives, we have

$$\|\boldsymbol{\phi}\| \leq M_B\ddot{q}_d + M_B \max(\text{eig}(\boldsymbol{\Lambda}))\|\boldsymbol{r}\| + F_B(\dot{q}_d + \|\boldsymbol{r}\|) + K_B + G_B + D_B + \epsilon_N \tag{6.88}$$

Similarly, the bound of $-\boldsymbol{r}^T\boldsymbol{K}_{\text{PD}}\boldsymbol{r} + r^T\boldsymbol{\phi}$, is given by

$$\begin{aligned} -\boldsymbol{r}^T\boldsymbol{K}_{\text{PD}}\boldsymbol{r} + r^T\boldsymbol{\phi} &\leq -K_{\text{PD}\min}\|\boldsymbol{r}\|^2 + \|\boldsymbol{\phi}\|\,\|\boldsymbol{r}\| \\ &\leq -K_{PD\min}\|\boldsymbol{r}\|^2 + M_B\ddot{q}_d\|\boldsymbol{r}\| + M_B \max(\text{eig}(\boldsymbol{\Lambda}))\|\boldsymbol{r}\|^2 \\ &\quad + F_B(\dot{q}_d + \|\boldsymbol{r}\|)\|\boldsymbol{r}\| + (K_B + G_B + D_B + \epsilon_N)\|\boldsymbol{r}\| \end{aligned} \tag{6.89}$$

where we have used the following well-known concept from linear algebra (Khalil, 1996):

$$\|\boldsymbol{x}\|^2 \min(\text{eig}(\boldsymbol{P})) \leq \boldsymbol{x}^T\boldsymbol{P}\boldsymbol{x} \leq \|\boldsymbol{x}\|^2 \max(\text{eig}(\boldsymbol{P})) \tag{6.90}$$

in which $K_{\text{PD}\min}$ is the minimum eigenvalue of the matrix $\boldsymbol{K}_{\text{PD}}$.

Equation (6.89) can be written as

$$-\boldsymbol{r}^T\boldsymbol{K}_{\text{PD}}\boldsymbol{r} + r^T\boldsymbol{\phi} \leq (A\|\boldsymbol{r}\| + B)\|\boldsymbol{r}\| \tag{6.91}$$

where A and B are given as follows:

$$A = -K_{\mathrm{PD\,min}} + M_B \max(\mathrm{eig}(\boldsymbol{\Lambda})) + F_B \tag{6.92}$$

$$B = M_B \ddot{q}_d + F_B \dot{q}_d + (K_B + G_B + D_B + \epsilon_N) \tag{6.93}$$

Observe that $\dot{\boldsymbol{M}} - 2\boldsymbol{V}_m$, is a skew-symmetric matrix (characteristics of robot manipulators). Hence, Eq. (6.94) is written in a more simplified way as

$$\dot{V} = -\boldsymbol{\Theta}^T(\boldsymbol{X}^T\boldsymbol{r} - \boldsymbol{F}\Theta) - \boldsymbol{r}^T\boldsymbol{K}_{\mathrm{PD}}\boldsymbol{r} + \boldsymbol{r}^T\boldsymbol{\phi} \tag{6.94}$$

If we ensure the following,

$$-\boldsymbol{r}^T\boldsymbol{K}_{\mathrm{PD}}\boldsymbol{r} + r^T\boldsymbol{\phi} < 0 \tag{6.95}$$

$$\boldsymbol{X}^T\boldsymbol{r} - \boldsymbol{F}\Theta = \boldsymbol{0} \tag{6.96}$$

then $\dot{V}$ in Eq. (6.94) will be negative definite.
First, if $(A\|\boldsymbol{r}\| + B)\|\boldsymbol{r}\| < 0$, then $-\boldsymbol{r}^T\boldsymbol{K}_{\mathrm{PD}}\boldsymbol{r} + r^T\boldsymbol{\phi} < 0$.

Thus

$$A\|\boldsymbol{r}\| + B < 0 \tag{6.97}$$

By making $A < 0$ and knowing that B is a positive constant, for $\forall\|\boldsymbol{r}\| > -B/A$, it is guaranteed to have $-\boldsymbol{r}^T\boldsymbol{K}_{\mathrm{PD}}\boldsymbol{r} + r^T\boldsymbol{\phi} < 0$.

The next condition to satisfy is called the "adaptive law," $\boldsymbol{X}^T\boldsymbol{r} - \boldsymbol{F}\Theta = \boldsymbol{0}$ and can be expressed as

$$\boldsymbol{\Theta} = \boldsymbol{F}^{-1}\boldsymbol{X}^T\boldsymbol{r} \tag{6.98}$$

where $\boldsymbol{F}$ is a design matrix (needs to be positive definite), $\boldsymbol{X}$ contains the control parameters, and $\boldsymbol{r}$ is the tracking error.

To satisfy Eq. (6.98), the initially chosen controller parameters, m and n, are kept fixed in the adaptation process and only the TSK parameters are adjusted ensuring $\dot{V} < 0$.

6.6.3 Application to Modular and Reconfigurable Robot Manipulators (MRR)

A modular and reconfigurable robot with two degrees of freedom is used for control. This robot consists of several modules that can be put together in different positions/orientations with respect to each other. Thus, the robot can assume multiple configurations and hence it is called reconfigurable. The robot in this experiment has a payload of $m = 6.80$ kg. The MRR is controlled by a MSK2812 DSP-based microcontroller via a controller area network (CAN) communication bus. The microcontroller is operated at 150 MHz. Figure 6.8 shows the experimental setup.

The design matrices are

$$\boldsymbol{F} \in \mathbb{R}^{18\times 18}$$

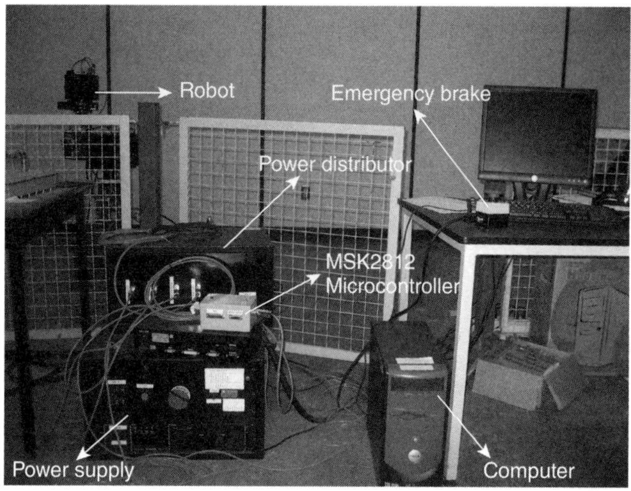

Figure 6.8 System hardware: computer, microcontroller, and robot (Biglarbegian et al., 2011; © 2011, IEEE).

TABLE 6.3 Controller Parameters

Controller	First Joint	Second Joint
PD	$K_{\text{PD}} = \begin{bmatrix} 0.025 & 0 \\ 0 & 0.008 \end{bmatrix}$	
Fixed Parameters	$m = 0.001$	$n = 0.001$

with $F_{ii} = 10$ $(i = 1, \ldots, 18)$, and $\Lambda_{\text{IT2}} = \text{diag}[0.001, 0.0019]$, and $\Lambda_{\text{T1}} = \text{diag}[0.0005, 0.0004]$. Control parameters are summarized in Table 6.3.

A sinusoidal trajectory is applied to each joint of the robot and control performance is compared in terms of mean-squared error (MSE) and percentage improvement (*PI*). Two different configurations of the robots are considered, designated as first and second, and the controller performance for both of them will be evaluated.

In the first configuration, the first and second joint axes are parallel to each other. In the second configuration, the second joint axis is perpendicular to the first joint axis. These two configurations are shown in Figs. 6.9 and 6.10.

The controllers' performance in terms of MSE and PI for the first and second configurations are shown in Tables 6.4 and 6.5, respectively. For the design of PD controller please see Lewis et al. (1999).

From the results, the superiority of the adaptive IT2 over other nonlinear controllers can be seen. The outcome verifies the potential of IT2 in handling uncertainties and varying dynamics that we cannot model properly.

The next section is dedicated to the design of robust IT2 TSK FLCs.

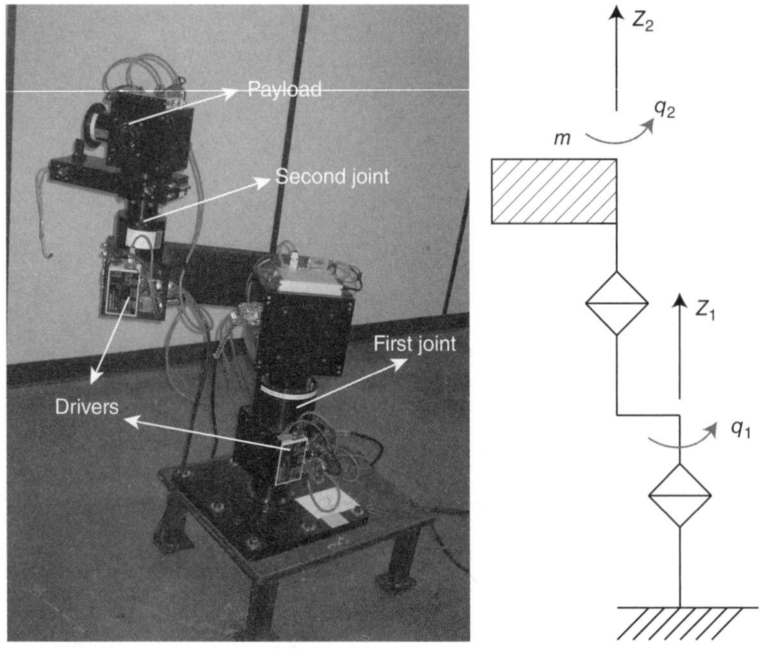

Figure 6.9 First configuration and its schematic (Biglarbegian et al., 2011; © 2011, IEEE).

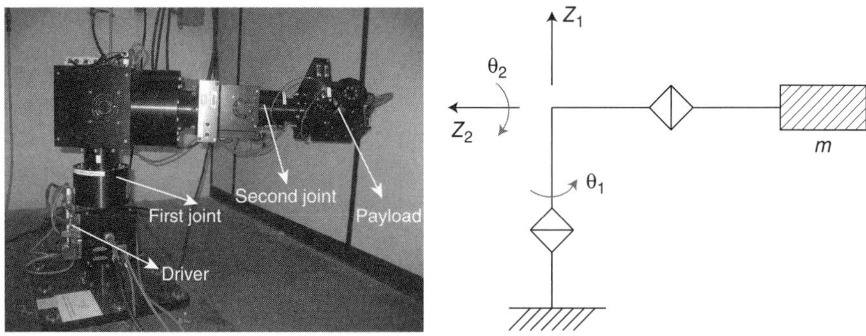

Figure 6.10 Second configuration and its schematic (Biglarbegian et al., 2011; © 2011, IEEE).

TABLE 6.4 First Configuration Results

	First Joint		Second Joint	
Controller	MSE	PI (%)	MSE	PI (%)
PD	1.2478	—	0.3323	—
T1 nonadaptive	0.7682	38.44	0.2572	22.61
T1 adaptive	0.7294	41.55	0.2313	30.41
IT2 nonadaptive	0.6098	51.13	0.1851	44.29
IT2 adaptive	0.5124	58.93	0.1724	48.12

TABLE 6.5 Second Configuration Results

	First Joint		Second Joint	
Controller	MSE	PI (%)	MSE	PI (%)
PD	1.1849	—	0.5709	—
T1 nonadaptive	0.7391	37.62	0.5303	7.11
T1 adaptive	0.7344	38.02	0.5129	8.34
IT2 nonadaptive	0.6302	46.82	0.5107	10.54
IT2 adaptive	0.5637	52.43	0.4181	26.77

6.7 ROBUST CONTROL DESIGN[8]

In this section, we develop robust IT2 TSK FLCs. So far, we have addressed important control aspects such as stability and adaptivity, but robustness remains to be discussed. Most of the times, particularly in dealing with real systems, we are faced with disturbances that are undesired and cannot be modeled accurately. Therefore, design of a robust control system that can suppress the disturbances is much needed. The specific robust control problem we look at is the disturbance rejection, which will be introduced later. Using the foundations built in earlier sections, we now are ready to design a robust IT2 FLC.

6.7.1 System Description

Similar to the stability analysis that was performed in Section 6.4, we assume the plant is modeled with a T1 TS model. To accommodate the disturbances, we first introduce the model structure of the plant that will be used throughout our design. The general structure of a plant described by a continuous T1 TS model is as follows:

$$R^s\text{: If } r_1 \text{ is } F_1^s \text{ and } r_2 \text{ is } F_2^s \text{ and } \cdots r_z \text{ is } F_z^s\text{, then} \tag{6.99}$$

$$R^s\text{: } \dot{\boldsymbol{x}}_s(t) = \boldsymbol{A}_s\boldsymbol{x}(t) + \boldsymbol{B}_s\boldsymbol{u}(t) + \boldsymbol{D}_s\boldsymbol{w}(t) \tag{6.100a}$$

$$\boldsymbol{y}_s(t) = \boldsymbol{C}_s\boldsymbol{x}(t) \tag{6.100b}$$

where r_i is the ith input to the T1 TSK FLC, $\boldsymbol{A}_s \in \mathbb{R}^{n\times n}$, $\boldsymbol{b}_s \in \mathbb{R}^{n\times m}$, $\boldsymbol{u}(t) \in \mathbb{R}^m$ (controller output), and F_i^s represents a T1 FS of the ith input state of rule s, $\boldsymbol{x}_s(t)$ is the output of each rule, $\boldsymbol{x}(t)$ is the state vector given by Eq. (6.2), and M is the number of rules, and the new vector introduced here to represent disturbances is $\boldsymbol{w}(t) \in \mathbb{R}^s$. The state and output vectors of the plant are, respectively, given by

$$\dot{\boldsymbol{x}}(t) = \frac{\sum_{s=1}^{M} f^s \dot{\boldsymbol{x}}_s(t)}{\sum_{s=1}^{M} f^s} \tag{6.101}$$

[8]Much of the material in this section is taken directly from Biglarbegian (2012; © 2012, IEEE).

$$y(t) = \frac{\sum_{s=1}^{M} f^s y_s(t)}{\sum_{s=1}^{M} f^s} \tag{6.102}$$

Note that we have used the same membership functions to obtain $\dot{x}(t)$ and $y(t)$, thus the firing intervals for both are the same, that is, f^s. Doing this will simplify the control design.

To simplify notations, we use f^s to represent $f^s(\boldsymbol{r})$ where $\boldsymbol{r}$ is the state vector, that is, $\boldsymbol{r} = [r_1, r_2, \cdots, r_p]^T$. Using the same controller defined earlier in Eq. (6.21), the IT2 FLC output is given by

$$\boldsymbol{u}(t) = m_c \frac{\sum_{s=1}^{M} \underline{g}^s(\boldsymbol{r})\boldsymbol{u}_s(t)}{\sum_{s=1}^{M} \underline{g}^s(\boldsymbol{r})} + n_c \frac{\sum_{s=1}^{M} \overline{g}^s(\boldsymbol{r})\boldsymbol{u}_s(t)}{\sum_{i=1}^{M} \overline{g}^s(\boldsymbol{r})} \tag{6.103}$$

where $\underline{g}^s$ and $\overline{g}^s$ are short for $\underline{g}^s(\boldsymbol{r})$ and $\overline{g}^s(\boldsymbol{r})$ and represent, respectively, the lower and upper firing strengths of the controller. Different notations for the firing strengths of the plant and controller are used.

Before introducing the control problem, let us simplify the closed-loop expression of the plant controller. It is easy to see that the expressions for the state vector and output are given as follows, respectively:

$$\dot{\boldsymbol{x}}(t) = \frac{\sum_{i=1}^{M} f^s[\boldsymbol{A}_s\boldsymbol{x}(t) + \boldsymbol{B}_s\boldsymbol{u}(t) + \boldsymbol{D}_s\boldsymbol{w}(t)]}{\sum_{s=1}^{M} f^s} \tag{6.104}$$

and

$$y(t) = \frac{\sum_{s=1}^{M} f^s \boldsymbol{C}_s\boldsymbol{x}(t)}{\sum_{s=1}^{M} f^s} \tag{6.105}$$

Using the feedback design matrix, that is, $\boldsymbol{F}_s\boldsymbol{x}(t)$, Eq. (6.103) can be written as

$$\boldsymbol{u}(t) = m_c \frac{\sum_{s=1}^{M} \underline{g}^s \boldsymbol{F}_s\boldsymbol{x}(t)}{\sum_{s=1}^{M} \underline{g}^s} + n_c \frac{\sum_{s=1}^{M} \overline{g}^s \boldsymbol{F}_s\boldsymbol{x}(t)}{\sum_{s=1}^{M} \overline{g}^s} \tag{6.106}$$

The index s in $\sum_{s=1}^{M} f^s$, $\sum_{s=1}^{M} \underline{g}^s$, and $\sum_{s=1}^{M} \overline{g}^s$ is a dummy index. and it is therefore true that

$$\sum_{s=1}^{M} f^s = \sum_{i=1}^{M} f^i \tag{6.107}$$

$$\sum_{s=1}^{M} \underline{g}^s = \sum_{i=1}^{M} \underline{g}^i \tag{6.108}$$

$$\sum_{s=1}^{M} \overline{g}^s = \sum_{i=1}^{M} \overline{g}^i \tag{6.109}$$

Define

$$h^i \equiv \frac{f^i}{\sum_{s=1}^{M} f^s} \qquad \underline{k}^i \equiv \frac{\underline{g}^i}{\sum_{s=1}^{M} \underline{g}^s} \qquad \overline{k}^i \equiv \frac{\overline{g}^i}{\sum_{s=1}^{M} \overline{g}^s} \tag{6.110}$$

Observe that $\sum_{i=1}^{M} h^i = 1$, and $\sum_{i=1}^{M} \underline{k}^i = \sum_{i=1}^{M} \overline{k}^i = 1$.

Using Eq. (6.110), we can express $\dot{\boldsymbol{x}}(t), \boldsymbol{y}(t)$, and $\boldsymbol{u}(t)$ in shorter and more compact forms as follows:

$$\dot{\boldsymbol{x}}(t) = \sum_{i=1}^{M} h^i [\boldsymbol{A}_i \boldsymbol{x}(t) + \boldsymbol{B}_i \boldsymbol{u}(t) + \boldsymbol{D}_i \boldsymbol{w}(t)] \tag{6.111}$$

$$\boldsymbol{y}(t) = \sum_{i=1}^{M} h^i \boldsymbol{C}_i \boldsymbol{x}(t) \tag{6.112}$$

$$\boldsymbol{u}(t) = m_c \sum_{i=1}^{M} \underline{k}^i \boldsymbol{F}_i \boldsymbol{x}(t) + n_c \sum_{i=1}^{M} \overline{k}^i \boldsymbol{F}_i \boldsymbol{x}(t) \tag{6.113}$$

To find an expression for the plant dynamics, we need to substitute $\boldsymbol{u}(t)$ in Eq. (6.113) into Eq. (6.111). Observe that both $\dot{\boldsymbol{x}}(t)$ and $\boldsymbol{u}(t)$ have i as their summation index. In order to combine the two summations, we need to change one of the dummy indices to another index other than i. Therefore, we change the dummy index i in Eq. (6.113) to j, making it true that

$$\boldsymbol{u}(t) = m_c \sum_{j=1}^{M} \underline{k}^j \boldsymbol{F}_j \boldsymbol{x}(t) + n_c \sum_{j=1}^{M} \overline{k}^j \boldsymbol{F}_j \boldsymbol{x}(t) \tag{6.114}$$

If the dummy index j in the second term of Eq. (6.114) is changed to l, it is also true that

$$\boldsymbol{u}(t) = m_c \sum_{j=1}^{M} \underline{k}^j \boldsymbol{F}_j \boldsymbol{x}(t) + n_c \sum_{l=1}^{M} \overline{k}^l \boldsymbol{F}_l \boldsymbol{x}(t) \tag{6.115}$$

The reason for expressing $\boldsymbol{u}(t)$ in Eq. (6.114) using two indices (j and l) is based on being able to combine the terms of the series in Eqs. (6.111) and (6.114) and factor out h^i, $\underline{k}^j$, $\overline{k}^l$. Thus, substituting $\boldsymbol{u}(t)$ in Eq. (6.114) into Eq. (6.111), we can express $\dot{\boldsymbol{x}}(t)$ as

$$\dot{\boldsymbol{x}}(t) = \sum_{i=1}^{M} \sum_{j=1}^{M} \sum_{l=1}^{M} h^i \underline{k}^j \overline{k}^l [\boldsymbol{A}_i \boldsymbol{x}(t) + m_c \boldsymbol{B}_i \boldsymbol{F}_j \boldsymbol{x}(t) + n_c \boldsymbol{B}_i \boldsymbol{F}_l \boldsymbol{x}(t) + \boldsymbol{D}_i \boldsymbol{w}(t)] \tag{6.116}$$

6.7.2 Disturbance Rejection Problem and Solution

The disturbance rejection problem is to ensure the following inequality holds (Biglarbegian, 2012):

$$\frac{\|y(t)\|_2}{\|w(t)\|_2} \leq \alpha \tag{6.117}$$

for $\|w(t)\|_2 \neq 0$; where $\|.\|$ denotes a 2-norm. In other words, we need to find the conditions for which $\|y(t)\|_2 \leq \alpha.\|w(t)\|_2$. This ensures robust disturbance rejection.

Choose the standard quadratic Lyapunov function given as follows:

$$V(t) = \boldsymbol{x}^T(t)\boldsymbol{P}\boldsymbol{x}(t) \tag{6.118}$$

$\dot{V}(t)$ can be calculated as

$$\dot{V}(t) = \dot{\boldsymbol{x}}^T(t)\boldsymbol{P}\boldsymbol{x}(t) + \boldsymbol{x}^T(t)\boldsymbol{P}\dot{\boldsymbol{x}}(t) \tag{6.119}$$

Substituting $\dot{\boldsymbol{x}}(t)$ in Eq.(6.116) into Eq. (6.119), we get

$$\begin{aligned}\dot{V}(t) = &\sum_{i=1}^{M}\sum_{j=1}^{M}\sum_{l=1}^{M} h^i \underline{k}^j \overline{k}^l [\boldsymbol{A}_i\boldsymbol{x}(t) + m_c\boldsymbol{B}_i\boldsymbol{F}_j\boldsymbol{x}(t)n_c\boldsymbol{B}_i\boldsymbol{F}_l\boldsymbol{x}(t) + \boldsymbol{D}_i\boldsymbol{w}(t)]^T\boldsymbol{P}\boldsymbol{x}(t)\\ &+ \boldsymbol{x}^T(t)\boldsymbol{P}\sum_{i=1}^{M}\sum_{j=1}^{M}\sum_{l=1}^{M} h^i \underline{k}^j \overline{k}^l [\boldsymbol{A}_i\boldsymbol{x}(t) + m_c\boldsymbol{B}_i\boldsymbol{F}_j\boldsymbol{x}(t) + n_c\boldsymbol{B}_i\boldsymbol{F}_l\boldsymbol{x}(t) + \boldsymbol{D}_i\boldsymbol{w}(t)]\end{aligned} \tag{6.120}$$

We now state the following theorem, which will help us derive the conditions for robustness:

THEOREM 6.2 If there exists a symmetric positive-definite matrix $\boldsymbol{P}$ such that the following condition holds

$$\dot{\boldsymbol{x}}^T(t)\boldsymbol{P}\boldsymbol{x}(t) + \boldsymbol{x}^T(t)\boldsymbol{P}\dot{\boldsymbol{x}}(t) + \boldsymbol{y}^T(t)\boldsymbol{y}(t) - \alpha^2\boldsymbol{w}^T(t)\boldsymbol{w}(t) \leq \boldsymbol{0} \tag{6.121}$$

then the closed-loop system is uniformly ultimately bounded (UUB) stable and also condition (6.117) is satisfied (Biglarbegian, 2012).

Proof. Define β as

$$\beta \equiv \dot{V}(t) + \boldsymbol{y}^T(t)\boldsymbol{y}(t) - \alpha^2\boldsymbol{w}^T(t)\boldsymbol{w}(t) \tag{6.122}$$

Note that $\beta \leq 0$ is the assumption of the theorem, that is, inequalities (6.121) and (6.122) are equivalent. If we make $\beta \leq 0$ for $\forall t$, then for any $t_f > 0$, $\int_0^{t_f} \beta\, dt \leq 0$, which in turn means

$$\int_0^{t_f} [\dot{V}(t) + y^T(t)y(t) - \alpha^2 w^T(t)w(t)]dt$$
$$= \int_0^{t_f} [y^T(t)y(t) - \alpha^2 w^T(t)w(t)]dt + V(t_f) - V(0) \leq 0 \tag{6.123}$$

Note that Eq. (6.123) can be rewritten as

$$V(t_f) + \int_0^{\tau} [y^T(t)y(t) - \alpha^2 w^T(t)w(t)]dt \leq 0 \tag{6.124}$$

and since $V(t_f) > 0$, we conclude that the following inequality must hold:

$$y^T(t)y(t) - \alpha^2 w^T(t)w(t) \leq 0 \tag{6.125}$$

If we derive the conditions for which $\beta \leq 0$, the control objective is met. To do so, we can express β as follows:

$$\beta = \dot{x}^T(t)Px(t) + x^T(t)P\dot{x}(t) + y^T(t)y(t) - \alpha^2 w^T(t)w(t) \tag{6.126}$$

Using the equivalent expression for $\dot{x}(t)$, we can write Eq. (6.126) as

$$\beta = \sum_{i=1}^{M}\sum_{j=1}^{M}\sum_{l=1}^{M} h^i \underline{k}^j \overline{k}^l [A_i x(t) + m_c B_i F_j x(t) + n_c B_i F_l x(t) + D_i w(t)]^T P x(t)$$
$$+ x^T(t)P\sum_{i=1}^{M}\sum_{j=1}^{M}\sum_{l=1}^{M} h^i \underline{k}^j \overline{k}^l [A_i x(t) + m_c B_i F_j x(t) + n_c B_i F_l x(t) + D_i w(t)]$$
$$+ y^T(t)y(t) - \alpha^2 w^T(t)w(t) \tag{6.127}$$

Substituting $y(t)$ in Eq. (6.112) into Eq. (6.127) we get

$$\beta = \sum_{i=1}^{M}\sum_{j=1}^{M}\sum_{l=1}^{M} h^i \underline{k}^j \overline{k}^l [A_i x(t) + m_c B_i F_j x(t) + n_c B_i F_l x(t) + D_i w(t)]^T P x(t)$$
$$+ x^T(t)P\sum_{i=1}^{M}\sum_{j=1}^{M}\sum_{l=1}^{M} h^i \underline{k}^j \overline{k}^l [A_i x(t) + m_c B_i F_j x(t) + n_c B_i F_l x(t) + D_i w(t)]$$
$$+ x^T(t)\sum_{i=1}^{M}\sum_{r=1}^{M} h^i h^r C_i^T C_r x(t) - \alpha^2 w^T(t)w(t) \tag{6.128}$$

Defining $s_{ijlr} \equiv h^i h^r \underline{k}^j \overline{k}^l$, we can express β as (Biglarbegian, 2012)

$$\begin{aligned}\beta = &\sum_{i,j,l,r=1}^{M} s_{ijlr}\boldsymbol{x}^T(t)[\boldsymbol{A}_i + \boldsymbol{B}_i(m_c\boldsymbol{F}_j + n_c\boldsymbol{F}_l)]^T\boldsymbol{P}\boldsymbol{x}(t)\\ &+ \sum_{i,j,l,r=1}^{M} s_{ijlr}\boldsymbol{x}^T(t)\boldsymbol{P}[\boldsymbol{A}_i + \boldsymbol{B}_i(m_c\boldsymbol{F}_j + n_c\boldsymbol{F}_l)]\\ &+ \sum_{i,j,l,r=1}^{M} s_{ijlr}\boldsymbol{w}^T(t)\boldsymbol{D}_i^T\boldsymbol{P}\boldsymbol{x}(t) + \boldsymbol{x}^T(t)\boldsymbol{P}\sum_{i,j,l,r=1}^{M} s_{ijlr}\boldsymbol{D}_i\boldsymbol{w}(t)\\ &+ \boldsymbol{x}^T(t)\sum_{i,j,l,r=1}^{M} s_{ijlr}\boldsymbol{C}_i^T\boldsymbol{C}_r\boldsymbol{x}(t) - \sum_{i,j,l,r=1}^{M} s_{ijlr}\alpha^2\boldsymbol{w}^T(t)\boldsymbol{w}(t)\end{aligned} \tag{6.129}$$

We can even express β more concisely as follows (Biglarbegian, 2012):

$$\beta = \begin{bmatrix}\boldsymbol{x}^T(t) & \boldsymbol{w}^T(t)\end{bmatrix}\boldsymbol{Z}\begin{bmatrix}\boldsymbol{x}(t)\\ \boldsymbol{w}(t)\end{bmatrix} \tag{6.130}$$

where $\boldsymbol{Z}$ is given by

$$\boldsymbol{Z} = \sum_{i,j,l,r=1}^{M} s_{ijlr}\begin{bmatrix}\boldsymbol{Z}'_{11} & \boldsymbol{Z}_{12}\\ \boldsymbol{Z}_{21} & \boldsymbol{Z}_{22}\end{bmatrix} \tag{6.131}$$

In which

$$\boldsymbol{Z}'_{11} = [\boldsymbol{A}_i + \boldsymbol{B}_i(m_c\boldsymbol{F}_j + n_c\boldsymbol{F}_l)]^T\boldsymbol{P} + \boldsymbol{P}[\boldsymbol{A}_i + \boldsymbol{B}_i(m_c\boldsymbol{F}_j + n_c\boldsymbol{F}_l)] + \boldsymbol{C}_i^T\boldsymbol{C}_r \tag{6.132}$$

$$\boldsymbol{Z}_{12} = \tfrac{1}{4}\boldsymbol{P}(\boldsymbol{D}_i + \boldsymbol{D}_j + \boldsymbol{D}_l + \boldsymbol{D}_r) \tag{6.133}$$

$$\boldsymbol{Z}_{21} = \tfrac{1}{4}(\boldsymbol{D}_i + \boldsymbol{D}_j + \boldsymbol{D}_l + \boldsymbol{D}_r)^T\boldsymbol{P} \tag{6.134}$$

$$\boldsymbol{Z}_{22} = -\alpha^2\mathbf{I} \tag{6.135}$$

If $\boldsymbol{Z}$ in Eq. (6.131) is negative definite, that is, $\boldsymbol{Z} \leq \boldsymbol{0}$, then the control objective is achieved. In the following, we further work on the expression for $\boldsymbol{Z}$ to obtain a simpler condition for satisfying the control objective.

Using Schur's complement (Boyd et al., 1994), $\boldsymbol{Z}$ can be simply written as

$$\boldsymbol{Z} = \begin{bmatrix}\left(\sum_{i,j,l=1}^{M} s_{ijlr}\boldsymbol{Z}_{11}\right) & \left(\sum_{i=1}^{M} h_i\boldsymbol{Z}_{12}\right)\\ \left(\sum_{i=1}^{M} h_i\boldsymbol{Z}_{21}\right) & \boldsymbol{Z}_{22}\end{bmatrix} + \begin{bmatrix}\left(\sum_{i=1}^{M} h_i\boldsymbol{C}_i^T\right)\\ \boldsymbol{0}\end{bmatrix}\begin{bmatrix}\left(\sum_{i=1}^{M} h_i\boldsymbol{C}_i\right) & \boldsymbol{0}\end{bmatrix} \tag{6.136}$$

where $Z_{11} = Z'_{11} - C_i^T C_r$. $Z \leq 0$ is equivalently expressed as

$$\sum_{i,j,l,r=1}^{M} s_{ijlr} \begin{bmatrix} Z_{11} & Z_{12} & -\frac{1}{2}(C_i + C_r)^T \\ Z_{21} & Z_{22} & 0 \\ -\frac{1}{2}(C_i + C_r) & 0 & -I \end{bmatrix} \leq 0 \tag{6.137}$$

Because $h_i, h_j, \underline{k}^l, \overline{k}^r \geq 0$, Eq. (6.137) is equivalent to the following LMI:

$$\begin{bmatrix} Z_{11} & Z_{12} & -\frac{1}{2}(C_i + C_r)^T \\ Z_{21} & Z_{22} & 0 \\ -\frac{1}{2}(C_i + C_r) & 0 & -I \end{bmatrix} \leq 0 \tag{6.138}$$

Earlier in Section 6.4 we introduced G_{ijl} to derive simpler stability criteria. Because of the symmetric properties of G_{ijl} under summation, the LMIs required for stability can be reduced to (a) when $i \leq j$ and $j = l$ and (b) when $j \leq l$ for $\forall i$. As a result, we can further reduce the number of LMIs. Therefore, the following two sets of LMIs are needed:

$$\begin{bmatrix} Z_{11}^1 & Z_{12}^1 & Z_{13}^1 \\ (Z_{12}^1)^T & Z_{22}^1 & 0 \\ (Z_{13}^1)^T & 0 & -I \end{bmatrix} \leq 0 \tag{6.139}$$

where $Z_{11}^1 \equiv -(G_{ijj}^T P + PG_{ijj})$, $Z_{12}^1 \equiv -\frac{1}{4}P(D_i + 2D_j + D_r)$, $Z_{13}^1 \equiv -\mathbf{T}_{12}$, and $Z_{22}^1 \equiv -\alpha^2 I$; for $i \leq j$.

And

$$\begin{bmatrix} Z_{11}^2 & Z_{12}^2 & Z_{13}^2 \\ (Z_{12}^2)^T & Z_{22}^2 & 0 \\ (Z_{13}^2)^T & 0 & -I \end{bmatrix} \leq 0 \tag{6.140}$$

where $Z_{11}^2 \equiv -(G_{ijl}^T P + PG_{ijl})$, $Z_{12}^2 \equiv -\frac{1}{4}P(D_i + D_j + D_l + D_r)$, $Z_{13}^2 \equiv -\mathbf{T}_{12}$, and $Z_{22}^2 \equiv -\alpha^2 I$; for $j \leq l$.

Note. Similar to Section 6.4, performing a parallel analysis produces the final LMIs for discrete systems that are given as

$$\begin{bmatrix} Z_{11}^1 & 0 & Z_{13}^1 & Z_{14}^1 \\ 0 & Z_{22}^1 & Z_{23}^1 & 0 \\ (Z_{13}^1)^T & (Z_{23}^1)^T & Z_{33}^1 & 0 \\ (Z_{14}^1)^T & 0 & 0 & -I \end{bmatrix} \leq 0 \tag{6.141}$$

where $\boldsymbol{Z}_{11}^1 \equiv -\boldsymbol{P}$, $\boldsymbol{Z}_{13}^1 \equiv -\frac{1}{2}(\boldsymbol{G}_{ijj} + \boldsymbol{G}_{jii})^T$, $\boldsymbol{Z}_{14}^1 \equiv -\frac{1}{2}(\boldsymbol{C}_i + \boldsymbol{C}_r)^T$, $\boldsymbol{Z}_{22}^1 \equiv -\alpha^2 \boldsymbol{I}$, $\boldsymbol{Z}_{23}^1 \equiv -\frac{1}{4}(\boldsymbol{D}_i + 2\boldsymbol{D}_j + \boldsymbol{D}_r)^T$, and $\boldsymbol{Z}_{33}^1 \equiv -\boldsymbol{P}^{-1}$ for $i \le j$.

As well

$$\begin{bmatrix} \boldsymbol{Z}_{11}^2 & \boldsymbol{0} & \boldsymbol{Z}_{13}^2 & \boldsymbol{Z}_{14}^2 \\ \boldsymbol{0} & \boldsymbol{Z}_{22}^2 & \boldsymbol{Z}_{23}^2 & \boldsymbol{0} \\ (\boldsymbol{Z}_{13}^2)^T & (\boldsymbol{Z}_{23}^2)^T & \boldsymbol{Z}_{33}^2 & \boldsymbol{0} \\ (\boldsymbol{Z}_{14}^2)^T & \boldsymbol{0} & \boldsymbol{0} & -\boldsymbol{I} \end{bmatrix} \le \boldsymbol{0} \tag{6.142}$$

where $\boldsymbol{Z}_{11}^2 \equiv -\boldsymbol{P}$, $\boldsymbol{Z}_{13}^2 \equiv -\frac{1}{2}(\boldsymbol{G}_{ijl} + \boldsymbol{G}_{ilj})^T$, $\boldsymbol{Z}_{14}^2 \equiv -\frac{1}{2}(\boldsymbol{C}_i + \boldsymbol{C}_r)^T$, $\boldsymbol{Z}_{22}^2 \equiv -\alpha^2 \boldsymbol{I}$, and $\boldsymbol{Z}_{23}^2 \equiv -\frac{1}{4}(\boldsymbol{D}_i + \boldsymbol{D}_j + \boldsymbol{D}_l + \boldsymbol{D}_r)^T$ for $j \le l$.

6.7.3 Robust Control Example

Example 6.3 ***Robust control (the code for this example can be found online on the Wiley website of the textbook).*** This section presents an example to demonstrate how the derived LMIs used for robust disturbance rejection can control a design problem. Assume an IT2 TSK FLC is modeled as follows (The membership functions are shown in Fig. 6.11):

Rule 2: If $x_1(k)$ is "around 0," then $\boldsymbol{x}(k+1) = \boldsymbol{A}_1\boldsymbol{x}(k) + \boldsymbol{B}_1\boldsymbol{u}(k) + \boldsymbol{D}_1\boldsymbol{w}(k)$.
Rule 2: If $x_1(t)$ is "around 1," then $\boldsymbol{x}(k+1) = \boldsymbol{A}_2\boldsymbol{x}(k) + \boldsymbol{B}_2\boldsymbol{u}(k) + \boldsymbol{D}_2\boldsymbol{w}(k)$.

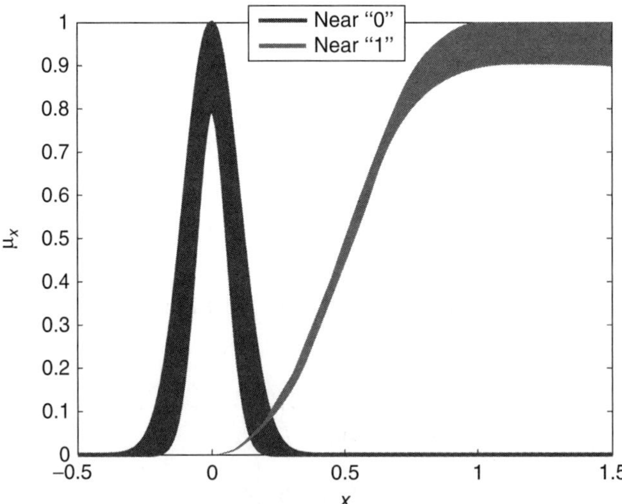

Figure 6.11 Membership functions (Biglarbegian, 2012; © 2012, IEEE).

where the matrices for the plant and controller are given by

$$A_1 = \begin{bmatrix} 0.9 & 0.636 \\ 0.737 & 0.1 \end{bmatrix} \qquad A_2 = \begin{bmatrix} 0.461 & 0.547 \\ 0.199 & 0.243 \end{bmatrix}$$

$$B_1 = \begin{bmatrix} 184 \\ 0.469 \end{bmatrix} \qquad B_2 = \begin{bmatrix} 0.164 \\ 0.988 \end{bmatrix} \qquad C_1 = \begin{bmatrix} 1 & 0 \\ 0 & 1 \end{bmatrix}$$

$$D_1 = \begin{bmatrix} 0.1 & 0.837 \\ 0.214 & 0.236 \end{bmatrix} \qquad C_1 = C_2, \quad D_1 = D_2$$

After solving the LMIs in Eqs. (6.141) and (6.142), the positive-definite matrix $\boldsymbol{P}$ is computed as

$$P = \begin{bmatrix} 0.2104 & -0.0264 \\ -0.0264 & 5.3830 \end{bmatrix}$$

The existence of $\boldsymbol{P}$ means LMIs in Eqs. (6.141) and (6.142) are satisfied. In other words, the LMIs are feasible, implying that the robust control objective given in Eq. (6.117) is met.

6.8 SUMMARY

In this chapter we used TSK model structure for the design of IT2 TSK FLCs. To design and analyze IT2 FLCs mathematically, a closed form is needed. We first introduced a novel inference engine that has a closed form. This inference engine has a simple structure and enables mathematical analysis of IT2 FLCs as well as easy real-time control implementations.

Starting with stability analysis, adaptive IT2 TSK FLCs for robotic arms were developed as well as systematic methods for robust control design. The benchmark examples and a real-time implementation on a modular and reconfigurable robotic system demonstrate how the developed theory can be effectively used in real applications. The mathematical methodologies presented in this chapter play a key role in analytical and systematic design as well as analyses of IT2 TSK FLCs. This design is theoretically sound and can be applied to nonlinear plants with uncertainty.

APPENDIX[9]

In this appendix, we show that $U_{m-n}(\boldsymbol{x})$ in Eq. (6.16) is a simplified version of the WM UBs. We prove this only for a continuous IT2 TSK model. To avoid repetition,

[9]Much of the material in this section is taken directly from Biglarbegian et al. (2010; © 2010, IEEE).

very similar analyses can be performed on the discrete case. Section 3.4 shows the general form of WM UBs is given by Eq. (3.88), that is,

$$U_{\mathrm{WM}}(\boldsymbol{x}) = \frac{1}{2}\left[\frac{\underline{u}_l(\boldsymbol{x}) + \overline{u}_l(\boldsymbol{x})}{2} + \frac{\underline{u}_r(\boldsymbol{x}) + \overline{u}_r(\boldsymbol{x})}{2}\right] \tag{6.143}$$

where $\underline{u}_l(\boldsymbol{x}), \overline{u}_l(\boldsymbol{x}), \underline{u}_r(\boldsymbol{x})$, and $\overline{u}_r(\boldsymbol{x})$ are given by Eqs. (3.83)–(3.86), respectively. We apply this general form of WM UBs to Eqs. (6.11)–(6.15).

Since the consequent part of the IT2 TSK are crisp numbers, that is, $u_l^s = u_r^s = u_s$, the boundaries defined by Eqs. (3.79)–(3.82) reduce to the following two equations:

$$u^{(0)}(\boldsymbol{x}) = \frac{\sum_{s=1}^{M} \underline{f}^s(\boldsymbol{x}) u_s}{\sum_{s=1}^{M} \underline{f}^s(\boldsymbol{x})} \tag{6.144}$$

$$u^{(M)}(\boldsymbol{x}) = \frac{\sum_{s=1}^{M} \overline{f}^s(\boldsymbol{x}) u_s}{\sum_{s=1}^{M} \overline{f}^s(\boldsymbol{x})} \tag{6.145}$$

Without loss of generality, assume $u^{(M)}(\boldsymbol{x}) > u^{(0)}(\boldsymbol{x})$ [$U_{\mathrm{WM}}(\boldsymbol{x})$ is invariant to $u^{(M)}(\boldsymbol{x}) > u^{(0)}(\boldsymbol{x})$]; therefore, Eqs. (3.83)–(3.86) can be written as

$$\overline{u}_l(\boldsymbol{x}) = u^{(0)}(\boldsymbol{x}) = \frac{\sum_{s=1}^{M} \underline{f}^s(\boldsymbol{x}) u_s}{\sum_{s=1}^{M} \underline{f}^s(\boldsymbol{x})} \tag{6.146}$$

$$\underline{u}_r(\boldsymbol{x}) = u^{(M)}(\boldsymbol{x}) = \frac{\sum_{s=1}^{M} \overline{f}^s(\boldsymbol{x}) u_s}{\sum_{s=1}^{M} \overline{f}^s(\boldsymbol{x})} \tag{6.147}$$

$$\begin{aligned} \underline{u}_l(\boldsymbol{x}) = {} & \frac{\sum_{s=1}^{M} \underline{f}^s(\boldsymbol{x}) u_s}{\sum_{s=1}^{M} \underline{f}^s(\boldsymbol{x})} \\ & - \left[\frac{\sum_{s=1}^{M}\left(\overline{f}^s(\boldsymbol{x}) - \underline{f}^s(\boldsymbol{x})\right)}{\sum_{s=1}^{M} \underline{f}^s(\boldsymbol{x}) . \sum_{s=1}^{M} \overline{f}^s(\boldsymbol{x})} \times \frac{\sum_{s=1}^{M} \underline{f}^s(\boldsymbol{x})(u_s - u_1) . \sum_{s=1}^{M} \overline{f}^s(\boldsymbol{x})(u_M - u_s)}{\sum_{s=1}^{M} \underline{f}^s(\boldsymbol{x})(u_s - u_1) + \sum_{s=1}^{M} \overline{f}^s(\boldsymbol{x})(u_M - u_s)}\right] \end{aligned} \tag{6.148}$$

$$\begin{aligned} \overline{u}_r(\boldsymbol{x}) = {} & \frac{\sum_{s=1}^{M} \overline{f}^s(\boldsymbol{x}) u_s}{\sum_{s=1}^{M} \overline{f}^s(\boldsymbol{x})} \\ & + \left[\frac{\sum_{s=1}^{M}\left(\overline{f}^s(\boldsymbol{x}) - \underline{f}^s(\boldsymbol{x})\right)}{\sum_{s=1}^{M} \underline{f}^s(\boldsymbol{x}) . \sum_{s=1}^{M} \overline{f}^s(\boldsymbol{x})} \times \frac{\sum_{s=1}^{M} \overline{f}^s(\boldsymbol{x})(u_s - u_1) . \sum_{s=1}^{M} \underline{f}^s(\boldsymbol{x})(u_M - u_s)}{\sum_{s=1}^{M} \overline{f}^s(\boldsymbol{x})(u_s - u_1) + \sum_{s=1}^{M} \underline{f}^s(\boldsymbol{x})(u_M - u_s)}\right] \end{aligned} \tag{6.149}$$

Using Eqs. (6.146)–(6.149), it is straightforward to show that $U_{\text{WM}}(\boldsymbol{x})$ in Eq. (6.143) can be expressed as

$$
\begin{aligned}
U_{\text{WM}}(\boldsymbol{x}) = {} & \frac{1}{2}\left(\frac{\sum_{s=1}^{M} \underline{f}^{s}(\boldsymbol{x})\, u_s}{\sum_{s=1}^{M} \underline{f}^{s}(\boldsymbol{x})} + \frac{\sum_{s=1}^{M} \overline{f}^{s}(\boldsymbol{x}) u_s}{\sum_{s=1}^{M} \overline{f}^{s}(\boldsymbol{x})}\right) \\
& - \frac{1}{4}\left[\frac{\sum_{s=1}^{M}(\overline{f}^{s}(\boldsymbol{x}) - \underline{f}^{s}(\boldsymbol{x}))}{\sum_{s=1}^{M} \underline{f}^{s}(\boldsymbol{x}) . \sum_{s=1}^{M} \overline{f}^{s}(\boldsymbol{x})} \times \frac{\sum_{s=1}^{M} \underline{f}^{s}(\boldsymbol{x})(u_s - u_1) . \sum_{s=1}^{M} \overline{f}^{s}(\boldsymbol{x})(u_M - u_s)}{\sum_{s=1}^{M} \underline{f}^{s}(\boldsymbol{x})(u_s - u_1) + \sum_{s=1}^{M} \overline{f}^{s}(\boldsymbol{x})(u_M - u_s)}\right] \\
& + \frac{1}{4}\left[\frac{\sum_{s=1}^{M}(\overline{f}^{s}(\boldsymbol{x}) - \underline{f}^{s}(\boldsymbol{x}))}{\sum_{s=1}^{M} \underline{f}^{s}(\boldsymbol{x}) . \sum_{s=1}^{M} \overline{f}^{s}(\boldsymbol{x})} \times \frac{\sum_{s=1}^{M} \overline{f}^{s}(\boldsymbol{x})(u_s - u_1) . \sum_{s=1}^{M} \underline{f}^{s}(\boldsymbol{x})(u_M - u_s)}{\sum_{s=1}^{M} \overline{f}^{s}(\boldsymbol{x})(u_s - u_1) + \sum_{s=1}^{M} \underline{f}^{s}(\boldsymbol{x})(u_M - u_s)}\right]
\end{aligned}
\tag{6.150}
$$

$U_{\text{WM}}(\boldsymbol{x})$ can be computed without having to perform TR, and therefore $U_{\text{WM}}(\boldsymbol{x})$ can be considered a viable alternative to using Eqs. (6.15) and (6.14).

We apply $U_{\text{WM}}(\boldsymbol{x})$ to $U_{\text{TSK/A2-C0}}(\boldsymbol{x})$ using the following model that appears in the consequent of rule s in Eq. (6.4):

$$
u_s = \sum_{i=1}^{p} c_i^s x_i \tag{6.151}
$$

It follows that

$$
u_s - u_1 = \sum_{i=1}^{p} (c_i^s - c_i^1) x_i \equiv \sum_{i=1}^{p} v_{s,p} c_i^s x_i \tag{6.152}
$$

$$
u_M - u_s = \sum_{i=1}^{p} (c_i^M - c_i^s) x_i \equiv \sum_{i=1}^{p} w_{s,p} c_i^s x_i \tag{6.153}
$$

where

$$
v_{s,p} \equiv \frac{c_i^s - c_i^1}{c_i^s} \tag{6.154}
$$

$$
w_{s,p} \equiv \frac{c_i^M - c_i^s}{c_i^s} \tag{6.155}
$$

Substituting Eqs. (6.151)–(6.153) into Eq. (6.150), $U_{\text{WM}}(\boldsymbol{x})$ can be expressed as

$$
U_{\text{WM}}(\boldsymbol{x}) = \frac{1}{2}\frac{\sum_{s=1}^{M} \underline{f}^{s}(\boldsymbol{x}) \left(\sum_{i=1}^{p} c_i^s x_i\right)}{\sum_{s=1}^{M} \underline{f}^{s}(\boldsymbol{x})} + \frac{1}{2}\frac{\sum_{s=1}^{M} \overline{f}^{s}(\boldsymbol{x}) \left(\sum_{i=1}^{p} c_i^s x_i\right)}{\sum_{s=1}^{M} \overline{f}^{s}(\boldsymbol{x})} + \alpha(\boldsymbol{x}) + \beta(\boldsymbol{x}) \tag{6.156}
$$

where

$$\alpha(\boldsymbol{x}) = -\frac{1}{4}\frac{\sum_{s=1}^{M}(\bar{f}^{s}(\boldsymbol{x}) - \underline{f}^{s}(\boldsymbol{x}))}{\sum_{s=1}^{M}\underline{f}^{s}(\boldsymbol{x}).\sum_{s=1}^{M}\bar{f}^{s}(\boldsymbol{x})}$$

$$\times \frac{\sum_{s=1}^{M}\underline{f}^{s}(\boldsymbol{x})\left(\sum_{i=1}^{p}\upsilon_{s,p}c_{i}^{s}x_{i}\right).\sum_{s=1}^{M}\bar{f}^{s}(\boldsymbol{x})\left(\sum_{i=1}^{p}w_{s,p}c_{i}^{s}x_{i}\right)}{\sum_{s=1}^{M}\underline{f}^{s}(\boldsymbol{x})\left(\sum_{i=1}^{p}\upsilon_{s,p}c_{i}^{s}x_{i}\right) + \sum_{s=1}^{M}\bar{f}^{s}(\boldsymbol{x})\left(\sum_{i=1}^{p}w_{s,p}c_{i}^{s}x_{i}\right)} \tag{6.157}$$

$$\beta(\boldsymbol{x}) = \frac{1}{4}\frac{\sum_{s=1}^{M}(\bar{f}^{s}(\boldsymbol{x}) - \underline{f}^{s}(\boldsymbol{x}))}{\sum_{s=1}^{M}\underline{f}^{s}(\boldsymbol{x}).\sum_{s=1}^{M}\bar{f}^{s}(\boldsymbol{x})}$$

$$\times \frac{\sum_{s=1}^{M}\bar{f}^{s}(\boldsymbol{x})\left(\sum_{i=1}^{p}\upsilon_{s,p}c_{i}^{s}x_{i}\right).\sum_{s=1}^{M}\underline{f}^{s}(\boldsymbol{x})\left(\sum_{i=1}^{p}w_{s,p}c_{i}^{s}x_{i}\right)}{\sum_{s=1}^{M}\bar{f}^{s}(\boldsymbol{x})\left(\sum_{i=1}^{p}\upsilon_{s,p}c_{i}^{s}x_{i}\right) + \sum_{s=1}^{M}\underline{f}^{s}(\boldsymbol{x})\left(\sum_{i=1}^{p}w_{s,p}c_{i}^{s}x_{i}\right)} \tag{6.158}$$

Using Eqs. (6.159) and (6.160), $\alpha(\boldsymbol{x})$ and $\beta(\boldsymbol{x})$ can be expressed as nonlinear functions of the upper and lower firing levels of each rule, as well as the input states, that is,

$$\alpha(\boldsymbol{x}) = g_1(\underline{f}^{s}(\boldsymbol{x}), \bar{f}^{s}(\boldsymbol{x}), \boldsymbol{x}, \upsilon_{s,p}, w_{s,p}) \times \frac{\sum_{s=1}^{M}\underline{f}^{s}(\boldsymbol{x})\left(\sum_{i=1}^{p}c_{i}^{s}x_{i}\right)}{\sum_{s=1}^{M}\underline{f}^{s}(\boldsymbol{x})} \tag{6.159}$$

$$\beta(\boldsymbol{x}) = g_2(\underline{f}^{s}(\boldsymbol{x}), \bar{f}^{s}(\boldsymbol{x}), \boldsymbol{x}, \upsilon_{s,p}, w_{s,p}) \times \frac{\sum_{s=1}^{M}\bar{f}^{s}(\boldsymbol{x})\left(\sum_{i=1}^{p}c_{i}^{s}x_{i}\right)}{\sum_{s=1}^{M}\bar{f}^{s}(\boldsymbol{x})} \tag{6.160}$$

where functions $g_1$and g_2 are given by[10]

$$g_1 = -\frac{1}{4}\frac{\sum_{s=1}^{M}(\bar{f}^{s}(\boldsymbol{x}) - \underline{f}^{s}(\boldsymbol{x}))}{\left[\sum_{s=1}^{M}\underline{f}^{s}(\boldsymbol{x})\left(\sum_{i=1}^{p}c_{i}^{s}x_{i}\right)\right]\sum_{s=1}^{M}\bar{f}^{s}(\boldsymbol{x})}$$

$$\times \frac{\sum_{s=1}^{M}\left[\underline{f}^{s}(\boldsymbol{x})\left(\sum_{i=1}^{p}\upsilon_{s,p}c_{i}^{s}x_{i}\right)\right]\sum_{s=1}^{M}\left[\bar{f}^{s}(\boldsymbol{x})\left(\sum_{i=1}^{p}w_{s,p}c_{i}^{s}x_{i}\right)\right]}{\sum_{s=1}^{M}\left[\underline{f}^{s}(\boldsymbol{x})\left(\sum_{i=1}^{p}\upsilon_{s,p}c_{i}^{s}x_{i}\right)\right] + \sum_{s=1}^{M}\left[\bar{f}^{s}(\boldsymbol{x})(\sum_{i=1}^{p}w_{s,p}c_{i}^{s}x_{i})\right]} \tag{6.161}$$

[10]In order to simplify the notation, in the rest of the derivation g_1 and g_2 are short for $g_1(\underline{f}^{s}(\boldsymbol{x}), \bar{f}^{s}(\boldsymbol{x}), \boldsymbol{x}, \upsilon_{s,p}, w_{s,p})$ and $g_2(\underline{f}^{s}(\boldsymbol{x}), \bar{f}^{s}(\boldsymbol{x}), \boldsymbol{x}, \upsilon_{s,p}, w_{s,p})$.

$$g_2 = \frac{1}{4} \frac{\sum_{s=1}^{M} \left(\overline{f}^s(\boldsymbol{x}) - \underline{f}^s(\boldsymbol{x}) \right)}{\left[\sum_{s=1}^{M} \overline{f}^s(\boldsymbol{x}) \left(\sum_{i=1}^{p} c_i^s x_i \right) \right] \sum_{s=1}^{M} \underline{f}^s(\boldsymbol{x})}$$
$$\times \frac{\sum_{s=1}^{M} \left[\overline{f}^s(\boldsymbol{x}) \left(\sum_{i=1}^{p} v_{s,p} c_i^s x_i \right) \right] \sum_{s=1}^{M} \left[\underline{f}^s(\boldsymbol{x}) \left(\sum_{i=1}^{p} w_{s,p} c_i^s x_i \right) \right]}{\sum_{s=1}^{M} \left[\overline{f}^s(\boldsymbol{x}) (\sum_{i=1}^{p} v_{s,p} c_i^s x_i) \right] + \sum_{s=1}^{M} \left[\underline{f}^s(\boldsymbol{x}) (\sum_{i=1}^{p} w_{s,p} c_i^s x_i) \right]} \qquad (6.162)$$

Using Eqs. (6.159)–(6.162), $U_{\mathrm{WM}}(\boldsymbol{x})$ in Eq. (6.156) can be written as

$$U_{\mathrm{WM}}(\boldsymbol{x}) = \frac{\sum_{s=1}^{M} \underline{f}^s(\boldsymbol{x}) \left(\frac{1}{2} \sum_{i=1}^{p} c_i^s x_i \right)}{\sum_{s=1}^{M} \underline{f}^s(\boldsymbol{x})} + \frac{\sum_{s=1}^{M} \overline{f}^s(\boldsymbol{x}) \left(\frac{1}{2} \sum_{i=1}^{p} c_i^s x_i \right)}{\sum_{s=1}^{M} \overline{f}^s(\boldsymbol{x})}$$
$$+ g_1 \times \frac{\sum_{s=1}^{M} \underline{f}^s(\boldsymbol{x}) \left(\sum_{i=1}^{p} c_i^s x_i \right)}{\sum_{s=1}^{M} \underline{f}^s(\boldsymbol{x})} + g_2 \times \frac{\sum_{s=1}^{M} \overline{f}^s(\boldsymbol{x}) \left(\sum_{i=1}^{p} c_i^s x_i \right)}{\sum_{s=1}^{M} \overline{f}^s(\boldsymbol{x})} \qquad (6.163)$$

Combining the first and third terms, and second and fourth terms of $U_{\mathrm{WM}}(\boldsymbol{x})$, Eq. (6.163) can be rewritten as

$$U_{\mathrm{WM}}(\boldsymbol{x}) = \frac{\sum_{s=1}^{M} \underline{f}^s(\boldsymbol{x}) \left(\sum_{i=1}^{p} \overbrace{\left(\frac{1}{2} + g_1 \right)}^{m} [c_i^s x_i] \right)}{\sum_{s=1}^{M} \underline{f}^s(\boldsymbol{x})} + \frac{\sum_{s=1}^{M} \overline{f}^s(\boldsymbol{x}) \left(\sum_{i=1}^{p} \overbrace{\left(\frac{1}{2} + g_2 \right)}^{n} [c_i^s x_i] \right)}{\sum_{s=1}^{M} \overline{f}^s(\boldsymbol{x})} \qquad (6.164)$$

Comparing Eqs. (6.164) and (6.16), remembering that u_s is given by Eq. (6.151), it can be seen that m and n correspond to $\left(\frac{1}{2} + g_1 \right)$ and $\left(\frac{1}{2} + g_2 \right)$, respectively, and hence $U_{\mathrm{WM}}(\boldsymbol{x})$ simplifies to $U_{m-n}(\boldsymbol{x})$.

CHAPTER 7

Looking into the Future

7.1 INTRODUCTION

Writing a multiauthored book about T2 FL control has been very worthwhile but has also been very challenging. It has been very worthwhile because all of the authors are in agreement that the time is right to bring different perspectives about T2 FL control together, as we have done in this book, so that the FL control community has all of this material in one place. It has been very challenging because there are different perspectives on T2 FL control, some very mathematically rigorous and others not so much.

This last chapter focuses on where we feel T2 FL control should be heading. Unlike all of the previous chapters, in this chapter the authors of each section are identified and have an opportunity to express their viewpoints on where they feel T2 FL control should be heading.

7.2 WILLIAM MELEK AND HAO YING LOOK INTO THE FUTURE

This book presents the foundations of IT2 FL control and provides several methodologies for modeling and control that utilize this advanced approach. From the six chapters, it should be clear to the reader by now that IT2 FLCs are (1) nonlinear controllers (with complicated input–output relations) and (2) more complex than their T1 counterparts in terms of the mathematical descriptions of their input–output relations and the number of their design parameters. Consequently, analyzing or designing an IT2 FLC system is substantially more challenging than analyzing or designing a T1 FLC system. With these facts in mind, we would like to point out the following research directions considered to be important for the future of IT2 FL control.

In the 1980s, the following question had to be faced by the fuzzy control community: *When should a T1 FLC be used instead of a conventional controller?* Because the advantages and disadvantages of T1 FL control, with respect to those of conventional control, were relatively easy to determine and understand, that question was not too difficult to be settled. The advantages include no need for the system's

Introduction to Type-2 Fuzzy Logic Control: Theory and Applications, First Edition.
Jerry M. Mendel, Hani Hagras, Woei-Wan Tan, William W. Melek, and Hao Ying.

mathematical model, as well as the ability to incorporate expert knowledge and experience in the form of fuzzy rules and fuzzy sets. The disadvantages include more difficult to construct and tune a T1 FLC, and more expensive to implement in hardware. Of course, T1 FLC does not and cannot replace conventional control, linear or nonlinear; instead, it complements conventional control rather nicely.

By the same token, it should not be difficult to understand that IT2 FL control will not, and cannot, replace either T1 FL control or conventional control. The three control methodologies are complementary. Arguably, one of the most important research directions is to develop a theory capable of determining whether or not an IT2 FLC should be used for any given practical control application, that is, a theory is needed that can be used ahead of time to determine whether an IT2 FLC should be employed as opposed to using a T1 FLC.

It is important that such a theory be simple and effective so that it can be used by a control practitioner who may know some thing about T1 FL control but has little knowledge about IT2 FL control (it is not very realistic to assume that someone knows nothing about T1 FL control and is considering to use IT2 FL control).

This theory should not be simulation based because a system's accurate mathematical model is, realistically speaking, always nonlinear and thus is very difficult to obtain in practice. This theory should also not be heavily reliant on trial-and-error efforts because it is not only costly but it is also risky to experiment with a real system (such as, e.g., systems in the nuclear industry and human physiological systems).

In practice, IT2 FL control may have to prove itself superior to both T1 FL control and conventional control for a particular control problem before it will actually be used. Because T1 FL control and conventional control are able to deliver satisfactory solutions for so many different practical control problems, defining the niche applications that require the distinct merits of IT2 FL control is a critically important but technically challenging area for study.

Another important factor that one has to keep in mind is that a real-world control application typically seeks the simplest and least expensive hardware/software solution that satisfies the technical specifications imposed by the customer or user. This is why PID control, with only three design parameters, all of which can be tuned manually in an intuitive manner, has become the most popular control strategy since its inception, dating back to the preelectronic period. It currently dominates about 90% of industrial processes worldwide (O'Dwyer, 2003) despite the availability of numerous more advanced and better (at least in theory) controllers developed in the past dozens of years (e.g., optimal controllers and robust controllers). A T2 FLC will not be used unless its added structural complexity and additional design parameters (as compared to a T1 FLC) can be reasonably justified by demonstrated *significant* gains in control performance (e.g., better transition control response and more robust performances in the presence of noise and disturbances to the system). Research has been under way to address the question of when IT2 FL control can be used for substantial performance improvement, and more and more publications are appearing about this.

As is evident by trends in the recent literature, another important research direction is to extend the analysis and design techniques that have been developed for various T1 FLCs and systems to IT2 FLCs and systems. Interestingly, methodologies available for analyzing and designing IT2 FLCs and systems are fundamentally the same as those utilized for designing T1 FLCs and systems. For example, the Lyapunov approach, which has been widely used for T1 FL control systems as well as for conventional nonlinear control systems, is the only general tool that has been used for analyzing the stability of or designing a stable IT2 FL control system. To date, there exists no other more effective stability approach for IT2 FL control systems. It is presently the most general and best technique available for IT2 FL controllers and systems, and we believe that it will play a crucial role in the development of future IT2 FL control theory. Note, however, that extending T1 results to IT2 results is very challenging because, generally speaking, an IT2 FL controller is a (much) more complicated nonlinear controller than is a T1 FL controller.

Among the many real-world application areas that may be appropriate for IT2 FL control is robotics. For example, in the field of mobile robots, IT2 FL control can be used to design motion controllers for mobile manipulators and mobile robot formations. The governing equations, which include the dynamics of robots, motors, and actuators, can be transformed and reformulated into a new space that includes nonholonomic constraints. Assuming only the upper bounds of uncertainties to be known, tracking error dynamics can be obtained for which a controller is designed to guarantee the stability of the error dynamics. IT2 FL control is a good candidate for this type of control problem because it has been shown to be very effective for handling uncertainties.

Biomedicine is another promising application area where IT2 FL control may be effectively utilized. For instance, IT2 FL control may be applied for control of drug dosage in chronic disease management such as diabetes and hypertension. This kind of controller is not necessarily of the PID type that has been the focus of some of the earlier chapters of this book. It is a feedback controller in a more general sense, that is, a different and more advanced and intelligent control form where IT2 FL can play an even more critical role by taking advantage of an expert's linguistically expressed knowledge and experience. IT2 FL supervisory control and hierarchal controllers are representatives of such IT2 FLCs; both have been studied in the context of T1 FL control many years ago.

For many disease treatments, effective drug dosages for condition management differ for different patients based on several factors such as preexisting conditions and other medications being taken, as well as age and ethnic background. The FOU of an IT2 FS can be used to represent a range of dosages considered for condition management, for example, to define a range of Coumadin dosage that can be prescribed for more effective control of international normalized ratio (INR) levels or as an indicator of the time it takes for blood to clot for patients with atrial fibrillation. The adjustment of Coumadin dosage has always been a challenge for such patients; too high of an INR value can result in internal bleeding, whereas too low of an INR value can result in blood clotting and an increased risk of a stroke. The target INR for most people with cardiac conditions is 2.5–3.5, and the Coumadin dosage to achieve this range varies among individuals based on several factors such as vitamin K intake levels, other medications, age, weight, and social habits. IT2 FL control methodologies similar to those presented in Chapters 3–6 can be applied offline to personalize medication dosages for individual patients in order to control their Coumadin intake based on target INR and other input factors such as vitamin intake, many of which introduce uncertainties in terms of their input ranges. Such uncertainties suggest that IT2 FL control may be very effective for this kind of individualized control.

7.3 HANI HAGRAS LOOKS INTO THE FUTURE[1]

The emphasis in Chapters 3–6 has been on the simplest kind of T2 FL control, namely IT2 FL control that uses singleton fuzzification. As control environments become more challenging (e.g., nonstationary noises and conditions of observation that affect the values obtained from sensor values), and as our understanding of capturing and modeling different types of uncertainty develops further, there will be a need for other kinds of more advanced T2 FLCs.

There are three directions in which such advances can occur: (1) nonsingleton IT2 FL control, (2) singleton general T2 FL control, and (3) nonsingleton general T2 FL control. This section focuses only on advances 1 and 2 because advance 3 is a mixture of advances 1 and 2.

7.3.1 Nonsingleton IT2 FL Control

A nonsingleton FLC is one whose inputs are modeled as fuzzy numbers. Recall that the inputs to a nonsingleton T1 FLC can only be modeled as T1 fuzzy numbers. On the other hand, the inputs to a nonsingleton IT2 FLC can be modeled either as T1 fuzzy numbers or as IT2 fuzzy numbers. Because the inputs to a nonsingleton IT2 FLC can be fuzzified in these two ways, each is considered separately next.

7.3.1.1 T1 Nonsingleton IT2 FLC A T1 nonsingleton IT2 FLC is described by the same diagram as is the singleton IT2 FLC that is depicted in Fig. 3.2. The rules of a T1 nonsingleton IT2 FLC are the same as for a singleton IT2 FLC. What are different is the fuzzifier, which treats the inputs as T1 FSs (i.e., a measured value is treated as signal plus stationary noise), and the effect of this on the inference block. The output of the inference block will again be an IT2 FS; so, the type-reducers and defuzzifiers that were described for a singleton IT2 FLC (Chapter 3) apply as well to a T1 nonsingleton IT2 FLC.

In a T1 nonsingleton IT2 FLC, measurement $x_i = x_i'$ is mapped into a T1 fuzzy number, which means that a T1 FS membership function (MF) is associated with it. More specifically, a T1 nonsingleton fuzzifier is one for which $(i = 1, \ldots, p)$:

[1]These statements reflect the joint opinions of Hani Hagras and Christian Wagner.

$\mu_{X_i}(x_i') = 1$ and $\mu_{X_i}(x_i)$ decreases from unity as x_i moves away from x_i' [contrast this with Eq. (3.10)]. Conceptually, the T1 nonsingleton fuzzifier implies that the given input value x_i' is the most likely value to be the correct one from all of the values in its immediate neighborhood.

Example 7.1 When the input is modeled as a Gaussian T1 fuzzy number whose standard deviation σ is proportional to the MF spread, then $\mu_{X_i}(x_i)$ can be expressed as (all secondary grades equal 1)

$$\mu_{\tilde{X}_i}(x_i) = 1/e^{-(x_i - x_i')^2/2\sigma^2} \qquad \forall x_i \in X_i \tag{7.1}$$

When the input is modeled as a triangular fuzzy number, as in Fig. 7.1, where c is the MF spread, then $\mu_{X_i}(x_i)$ can be expressed as

$$\mu_{\tilde{X}_i}(x_i) = 1 \Bigg/ \begin{cases} \dfrac{x_i - (x_i' - c/2)}{c/2} & \left(x_i' - \frac{c}{2}\right) \le x_i \le x_i' \\ \dfrac{(x_i' + c/2) - x_i}{c/2} & x_i' \le x_i \le \left(x_i' + \frac{c}{2}\right) \end{cases} \tag{7.2}$$

Turning to the effects of T1 nonsingleton fuzzification on the inference block, when the inputs to the IT2 FLC are T1 FSs, then the computations for the firing interval are more complicated than they are in the singleton fuzzification case.

Because all of the Chapter 3 derivations for an IT2 FLC used T1 FS mathematics, this would be a good time for the reader to review the sup-star composition formulas for a nonsingleton T1 FLC that are given in Eqs. (3.9)–(3.12). Observe from these equations that it is the computations of

$$x^s_{m,\max} = \arg \sup_{x_m \in X_m} \underbrace{\mu_{X_m}(x_m) \star \mu_{F^s_m}(x_m)}_{\mu_{Q^s_m}(x_m)} \tag{7.3}$$

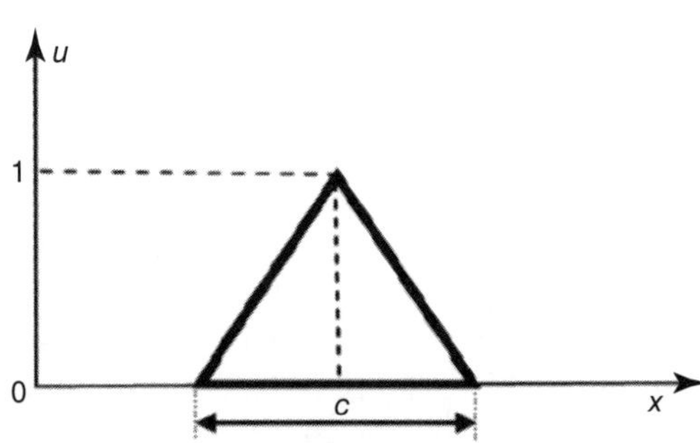

Figure 7.1 Primary membership of a triangular T1 nonsingleton fuzzy number in the x-u domain.

followed by

$$\mu_{Q_m^s}(x_{m,\max}^s) = \mu_{X_m}(x_{m,\max}^s) \star \mu_{F_m^s}(x_{m,\max}^s) \tag{7.4}$$

that are challenging.

In Section 3.3 we showed that all computations for an IT2 FLC involve using only the lower and upper MFs of the IT2 FSs. Because a T1 FS can be interpreted as an IT2 FS whose lower and upper MFs are both equal to the T1 MF, it is intuitively obvious that for T1 nonsingleton fuzzification, Corollary 3.1, for computing the firing interval, becomes [for derivations of these equations, see Mendel (2001, Chapter 11) and Mendel et al. (2006)]:

$$F(\mathbf{x}') \equiv [\underline{f}(\mathbf{x}'), \overline{f}(\mathbf{x}')] \tag{7.5}$$

$$\underline{f}(\mathbf{x}') \equiv T_{m=1}^{p} \underline{\mu}_{Q_m}(\underline{x}_{m,\max}^s) \tag{7.6}$$

$$\overline{f}(\mathbf{x}') \equiv T_{m=1}^{p} \overline{\mu}_{Q_m}(\overline{x}_{m,\max}^s) \tag{7.7}$$

where

$$\underline{x}_{m,\max}^s = \arg\sup_{x_m \in X_m} \underbrace{\mu_{X_m}(x_m) \star \underline{\mu}_{F_m^s}(x_m)}_{\underline{\mu}_{Q_m^s}(x_m)} \tag{7.8}$$

$$\overline{x}_{m,\max}^s = \arg\sup_{x_m \in X_m} \underbrace{\mu_{X_m}(x_m) \star \overline{\mu}_{F_m^s}(x_m)}_{\overline{\mu}_{Q_m^s}(x_m)} \tag{7.9}$$

$$\underline{\mu}_{Q_m^s}(\underline{x}_{m,\max}^s) = \mu_{X_m}(\underline{x}_{m,\max}^s) \star \underline{\mu}_{F_m^s}(\underline{x}_{m,\max}^s) \tag{7.10}$$

$$\overline{\mu}_{Q_m^s}(\overline{x}_{m,\max}^s) = \mu_{X_m}(\overline{x}_{m,\max}^s) \star \overline{\mu}_{F_m^s}(\overline{x}_{m,\max}^s) \tag{7.11}$$

The next example illustrates the computations in Eqs. (7.8)–(7.11).

Example 7.2 Figure 7.2 depicts the measurement $x_i = x_i' = 14$ modeled as a triangular T1 fuzzy number, where $c = 4$, so that it spans from 12 to 16. Observe that this fuzzy set overlaps both Low and Medium, which are the FOUs of rule antecedents.

We shall now provide each of the computations in Eqs. (7.8)–(7.11) with a geometrical interpretation, when the minimum is used for the t-norm and the maximum is used for the supremum. For notational simplicity we drop the s superscripts. In Eq. (7.8), $\underline{x}_{\text{Low},\max} = \arg\sup_{x_m \in X_m} \min(\mu_{X_m}(x_m), \underline{\mu}_{\text{Low}}(x_m))$ is found at one of the intersections of the LMF for Low and the triangle for $\mu_{X_m}(x_m)$ that is centered about 14. Observe, from Fig. 7.2, that there are two such intersections, but it is at the intersection of the LMF for Low and the left leg of the triangle for $\mu_{X_m}(x_m)$ at which u

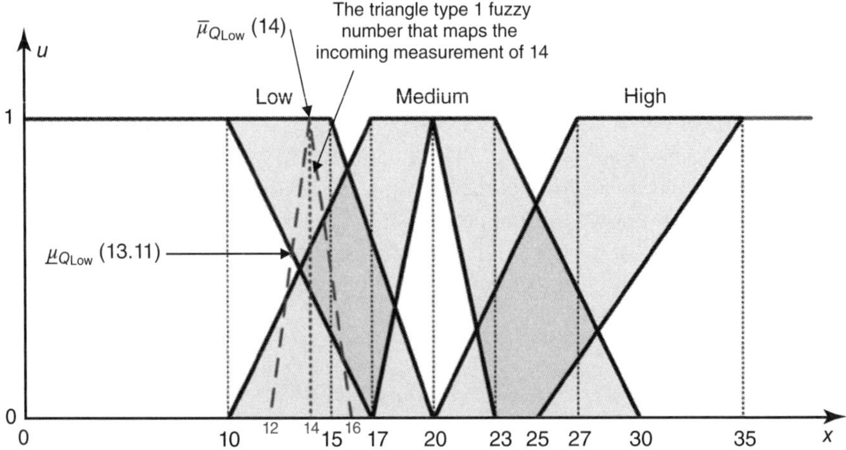

Figure 7.2 Example of T1 nonsingleton fuzzification.

has its maximum value; hence, $\underline{x}_{\text{Low,max}} = 13.11$, and

$$\begin{aligned}\underline{\mu}_{Q_{\text{Low}}}(13.11) &= \min(\mu_{X_m}(13.11), \underline{\mu}_{\text{Low}}(13.11)) \\ &= \mu_{X_m}(13.11) = \underline{\mu}_{\text{Low}}(13.11) = 0.557\end{aligned} \tag{7.12}$$

Of course, in order to obtain the actual numerical values for $\underline{x}_{\text{Low,max}}$ and $\underline{\mu}_{Q_{\text{Low}}}(13.11)$, we wrote equations for the LMF for Low and the left leg of the triangle for $\mu_{X_m}(x_m)$ that is centered about 14, and used some algebra.

In Eq. (7.9), $\bar{x}_{\text{Low,max}} = \arg\sup_{x_m \in X_m} \min(\mu_{X_m}(x_m), \bar{\mu}_{\text{Low}}(x_m))$ is found at the single intersection of the UMF for Low and the apex of the triangle for $\mu_{X_m}(x_m)$ that is centered about 14, that is, $\bar{x}_{\text{Low,max}} = 14$; hence, $\bar{\mu}_{Q_{\text{Low}}}(14) = 1$.

We leave it to the reader to follow the same steps to compute $\underline{x}_{\text{Medium,max}}$, $\underline{\mu}_{Q_{\text{Low}}}(\underline{x}_{\text{Medium,max}})$, $\bar{x}_{\text{Medium,max}}$ and $\bar{\mu}_{Q_{\text{Medium}}}(\bar{x}_{\text{Medium,max}})$, for example, (see Fig. 7.2), $\underline{\mu}_{Q_{\text{Low}}}(\underline{x}_{\text{Medium,max}}) = 0$.

Mendel (2001, Chapter 12, Examples 12-2 and 12-3) shows that by using T1 nonsingleton fuzzification the firing interval is increased over the firing interval obtained using singleton fuzzification. This makes physical sense in that uncertainty about the measured inputs must be protected against, and the way to do this is to increase the firing interval.

It is only the computations of the firing interval for a T1 nonsingleton IT2 FLC that change. All of the remaining computations, such as the ones that are illustrated in the comprehensive example in Section 3.3.2.6, remain the same.

7.3.1.2 T2 Nonsingleton IT2 FLC A T2 nonsingleton IT2 FLC is also described by the same diagram as is a singleton IT2 FLC (Fig. 3.4). The rules of a T2 nonsingleton IT2 FLC are also the same as for a singleton IT2 FLC. What are

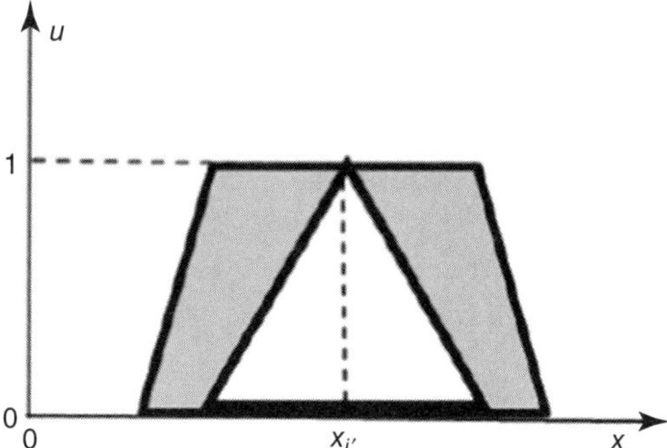

Figure 7.3 Primary membership of a T2 nonsingleton fuzzy number in the *x*-*u* domain.

different is the fuzzifier, which treats the inputs now as IT2 FSs (i.e., a measured value is treated as signal plus nonstationary noise), and also the effect of this on the inference block. The output of the inference block will again be an IT2 FS; so, again the type-reducers and defuzzifiers that were described for a singleton IT2 FLC (Chapter 3) apply as well to the T2 nonsingleton IT2 FLC.

In a T2 nonsingleton IT2 FLC, measurement $x_i = x_i'$ is mapped into a T2 fuzzy number, which means that an FOU is associated with it. Possible MFs are: Gaussian with uncertain standard deviation whose mean is located at x_i'; Gaussian with uncertain mean that is located in some range about x_i'; Gaussian with uncertain mean and standard deviation, where the mean is also located in some range about x_i'; and so forth. Conceptually, the T2 nonsingleton fuzzifier implies that the given input value x_i' is the most likely value to be the correct one from all values in its immediate neighborhood (Mendel, 2001). Because the incoming input signal is corrupted by nonstationary noise and uncertainty, neighboring points are also likely to be a correct value, but to a lesser degree.

Example 7.3 The shape of the incoming measurement x_i' modeled as a triangular nonsingleton fuzzy number is shown in Fig. 7.3.

Turning to the effects of T2 nonsingleton fuzzification on the inference block, when the inputs to the IT2 FLC are T2 FSs, then the computations for the firing interval are even more complicated than they were in the T1 fuzzification case.

Because an IT2 FS is described by its lower and upper MFs, Corollary 2.2, for computing the firing interval, becomes [for derivations of these equations, see Mendel (2001, Chapter 12) and Mendel et al. (2006)]: (1) Equations (7.5)–(7.7) are unchanged, but Eqs. (7.8)–(7.11) are changed to

$$\underline{x}^s_{m,\max} = \arg\sup_{x_m \in X_m} \underbrace{\underline{\mu}_{X_m}(x_m) \star \underline{\mu}_{F^s_m}(x_m)}_{\underline{\mu}_{Q^s_m}(x_m)} \tag{7.13}$$

$$\overline{x}^s_{m,\max} = \arg\sup_{x_m \in X_m} \underbrace{\overline{\mu}_{X_m}(x_m) \star \overline{\mu}_{F^s_m}(x_m)}_{\overline{\mu}_{Q^s_m}(x_m)} \tag{7.14}$$

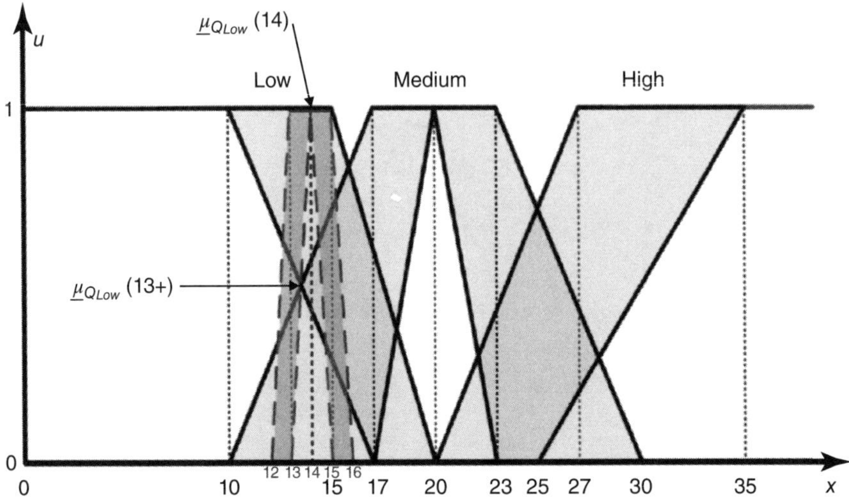

Figure 7.4 Example of nonsingleton T2 fuzzification for an IT2 Mamdani FLC.

$$\underline{\mu}_{Q^s_m}(\underline{x}^s_{m,\max}) = \underline{\mu}_{X_m}(\underline{x}^s_{m,\max}) \star \underline{\mu}_{F^s_m}(\underline{x}^s_{m,\max}) \tag{7.15}$$

$$\overline{\mu}_{Q^s_m}(\overline{x}^s_{m,\max}) = \overline{\mu}_{X_m}(\overline{x}^s_{m,\max}) \star \overline{\mu}_{F^s_m}(\overline{x}^s_{m,\max}) \tag{7.16}$$

The next example illustrates the computations in Eqs. (7.13)–(7.16).

Example 7.4 This example begins with the antecedent word FOUs that were used in Example 7.2 (Fig. 7.2) and models the measured value of the input at $x_i = x_i' = 14$ by a trapezoidal FOU. This FOU also overlaps both Low and Medium.

As we did in Example 7.2, we shall provide each of the computations in Eqs. (7.13)–(7.16) with a geometrical interpretation, but with less detail than in that example. As in Example 7.2, the minimum is again used for the t-norm and the maximum is used for the supremum. In Eq. (7.13), $\underline{x}_{\text{Low,max}} = \arg\sup_{x_m \in X_m} \min(\underline{\mu}_{X_m}(x_m), \underline{\mu}_{\text{Low}}(x_m))$ is found at one of the intersections of the LMFs for Low and $\underline{\mu}_{X_m}(x_m)$ that is centered about 14. Observe, from Fig. 7.4, that there are two such intersections, but it is again at the intersection of the LMF for Low and the left leg of the triangle for $\underline{\mu}_{X_m}(x_m)$ at which u has its maximum value. Observe also that $\underline{x}_{\text{Low,max}}$ appears to be a bit to the right of 13, say 13+; and $\underline{\mu}_{Q_{\text{Low}}}(13+)$ appears to be a bit below 0.5, say 0.5 –, that is,

$$\underline{\mu}_{Q_{\text{Low}}}(13+) = \min(\underline{\mu}_{X_m}(13+), \underline{\mu}_{\text{Low}}(13+)) = \underline{\mu}_{X_m}(13+) = \underline{\mu}_{\text{Low}}(13+) = 0.5- \tag{7.17}$$

Of course, in order to obtain the actual numerical values for $\underline{x}_{\text{Low,max}}$ and $\underline{\mu}_{Q_{\text{Low}}}(13+)$, we need to find equations for the LMFs of Low and $\underline{\mu}_{X_m}(x_m)$ that are centered about 14, and use some algebra.

In Eq. (7.14), $\overline{x}_{\text{Low,max}} = \arg\sup_{x_m \in X_m} \min(\overline{\mu}_{X_m}(x_m), \overline{\mu}_{\text{Low}}(x_m))$ is found at one of the intersections of the UMFs for Low and $\overline{\mu}_{X_m}(x_m)$ that is centered about 14. Observe from Fig. 7.4 that such intersections occur where the top portion of the UMFs for Low and $\overline{\mu}_{X_m}(x_m)$ intersect, where u has its maximum value of 1. Because $\overline{\mu}_{Q_{\text{Low}}} = 1$ for all of these x values, we can associate any one of them with $\overline{\mu}_{Q_{\text{Low}}}$, for example, $x_m = 14$.

We leave it to the reader to follow the same steps to compute $\underline{x}_{\text{Medium,max}}$, $\underline{\mu}_{Q_{\text{Low}}}(\underline{x}_{\text{Medium,max}})$, $\overline{x}_{\text{Medium,max}}$ and $\overline{\mu}_{Q_{\text{Medium}}}(\overline{x}_{\text{Medium,max}})$.

Mendel (2001, Chapter 12, Examples 12-2 and 12-3) shows that by using T2 nonsingleton fuzzification the firing interval is further increased over the firing intervals obtained by using either singleton or T1 fuzzification. This again makes physical sense because the increased uncertainty about the measured inputs, modeled now by FOUs, must be protected against, and the way to do this is to further increase the firing interval.

7.3.1.3 Comments All of the above results for nonsingleton fuzzification have been known for a long time. The research challenge is to figure out how to perform the much more complicated firing interval computations so that nonsingleton fuzzification can be used in real-time FLC. Another research challenge is to quantify the relationships between nonsingleton fuzzification and control measures such as robustness.

7.3.2 zSlices-Based Singleton General T2 FL Control[2]

Mamdani IT2 FLCs as discussed throughout most of this book are based on IT2 FSs. As detailed in Section 2.3, IT2 FSs are a simplification of what is now commonly referred to as general type 2 (GT2) FSs. Whereas for GT2 FSs the degree of membership of a given point is modeled as a distribution (specifically a T1 FS), the degree of membership in IT2 sets is modeled as an interval, sacrificing the additional degree of freedom but greatly reducing the complexity in the underlying theory, implementation, and computation.

A detailed overview of the background on GT2 FSs as well as the slices-based representation (through zSlices or α planes) is included in Section 2.4; here we provide a brief reminder of the development and types of representations of GT2 FSs, followed by a detailed description of how GT2-FS-based FLCs can be implemented based on IT2 FLCs by leveraging the zSlice representation.

Although IT2 FLCs have been employed successfully in many applications, the potential of employing GT2 FLCs has driven research over the last 10 years (e.g., Coupland and John, 2007; Liu, 2008; Wagner and Hagras, 2008) to develop different representations for GT2 FSs that enable the harnessing of the modeling power of GT2 FLCs while avoiding the computation complexities that traditionally prevented their real-world application. Most recently, the α-plane (Liu,

[2]The material in this section is taken from Wagner and Hagras (2010; © 2010, IEEE).

2008) and zSlice (Wagner and Hagras, 2010) representations for GT2 FSs have enabled the representation of GT2 FSs as a series of modified IT2 FS "slices," which enables the real-world application of GT2 FLCs with minimal implementation and design effort (beyond that involved in IT2 FLCs) and manageable computational complexity.

Fundamentally, the α-plane and zSlice representations are based on identical concepts and ideas (see Table 2.6). The difference in naming originally arose from the independent development from different starting points of both representations around the same time and both names have been used interchangeably since their introduction in 2008. From a theoretical point of view, the α-plane representation lends itself in terms of notation and naming to the seamless extension of fuzzy logic theory (e.g., α-cuts) and thus it has been the focus of Section 2.4. In practical terms, the zSlice representation provides the advantage of enabling the intuitive working in three dimensions (used for GT2 FSs), in particular, for practitioners with a mathematical background who are used to working in the x-y-z domain (see Table 2.6). Thus, in the following sections, we will focus on the zSlice representation introduced in Wagner and Hagras (2008, and 2010).

For illustration purposes,[3] Fig. 7.5 details the representation of a GT2 FS based on the zSlice representation. This figure shows the transition from a standard GT2 FS, (i.e., one with a continuous third dimension) to a zGT2 FS with three zSlices. In practice, zGT2 FSs are frequently generated directly as a series of zSlices, which together form the GT2 set.

In the rest of this section we provide a brief introduction to zSlices and zSlices-based GT2 (zGT2) FSs in Sections 7.3.2.1 and 7.3.2.2, followed by an in-depth review of the zGT2 FLC in Section 7.3.2.3 and some observations in

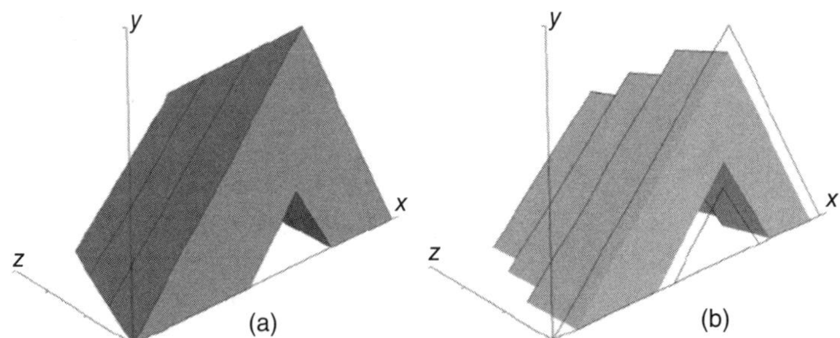

Figure 7.5 (a) Side view of a GT2 FS, indicating three zLevels on the third dimension and (b) side view of the zSlices representation of the same set (using three zSlices and indicating zLevel 0).

[3]Note that GT2 FSs (zSlice based or not) are not "filled-in," an impression one may get from figures such as Fig. 7.5. The set is defined through its primary/secondary membership combinations over its domain as laid out in Section 2.4 and recapitulated in Section 7.3.2.1. The figures employed in this chapter and other parts of the book are solely "filled-in" to facilitate the visualization of the GT2 FSs.

Section 7.3.2.4. Finally, for the reader interested in the implementation of zGT2 FLCs, see the book's Appendix A.

7.3.2.1 *From IT2 FSs to zSlices* zSlices are at the core of the zSlice-based representation for GT2 FSs. As noted in Wagner and Hagras (2010) (see, also, Section 2.4.1), a zSlice is formed by slicing a GT2 FS in the third dimension (z) at level z_i. This slicing action results in an interval set in the third dimension with height/depth z_i. As such, a zSlice $\widetilde{Z}_i$ is equivalent to an IT2 FS with the exception that its membership grade $\mu_{\widetilde{Z}_i}(x, y)$ in the third dimension is not fixed to 1 but is equal to z_i where $0 \leq z_i \leq 1$. Thus, following a vertical slice representation, the zSlice $\widetilde{Z}_i$ can be written as

$$\widetilde{Z}_i = \int_{x \in X} \int_{y \in \overline{y}_{i_x}} z_i/(x, y) \tag{7.18}$$

At each x value (as shown in Fig. 7.6a), zSlicing creates an interval set (i.e., a rectangular well) with height z_i and support $\overline{y}_{i_x}$ that ranges from l_i to r_i as shown in Fig. 7.6b. Note that, for notational simplicity, we have dropped x as an index on l_i and r_i. Further $1 \leq i \leq I$, where I is number of zSlices (excluding $\widetilde{Z}_0$) and generally $z_i = i/I$. Thus Eq. (7.18) can be written as

$$\widetilde{Z}_i = \int_{x \in X} \int_{y \in [l_i, r_i]} z_i/(x, y) \tag{7.19}$$

Observe that

$$\widetilde{Z}_0 = \int_{x \in X} \int_{y \in \overline{y}_{0_x}} 0/(x, y) \tag{7.20}$$

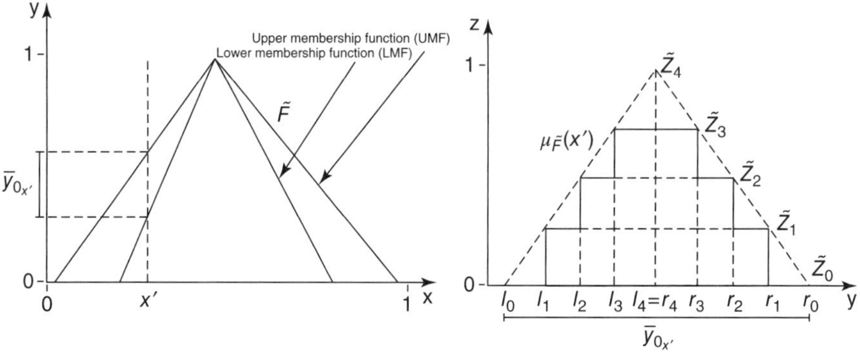

Figure 7.6 (a) Front view of a GT2 FS $\widetilde{F}$ and (b) vertical slice at x' of zSlice-based T2 FS with $I = 4$ (Wagner and Hagras, 2010; © 2010, IEEE).

Because all secondary memberships of $\widetilde{Z}_0$ are zero, it does not contribute anything to any computations; hence, it is not counted as part of the number of zSlices making up a complete zGT2 FS.

7.3.2.2 *zSlices-Based GT2 FSs* Analogous to the α-plane representation of a GT2 FS [see Section 2.4 and specifically Eq. (2.84)], a GT2 FS $\widetilde{F}$ is seen to be equivalent to the collection of an infinite number of zSlices, that is,

$$\widetilde{F} = \int_{0 \le i \le I} \widetilde{Z}_i \quad I \to \infty \tag{7.21}$$

In a discrete universe of discourse Eq. (7.21) can be expressed as

$$\widetilde{F} = \sum_{i=0}^{I} \widetilde{Z}_i \tag{7.22}$$

We will be referring to the discrete version of a GT2 FS, in Eq. (7.22), throughout the rest of this section. It should be noted that the summation sign in Eq. (7.22) does not denote arithmetic addition but, instead, the union set-theoretic operation, as discussed just below Eq. (2.2). We have employed the *max* operation to represent the union; hence, whenever a y value is attached to more than one z_i value, the maximum of those z_i values is chosen and attached to the given y value. Note that, analogous to Theorem 2.4 expressed for α-planes, the membership function $\mu_{\widetilde{F}}(x')$ at x' of the zSlices-based GT2 FS $\widetilde{F}$ shown Fig. 7.6b can be expressed as

$$\mu_{\widetilde{F}}(x') = \sum_{i=1}^{I} \sum_{y \in [l_i, r_i]} z_i / y \tag{7.23}$$

where $0 \le i \le I$. It is worth reminding the reader that, at x', $\mu_{\widetilde{F}}(x')$ is a type 1 FS.

Having described the structure of both zSlices and zGT2 FSs, we proceed in the following section to detail the construction of zGT2-based FLCs. For more detailed information about the nature of and set-theoretic operations on zGTS FSs, see Section 2.4.2 (Wagner and Hagras, 2008, 2010).

7.3.2.3 *Mamdani zSlice-Based GT2 FL Control* The structure of a zSlice-based GT2 FLC (zFLC) shown in Fig. 7.7 is very similar to that of the IT2 FLC depicted in Fig. 3.4, that is, it is composed of a fuzzifier, an inference engine, and rule base as well as a type-reducer and defuzzifier. Note that the type-reducer and defuzzifier have been combined and are shown as a single defuzzifier component in Fig. 7.7 (and also in Fig. 7.8). The difference between the FLCs is in the nature of the type-2 sets, which are zSlice-based GT2 FSs in the case of the zFLC and IT2 FSs in the case of the IT2 FLC.

A zFLC can be expressed directly as the weighted combination of the IT2 FLCs computed for each zLevel, where the secondary membership of each zLevel acts

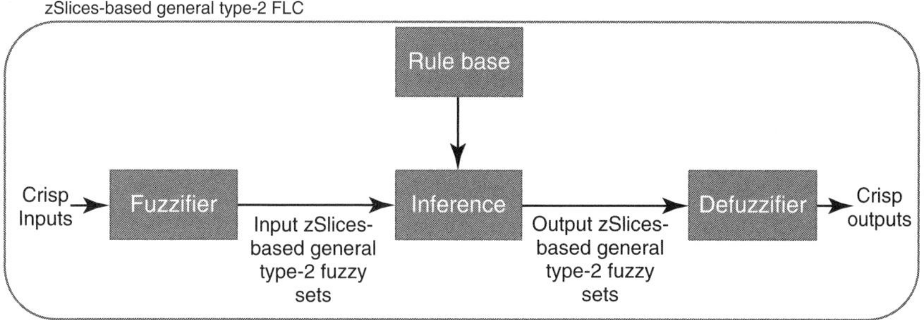

Figure 7.7 Standard zSlices-based GT2 FLC.

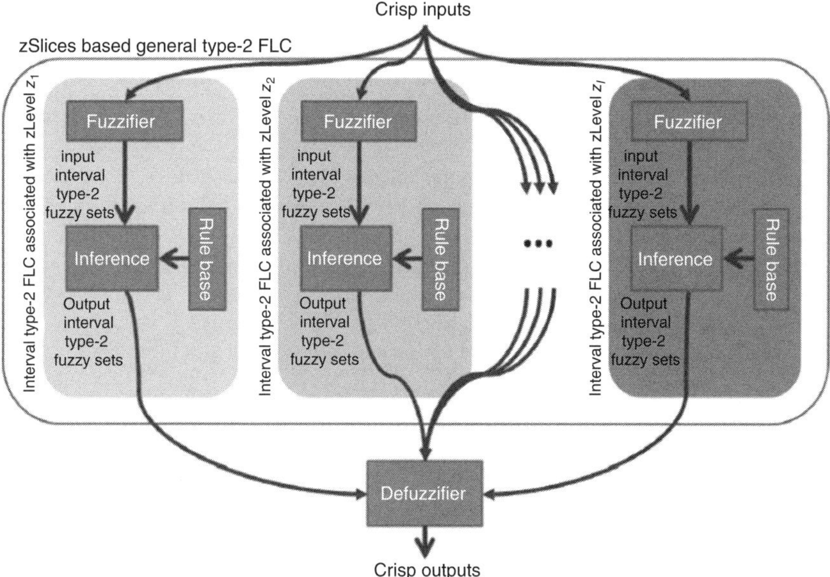

Figure 7.8 zSlices-based GT2 FLC implemented as a collection of IT2 FLCs, where each IT2 FLC is associated with a given zLevel and all outputs are fused in the defuzzification stage of the FLC. Note that darker shading from left to right reflects higher zLevels.

as the weight; thus, all operations of the zFLC can be directly computed using zSlices-based GT2 FSs as described in Section 2.4.2. Consequently, the same FLC can be implemented by constructing a collection of IT2 FLCs where each of these FLCs is associated with a given zLevel. Figure 7.8 provides an illustration of this process.

zFLCs enable a straightforward defuzzification based on the individual centroids generated by each "contained IT2 FLC" (see Fig. 7.8). As shown in Section 3.3.2.4, the centroid $C_{\widetilde{B}}$ of an output set $\widetilde{B}$ of an IT2 FLC is an IT1 FS defined by its left and right end point, that is,

$$C_{\widetilde{B}}(\mathbf{x}) = 1/[u_l(\mathbf{x}), u_r(\mathbf{x})] \tag{7.24}$$

For a given zLevel z_i, the centroid of the output set $\widetilde{B}_i$ of the modified IT2 FS can be expressed as

$$C_{\widetilde{B}_i}(\mathbf{x}) = z_i/[u_{l_i}(\mathbf{x}), u_{r_i}(\mathbf{x})] \tag{7.25}$$

Consequently, the overall centroid C of a zFLC can be expressed as the union of the centroids of all "contained IT2" FLCs (associated with their respective zLevel) and written as

$$C(\mathbf{x}) = \bigcup_{i=1}^{I} C_{\widetilde{B}_i}(\mathbf{x}) \tag{7.26}$$

Figure 7.9 shows an example centroid of a zFLC with three zLevels. This figure further clarifies the result of the union of the individual IT1 output sets each with a height of z_i.

Considering the centroid in Eq. (7.26), the resulting defuzzified value of a zFLC can be computed either as the weighted average of the averages of the centroids $C_{\widetilde{B}_i}(\mathbf{x})$ of each given zLevel or by computing the centroid defuzzifier for the overall centroid $C(\mathbf{x})$.

Applying the centroid defuzzifier to the averages of the centroids of each zLevel, we obtain

$$y_C(\mathbf{x}) = \frac{z_1\left[\frac{u_{l_1}(\mathbf{x})+u_{r_1}(\mathbf{x})}{2}\right] + z_2\left[\frac{u_{l_2}(\mathbf{x})+u_{r_2}(\mathbf{x})}{2}\right] + \cdots + z_I u_I(\mathbf{x})}{z_1 + z_2 + \cdots + z_I} \tag{7.27}$$

Note that, as mentioned above, we have excluded the values associated the zSlice $\widetilde{Z}_0$ because they will not have any impact on the output of the zFLC [when $z_0 = 0$, its terms in the numerator and denominator of Eq. (7.27) are 0].

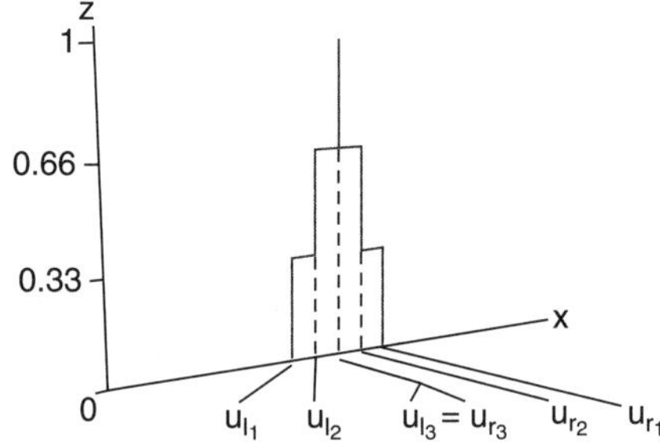

Figure 7.9 Example centroid of a zFLC with $I = 3$ (Wagner and Hagras, 2010; © 2010, IEEE).

From Eq. (7.27), one can easily see that the crisp output of the zFLC is a weighted average of the outputs of the different zLevel-induced IT2 FLCs (where the output of a given IT2 FLC for a given zLevel z_i is equal to the average of the left and right end points of the type-reduced set for that zLevel z_i, that is, $[u_l(\mathbf{x}) + u_r(\mathbf{x})]/2$, each associated with its specific zLevel. Thus, the output of the zFLC is an aggregation of the outputs of several IT2 FLCs, each associated with a specific zLevel. It is worthwhile to repeat here that this also implies that all the underlying IT2 FLCs can be processed independently and are only recombined during the defuzzification stage.

To date, due to the relative newness of the α -plane/zSlice representation, no publications have appeared about the automatic (i.e., non-expert-based) design of Mamdani zFLCs; however, because their structure uses IT2 FSs raised to specific zLevels, the same approaches to designing IT2 Mamdani FLCs as detailed in Section 3.6 can be applied, subject to addressing the connection and correct structure of the individual zSlice-based FSs. In other words, care needs to be taken to ensure that the FOU of zSlices $\widetilde{Z}_i$ with a higher zLevel is contained within the zSlices of the lower levels, that is,

$$\mathrm{FOU}(\widetilde{Z}_1) \subseteq \mathrm{FOU}(\widetilde{Z}_2) \subseteq \cdots \subseteq \mathrm{FOU}(\widetilde{Z}_I) \tag{7.28}$$

The actual rule-based design of a Mamdani zFLC is identical to that of a T1 or IT2 FLC and thus the same approaches for learning it from data, as laid out in Section 3.6, apply to it.

7.3.2.4 Observations We conclude this section with some observations about zGT2 FLCs that suggest such FLCs are worthy of further study:

- The fact that the complex operations on GT2 FSs can be reduced to common IT2 FS operations significantly reduces the design and implementation complexity and thus should facilitate the use of GT2 FLCs.
- The property of zFLCs that allows the computation of each zLevel independently also allows for a high degree of parallel computation. In fact, all zSlices levels can be computed simultaneously on separate processors followed only by the very simple defuzzification stage, which is done centrally and the output of which is fed to the controlled system. This offers great potential with minimal implementation effort and should allow the use of GT2 FSs not only for FLCs but also for a variety of other applications.
- Current IT2 theory can be reused and only very small modifications are necessary to use current IT2 implementations to compute zFLCs.
- When computing the centroid of a zGT2 FS as done during the type reduction stage, the resulting type-reduced T1 FS still (as for standard GT2 FLCs) gives an indicative model of the amount of uncertainty contained within the current iteration of the zFLC (as shown in Fig. 7.9).

- The use of zFLCs may let one achieve real-time performance for GT2 FLCs as a result of significantly simplifying the computational complexity associated with the deployment of GT2 FLCs.

7.4 WOEI WAN TAN LOOKS INTO THE FUTURE

The preceding chapters have demonstrated that IT2 FL control is a diverse field of rich promise; however, a central challenge is to increase the number of applications and users in order to bring IT2 FL control into the mainstream of automation and control.

Because control theory is firmly grounded in mathematical theory, development of a set of convenient theoretical tools may bring about new possibilities. The analytical structures described in Chapters 4 and 5 are a first step.

By now it should be clear to the reader that type reduction presents the major hurdle to a theoretical framework of IT2 FL control.

Nie and Tan (2008) proposed a type reduction algorithm in which the output of an IT2 FLC (that is obtained by first aggregating the fired rule output sets by using the union operation and then type reducing that IT2 FS) is expressed as a closed-form equation. This has been referred to as the *Nie–Tan (NT) method* (e.g., Mendel, 2013) and is based on the vertical-slice representation of an IT2 FS. Recall that, in the vertical-slice representation [see Eqs. (2.29) and (2.30)], an IT2 FS is considered as the union of the vertical slices for all the values of the primary variable, which are intervals.

The main idea behind the NT method is to average the vertical slices of an IT2 FS $\widetilde{B}$ for all the values of the primary variable, producing a T1 FS, and to then compute the centroid (COG) of this T1 FS. Mathematically, the defuzzified output of an IT2 FS $\widetilde{B}$ using the NT method can be expressed as

$$u_{\text{NT}}(\mathbf{x}') = \text{COG}\left\{\frac{1}{2}\left[\underline{\mu}_{\widetilde{B}}(u|\mathbf{x}') + \overline{\mu}_{\widetilde{B}}(u|\mathbf{x}')\right]\right\} = \frac{\sum_{i=1}^{N} u_i[\underline{\mu}_{\widetilde{B}}(u_i|\mathbf{x}') + \overline{\mu}_{\widetilde{B}}(u_i|\mathbf{x}')]}{\sum_{i=1}^{N}[\underline{\mu}_{\widetilde{B}}(u_i|\mathbf{x}') + \overline{\mu}_{\widetilde{B}}(u_i|\mathbf{x}')]} \tag{7.29}$$

Mendel and Liu (2013) proved that $u_{\text{NT}}(\mathbf{x}')$ is a first-order approximation to the actual defuzzified value of $\widetilde{B}$, $m_{\widetilde{B}}(\mathbf{x}')$, where

$$m_{\widetilde{B}}(\mathbf{x}') = \frac{c_l(\widetilde{B}|\mathbf{x}') + c_r(\widetilde{B}|\mathbf{x}')}{2} \tag{7.30}$$

and $c_l(\widetilde{B}|\mathbf{x}')$ and $c_r(\widetilde{B}|\mathbf{x}')$ are the left and right end points of the actual centroid (computed by using KM algorithms). More specifically, they proved

$$u_{\text{NT}}(\mathbf{x}') \approx m_{\widetilde{B}}(\mathbf{x}') + \delta(\widetilde{B}|\mathbf{x}') \tag{7.31}$$

where a formula for $\delta(\widetilde{B}|\mathbf{x}')$ is given in their paper but is not needed for the present section. Examples given in Mendel and Liu (2013) show that $\delta(\widetilde{B}|\mathbf{x}')$ is quite small, so a very reasonable approximation to $m_{\widetilde{B}}(\mathbf{x}')$ is $u_{\mathrm{NT}}(\mathbf{x}')$.

Compared with the KM method, the advantages of the NT method are the simplicity in its implementation and its closed-form formula that should make theoretical analyses of an IT2 FLC possible. Such analyses (e.g., like the ones in Chapter 6) remain to be performed.

Because the NT method is only a first-order approximation to the KM approach, a future research direction is to study the ability of the NT method to handle large uncertainties for control problems in which there are uncertain environments. Another possible future direction is to study the characteristics of IT2 FLCs that use the NT method. The input–output relationship of IT2 FLCs using the NT method and their robustness may be the focus of such a study.

7.5 JERRY MENDEL LOOKS INTO THE FUTURE

Control occurs in real time, and therefore computational complexity is a very important issue for it. For T2 FLCs to be used in real-world applications they must be simplified. Ideally, the simplified T2 FLC should also lend itself to mathematical analyses.

7.5.1 IT2 FLC

7.5.1.1 Simplified Architectures Section 7.4 has already described an important simplification for type reduction in an IT2 FLC, the Nie–Tan method. That method is presently limited by its requirement that the fired rule output sets must be aggregated (by, e.g., using the union operation) so as to obtain a composite IT2 FS. In many FLCs such an aggregation is avoided because it is time consuming. That is the main reason for using COS type reduction instead of centroid type reduction.

The m-n TSK IT2 FLC, described in Chapter 6, is another way to greatly simplify an FLC. It replaces COS type reduction by the simpler formula:

$$u_{mn}(\mathbf{x}') \equiv m\frac{\sum_{s=1}^{M} u_s \underline{f}^s(\mathbf{x}')}{\sum_{s=1}^{M} \underline{f}^s(\mathbf{x}')} + n\frac{\sum_{s=1}^{M} u_s \overline{f}^s(\mathbf{x}')}{\sum_{s=1}^{M} \overline{f}^s(\mathbf{x}')} \tag{7.32}$$

One of the important design features of Eq. (7.32) is its two design parameters (degrees of freedom) m and n. Presently, the Nie–Tan method does not have comparable design degrees of freedom; however, there is an interesting connection between the Nie–Tan and m-n TSK IT2 FLCs, as explained next.

It is straightforward (Mendel, 2013), by using some simple algebra, to reexpress $u_{\text{NT}}(\mathbf{x}')$ in Eq. (7.29) as

$$u_{\text{NT}}(\mathbf{x}') = \frac{\sum_{i=1}^{N} \underline{\mu}_{\tilde{B}}(u_i|\mathbf{x}')}{\sum_{i=1}^{N} [\underline{\mu}_{\tilde{B}}(u_i|\mathbf{x}') + \overline{\mu}_{\tilde{B}}(u_i|\mathbf{x}')]} \times \frac{\sum_{i=1}^{N} u_i \underline{\mu}_{\tilde{B}}(u_i|\mathbf{x}')}{\sum_{i=1}^{N} \underline{\mu}_{\tilde{B}}(u_i|\mathbf{x}')}$$

$$+ \frac{\sum_{i=1}^{N} \overline{\mu}_{\tilde{B}}(u_i|\mathbf{x}')}{\sum_{i=1}^{N} [\underline{\mu}_{\tilde{B}}(u_i|\mathbf{x}') + \overline{\mu}_{\tilde{B}}(u_i|\mathbf{x}')]} \times \frac{\sum_{i=1}^{N} u_i \overline{\mu}_{\tilde{B}}(u_i|\mathbf{x}')}{\sum_{i=1}^{N} \overline{\mu}_{\tilde{B}}(u_i|\mathbf{x}')}$$

$$\equiv m'_{\tilde{B}}(\mathbf{x}') \frac{\sum_{i=1}^{N} u_i \underline{\mu}_{\tilde{B}}(u_i|\mathbf{x}')}{\sum_{i=1}^{N} \underline{\mu}_{\tilde{B}}(u_i|\mathbf{x}')} + n'_{\tilde{B}}(\mathbf{x}') \frac{\sum_{i=1}^{N} u_i \overline{\mu}_{\tilde{B}}(u_i|\mathbf{x}')}{\sum_{i=1}^{N} \overline{\mu}_{\tilde{B}}(u_i|\mathbf{x}')} \tag{7.33}$$

where

$$m'_{\tilde{B}}(\mathbf{x}') = \frac{\sum_{i=1}^{N} \underline{\mu}_{\tilde{B}}(u_i|\mathbf{x}')}{\sum_{i=1}^{N} [\underline{\mu}_{\tilde{B}}(u_i|\mathbf{x}') + \overline{\mu}_{\tilde{B}}(u_i|\mathbf{x}')]} \tag{7.34}$$

$$n'_{\tilde{B}}(\mathbf{x}') = \frac{\sum_{i=1}^{N} \overline{\mu}_{\tilde{B}}(u_i|\mathbf{x}')}{\sum_{i=1}^{N} [\underline{\mu}_{\tilde{B}}(u_i|\mathbf{x}') + \overline{\mu}_{\tilde{B}}(u_i|\mathbf{x}')]} \tag{7.35}$$

Observe that Eq. (7.33) resembles Eq. (7.32), although m and n in Eq. (7.32) are arbitrary constants, whereas $m'_{\tilde{B}}(\mathbf{x}')$ and $n'_{\tilde{B}}(\mathbf{x}')$ in Eq. (7.33) are not, that is, they are computed by means of Eqs. (7.34) and (7.35).

If, hypothetically, Eq. (7.32) were to be used for centroid TR, then it could be re-expressed as (Mendel, 2013)

$$C_{mn}(\mathbf{x}') = m \frac{\sum_{i=1}^{N} u_i \underline{\mu}_{\tilde{B}}(u_i|\mathbf{x}')}{\sum_{i=1}^{N} \underline{\mu}_{\tilde{B}}(u_i|\mathbf{x}')} + n \frac{\sum_{i=1}^{N} u_i \overline{\mu}_{\tilde{B}}(u_i|\mathbf{x}')}{\sum_{i=1}^{N} \overline{\mu}_{\tilde{B}}(u_i|\mathbf{x}')} \tag{7.36}$$

Comparing Eqs. (7.33) and (7.36), we see they are the same when $m = m'_{\tilde{B}}$ and $n = n'_{\tilde{B}}$. This provides an interesting new connection between the *m-n* IT2 TSK FLC formula and the Nie–Tan formula, a connection that needs further strengthening (research) since the *m-n* IT2 TSK FLC formula has not actually been suggested before as a way to replace the centroid computation when an FOU is present, and the Nie–Tan formula has not actually been suggested before as a way to replace a TR computation when an FOU is not present.

7.5.1.2 *Continuous IT2 FLC* Recently, Wu and Mendel (2011) examined the continuity of T1 and IT2 FLCs using rigorous mathematics. They reached the following guidelines for practitioners who want to design continuous IT2 FLCs:

1. To guarantee a continuous input–output mapping regardless of which type reduction and defuzzification method is used, Gaussian IT2 FSs should be employed for all rule antecedents.
2. When triangular and/or trapezoidal IT2 FSs are used, to guarantee a continuous input–output mapping, *the LMFs should cover every input domain for all rule antecedents*. This implies that the UMFs must also cover every input domain.

The fact that antecedent LMFs cannot have gaps between them is a somewhat surprising result, and, unless Gaussian MFs are used during an optimal design of an FLC, this imposes additional constraints on optimal designs when piecewise-linear antecedent LMFs are used. How such constraints can be enforced during designs of optimal FLCs, as described in Section 3.6, is an open area for research.

7.5.2 GT2 FLC

7.5.2.1 *Simplified Architectures* Both the Nie–Tan and *m-n* IT2 FLCs can be extended to GT2 FLCs.

The gist of how the NT method can be used for a GT2 FLS is (Mendel, 2014):

1. For each of the M rules, compute its firing interval $F_\alpha^s(\mathbf{x}')$ for level α, that is, compute
$$F_\alpha^s(\mathbf{x}') \equiv [\underline{f}_\alpha^s(\mathbf{x}'), \overline{f}_\alpha^s(\mathbf{x}')] \tag{7.37}$$
where $\underline{f}_\alpha^s(\mathbf{x}')$ and $\overline{f}_\alpha^s(\mathbf{x}')$ are computed using α-cuts in Eqs. (3.51) and (3.52), respectively.
2. For each of the M rules compute $\mathrm{FOU}(\widetilde{B}_\alpha^s)$, that is,
$$\mathrm{FOU}(\widetilde{B}_\alpha^s) = [\underline{\mu}_{\widetilde{B}_\alpha^s}(u|\mathbf{x}'), \overline{\mu}_{\widetilde{B}_\alpha^s}(u|\mathbf{x}')] \quad \forall u \in U \tag{7.38}$$
where $\underline{\mu}_{\widetilde{B}_\alpha^s}(u|\mathbf{x}')$ and $\overline{\mu}_{\widetilde{B}_\alpha^s}(u|\mathbf{x}')$ are computed using α-cuts in Eq. (3.53).
3. Compute the aggregated output horizontal slice $\mathrm{FOU}(\widetilde{B}_\alpha)$, that is,
$$\mathrm{FOU}(\widetilde{B}_\alpha) = [\underline{\mu}_{\widetilde{B}_\alpha}(u|\mathbf{x}'), \overline{\mu}_{\widetilde{B}_\alpha}(u|\mathbf{x}')] \quad \forall u \in U \tag{7.39}$$
where $\underline{\mu}_{\widetilde{B}_\alpha}(u|\mathbf{x}')$ and $\overline{\mu}_{\widetilde{B}_\alpha}(u|\mathbf{x}')$ are computed using α-cuts in Eqs. (3.56) and (3.57), respectively.

4. Compute α - level NT outputs, $u_{\text{NT},\alpha}(\mathbf{x}')$ $(\alpha = \alpha_1, \ldots, \alpha_k)$ by using Eq. (7.29) at each α-level.
5. Defuzzify the $u_{\text{NT},\alpha}(\mathbf{x}')$ $(\alpha = \alpha_1, \ldots, \alpha_k)$ by creating a spike of amplitude α_j for $u_{\text{NT},\alpha_j}(\mathbf{x}')$, and computing $u_{\text{NT}}(\mathbf{x}')$ as

$$u_{\text{NT}}(\mathbf{x}') = \frac{\sum_{j=1}^{k} \alpha_j u_{\text{NT},\alpha_j}(\mathbf{x}')}{\sum_{j=1}^{k} \alpha_j} \tag{7.40}$$

The gist of how the *m-n* TSK architecture can be used for a TSK GT2 FLS is (Mendel, 2014):

1. For each of the *M* rules, compute its firing interval $F_\alpha^s(\mathbf{x}')$ for level α, exactly as in Eq. (7.37).
2. Decide if the same or different values of m, n, and u_s will be used for each α.
3. Compute α - level *m-n* outputs, $u_{mn,\alpha_j}(\mathbf{x}')(\alpha = \alpha_1, \ldots, \alpha_k)$ by using Eq. (7.32) at each α-level.
4. Defuzzify the $u_{mn,\alpha_j}(\mathbf{x}')$ $(\alpha = \alpha_1, \ldots, \alpha_k)$ by creating a spike of amplitude α_j for $u_{mn,\alpha_j}(\mathbf{x}')$, and computing $u_{mn}(\mathbf{x}')$, as

$$u_{mn}(\mathbf{x}') = \frac{\sum_{j=1}^{k} \alpha_j u_{mn,\alpha_j}(\mathbf{x}')}{\sum_{j=1}^{k} \alpha_j} \tag{7.41}$$

Doing what has just been proposed, as well as studying whether or not Eqs. (7.40) and/or (7.41) lend themselves to mathematical analyses, such as the analyses given in Chapter 6, remains to be explored.

***7.5.2.2 Parameterization*[4]** We have seen (Chapter 2) that a GT2 FS can be represented in four different ways: points, wavy slices, zSlices (i.e., horizontal slices), and vertical slices. While each of the latter three plays an important role for theoretical aspects of GT2 FSs, we need to know *which of these representations should be used during an optimal design of a GT2 FLC.*

The basic premise of this subsection is that *one should use a parsimonious parametric representation of a GT2 FS during an optimal design of a GT2 FLC*. Such a representation is one that is described by as few parameters as possible.

Clearly, the point representation of a GT2 FS is not parsimonious because it is not even a parametric representation. Similarly, the wavy-slice representation in which the wavy slices must be enumerated from all of the points of the GT2 FS is also a non-parametric representation. Neither of these representations is useful for an optimal design of a GT2 FLC.

[4]The material in this section is taken from Mendel (2014).

The horizontal-slice representation needs to be made parametric for it to be useful for an optimal design of a GT2 FLC. This representation requires choosing the number of horizontal slices and then parameterizing each of them. If, for example, each horizontal slice needs n_h parameters to describe it (i.e., assume that all of the horizontal slices have the same shape, and a higher horizontal slice is a squished version of the one just below it), and there are k horizontal slices, then this representation will require kn_h parameters, which is not parsimonious. Additionally, a change in the numerical value of k from k_1 to k_2, where $k_2 > k_1$, leads to $(k_2 - k_1)n_h$ additional parameters. Finally, there does not seem to be a simple way to map (squish) a horizontal slice at level α_1 into a horizontal slice at level α_2, for example, I do not know of a simple way to shrink the horizontal slice at level α_1 into a horizontal slice at level α_2 such that the latter horizontal slice is contained (nested) within the former horizontal slice [see Eq. (7.28)], as would be the case if, for example, all of the secondary MFs were triangles or trapezoids. Even if one could find such a mathematical transformation, it would have to be parsimonious, and then the quantification of the nesting of successive horizontal slices would have to be included as constraints during the optimization of a performance objective function as part of the optimal design of a GT2 FLC. Such a constrained optimization problem (although it does not yet exist) would be very challenging to solve, to say the least. Based on these arguments I conclude (somewhat surprisingly) that *the horizontal-slice representation of a GT2 FS is not useful for an optimal design of a GT2 FLC.*

This leaves only the vertical-slice representation of a GT2 FS. I will now demonstrate that this is a very flexible and parsimonious representation of such a FS.

There can be different ways to parameterize vertical slices. The worst way to do this is to parameterize each of them separately; such a representation would suffer from the same nonparsimony that the horizontal-slice representation does.

My suggestion for parameterizing the vertical-slice representation of a GT2 FS is to: (1) parameterize its FOU exactly as one presently parameterizes the FOU of an IT2 FS, and (2) parameterize the secondary MFs by choosing a fairly simple function that introduces only one new parameter.[5] Because the secondary MFs are vertical slices, they are always anchored on the already parameterized FOU. Examples of such secondary MFSs are:

1. *Triangle* The base of each triangle equals $\overline{\mu}_{\tilde{A}}(x) - \underline{\mu}_{\tilde{A}}(x)$, and its apex location, Apex(x), is parameterized as ($w \in [0, 1]$) (Liu, 2008; Mendel et al., 2009)

$$\text{Apex}(x) = \underline{\mu}_{\tilde{A}}(x) + w[\overline{\mu}_{\tilde{A}}(x) - \underline{\mu}_{\tilde{A}}(x)] \tag{7.42}$$

When $w = 0$, the secondary MF is a right triangle whose right angle is perpendicular to $\underline{\mu}_{\tilde{A}}(x)$; when $w = \frac{1}{2}$, the secondary MF is an isosceles triangle;

[5] Begin with secondary MFs that can be described by using only one new parameter; if performance is not acceptable, then use secondary MFs that can be described by using two new parameters; and so forth.

and, when $w = 1$, the secondary MF is a right triangle whose right angle is perpendicular to $\overline{\mu}_{\tilde{A}}(x)$. And w is treated as a design parameter, but it is the *same* for all of the vertical slices.

2. *Symmetrical Trapezoid* The base of each trapezoid equals $\overline{\mu}_{\tilde{A}}(x) - \underline{\mu}_{\tilde{A}}(x)$ and its top is defined by left and right end points, $\text{EP}_l(x)$ and $\text{EP}_r(x)$, which are parameterized as ($w \in [0, 1]$) (Liu, 2008; Mendel et al., 2009)

$$\text{EP}_l(x) = \underline{\mu}_{\tilde{A}}(x) + \frac{1}{2}w[\overline{\mu}_{\tilde{A}}(x) - \underline{\mu}_{\tilde{A}}(x)] \tag{7.43}$$

$$\text{EP}_r(x) = \overline{\mu}_{\tilde{A}}(x) - \frac{1}{2}w[\overline{\mu}_{\tilde{A}}(x) - \underline{\mu}_{\tilde{A}}(x)] \tag{7.44}$$

When $w = 0$, the trapezoid reduces to a square well, and *the GT2 FS reduces to an IT2 FS*; and, when $w = 1$, $\text{EP}_l(x) = \text{EP}_r(x)$, so that the trapezoid reduces to an isosceles triangle. Again w is treated as a design parameter, but it is the *same* for all of the vertical slices.

Another choice for a secondary MF is a nonsymmetrical trapezoid; however, it requires two parameters to define it, and so we leave its formulas to the reader [see Mendel et al. (2009) for an example].

Examples of the number of parameters that would have to be optimized during an optimal design of a GT2 FLC, for each of its GT2 FSs, are:

1. *FOU is Gaussian with uncertain mean* $m \in [m_1, m_2]$ *and/or standard deviation* $\sigma \in [\sigma_1, \sigma_2]$, *and triangle or symmetrical trapezoid secondary MFs*: four or five parameters per GT2 FS, namely $\theta = \{m_1, m_2, \sigma, w\}$, or $\theta = \{m, \sigma_1, \sigma_2, w\}$, or $\theta = \{m_1, m_2, \sigma_1, \sigma_2, w\}$.
2. *FOU is trapezoidal with a normal (not-necessarily symmetrical) trapezoid UMF* $\{a, b, c, d\}$ *and a subnormal (not-necessarily symmetrical) triangle LMF* $\{e, f, g, h\}$ [*h is the height of the LMF; if the LMF is normal, then* $h = 1$ *and it is described by* $\{e, f, g\}$], *and triangle or symmetrical trapezoid secondary MFs*: nine or eight parameters per GT2 FS, namely $\theta = \{a, b, c, d, e, f, g, h, w\}$ or $\theta = \{a, b, c, d, e, f, g, w\}$.

Clearly, a Gaussian FOU is more parsimonious than a trapezoidal FOU.[6]

My conclusion is that *for the optimal design of a GT2 FLC one should use the vertical-slice representation of its GT2 FSs.*

7.5.2.3 Performance Improvement[7]

We have seen, in Section 3.6.2, that sometimes the parameters of an IT2 FLC are optimized (tuned, learned) during

[6]Gaussian FOUs are also necessary for continuous GT2 FLCs (see Section 7.5.1.2).
[7]The material in this section is taken from Mendel (2014).

its design phase.[8] The optimized parameters are then fixed during its operational phase, unless continued adaptation is required, in which case online changes to parameters take place. Consequently, the parameters of a GT2 FLC can also be optimized during its design phase. The parameters of a GT2 FLC are in the antecedents, consequent, and (possibly) defuzzifier.

Recall, from Section 3.6.2, that during the optimal design of a GT2 FLC one sets up a mathematical objective function, $J(\boldsymbol{\phi})$, that depends upon the design parameters, $\boldsymbol{\phi}$. For a GT2 FLC, the elements of $\boldsymbol{\phi}$ include all of the antecedent and consequent MF parameters as well as any defuzzification parameters (if there are any), and $J(\boldsymbol{\phi})$ is again a nonlinear function of $\boldsymbol{\phi}$ and so some sort of mathematical programming approach has to be used to optimize it.

If computing derivatives was very difficult for IT2 FLCs, it will be considerably more difficult for GT2 FLCs. This is because, not only must such derivatives be computed for each horizontal slice, where the same secondary parameter (w) appears in all of the horizontal slices, but they must also be computed for the defuzzification method, in which the horizontal slices become coupled.

In conclusion, my recommendation (as it was for IT2 FLCs) is not to use gradient-based optimization algorithms for the designs of GT2 FLCs.

I maintain that using GT2 FSs in a FLC has the potential to provide better (and certainly no worse) performance for an FLC than using IT2 FSs. For this to be true one needs to use an optimization method that ensures that this happens. Using QPSO (see Section 3.6.2) lets us do this by using the following *design procedure*.

1. Design a T1 FLC by optimizing its parameters using QPSO.
2. Design an IT2 FLC by optimizing its parameters using QPSO in which one particle is associated with the just designed T1 FLC.
3. Design a GT2 FLC by optimizing its parameters using QPSO in which one particle is associated with the just designed IT2 FLC.

Mendel (2014) has proven that by virtue of the QPSO algorithm, the performance of an optimized GT2 FLC cannot be worse than that of an optimized IT2 FLC. This does not mean that the performance of the optimized GT2 FLC will be *significantly better* than that of the optimized IT2 FLC. As for the case of the IT2 FLC, there is no analysis that is available to date that focuses on such relative performance improvements.

Example 7.5 In step 3 of the above design procedure, one particle is associated with a just designed IT2 FLC. Suppose, for example, that the GT2 FLC is Mamdani + COS TR, the antecedent and consequent FOUs are Gaussian with uncertain mean $m \in [m_1, m_2]$, certain standard deviation σ, and have symmetrical trapezoid secondary MFs, and Eq. (7.27) is used for defuzzification. The structure of a particle

[8]In this Section I am focusing only on the optimization of MF parameters. Choosing the number of rules (M), MF shapes and t-norms are other important design issues but are not covered herein.

for such a GT2 FLC is[9]

$$\boldsymbol{\phi}_{\mathrm{GT2}} = \mathrm{col}(\underbrace{m_{11}^{1}, m_{12}^{1}, \sigma_{1}^{1}, w_{1}^{1}, \ldots, m_{p1}^{1}, m_{p2}^{1}, \sigma_{p}^{1}, w_{p}^{1}, mc_{1}^{1}, mc_{2}^{1}, \sigma c_{p}^{1}, wc_{p}^{1}}_{\text{Rule 1}}; \ldots; \underbrace{m_{11}^{M}, m_{12}^{M}, \sigma_{1}^{M}, w_{1}^{M}, \ldots, m_{p1}^{M}, m_{p2}^{M}, \sigma_{p}^{M}, w_{p}^{M}, mc_{1}^{M}, mc_{2}^{M}, \sigma c_{p}^{M}, wc_{p}^{M}}_{\text{Rule } M}) \quad (7.45)$$

The IT2 particle must be of the same length as this GT2 particle, begins with Eq. (7.45), and can be expressed as

$$\boldsymbol{\phi}_{\mathrm{IT2}} = \mathrm{col}(\underbrace{m_{11}^{1}, m_{12}^{1}, \sigma_{1}^{1}, 0, \ldots, m_{p1}^{1}, m_{p2}^{1}, \sigma_{p}^{1}, 0, mc_{1}^{1}, mc_{2}^{1}, \sigma c_{p}^{1}, 0}_{\text{Rule 1}}; \ldots; \underbrace{m_{11}^{M}, m_{12}^{M}, \sigma_{1}^{M}, 0, \ldots, m_{p1}^{M}, m_{p2}^{M}, \sigma_{p}^{M}, 0, mc_{1}^{M}, mc_{2}^{M}, \sigma c_{p}^{M}, 0}_{\text{Rule } M}) \quad (7.46)$$

Observe in Eq. (7.46), that:

- All of the MF parameters are taken from the optimized IT2 FLC design.
- By setting all of the secondary MF w parameters equal to 0, an IT2 FS is embedded into a GT2 FS (the secondary MFs become rectangular wells).
- In this way it is straightforward to embed an IT2 particle into a GT2 particle.

Optimally designing GT2 FLCs for guaranteed control system performance improvement is an important area for future research.

[9]In order to distinguish between the antecedent and consequent parameters, the consequent parameters are identified by using an additional letter c, for example, mc, σc and wc.

APPENDIX A

T2 FLC Software: From Type-1 to zSlices-Based General Type-2 FLCs

A.1 INTRODUCTION

This appendix is aimed at demonstrating the transition from T1 to IT2 FLCs and finally to zGT2 FLCs in terms of actual software implementations of the FLCs. In order to make it as useful as possible, this appendix is written as a *tutorial* and employs the freely available software framework Juzzy (Wagner, 2013) (available at http://juzzy.wagnerweb.net) written in Java that supports T1, IT2, and GT2 FLCs. Some of the relevant source code is included directly in the tutorial. It is expected that this will be helpful to the reader for the implementation of FLCs in Java and other programming languages, in particular those that are object oriented.

Section A.2 briefly reviews the details of the sample FLC that is developed throughout the tutorial, followed by individual subsections on the T1, IT2, and GT2 FLC implementations.

A.2 FLC FOR RIGHT-EDGE FOLLOWING

This tutorial is set around the same robotic application as detailed in Example 3.3 in Chapter 3, that is, the implementation of a right-edge (wall) following behavior. In control terms, we have a two-input (front and back sonar sensors on the right side of the robot) and single-output (steering angle) control problem. Both sonar sensors are modeled as variables with domain [0, 500] (e.g., in millimeters), and the steering angle is modeled as a variable with domain [−180, 180] (degrees).

In the sections below, we show and discuss the details of the FLC implementations for this problem based on T1, IT2, and zSlices-based GT2 FLCs. In particular, we highlight the similarities and differences as transitions are made from the T1 to the GT2 case and provide suggestions for a straightforward implementation of the FLCs and some general FLC design guidelines. Finally, we provide a brief discussion of the different FLCs and their outputs.

Note that we are not focusing on "control performance" or developing the "best" controller but on illustrating the different kinds of FLCs. In this context, the fuzzy

Introduction to Type-2 Fuzzy Logic Control: Theory and Applications, First Edition.
Jerry M. Mendel, Hani Hagras, Woei-Wan Tan, William W. Melek, and Hao Ying.

sets and the rule bases used are only examples and no optimization techniques have been applied to them. In a real application it is expected that the FLC would be designed so as to achieve desirable control system performance properties, using either expert knowledge, consensus, or one of the training techniques discussed earlier, for example, BP, QPSO, or GA.

A.3 TYPE-1 FLC SOFTWARE

In order to set up the T1 FLC we take a step-by-step approach that is later mirrored in the IT2 and zGT2 FLCs. Each step is detailed below.

A.3.1 Define and Set Up T1 FLC Inputs

We define both of the inputs (right-front and right-back sonar sensors) as numbers between 0 and 500. Figure A.1 depicts the definition of input objects in Java using the Juzzy software package. The input objects serve a twofold purpose: They define the domain for each input, and they allow one to directly associate the antecedent and the consequent objects of the rules, as is illustrated below.

A.3.2 Define T1 FSs That Quantify Each Variable

In order to maintain a low level of complexity in the fuzzy system (for illustration purposes), we model both sonar inputs using two trapezoidal fuzzy sets, corresponding to the linguistic labels Near and Far. Additionally, the steering angle output is modeled using three trapezoidal fuzzy sets corresponding to the linguistic labels Left, Zero, and Right. All MFs/FSs are shown in Figs. A.2–A.4, respectively.

Figure A.5 illustrates the setup of the T1 FSs in Java, based on the Juzzy fuzzy systems package. It shows how the individual fuzzy sets, shown in Figs. A.2–A.4, are created by the user, providing a label for each set as well as the four parameters required to define a trapezoidal membership function for a type 1 fuzzy set.

Note that the fuzzy sets for the right-front sensor (Fig. A.2) differ from those designed for the back sensor (Fig. A.3) because we are modeling the fact that the front sensor points at the wall at a $<90°$ angle, whereas the back sensor is perpendicular to the wall. Note further that the sets in Fig. A.4 do not overlap because they are output sets that are meant to capture the different ranges of the steering-angle output domain, that is, the labels Left, Zero, and Right.

```
//Define the inputs
rfs = new input ("Right Front Sonar", new Tuple(0,500));
rbs = new input ("Right Back Sonar", new Tuple(0,500));
```

Figure A.1 Defining FLC inputs in Java using the Juzzy package.

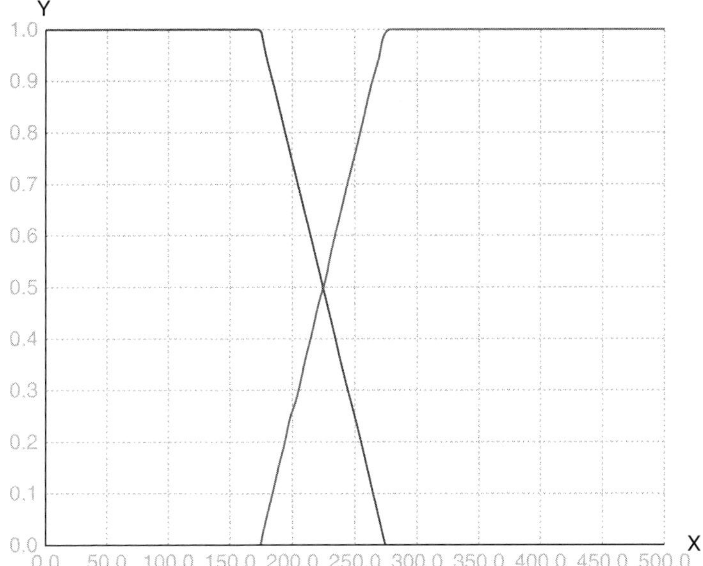

Figure A.2 Trapezoidal T1 FSs for Near and Far for the right-front sonar (RFS).

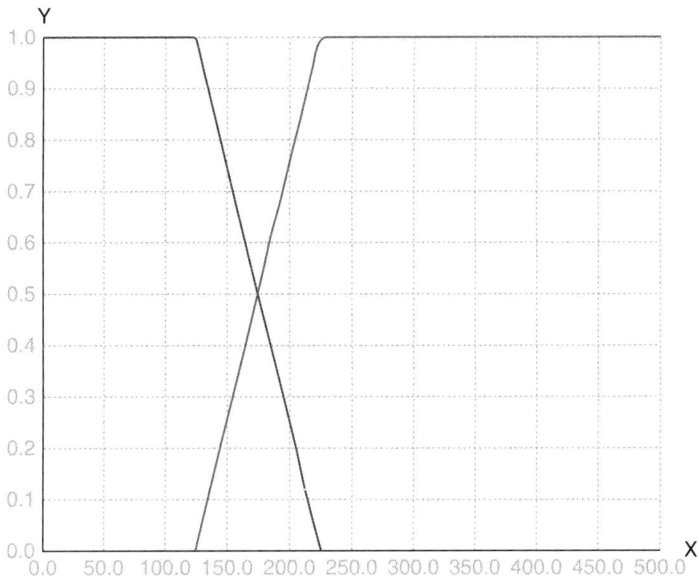

Figure A.3 Trapezoidal T1 FSs sets for Near and Far for the right-back sonar (RBS).

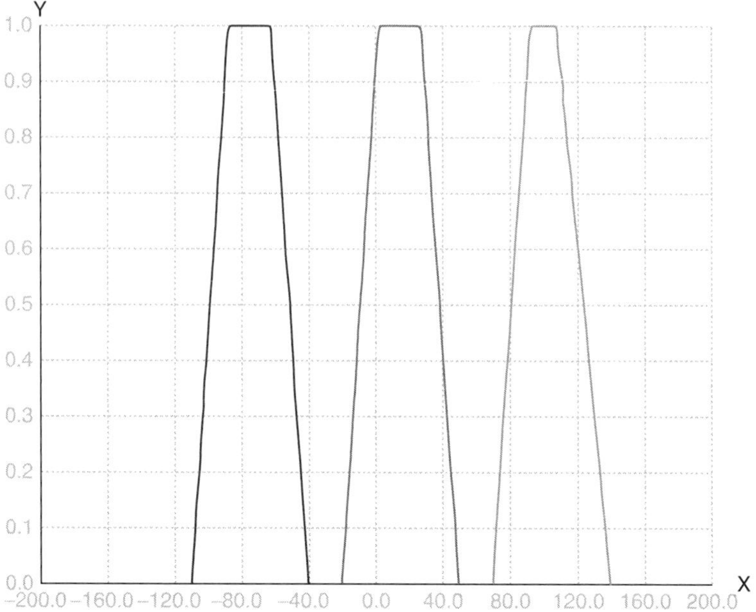

Figure A.4 Trapezoidal T1 FSs for Left, Zero, and Right steering angle.

A.3.3 Define Logical Antecedents and Consequents for the FL Rules

In preparation for designing the rule base for the T1 FLC, the previously created MFs are associated with antecedent and consequent terms. Note that each antecedent is labeled (the labels are later used to create a list of rules for the output) and that all antecedents are associated with the respective input defined previously (i.e., right front and back sonars). The source code for this step is illustrated in Fig. A.6.

A.3.4 Define Rule Base of T1 FLC

Having set up the antecedents and consequents of the T1 FLC, it is straightforward to create a new rule base and to add rules. Figure A.7 illustrates the set up of a new rule base, one with four rules, as well as the actual definition and inclusion of each rule. Note how the antecedents of the rules are specified as an array, followed by the respective consequent for each rule.

A rule base can be printed using the rule base object's toString() method. The resulting printout (which is human readable) for the rule base created in Fig. A.7 is given in Fig. A.8.

After specifying the rule base, the T1 FLC is ready for use. The control surface for the T1 FLC, based on minimum inference, combining all individual rule outputs through the union operation and applying centroid defuzzification, is shown in Fig. A.9 (both inputs are discretized into 100 steps). The method employed using

```
T1MF_Trapezoidal rfsNearMF = new T1MF_Trapezoidal("MF for Near distance of the RFS", new double[]{0.0, 0.0, 125.0, 225.0});
T1MF_Trapezoidal rfsFarMF = new T1MF_Trapezoidal("MF for Far distance for the RFS", new double[]{125.0, 225.0, 500.0, 500.0});

T1MF_Trapezoidal rbsNearMF = new T1MF_Trapezoidal("MF for Near distance of the RBS", new double[]{0.0, 0.0, 175.0, 275.0});
T1MF_Trapezoidal rbsFarMF = new T1MF_Trapezoidal("MF for Far distance for the RBS", new double[]{175.0, 275.0, 500.0, 500.0});

T1MF_Trapezoidal leftMF = new T1MF_Trapezoidal("MF for Left Steering Angle", new double[]{-110.0, -87.5, -62.5, -40.0});
T1MF_Trapezoidal zeroMF = new T1MF_Trapezoidal("MF for Zero Steering Angle", new double[]{-20.0, 2.5, 27.5, 50.0});
T1MF_Trapezoidal rightMF = new T1MF_Trapezoidal("MF for Right Steering Angle", new double[]{70.0, 92.5, 107.5, 140.0});
```

Figure A.5 Source code snippet showing the setup of the FSs/MFs for the T1 FLC.

```
T1_Antecedent rfsNear = new T1_Antecedent("Near distance of the RFS",rfsNearMF, rfs);
T1_Antecedent rfsFar = new T1_Antecedent("Far distance of the RFS",rfsFarMF, rfs);

T1_Antecedent rbsNear = new T1_Antecedent("Near distance of the RBS",rbsNearMF, rbs);
T1_Antecedent rbsFar = new T1_Antecedent("Far distance of the RBS",rbsFarMF, rbs);

T1_Consequent left = new T1_Consequent("Left", leftMF);
T1_Consequent zero = new T1_Consequent("Zero", zeroMF);
T1_Consequent right = new T1_Consequent("Right", rightMF);
```

Figure A.6 Definitions of antecedent and consequent terms in Juzzy.

```
rulebase = new T1_Rulebase(4);
rulebase.addRule(new T1_Rule(new T1_Antecedent[]{rfsNear, rbsNear}, left));
rulebase.addRule(new T1_Rule(new T1_Antecedent[]{rfsNear, rbsFar}, left));
rulebase.addRule(new T1_Rule(new T1_Antecedent[]{rfsFar, rbsNear}, zero));
rulebase.addRule(new T1_Rule(new T1_Antecedent[]{rfsFar, rbsFar}, right));
```

Figure A.7 Creation of a new T1 rule base in Juzzy.

```
IF Near distance of the RFS AND Near distance of the RBS THEN Left
IF Near distance of the RFS AND Far distance of the RBS THEN Left
IF Far distance of the RFS AND Near distance of the RBS THEN Zero
IF Far distance of the RFS AND Far distance of the RBS THEN Right
```

Figure A.8 Printout of T1 FLC rule base (the structure of this printout can be changed within the Juzzy software to accommodate other rule formats, such as, for example, "If distance of the RFS is Near AND distance of the RBS is Near THEN Output is Left").

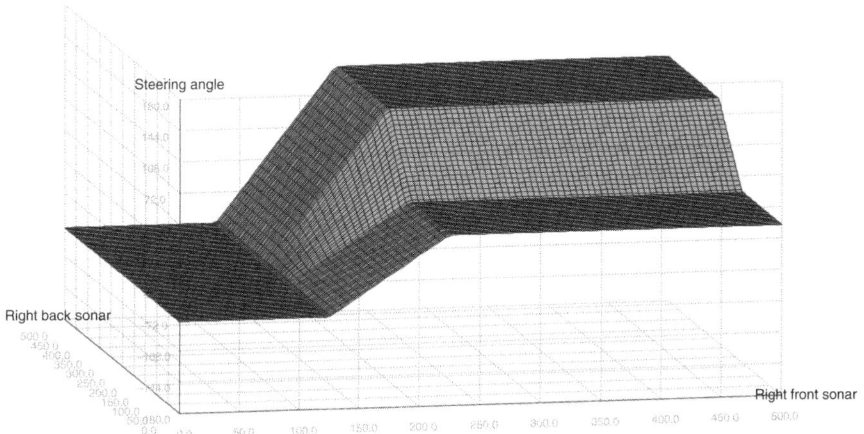

Figure A.9 Control surface for the T1 FLC.

the Juzzy framework to create the control surface is included in Fig. A.10. Note that the control surface in Fig. A.9 is rather "angular," that is, the control signal changes abruptly at different points over the control surface. A smoother control surface is generally expected for T2 FLCs, and this expectation will be demonstrated in the subsequent sections.

A.4 INTERVAL T2 FLC SOFTWARE

Similar to the just described approach for constructing a T1 FLC, in this section we proceed to set up an IT2 Mamdani FLC step by step, highlighting in particular the similarities and differences between both kinds of FLCs.

```
private void plotControlSurface(boolean useCentroidDefuzzification, int input1Discs, int input2Discs)
{
    double output;
    double[] x = new double[input1Discs];
    double[] y = new double[input2Discs];
    double[][] z = new double[y.length][x.length];
    double incrX, incrY;
    incrX = rfs.getDomain().getSize()/(input1Discs-1.0);
    incrY = rbs.getDomain().getSize()/(input2Discs-1.0);

    //first, get the discretisation values
    for(int currentX=0; currentX<input1Discs; currentX++)
    {
        x[currentX] = currentX * incrX;
    }
    for(int currentY=0; currentY<input2Discs; currentY++)
    {
        y[currentY] = currentY * incrY;
    }

    for(int currentX=0; currentX<input1Discs; currentX++)
    {
        rfs.setInput(x[currentX]); //set the front sonar input
        for(int currentY=0; currentY<input2Discs; currentY++)
        {
            rbs.setInput(y[currentY]); //set the back sonar input
            if(useCentroidDefuzzification)
                output = rulebase.evaluate(1, 100);
            else
                output = rulebase.evaluate(0, 0);
            z[currentY][currentX] = output;
        }
    }

    //now do the plotting (relies on JMathPlot library: http://code.google.com/p/jmathplot/ (Sep. 2012))
    JMathPlotter plotter = new JMathPlotter(17, 17, 14);
    plotter.plotControlSurface("Type-1 Control Surface",
            new String[]{rfs.getName(), rbs.getName(), "Steering Angle"}, x, y, z, new Tuple(-180,180.0));
   plotter.show("Type-1 Fuzzy Logic System Control Surface for Robot Control Example");
}
```

Figure A.10 Code to execute the FLC with the purpose of generating the control surface.

A.4.1 Define and Set Up FLC Inputs

Because the actual inputs to the IT2 FLC are identical to those used in the T1 FLC case, the setup of the IT2 FLC inputs is identical to that of the T1 case.

A.4.2 Define IT2 FSs That Quantify Each Variable

All the sets in our IT2 FLC are IT2 FSs, which are based on blurred (see Section 3.6.1) T1 FSs. Consequently, we chose to model the inputs as trapezoidal IT2 FSs, as shown in Figs. A.11 and A.12. Additionally, the steering angle is modeled using three trapezoidal fuzzy sets, as shown in Fig. A.13. The rule base of the IT2 FLC is identical to that of the T1 FLC because the structure of a rule base does not change; it is only the FSs that are used to model the antecedents and consequents that change.

Figure A.14 illustrates the setup of the IT2 FSs in Java, based on the Juzzy fuzzy systems package. Observe how each IT2 FS is defined by using both an upper and a lower T1 MF.

A.4.3 Define Logical Antecedents and Consequents for the FL Rules

After all FSs have been defined, all of them are associated with antecedents and consequents (as in the T1 case) in preparation for the construction of FL rules. Figure A.15 shows the construction of the respective antecedents and consequents in Juzzy for the IT2 FLC.

A.4.4 Define Rule Base of the IT2 FLC

The example four-rule rule base is specified as for the T1 case, and is shown in Fig. A.16.

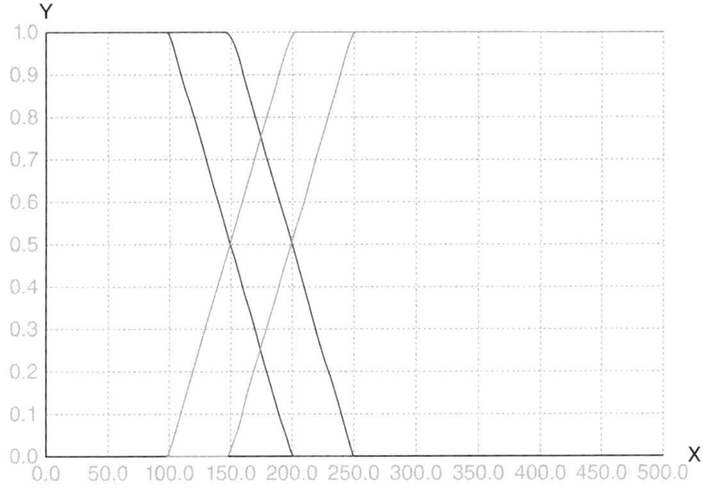

Figure A.11 Trapezoidal IT2 FSs for Near and Far for the right-front sonar (RFS).

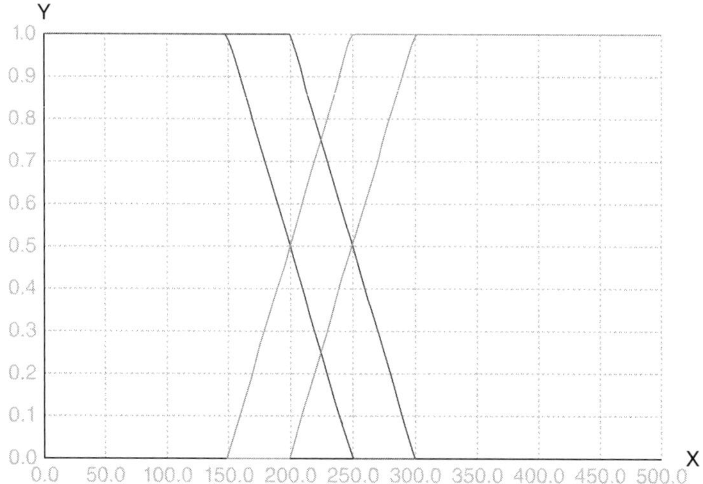

Figure A.12 Trapezoidal IT2 FSs for Near and Far for the right-back sonar (RBS).

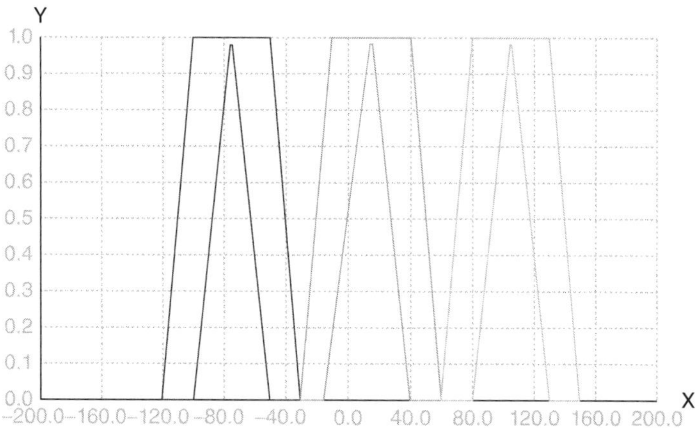

Figure A.13 Trapezoidal IT2 FSs for Left, Zero, and Right steering angle.

Having specified the rule base, the resulting IT2 FLC is ready for use. For illustrative purposes, the resulting rule base that uses minimum inference, KM center-of-sets type reduction [Eqs. (3.64)–(3.66)] and 100 discretization steps for both inputs is shown in Fig. A.17. The method employed using the Juzzy framework to create the control surface is included in Fig. A.18.

Comparing the IT2 FLC control surface (Fig. A.17) with that of the T1 FLC control surface (Fig. A.9) highlights the effects of using IT2 FSs, that is, the control surface of the IT2 FLC seems smoother overall than that of the T1 FLC. In other words, the IT2 FL control output changes less abruptly as the input measurements to the FLC change.

```
T1MF_Trapezoidal rfsNearUMF = new T1MF_Trapezoidal("Upper MF for Near distance of the RFS",new double[]{0.0, 0.0, 150.0, 250.0});
T1MF_Trapezoidal rfsNearLMF = new T1MF_Trapezoidal("Lower MF for Near distance of the RFS",new double[]{0.0, 0.0, 100.0, 200.0});
IntervalT2MF_Trapezoidal rfsNearMF = new IntervalT2MF_Trapezoidal("IT2MF for Near distance of the RFS",rfsNearUMF,rfsNearLMF);

T1MF_Trapezoidal rfsFarUMF = new T1MF_Trapezoidal("Upper MF for Far distance of the RFS",new double[]{100.0, 200.0, 500.0, 500.0});
T1MF_Trapezoidal rfsFarLMF = new T1MF_Trapezoidal("Lower MF for Far distance of the RFS",new double[]{150.0, 250.0, 500.0, 500.0});
IntervalT2MF_Trapezoidal rfsFarMF = new IntervalT2MF_Trapezoidal("IT2MF for Far distance of the RFS",rfsFarUMF,rfsFarLMF);

T1MF_Trapezoidal rbsNearUMF = new T1MF_Trapezoidal("Upper MF for Near distance of the RBS",new double[]{0.0, 0.0, 200.0, 300.0});
T1MF_Trapezoidal rbsNearLMF = new T1MF_Trapezoidal("Lower MF for Near distance of the RBS",new double[]{0.0, 0.0, 150.0, 250.0});
IntervalT2MF_Trapezoidal rbsNearMF = new IntervalT2MF_Trapezoidal("IT2MF for Near distance of the RBS",rbsNearUMF,rbsNearLMF);

T1MF_Trapezoidal rfsFarUMF = new T1MF_Trapezoidal("Upper MF for Far distance of the RBS",new double[]{150.0, 250.0, 500.0, 500.0});
T1MF_Trapezoidal rfsFarLMF = new T1MF_Trapezoidal("Lower MF for Far distance of the RBS",new double[]{200.0, 300.0, 500.0, 500.0});
IntervalT2MF_Trapezoidal rbsFarMF = new IntervalT2MF_Trapezoidal("IT2MF for Far distance of the RBS",rbsFarUMF,rbsFarLMF);

T1MF_Trapezoidal leftUMF = new T1MF_Trapezoidal("Upper MF for Left Steering Angle",new double[]{-120.0, -100.0, -50.0, -30.0});
T1MF_Trapezoidal leftLMF = new T1MF_Trapezoidal("Lower MF for Left Steering Angle",new double[]{-100.0, -75.0, -75.0, -50.0});
IntervalT2MF_Trapezoidal leftMF = new IntervalT2MF_Trapezoidal("IT2MF for Left Steering Angle",leftUMF,leftLMF);

T1MF_Trapezoidal zeroUMF = new T1MF_Trapezoidal("Upper MF for Zero Steering Angle",new double[]{-30.0, -10.0, 40.0, 60.0});
T1MF_Trapezoidal zeroLMF = new T1MF_Trapezoidal("Lower MF for Zero Steering Angle",new double[]{-15.0, 15.0, 15.0, 40.0});
IntervalT2MF_Trapezoidal zeroMF = new IntervalT2MF_Trapezoidal("IT2MF for Zero Steering Angle",zeroUMF,zeroLMF);

T1MF_Trapezoidal rightUMF = new T1MF_Trapezoidal("Upper MF for Right Steering Angle",new double[]{60.0, 80.0, 130.0, 150.0});
T1MF_Trapezoidal rightLMF = new T1MF_Trapezoidal("Lower MF for Right Steering Angle",new double[]{80.0, 105.0, 105.0, 130.0});
IntervalT2MF_Trapezoidal rightMF = new IntervalT2MF_Trapezoidal("IT2MF for Right Steering Angle",rightUMF,rightLMF);
```

Figure A.14 Definitions of IT2 FSs in Juzzy; each IT2 FS is based on an upper and a lower T1 MF.

```
IT2_Antecedent rfsNear = new IT2_Antecedent("Near distance of the RFS",rfsNearMF, rfs);
IT2_Antecedent rfsFar = new IT2_Antecedent("Far distance of the RFS",rfsFarMF, rfs);

IT2_Antecedent rbsNear = new IT2_Antecedent("Near distance of the RBS",rbsNearMF, rbs);
IT2_Antecedent rbsFar = new IT2_Antecedent("Far distance of the RBS",rbsFarMF, rbs);

IT2_Consequent left = new IT2_Consequent("Left", leftMF);
IT2_Consequent zero = new IT2_Consequent("Zero", zeroMF);
IT2_Consequent right = new IT2_Consequent("Right", rightMF);
```

Figure A.15 Antecedent and consequent definitions for the IT2 FLC in Juzzy.

```
rulebase = new IT2_Rulebase(4);
rulebase.addRule(new IT2_Rule(new IT2_Antecedent[]{rfsNear, rbsNear}, left));
rulebase.addRule(new IT2_Rule(new IT2_Antecedent[]{rfsNear, rbsFar}, left));
rulebase.addRule(new IT2_Rule(new IT2_Antecedent[]{rfsFar, rbsNear}, zero));
rulebase.addRule(new IT2_Rule(new IT2_Antecedent[]{rfsFar, rbsFar}, right));
```

Figure A.16 Creation of a new IT2 rule base in Juzzy.

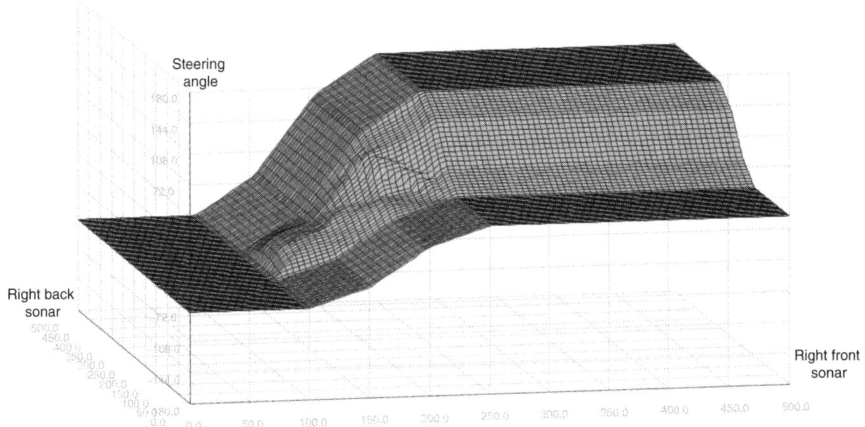

Figure A.17 Control surface for the IT2 FLC.

A.5 zSLICES-BASED GENERAL TYPE-2 FLC SOFTWARE

Building directly on the just-described implementation of an IT2 FLC, we proceed in this section to design and implement a zSlices-based GT2 FLC. In order to maintain tractability of the different zLevels, especially in the figures, we use only four zLevels for the zFLC; hence, all zSlices-based sets will be modeled using four zSlices. In the rest of this section, we provide step-by-step details for the implementation of the zFLC, analogous to previous sections on the T1 and IT2 FLCs.

A.5.1 Define and Set Up FLC Inputs

Because the actual inputs to the zFLC are identical to those for the T1 and IT2 FLC cases, the setup of the zFLC inputs is identical to that of the T1 and IT2 cases.

A.5.2 Define zSlices-Based GT2 FSs That Quantify Each Variable

All zSlices-based GT2 FSs are based on the IT2 FSs from the previous section. The third dimension for each zSlices-based GT2 FS is formed by creating four zSlices that are evenly distributed within the FOU of each of the original IT2 FSs. Consequently, both inputs are modeled using trapezoidal zSlices-based fuzzy sets as shown in Figs. A.19 and A.20. Additionally,

```
private void plotControlSurface(boolean useCentroidDefuzzification, int input1Discs, int input2Discs)
{
    double output;
    double[] x = new double[input1Discs];
    double[] y = new double[input2Discs];
    double[][] z = new double[y.length][x.length];
    double incrX, incrY;
    incrX = rfs.getDomain().getSize()/(input1Discs-1.0);
    incrY = rbs.getDomain().getSize()/(input2Discs-1.0);

    //first, get the discretisation values
    for(int currentX=0; currentX<input1Discs; currentX++)
    {
        x[currentX] = currentX * incrX;
    }
    for(int currentY=0; currentY<input2Discs; currentY++)
    {
        y[currentY] = currentY * incrY;
    }

    for(int currentX=0; currentX<input1Discs; currentX++)
    {
        rfs.setInput(x[currentX]);
        for(int currentY=0; currentY<input2Discs; currentY++)
        {
            rbs.setInput(y[currentY]);
            if(useCentroidDefuzzification)
                output = rulebase.evaluate(1, 100);
            else
                output = rulebase.evaluate(0,0);
            z[currentY][currentX] = output;
        }
    }

    //now do the plotting (relies on JMathPlot library: http://code.google.com/p/jmathplot/ (Sep. 2012))
    JMathPlotter plotter = new JMathPlotter(17, 17, 14);
    plotter.plotControlSurface("Internal Type-2 Control Surface",
            new String[]{rfs.getName(), rbs.getName(), "Steering Angle"}, x, y, z, new Tuple(-180,180.0));
   plotter.show("Interval Type-2 Fuzzy Logic System Control Surface for Robot Control Example");
}
```

Figure A.18 Code to execute the IT2 FLC for the purpose of generating the control surface.

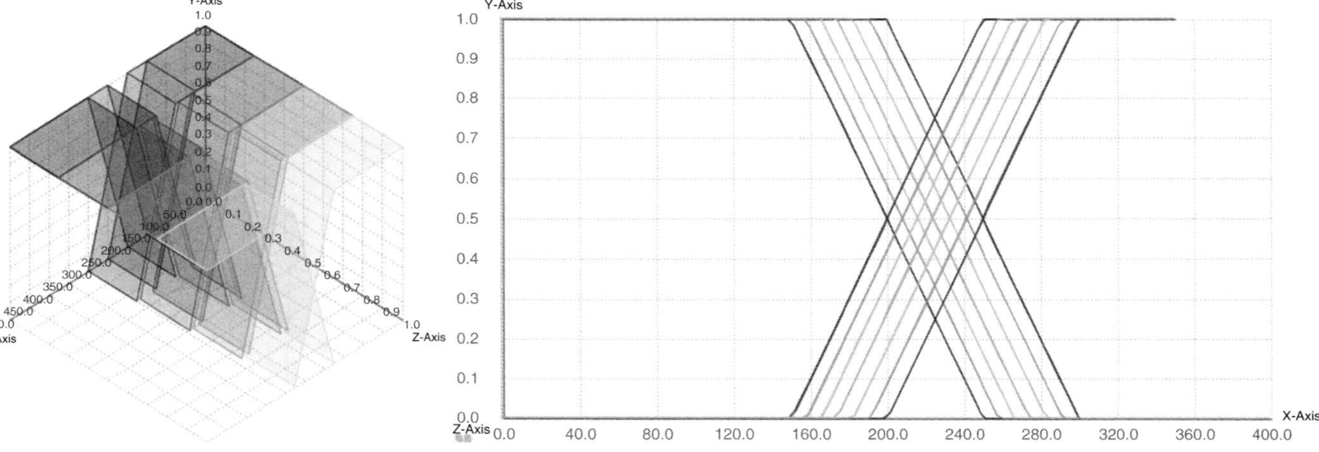

Figure A.19 Two views of the same trapezoidal zSlices-based GT2 FSs for Near and Far for the right-front sonar (RFS) (four zLevels).

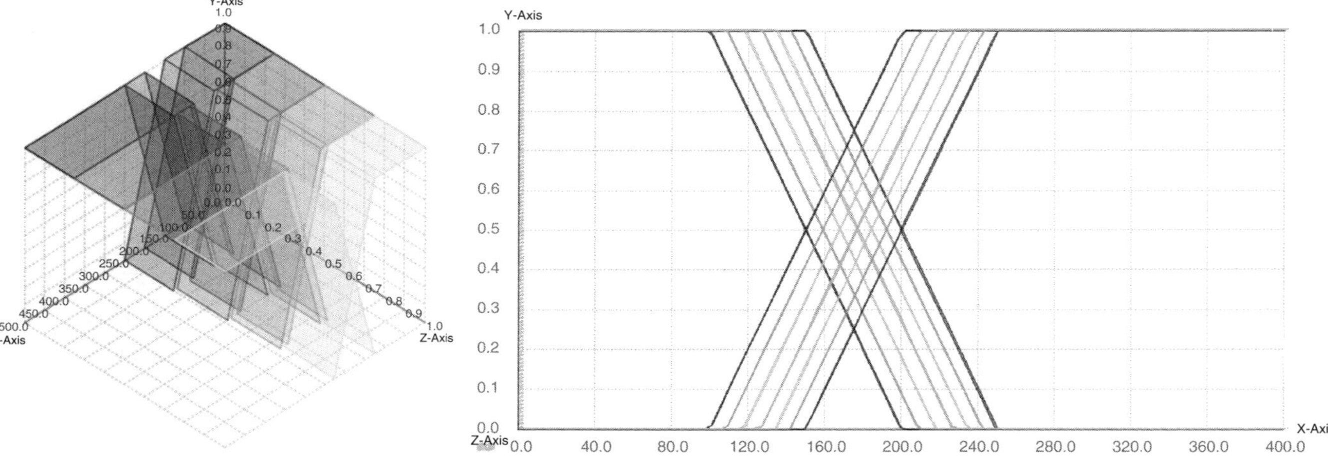

Figure A.20 Two views of the same trapezoidal zSlices-based GT2 FSs for Near and Far for the right-back sonar (RBS) (four zLevels).

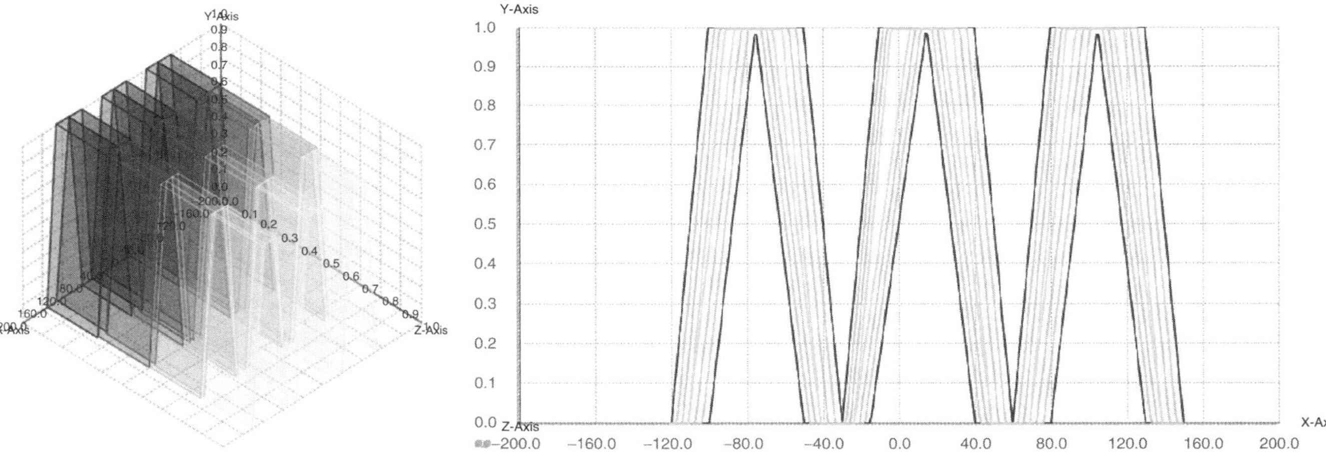

Figure A.21 Two views of the same trapezoidal zSlices-based GT2 FSs for Left, Zero, and Right steering angle (four zLevels).

```
//Set up the lower and upper membership functions (MFs) making up the
//overall Interval Type-2 Fuzzy Sets for each input and output
T1MF_Trapezoidal rfsNearUMF = new T1MF_Trapezoidal("Upper MF for Near distance of the RFS",new double[]{0.0, 0.0, 150.0, 250.0});
T1MF_Trapezoidal rfsNearLMF = new T1MF_Trapezoidal("Lower MF for Near distance of the RFS",new double[]{0.0, 0.0, 100.0, 200.0});
IntervalT2MF_Trapezoidal rfsNearIT2MF = new IntervalT2MF_Trapezoidal("IT2MF for Near distance of the RFS",rfsNearUMF,rfsNearLMF);
//now spawn a basic zSlices-based set with 4 zLevels
GenT2zMF_Trapezoidal rfsNearMF = new GenT2zMF_Trapezoidal("zGT2MF for Near distance of the RFS", rfsNearIT2MF, numberOfzLevels);

T1MF_Trapezoidal rfsFarUMF = new T1MF_Trapezoidal("Upper MF for Far distance of the RFS",new double[]{100.0, 200.0, 500.0, 500.0});
T1MF_Trapezoidal rfsFarLMF = new T1MF_Trapezoidal("Lower MF for Far distance of the RFS",new double[]{150.0, 250.0, 500.0, 500.0});
IntervalT2MF_Trapezoidal rfsFarIT2MF = new IntervalT2MF_Trapezoidal("IT2MF for Far distance of the RFS",rfsFarUMF,rfsFarLMF);
GenT2zMF_Trapezoidal rfsFarMF = new GenT2zMF_Trapezoidal("zGT2MF for Far distance of the RFS", rfsFarIT2MF, numberOfzLevels);

T1MF_Trapezoidal rbsNearUMF = new T1MF_Trapezoidal("Upper MF for Near distance of the RBS",new double[]{0.0, 0.0, 200.0, 300.0});
T1MF_Trapezoidal rbsNearLMF = new T1MF_Trapezoidal("Lower MF for Near distance of the RBS",new double[]{0.0, 0.0, 150.0, 250.0});
IntervalT2MF_Trapezoidal rbsNearIT2MF = new IntervalT2MF_Trapezoidal("IT2MF for Near distance of the RBS",rbsNearUMF,rbsNearLMF);
GenT2zMF_Trapezoidal rbsNearMF = new GenT2zMF_Trapezoidal("zGT2MF for Near distance of the RBS", rbsNearIT2MF, numberOfzLevels);

T1MF_Trapezoidal rbsFarUMF = new T1MF_Trapezoidal("Upper MF for Far distance of the RBS",new double[]{150.0, 250.0, 500.0, 500.0});
T1MF_Trapezoidal rbsFarLMF = new T1MF_Trapezoidal("Lower MF for Far distance of the RBS",new double[]{200.0, 300.0, 500.0, 500.0});
IntervalT2MF_Trapezoidal rbsFarIT2MF = new IntervalT2MF_Trapezoidal("IT2MF for Far distance of the RBS",rbsFarUMF,rbsFarLMF);
GenT2zMF_Trapezoidal rbsFarMF = new GenT2zMF_Trapezoidal("zGT2MF for Far distance of the RBS", rbsFarIT2MF, numberOfzLevels);

T1MF_Trapezoidal leftUMF = new T1MF_Trapezoidal("Upper MF for Left Steering Angle",new double[]{-120.0, -100.0, -50.0, -30.0});
T1MF_Trapezoidal leftLMF = new T1MF_Trapezoidal("Lower MF for Left Steering Angle",new double[]{-100.0, -75.0, -75.0, -50.0});
IntervalT2MF_Trapezoidal leftIT2MF = new IntervalT2MF_Trapezoidal("IT2MF for Left Steering Angle",leftUMF,leftLMF);
GenT2zMF_Trapezoidal leftMF = new GenT2zMF_Trapezoidal("zGT2MF for Left Steering Angle", leftIT2MF, numberOfzLevels);

T1MF_Trapezoidal zeroUMF = new T1MF_Trapezoidal("Upper MF for Zero Steering Angle",new double[]{-30.0, -10.0, 40.0, 60.0});
T1MF_Trapezoidal zeroLMF = new T1MF_Trapezoidal("Lower MF for Zero Steering Angle",new double[]{-15.0, 15.0, 15.0, 40.0});
IntervalT2MF_Trapezoidal zeroIT2MF = new IntervalT2MF_Trapezoidal("IT2MF for Zero Steering Angle",zeroUMF,zeroLMF);
GenT2zMF_Trapezoidal zeroMF = new GenT2zMF_Trapezoidal("zGT2MF for Zero Steering Angle", zeroIT2MF, numberOfzLevels);

T1MF_Trapezoidal rightUMF = new T1MF_Trapezoidal("Upper MF for Right Steering Angle",new double[]{60.0, 80.0, 130.0, 150.0});
T1MF_Trapezoidal rightLMF = new T1MF_Trapezoidal("Lower MF for Right Steering Angle",new double[]{80.0, 105.0, 105.0, 130.0});
IntervalT2MF_Trapezoidal rightIT2MF = new IntervalT2MF_Trapezoidal("IT2MF for Right Steering Angle",rightUMF,rightLMF);
GenT2zMF_Trapezoidal rightMF = new GenT2zMF_Trapezoidal("zGT2MF for Right Steering Angle", rightIT2MF, numberOfzLevels);
```

Figure A.22 Definition of zSlices-based GT2 FSs in Juzzy; each zGT2 set is based on an IT2 FS (itself based on an upper and lower T1 FSs) and is constructed using four zSlices.

```
GenT2Z_Antecedent rfsNear = new GenT2z_Antecedent("Near distance of the RFS",rfsNearMF, rfs);
GenT2Z_Antecedent rfsFar = new GenT2z_Antecedent("Far distance of the RFS",rfsFarMF, rfs);

GenT2Z_Antecedent rbsNear = new GenT2z_Antecedent("Near distance of the RBS",rbsNearMF, rbs);
GenT2Z_Antecedent rbsFar = new GenT2z_Antecedent("Far distance of the RBS",rbsFarMF, rbs);

GenT2ZEngine_Defuzzification gT2zED = new GenT2zEngine_Defuzzification(100);
GenT2Z_Consequent left = new GenT2z_Consequent("Left", leftMF, gT2zED);
GenT2Z_Consequent zero = new GenT2z_Consequent("Zero", zeroMF, gT2zED);
GenT2Z_Consequent right = new GenT2z_Consequent("Right", rightMF, gT2zED);
```

Figure A.23 Antecedent and consequent definitions for the zSlices-based GT2 FLC in Juzzy.

```
rulebase = new GenT2z_Rulebase(4);
rulebase.addRule(new GenT2z_Rule(new GenT2z_Antecedent[]{rfsNear, rbsNear}, left));
rulebase.addRule(new GenT2z_Rule(new GenT2z_Antecedent[]{rfsNear, rbsFar}, left));
rulebase.addRule(new GenT2z_Rule(new GenT2z_Antecedent[]{rfsFar, rbsNear}, zero));
rulebase.addRule(new GenT2z_Rule(new GenT2z_Antecedent[]{rfsFar, rbsFar}, right));
```

Figure A.24 Creation of zSlices-based GT2 rule base in Juzzy.

the steering output is modeled using the three trapezoidal zSlices-based GT2 FSs shown in Fig. A.21. Each of these figures includes both a rear/side view as well as a front view of the respective sets so as to facilitate their visualization. Note, also, that in all zSlices-based set figures, the right-hand figure indicates the upper and lower membership functions of each zSlice for the Near and Far sets, similar to the common representation of IT2 FSs. The left-hand figure provides a 3D view in which each zSlice is shown in a different shade of gray. In reality, each zSlice "protrudes" from its zLevel to $z = 0$ but is only shown from its zLevel to the next zLevel below it to facilitate visualization (e.g., z2 is shown to protrude from $z = z2$ to $z = z1$).

The process of creating the zSlices-based GT2 FSs is illustrated by the source code depicted in Fig. A.22, obtained from the Juzzy package.

Note, as stated earlier, that it is a requirement for all zSlices-based GT2 FSs in a given GT2 FLC to have the same number of zLevels because this enables the computation of each zLevel in isolation (and a later recombination of the results).

A.5.3 Define Logical Antecedents and Consequents for the FL Rules

As in the IT2 case, after all FSs have been defined, all of them are associated with antecedents and consequents, in preparation for the construction of FL rules.

Figure A.23 shows the construction of the respective antecedents and consequents in Juzzy for the GT2 FLC.

A.5.4 Define Rule Base of the GT2 FLC

Because the antecedents and consequents are the same as in the IT2 case, the example four-rule rule base of the zFLC is identical to that of the IT2 FLC (only the underlying FSs differ). The code for the construction (in Juzzy) of the rule base for the zFLC is provided in Fig. A.24.

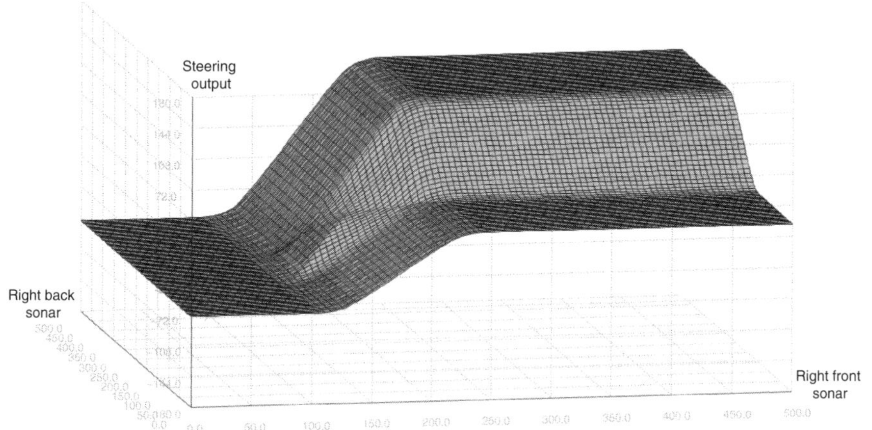

Figure A.25 Control surface for zSlices-based GT2 FLC.

```
private void plotControlSurfaceMC(int input1Discs, int input2Discs)
{
    double output;
    double[] x = new double[input1Discs];
    double[] y = new double[input2Discs];

    double[][] z = new double[y.length][x.length];
    double incrX, incrY;
    incrX = rfs.getDomain().getSize()/(input1Discs-1.0);
    incrY = rbs.getDomain().getSize()/(input2Discs-1.0);

    //first, get the values
    for(int currentX=0; currentX<input1Discs; currentX++)
    {
        x[currentX] = currentX * incrX;
    }
    for(int currentY=0; currentY<input2Discs; currentY++)
    {
        y[currentY] = currentY * incrY;
    }

    for(int currentX=0; currentX<input1Discs; currentX++)
    {
        rfs.setInput(x[currentX]); //set the front sonar input
        for(int currentY=0; currentY<input2Discs; currentY++)
        {
            rbs.setInput(y[currentY]); //set the back sonar input
            output = factory.runFactory();

            z[currentY][currentX] = output;
        }
    }

    //now do the plotting (relies on JMathPlot library: http://code.google.com/p/jmathplot/ (Sep. 2012))
    JMathPlotter plotter = new JMathPlotter();
    plotter.plotControlSurface("Control Surface",
            new String[]{rfs.getName(), rbs.getName(), "Steering Output"}, x, y, z, new Tuple(-180.0, 180.0));
    plotter.show("zSlices based General Type-2 Fuzzy Logic System Control Surface for Tipping Example");
}
```

Figure A.26 Code to execute the FLC with the purpose of generating the control surface.

After specifying the rule base, the zFLC is ready for execution. During execution, each zLevel (i.e., four in our case) is computed in parallel (subject to computational resources) and the output of each individual zLevel is recombined to produce the overall result, as explained in Chapter 7. The control surface for the zFLC is shown in Fig. A.25. The method employed using the Juzzy framework to create the control surface is included in Fig. A.26.

Similar to Section A.4.4, direct comparisons of the T1 FLC, IT2 FLC, and zFLC, in terms of absolute performance, are not provided here because they are highly dependent on the robot, its settings, and the like.

Comparing the zFLC control surface (Fig. A.25) with that of the IT2 FLC control surface (Fig. A.17) highlights the impact of the underlying zSlices-based GT2 FSs. Observe that the control surface of the zFLC is smoother overall than that of the IT2 FLC, so its control output will change less drastically than that of the IT2 FLC as its input measurements change.

REFERENCES

J. Aisbett, J. T. Rickard, and D. G. Morgenthaler, "Type-2 fuzzy sets as functions on spaces," *IEEE Trans. on Fuzzy Systems*, vol. 18, pp. 841–844, Aug. 2010.

R. A. Aliev, W. Pedrycz, B. G. Guirimov, R. R. Aliev, U. Ilhan, M. Babagil, and S. Mammadli, "Type-2 fuzzy neural networks with fuzzy clustering and differential evolution optimization," *Information Sciences*, vol. 181, pp. 1591–1608, 2011.

M. Biglarbegian, "On the design of robust intelligent controllers with application to mobile robot tracking," Proceedings of 2012 IEEE American Control Conf., Montreal, Canada, pp. 4879–4884, 2012.

M. Biglarbegian, W. W. Melek, and J. M. Mendel, "Parametric design of stable type-2 TSK fuzzy systems," Proceedings of 2008 North American Information Processing Society Conf., New York, pp. 1–6, 2008.

M. Biglarbegian, W. W. Melek, and J. M. Mendel, "A practical approach for design of PD and PI-like interval type-2 fuzzy controllers," Proceedings of 2009 IEEE Conf. on Systems, Man and Cybernetics, San Antonio, TX, pp. 255–261, 2009.

M. Biglarbegian, W. W. Melek, and J. M. Mendel, "On the stability of interval type-2 TSK fuzzy logic control systems," *IEEE Trans. on Systems, Man and Cybernetics—Part B: Cybernetics*, vol. 40, pp. 798–818, June 2010.

M. Biglarbegian, W. W. Melek, and J. M. Mendel, "Design of novel interval type-2 fuzzy controllers for modular and reconfigurable robots: Theory and experiments," *IEEE Trans. on Industrial Electronics*, vol. 58, pp. 1371–1384, April 2011.

S. P. Boyd, L. E. Ghaoui, E. Feron, and V. Balakrishnan, *Linear Matrix Inequalities in System and Control Theory*, Philadelphia: SIAM, 1994.

British Standards, *BS 5514 Reciprocating Internal Combustion Engines: Speed Governing*, 2nd ed., accessed at http://www.bsonline.bsiglobal.com/server/index.jsp.

O. Castillo and P. Melin, "A review on the design and optimization of interval type-2 fuzzy controllers," *Applied Soft Computing*, vol. 12, no. 4, pp. 1267–1278, 2012.

J. Castro, "Fuzzy logic controllers are universal approximators," *IEEE Trans. on Systems, Man and Cybernetics*, vol. 25, pp. 629–635, Apr. 1995.

J. R. Castro, O. Castillo, P. Melin, and A. R.-Diaz, "A hybrid learning algorithm for a class of interval type-2 fuzzy neural networks," *Information Sciences*, vol. 179, pp. 2175–2193, 2009.

L. Chua, M. Komuro, and T. Matsumoto, "The double scroll family," *IEEE Trans. on Circuits and Systems*, vol. 33, no. 11, pp. 1073–1118, 1986.

Introduction to Type-2 Fuzzy Logic Control: Theory and Applications, First Edition.
Jerry M. Mendel, Hani Hagras, Woei-Wan Tan, William W. Melek, and Hao Ying.

L. Chua, C. W. Wu, A. Hunang, and G. Q. Zhong, "A universal circuit for studying and generating chaos-part I: Routes to chaos," *IEEE Trans. on Circuits and Systems-I: Fundamental Theory and Appl.*, vol. 40, no. 10, pp. 732–744, 1993.

S. Chopra, R. Mitra, and V. Kumar, "Fuzzy controller: Choosing an appropriate and smallest rules set," *J. of Computational Cognition*, vol. 3, no. 4, pp. 73–79, 2005.

S. Coupland and R. John, "Geometric type-1 and type-2 fuzzy logic systems," *IEEE Trans. on Fuzzy Systems*, vol. 15, pp. 3–15, Feb. 2007.

T. M. Cover and J. A. Thomas, *Elements of Information Theory*, New York: Wiley, 1991.

X. Du and H. Ying, "Deriving analytical structure of a type-2 fuzzy PD/PI controller," Proceedings of 2008 North American Information Processing Society Conf., New York, pp. 1–6, 2008.

X. Du and H. Ying, "Derivation and analysis of the analytical structures of the interval type-2 fuzzy-PI and PD controllers," *IEEE Trans. on Fuzzy Systems*, vol. 18, pp. 802–814, Aug. 2010.

K. Duran, H. Bernal, and M. Melgarejo, "Improved iterative algorithm for computing the generalized centroid of an interval type-2 fuzzy set," Proceedings of 2008 North American Information Processing Society Conf., New York, Paper 50056, 2008.

G. Feng, "A survey on analysis and design of model-based fuzzy control systems," *IEEE Trans. on Fuzzy Systems*, vol. 14, pp. 676–697, Oct. 2006.

J. Figueroa, J. Posada, J. Soriano, M. Melgarejo, and S. Roj, "A type-2 fuzzy logic controller for tracking mobile objects in the context of robotic soccer games," Proceedings of 2005 IEEE Int'l. Conf. on Fuzzy Systems, Reno, NV, pp. 359–364, 2005.

D. E. Goldberg, *Genetic Algorithms in Search, Optimization and Machine Learning*, Reading, MA: Addison Wesley, 1989.

M. Grant and S. Boyd, "Graph implementations for non-smooth convex programs," in V. Blondel, S. Boyd, and H. Kimura (Eds.), *Recent Advances in Learning and Control, Lecture Notes in Control and Information Sciences*, pp. 95–110, London: Springer-Verlag Limited, 2008.

M. Grant and S. Boyd, "CVX: Matlab software for disciplined convex programming, version 1.21"; accessed at http://stanford.edu/ boyd/ cvx, 2011.

H. Hagras, "A hierarchical type-2 fuzzy logic control architecture for autonomous mobile robots," *IEEE Trans. on Fuzzy Systems*, vol. 12, pp. 524–539, Aug. 2004.

H. Hagras, "Comments on dynamical optimal training for interval type-2 fuzzy neural network" *IEEE Trans. on Systems, Man and Cybernetics, Part B: Cybernetics*, Vol. 36, pp. 1206–1209, Oct. 2006.

H. Hagras, "Type-2 FLCs: A new generation of fuzzy controllers," *IEEE Computational Intelligence Mag.*, vol. 2, pp. 30–43, Feb. 2007.

H. Hagras and C. Wagner, "Towards the wide spread use of type-2 fuzzy logic systems in real world applications," *IEEE Computational Intelligence Mag.*, vol. 7, pp. 14–24, Aug. 2012.

H. Hagras, F. Doctor, A. Lopez, and V. Callaghan, "An incremental adaptive life long learning approach for type-2 fuzzy embedded agents in ambient intelligent environments," *IEEE Trans. on Fuzzy Systems*, vol. 15, pp 41–55, Feb. 2007.

D. Hidalgo, P. Melin, and O. Castillo, "An optimization method for designing type-2 fuzzy inference systems based on the footprint of uncertainty using genetic algorithms," *Expert Systems Applications*, vol. 39, no. 4, pp. 4590–4598, 2012.

K. Hirota, "History of industrial applications of fuzzy logic in Japan," in J. Yen, R. Langari, and L. A. Zadeh (Eds.), *Industrial Applications of Fuzzy Logic and Intelligent Systems*, Piscataway, NJ: IEEE Press, pp. 43–54, 1995.

J. C. Holland, *Adaptation in Natural and Artificial Systems*, Ann Arbor, MI: Univ. of Michigan Press, 1975 (as of 1992, available from the MIT Press).

L. Holmblad and I. Ostergaard, "Control of a cement kiln by fuzzy logic," in *Fuzzy Information and Decision-Processes*, M. M. Gupta and E. Sanchez (Eds.), Amsterdam: North-Holland, pp. 389–399, 1982.

S. Horikawa, T. Furahashi, and Y. Uchikawa, "On fuzzy modeling using fuzzy neural networks with back-propagation slgorithm," *IEEE Trans. on Neural Networks*, vol. 3, pp. 801–806, Sept., 1992.

H. Hu, Y. Wang, and Y. Cai, "Advantages of the enhanced opposite direction searching algorithm for computing the centroid of an interval type-2 fuzzy set," *Asian J. of Control*, vol. 14, no. 6, pp. 1–9, Nov. 2012

J.-S. R. Jang, "Self-learning fuzzy controllers based on temporal back-propagation," *IEEE Trans. on Neural Networks*, vol. 3, pp. 714–723, Sept., 1992.

C.-F. Juang and Y.-W. Tsao, "A type-2 self-organizing neural fuzzy system and its FPGA implementation," *IEEE Trans. on Systems, Man, and Cybernetics—Part B: Cybernetics*, vol. 38, no. 6, pp. 1537–1548, 2008.

N. N. Karnik and J. M. Mendel, "Centroid of a type-2 fuzzy set," *Information Sciences*, vol. 132, pp. 195–220, 2001a.

N. N. Karnik and J. M. Mendel, "Operations on type-2 fuzzy sets," *Fuzzy Sets and Systems*, vol. 122, pp. 327–348, 2001b.

N. N. Karnik, J. M. Mendel, and Q. Liang "Type-2 fuzzy logic systems," *IEEE Trans. on Fuzzy Systems*, vol. 7, pp. 643–658, Dec. 1999.

J. Kennedy and R. Eberhart, "Particle swarm optimization," Proc. of IEEE Int'l. Conf. on Neural Networks, pp. 1942–1948, 1995.

K. H. Khalal, *Nonlinear Systems*, 2nd, ed., Englewood Cliffs, NJ: Prentice-Hall, 1996.

G. J. Klir and B. Yuan, *Fuzzy Sets and Fuzzy Logic: Theory and Applications* (Chapter 1), Englewood Cliffs, NJ: Prentice-Hall, 1995.

B. Kosko, "Fuzzy systems are universal approximators," *IEEE Trans. on Computers*, vol. 43, pp. 1329–1333, Nov. 1994.

V. Kreinovich, G. C. Mouzouris, and H. T. Nguyen, "Fuzzy rule based modeling as a universal approximation tool," in H. T. Nguyen and M. Sugeno (Eds.), *Fuzzy Systems: Modeling and Control*, Boston: Kluwer Academic, 1998.

P. M. Larsen, "Industrial applications of fuzzy logic control," *Int. J. Man, Mach. Studies*, vol. 12, no. 1, pp. 3–10, 1980.

C. C. Lee, "Fuzzy logic in control systems: Fuzzy logic controller-Part I," *IEEE Trans. on Systems, Man, and Cybernetics*, vol. 20, no. 2, pp. 404–418, 1990.

F. L. Lewis, S. Jagannathan, and A. Yesildirek, *Neural Network Control of Robot Manipulators and Nonlinear Systems*, New York: Taylor & Francis, 1999.

Q. Liang and J. M. Mendel, "Interval type-2 fuzzy logic systems: theory and design," *IEEE Trans. on Fuzzy Systems*, vol. 8, pp. 535–550, Oct. 2000.

Q. Liang and J. M. Mendel, "MPEG VBR video traffic modeling and classification using fuzzy techniques," *IEEE Trans. on Fuzzy Systems*, vol. 9, pp. 183–193, Feb. 2001.

C.-T. Lin and C. S. G. Lee, *Neural Fuzzy Systems*, Englewood Cliffs, NJ: Prentice-Hall, 1996.

P. Lin, C. Hsu, and T. Lee, "Type-2 fuzzy logic controller design for buck DC-DC converters," Proceedings of 2005 IEEE Int'l. Conf. on Fuzzy Systems, Reno, NV, pp. 365–370, 2005.

T.-C. Lin and M.-C. Chen, "Adaptive hybrid type-2 intelligent sliding model control for uncertain nonlinear multivariable dynamical systems," *Fuzzy Sets and Systems*, vol. 171, pp. 44–71, 2011.

O. Linda and M. Manic, "Monotone centroid flow algorithm for type-reduction of general type-2 fuzzy sets," *IEEE Trans. on Fuzzy Systems*, vol. 20, pp. 805–819, Oct. 2012.

F. Liu, "An efficient centroid type-reduction strategy for general type-2 fuzzy logic system," *Information Sciences*, vol. 178, pp. 2224–2236, 2008.

F. Liu and J. M. Mendel, "Aggregation using the fuzzy weighted average, as computed by the KM algorithms," *IEEE Trans. on Fuzzy Systems*, vol. 16, pp. 1–12, Feb. 2008.

X. Liu and J. M. Mendel, "Connect Karnik–Mendel algorithms to root-finding for computing the centroid of an interval type-2 fuzzy set," *IEEE Trans. on Fuzzy Systems*, vol. 19, pp. 652–665, Aug. 2011.

C. Lynch, H. Hagras, and V. Callaghan, "Embedded type-2 FLC for real-time speed control of marine and traction diesel engines," Proceedings of 2005 IEEE Int'l. Conf. on Fuzzy Systems, Reno, NV, pp. 347–352, 2005.

C. Lynch. H. Hagras, and V. Callaghan, "Embedded interval type-2 neuro-fuzzy speed controller for marine diesel engines," Proceedings of 2006 Information Processing and Management of Uncertainty in Knowledge-Based Systems Conf., Paris, France, pp. 1340–1347, 2006a.

C. Lynch, H. Hagras, and V. Callaghan, "Using uncertainty bounds in the design of embedded real-time type-2 neuro-fuzzy speed controller for marine diesel engines," Proceedings of 2006 IEEE Int'l. Conf. on Fuzzy Systems, Vancouver, Canada, pp. 7217–7224, 2006b.

E. H. Mamdani, "Applications of fuzzy algorithms for simple dynamic plant," *Proc. IEEE*, vol. 121, pp. 1585–1588, 1974.

E. H. Mamdani, "Fuzzy control—a misconception of theory and application," *IEEE Expert-A Fuzzy Logic Symposium*, vol. 9, no. 4, pp. 27–28, 1994.

E. H. Mamdani and S. Assilian, "An experiment in linguistic synthesis with a fuzzy logic controller," *Int. J. of Machine Studies*, vol. 7, pp. 1–13, 1975.

R. Martínez, O. Castillo, and L. T. Aguilar, "Optimization with genetic algorithms of interval type-2 fuzzy logic controllers for an autonomous wheeled mobile robot: A comparison under different kinds of perturbations," Proceedings of 2008 IEEE Int'l. Conf. on Fuzzy Systems, Paper FS0225, Hong Kong, China, June 2008.

R. Martínez, O. Castillo, and L. T. Aguilar, "Optimization of interval type-2 fuzzy logic controllers for a perturbed autonomous wheeled mobile robot using genetic algorithms," *Information Sciences*, vol. 179, pp. 2158–2174, 2009.

M. C. A. Melgarejo, "A fast recursive method to compute the generalized centroid of an interval type-2 fuzzy set," Proceedings of 2007 North American Fuzzy Info. Processing Society Conf., San Diego, CA, pp. 190–194, 2007.

P. Melin, L. Astudillo, O. Castillo, F. Valdez, and M. Garcia "Optimal design of type-2 and type-1 fuzzy tracking controllers for autonomous mobile robots under perturbed torques using a new chemical optimization paradigm," *Expert Systems Applications*, vol. 40, no. 8, pp. 3185–3195, 2013.

J. M. Mendel, "Fuzzy logic systems for engineering: a tutorial," *IEEE Proc.*, Vol. 83, pp. 345–377, Mar. 1995.

J. M. Mendel, *Uncertain Rule-Based Fuzzy Logic Systems: Introduction and New Directions*, Upper-Saddle River, NJ: Prentice-Hall, 2001.

J. M. Mendel, "Computing derivatives in interval type-2 fuzzy logic systems," *IEEE Trans. on Fuzzy Systems*, vol. 12, pp. 84–98, Feb. 2004

J. M. Mendel, "Type-2 fuzzy sets and systems: an overview," *IEEE Computational Intelligence Mag.*, vol. 2, pp. 20–29, Feb. 2007.

J. M. Mendel, "Comments on 'α-plane representation for type-2 fuzzy sets: theory and applications," *IEEE Trans. on Fuzzy Systems*, vol. 18, pp. 229–230, Feb. 2010.

J. M. Mendel, "Plotting 2-1/2 D figures for general type-2 fuzzy sets by hand or by PowerPoint," Proceedings of 2012 IEEE Int'l. Conf. on Fuzzy Systems, Brisbane, Australia, pp. 1490–1497, June 2012.

J. M. Mendel, "On KM algorithms for solving type-2 fuzzy set problems," *IEEE Trans. on Fuzzy Systems*, vol. 21, pp. 426–446, June 2013.

J. M. Mendel, "General type-2 fuzzy logic systems made simple: a tutorial," *IEEE Trans. on Fuzzy Systems*, vol. 22, 2014.

J. M. Mendel and R. I. John, "Type-2 fuzzy sets made simple," *IEEE Trans. on Fuzzy Systems*, vol. 10, pp. 117–127, Apr. 2002a.

J. M. Mendel and R. I. John, "Footprint of uncertainty and its importance to type-2 fuzzy sets," Proceedings of 6th IASTED Int'l. Conf. on Artificial Intelligence and Soft Computing, Banff, Alberta, Canada, pp. 587–592, July 2002b.

J. M. Mendel and F. Liu, "Super-exponential convergence of the Karnik-Mendel algorithms for computing the centroid of an interval type-2 fuzzy set," *IEEE Trans. on Fuzzy Systems*, vol. 15, pp. 309–320, Apr. 2007.

J. M. Mendel and X. Liu, "Simplified interval type-2 fuzzy logic systems," *IEEE Trans. on Fuzzy Systems*, vol. 21, pp. 1056–1069, Dec. 2013.

J. M. Mendel and D. Wu, *Perceptual Computing: Aiding People in Making Subjective Judgments*, Hoboken, NJ: Wiley and IEEE Press, 2010.

J. M. Mendel and H. Wu, "Type-2 fuzzistics for symmetric interval type-2 fuzzy sets: Part 1, forward problems," *IEEE Trans. on Fuzzy Systems*, vol. 14, pp. 781–792, Dec. 2006.

J. M. Mendel and H. Wu, "Type-2 fuzzistics for non-symmetric interval type-2 fuzzy sets: forward problems," *IEEE Trans. on Fuzzy Systems*, vol. 15, pp. 916–930, Oct., 2007a.

J. M. Mendel and H. Wu, "New results about the centroid of an interval type-2 fuzzy set, including the centroid of a fuzzy granule," *Information Sciences*, vol. 177, pp. 360–377, 2007b.

J. M. Mendel, R. I. John, and F. Liu, "Interval type-2 fuzzy logic systems made simple," *IEEE Trans. on Fuzzy Systems*, vol. 14, pp. 808–821, Dec. 2006.

J. M. Mendel, F. Liu, and D. Zhai, "Alpha-plane representation for type-2 fuzzy sets: Theory and applications," *IEEE Trans. on Fuzzy Systems*, vol. 17, pp. 1189–1207, Oct. 2009.

G. Mendez, L. Leduc-Lezama, R. Colas, G. Murillo-Perez, J. Ramirez-Cuellar and J. Lopez, "Modelling and control of coiling entry temperature using interval type-2 fuzzy logic systems," *J. of Iron Making and Steel Making*, vol. 37, pp. 126–134, Feb. 2010.

Y. Moldonado, O. Castillo, and P. Melin, "Particle swarm optimization of interval type-2 fuzzy systems for FPGA applications," *Applied Soft Computing*, vol. 13, no. 1, pp. 496–508, 2013.

T. Munakata and Y. Jani, "Fuzzy systems: An overview," *Communications of the ACM*, vol. 37, pp. 69–96, Mar. 1994.

M. Nie and W. W. Tan, "Towards an efficient type-reduction method for interval type-2 fuzzy logic systems," Proceeding of 2008 IEEE Int'l. Conf. on Fuzzy Systems, Hong Kong, China, Paper FS0339, 2008.

M. Nie and W. W. Tan, "Derivation of the analytical structure of symmetrical IT2 fuzzy PD and PI controllers," Proceedings of 2010 IEEE Int'l. Conf. on Fuzzy Systems, Barcelona, Spain, pp. 1–8, July 2010.

M. Nie and W. W. Tan, "Analytical structure and characteristics of symmetric Karnik-Mendel type-reduced interval type-2 fuzzy PI and PD controllers," *IEEE Trans. on Fuzzy Systems*, vol. 20, pp. 416–430, 2012.

A. O'Dwyer, *Handbook of PI and PI Controller Tuning Rules*, London: Imperial College Press, 2003.

R. Palm, "Sliding mode fuzzy control," Proceedings of 1992 IEEE Int'l. Conf. on Fuzzy Systems, San Diego, CA, pp. 519–526, 1992.

Q. Ren, L. Baron, K. Jemielniak, and M. Balazinski, "Modeling of dynamic micro-milling cutting forces using type-2 fuzzy rule-based system," Proceedings of 2010 IEEE Int'l. Conf. on Fuzzy Systems, Barcelona, Spain, pp. 1–7, 2010.

P. C. Shill, M. F. Amin, M. A. H. Akhand, and K. Murase, "Optimization of interval type-2 fuzzy logic controller using quantum genetic algorithms," Proc. FUZZ-IEEE 2012, Brisbane, Australia, pp. 1027–1034, June 2012.

M. Sugeno, "An introductory survey of fuzzy control," *Information Sciences*, vol. 36, pp. 59–83, July–Aug. 1985.

T. Takagi and M. Sugeno, "Fuzzy identification of systems and its application to modeling and control," *IEEE Trans. on Systems, Man and Cybernetics*, vol. 15, pp. 116–132, Jan/Feb. 1985.

K. Tanaka and M. Sano, "Trajectory stabilization of a model car via fuzzy control," *Fuzzy Sets and Systems*, vol. 70, no. 2–3, pp. 155–170, 1995.

K. Tanaka and M. Sugeno, "Introduction to fuzzy modeling," in H. T. Nguyen and M. Sugeno (Eds.), *Fuzzy Systems Modeling and Control*, Boston: Kluwer Academic, 1998, pp. 63–89.

K. Tanaka, M. Sano, and H. Watanabe, "Modeling and control of carbon monoxide concentration using a neuro-fuzzy technique," *IEEE Trans. Fuzzy Systems*, vol. 3, pp. 271–279, Aug. 1995.

L. Teo, M. Khalid, and R. Yusof, "Self-tuning neuro-fuzzy control by genetic algorithms with an application to a coupled-tank liquid-level control system," *Eng. Appl. of Artificial Intelligence*, vol. 11, pp. 517–529, Aug. 1998.

M. Tripathy and S. Mishra, "Interval type-2-based thyristor controlled series capacitor to improve power system stability," *Generation, Transmission & Distribution*, vol. 5. pp. 209–222, Feb. 2011.

C. Wagner, "Juzzy–A Java based toolkit for type-2 fuzzy logic," Proc. of IEEE Symposium on Advances in Type-2 Fuzzy Logic Systems, Singapore, Apr. 2013.

C. Wagner and H. Hagras, "A genetic algorithm based architecture for evolving type-2 fuzzy logic controllers for real world autonomous mobile robots," Proceedings of 2007 IEEE Int'l. Conf. on Fuzzy Systems, pp. 193–198, London, July 2007.

C. Wagner and H. Hagras, "zSlices—towards bridging the gap between interval and general type-2 fuzzy logic," Proceedings of 2008 IEEE Int'l. Conf. on Fuzzy Systems, Hong Kong, pp. 489–497, 2008.

C. Wagner and H. Hagras, "Towards general type-2 fuzzy logic systems based on zSlices," *IEEE Trans. on Fuzzy Systems*, vol. 18, pp. 637–660, Aug. 2010.

C. Wang, C. Cheng, and T. Lee, "Dynamical optimal training for interval type-2 fuzzy neural network (T2FNN)," *IEEE Trans. on Systems, Man and Cybernetics Part B: Cybernetics*, vol. 34, pp. 1462–1477, June 2004.

H. O. Wang and K. Tanaka, "An LMI-based stable fuzzy control of nonlinear systems and its application to control of chaos," Proceedings of 1996 IEEE Int'l. Conf. on Fuzzy Systems, New Orleans, pp. 1433–1438, 1996.

L.-X. Wang, "Fuzzy systems are universal approximators," Proc. International Conference on Fuzzy Systems, San Diego, CA, 1992.

L.-X. Wang, *A Course in Fuzzy Systems and Control*, Englewood Cliffs, NJ: Prentice-Hall, 1997.

L.-X. Wang and J. M. Mendel, "Fuzzy basis functions, universal approximation, and orthogonal least squares learning," *IEEE Trans. on Neural Networks*, vol. 3, pp. 807–813, Sept. 1992a.

L.-X. Wang and J. M. Mendel, "Back-propagation of fuzzy systems as non-linear dynamic system identifiers," Proceedings of 1992 IEEE Int'l. Conference on Fuzzy Systems, San Diego, CA, pp. 1409–1418, 1992b.

X. Wang, Y. He, L. Dong, and H. Zhao, "Particle swarm optimization for determining fuzzy measures from data," *Information Sciences*, vol. 181, pp. 4230–4252, Oct. 2011.

F. Wei, S. Jun, X. Z.-Ping, and W.-B. Xu, "Convergence analysis of quantum-behaved particle swarm optimization algorithm and study on its control parameter," *Acta Phys. Sin.*, vol. 59, no. 6, pp. 3686–3694, 2010.

D. Wu, "An interval type-2 fuzzy logic system cannot be implemented by traditional type-1 fuzzy logic systems," Proceedings of 2011 World Conference on Soft Computing, San Francisco, 2011.

D. Wu, "Approaches for reducing the computational cost of interval type-2 fuzzy logic systems: overview and comparisons," *IEEE Trans. on Fuzzy Systems*, vol. 21, pp. 80–99, Feb. 2013.

D. Wu, "On the fundamental differences between interval type-2 and type-1 fuzzy logic controllers," *IEEE Transactions on Fuzzy Systems*, Vol. 20, pp. 832–848, Oct. 2012.

D. Wu and J. M. Mendel, "Enhanced Karnik-Mendel algorithms," *IEEE Trans. on Fuzzy Systems*, vol. 17, pp. 923–934, Aug. 2009.

D. Wu and J. M. Mendel, "On the continuity of type-1 and interval type-2 fuzzy logic systems," *IEEE Trans. on Fuzzy Systems*, vol. 19, pp. 179–192, Feb. 2011.

D. Wu and M. Nie, "Comparison and practical implementations of type-reduction algorithms for type-2 fuzzy sets and systems," Proceedings of 2011 IEEE Int'l. Conf. on Fuzzy Systems, Taipei, Taiwan, pp. 2131–2138, 2011.

D. Wu and W. Tan, "Type-2 fuzzy logic controller for the liquid-level process," Proceedings of 2004 IEEE Int'l. Conf. on Fuzzy Systems, Budapest, Hungary, pp. 248–253, 2004.

D. Wu and W. Tan, "Type-2 FLS modeling capability analysis," Proceedings of 2005 IEEE Int'l. Conf. on Fuzzy Systems, Reno, NV, pp. 242–247, 2005.

D. R. Wu and W. W. Tan, "Genetic learning and performance evaluation of type-2 fuzzy logic controllers," *Engineering Applications of Artificial Intelligence*, vol. 19, no. 8, pp. 829–841, 2006.

D. Wu and W. W. Tan, "Interval type-2 fuzzy PI controllers: Why they are more robust," Proceedings of 2010 IEEE Int'l. Conf. on Granular Computing, San Jose, CA, pp. 802–807, 2010.

H. Wu and J. M. Mendel, "Uncertainty bounds and their use in the design of interval type-2 fuzzy logic systems," *IEEE Trans. on Fuzzy Systems*, vol. 10, pp. 622–639, Oct. 2002.

R. R. Yager and D. P. Filev, *Essentials of Fuzzy Modeling and Control*, New York: Wiley, 1994.

S. Yasunobu and S. Miyamoto, "Automatic train operation by fuzzy predictive control," in M. Sugeno (Ed.), *Industrial Applications of Fuzzy Control*, Amsterdam: North Holland, 1985.

C.-Y. Yeh, W.-H. Roger Jeng, and S.-J. Lee, "An enhanced type-reduction algorithm for type-2 fuzzy sets," *IEEE Trans. on Fuzzy Systems*, vol. 19, pp. 227–240, Apr. 2011.

J. Yen and R. Langari, *Fuzzy Logic: Intelligence, Control and Information*, Upper Saddle River, NJ: Prentice-Hall, 1999.

H. Ying, "The simplest fuzzy controllers using different inference methods are different nonlinear proportional–integral controllers with variable gains," *Automatica*, vol. 29, pp. 1579–1589, March 1993a.

H. Ying, "A fuzzy controller with linear rules is the sum of a global two-dimensional multilevel relay and a local nonlinear proportional-integral controller," *Automatica*, vol. 29, pp. 499–505, March 1993b.

H. Ying, "Sufficient conditions on general fuzzy systems as function approximators," *Automatica*, vol. 30, pp. 521–525, March 1994a.

H. Ying, "Practical design of nonlinear fuzzy controllers with stability analysis for regulating processes with unknown mathematical models," *Automatica*, vol. 30, pp. 1185–1195, July 1994b.

H. Ying, "Sufficient conditions on uniform approximation of multivariate functions by general Takagi-Sugeno fuzzy systems with linear rule consequent," *IEEE Trans. on Systems, Man, and Cybernetics*, vol. 28, pp. 515–520, July 1998a.

H. Ying, "The Takagi-Sugeno fuzzy controllers using the simplified linear control rules are nonlinear variable gain controllers," *Automatica*, vol. 34, pp. 157–167, February 1998b.

H. Ying, "Constructing nonlinear variable gain controllers via the Takagi-Sugeno fuzzy control," *IEEE Trans. on Fuzzy Systems*, Vol. 6, pp. 226–234, May 1998c.

H. Ying, *Fuzzy Control and Modeling: Analytical Foundations and Applications*, Piscataway, NJ: IEEE Press, 2000.

H. Ying, "Deriving analytical input–output relationship for fuzzy controllers using arbitrary input fuzzy sets and Zadeh fuzzy AND operator," *IEEE Trans. on Fuzzy Systems*, vol. 14, pp. 654–662, October 2006.

H. Ying, "General interval type-2 Mamdani fuzzy systems are universal approximators," Proceedings of 2008 North American Fuzzy Information Processing Society Conf., New York, pp. 1–6, 2008.

H. Ying, "Interval type-2 Takagi-Sugeno fuzzy systems with linear rule consequent are universal approximators," Proceedings of 2009 North American Fuzzy Information Processing Society Conf., Cincinnati, OH, pp. 1–5, 2009.

H. Ying, M. McEachern, D. Eddleman, and L. C. Sheppard, "Fuzzy control of mean arterial pressure in postsurgical patients with sodium nitroprusside infusion," *IEEE Trans. on Biomedical Engineering*, vol. 39, pp. 1060–1070, Oct. 1992.

H. Ying, W. Siler, and J. J. Buckley, "Fuzzy control-theory—a nonlinear case," *Automatica*, vol. 26, pp. 513–520, May 1990.

L. A. Zadeh, "Outline of a new approach to the analysis of complex systems and decision processes," *IEEE Trans. on Systems, Man, and Cybernetics*, vol. SMC-3, pp. 28–44, Jan. 1973.

L. A. Zadeh, "The concept of a linguistic variable and its application to approximate reasoning–1," *Information Sciences*, vol. 8, pp. 199–249, 1975.

S. Zaheer and J. Kim, "Type-2 fuzzy airplane altitude control: A comparative study," Proceedings of 2011 IEEE Int'l. Conf. on Fuzzy Systems, Taipei, Taiwan, pp. 2170–2176, 2011.

D. Zhai and J. M. Mendel, "Computing the centroid of a general type-2 fuzzy set by means of the centroid flow algorithm," *IEEE Trans. on Fuzzy Systems*, vol. 19, pp. 401–422, June 2011a.

D. Zhai and J. M. Mendel, "Uncertainty measures for general type-2 fuzzy sets," *Information Sciences*, vol. 181, pp. 503–518, 2011b.

D. Zhai and J. M. Mendel, "Enhanced centroid-flow algorithm for computing the centroid of general type-2 fuzzy sets," *IEEE Trans. on Fuzzy Systems*, vol. 20, pp. 939–956, Oct. 2012.

H. Zhou and H. Ying, "Deriving the input–output mathematical relationship for a class of interval type-2 Mamdani fuzzy controllers," Proceedings of 2011 IEEE Int'l. Conf. on Fuzzy Systems, Taipei, Taiwan, pp. 2589–2593, 2011.

H. Zhou and H. Ying, "A technique for deriving analytical structure of a general class of interval type-2 TS fuzzy controllers," Proceedings of 2012 North American Fuzzy Information Processing Society Conf., Berkeley, CA, pp. 1–6, 2012.

H. Zhou and H. Ying, "A method for deriving the analytical structure of a broad class of typical interval type-2 Mamdani fuzzy controllers," *IEEE Trans. on Fuzzy Systems*, Vol. 21, pp. 447–458, June 2013.

INDEX

Introduction to Type-2 Fuzzy Logic Control: Theory and Applications, First Edition.
Jerry M. Mendel, Hani Hagras, Woei-Wan Tan, William W. Melek, and Hao Ying.

IEEE Press Series on
COMPUTATIONAL INTELLIGENCE

Series Editor, **David B. Fogel**

The IEEE Press Series on Computational Intelligence includes books on neural, fuzzy, and evolutionary computation, and related technologies, of interest to the engineering and scientific communities. Computational intelligence focuses on emulating aspects of biological systems to construct software and/or hardware that learns and adapts. Such systems include neural networks, our use of language to convey complex ideas, and the evolutionary process of variation and selection. The series highlights the most-recent and groundbreaking research and development in these areas, as well as the important hybridization of concepts and applications across these areas. The audiences for books in the series include undergraduate and graduate students, practitioners, and researchers in computational intelligence.

Computational Intelligence: The Experts Speak. Edited by David B. Fogel and Charles J. Robinson. 2003. 978-0-471-27454-4

Handbook of Learning and Approximate Dynamic Programming. Edited by Jennie Si, Andrew G. Barto, Warren B. Powell, and Donald Wunsch II. 2004. 978-0471-66054-X

Computationally Intelligent Hybrid Systems. Edited by Seppo J. Ovaska. 2005. 978-0471-47668-4

Evolutionary Computation: Toward a New Philosophy of Machine Intelligence, Third Edition. David B. Fogel. 2006. 978-0471-66951-7

Emergent Information Technologies and Enabling Policies for Counter-Terrorism. Edited by Robert L. Popp and John Yen. 2006. 978-0471-77615-4

Introduction to Evolvable Hardware: A Practical Guide for Designing Self-Adaptive Systems. Garrison W. Greenwood and Andrew M. Tyrrell. 2007. 978-0471-71977-9

Computational Intelligence in Bioinformatics. Edited by Gary B. Fogel, David W. Corne, and Yi Pan. 2008. 978-0470-10526-9

Computational Intelligence and Feature Selection: Rough and Fuzzy Approaches. Richard Jensen and Qiang Shen. 2008. 978-0470-22975-0

Clustering. Rui Xu and Donald C. Wunsch II. 2009. 978-0470-27680-8

Biometrics: Theory, Methods, and Applications. Edited by: N.V. Boulgouris, Konstantinos N. Plataniotis, and Evangelia Micheli-Tzanakou. 2009. 978-0470-24782-2

Evolving Intelligent Systems: Methodology and Applications. Edited by Plamen Angelov, Dimitar P. Filev, and Nikola Kasabov. 2010. 978-0470-28719-4

Perceptual Computing: Aiding People in Making Subjective Judgments. Jerry M. Mendel and Dongrui Wu. 2010. 978-0470-47876-9

Reinforcement Learning and Approximate Dynamic Programming for Feedback Control. Edited by Frank L. Lewis and Derong Liu. 2012. 978-1118-10420-0

Complex-Valued Neural Networks: Advances and Applications. Edited by Akira Hirose. 2013. 978-1118-34460-6

Unsupervised Learning: A Dynamic Approach. Matthew Kyan, Paisarn Muneesawang, Kambiz Jarrah, Ling Guan. 2014. 978-0470-27833-8

Introduction to Type-2 Fuzzy Logic Control: Theory and Applications. Jerry M. Mendel, Hani Hagras, Woei-Wan Tan, William W. Melek, Hao Ying. 2014 978-1118-278291